Conversion Factors

1 in.	= 2.54 cm
1 mi	= 1.609 km
1 lb	= 0.454 kg
1 AMU	= 1.66×10^{-27} kg
30 mi/h	= 44 ft/s
	= 13.4 m/s
1 eV	= 1.60×10^{-19} J
1 MeV	= 10^6 eV = 1.60×10^{-13} J
1 Cal	= 4186 J
1 Å	= 10^{-10} m

Greek Letters Used in the Text

LETTER	NAME	USE
α	alpha	Symbol for alpha particle (helium nucleus)
β	beta	Symbol for beta particle (electron)
γ	gamma	Symbol for gamma ray (nuclear photon)
Δ	delta	An increment of a quantity
θ	theta	Angle
λ	lambda	Wavelength
μ	mu	Symbol for muon
ν	nu	Symbol for neutrino; frequency
π	pi	Symbol for pion; $\pi = 3.14159 \cdots$
ρ	rho	Density
τ	tau	Period
ϕ	phi	Angle

Physics
in the
Modern
World

Physics in the Modern World

Jerry B. Marion

*University of Maryland
College Park*

Academic Press   New York / San Francisco / London

A Subsidiary of Harcourt Brace Jovanovich, Publishers

ACADEMIC PRESS, INC.
111 Fifth Avenue, New York, New York 10003

United Kingdom Edition published by
ACADEMIC PRESS, INC. (LONDON) LTD.
24/28 Oval Road, London NW1

Library of Congress Cataloging in Publication Data

Marion, Jerry B.
   Physics in the modern world.

      Bibliography: p.
      Includes index.
      1.   Physics.   I.   Title.
QC23.M36   530   75-13104
ISBN 0-12-472277-6

PRINTED IN THE UNITED STATES OF AMERICA

In addition to the acknowledgments expressed in credit lines accompanying
photographs in the text, permission to reproduce illustrations on the following
pages from the sources listed is gratefully acknowledged: 29, C. F. Powell,
P. H. Fowler, and D. H. Perkins (eds.), *The Study of Elementary Particles by
the Photographic Method* (Pergamon Press, Oxford, 1959); 48, 53, 275, 283,
285, 307, 310, *PSSC Physics* (Heath, Lexington, Massachusetts); 123, (1) NASA,
E. C. Goddard, (2) Novosti from Sovfoto; 187, J. N. Pitts and R. L. Metcalf,
Advances in Environmental Sciences, Vol. 1 (Wiley, New York, 1969); 247,
The University of Maryland, Richard Farkas; 324, J. B. Marion, *Physics and
the Physical Universe* (Wiley, New York, 1971); 326, Educational Development
Center; 366, British Crown Copyright, Science Museum, London; 387, 403,
ERDA, Lotte Jacobi; 415, Harbrace; 444, Lent to Science Museum, London
by Sir Lawrence Bragg, F.R.S.

Cover photo by Fritz Goro, Time/Life Agency, © Time Inc.

CONTENTS

PREFACE

This is a text for a one-year introductory course in physics for students who are specializing in other disciplines. In these chapters you will find a survey of topics in physics with emphasis on those aspects of current interest. No mathematics beyond basic trigonometry is required to follow the discussions.

Today we live in a world that is dominated by technology. Morever, the impact of technology on society, already enormous, will almost certainly continue to increase. In order to cope with the problems of a highly technical world, it is necessary to appreciate some of the basic scientific ideas that are the foundation stones upon which our modern technology is built. Lacking this understanding, we would find it increasingly difficult to contribute effectively to the complex decisions that affect our everyday lives. It is the purpose of this book to put forward these fundamental ideas as clearly as possible and to draw attention to the way that basic physical principles are applied in our technological world.

In *Physics in the Modern World* you will see that physical principles bring a pattern of simplicity and continuity to the diverse natural and technological world around us. To show the many ways that physical ideas are manifest in everyday situations, numerous short essays on various kinds of applications have been included. In these special sections the reader will learn about the operation of rockets and cameras, and about the principles at work in space travel and X-ray photography. Discussions of automobile air bags, drag racing, artificial gravity, pollution control, appliance economics, musical instruments, radar, and other modern phenomena and devices emphasize the way that physical principles are applied in today's world. Historical sketches of individual scientists detail their important contributions to our present knowledge and technology. Physics is not an abstract subject. Physical principles form the basis of the world in which we live, and they constitute a vital part of the knowledge we must have to understand and appreciate that world.

An important part of learning about physics is becoming familiar with some of the quantitative aspects of the subject by solving simple problems. However, the main thrust of this book is not concerned with problem-solving techniques. Instead, the emphasis here is on the basic concepts and principles. To be sure, these ideas are reinforced through examples and exercises. But it is much more important to understand the physical basis of an event or situation than to be able to substitute numbers into some formula.

A serious effort has been made to present each topic in the clearest possible terms. The explanations are developed carefully and in depth, frequently including a detailed example. Accordingly, this book is more than a source of questions for the instructor to answer. It is a book that the student can *read*.

To enhance the value of this text as a learning tool, a supplementary student guide is available. In this guide the student will find a short summary of the important ideas in each chapter, some additional worked examples, suggestions for outside reading, and a list of questions and problems (with answers) to test his comprehension of the material.

I hope that you, the reader, will enjoy this book as much as I have enjoyed writing it!

JERRY B. MARION

College Park, Maryland

1

INTRODUCTION TO PHYSICAL IDEAS

From his home on the Earth, Man can look through a telescope into the vast reaches of space. Or he can look through a microscope into the miniature world of cells and molecules. The scale of things that Man has been able to observe and study truly staggers the imagination. Roughly speaking, the Universe is as many times larger than the Earth as the Earth is larger than an atom. Thus, Man stands in a middle position, privileged to view the immensely large Universe populated with an incalculable number of stars and galaxies as well as the microscopic domain of incredibly tiny atoms and molecules.

Man has reached out from his position between the large and the small of the Universe and he has uncovered at least some of the rules by which Nature governs the *microscopic* (or small-scale) world of atoms and the *macroscopic* (or large-scale) realm of everyday objects, the Earth, planets, and stars. In this book we will examine some of these discoveries, and we will see how they are used to describe the world around us.

DESCRIBING AND MEASURING THINGS

The Basic Concepts

Progress is made in understanding our physical surroundings through *observation* and *measurement* coupled with *logic* and *reason*. In order to describe our observations and to record our measurements, we must agree on the language and the terms that we will use. Our intuitive ideas concerning physical concepts will serve as the starting points for most of our discussions of the world around us. One of the important aspects of measurements of any type is the existence of a set of *standards*. Unless we all agree on the meaning of terms such as *one quart* or *one acre* or *one hour*, it will be impossible to give a precise interpretation to any measurement. The necessity for standards of various kinds has given rise to an enormous number of measuring units. Many of these measuring units have very specialized applications—for example, the *tablespoon* in cooking or the *rod* in surveying or the *carat* in gemmology. Fortunately, in scientific matters a restricted set of measuring units is used.

The fundamental units of measure in science are those of *length, time,* and *mass*. These are familiar concepts, but because they are so basic to the description of physical events and phenomena, we will briefly discuss each of these units in turn.

 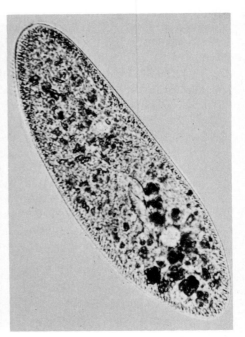

HALE OBSERVATORIES

The large and the small of the Universe. At the left is a telescopic photograph of the great galaxy in the constellation Andromeda and at the right is a microscopic photograph of a *paramecium*, a one-celled animal. The diameter of Andromeda is approximately 1 000 000 000 000 000 000 000 m (10^{21} m) whereas the size of the paramecium is approximately 0.000 1 m (10^{-4} m).

Length

Most Americans are accustomed to measuring distance in terms of inches, feet, yards, and miles, which are length units in the so-called *British* system of units. These length units are derived from a variety of sources, dating back hundreds or thousands of years to periods when there were only the crudest of standards for the measurement of length. Today, the scientific community universally uses the *metric system* of measure. Indeed, even for everyday matters, most of the world (with the primary exception of the United States) uses metric measure. In order to preserve our position in world trade, the United States will eventually change over from its archaic system to metric units. But it will probably be many years until we will have foregone completely our present system.

The standard of length in the metric system is the *meter* (m). Compared to the length units in the British system, the meter has the following values:

$$1 \text{ m} = 39.37 \text{ in.} = 3.281 \text{ ft} = 1.094 \text{ yd}$$

That is, a *meter* is about 10 percent longer than a *yard*.

Until 1961 the meter was defined as the distance between two finely drawn lines on a certain metal bar housed in the International Bureau of Weights and Measures, near Paris. Copies of this bar were distributed to national standards laboratories throughout the world. However, in 1961 an international agreement was made to define the meter in terms of the wavelength of the orange light emitted by krypton atoms. Thus, we now have an *atomic* standard for length. Because all atoms of krypton are exactly alike,* a length standard can be established in any laboratory where it is required, and it is guaranteed that all such krypton standards will be absolutely identical. Not only does the adoption of an atomic standard for length eliminate the necessity of relying on the inconvenient standard meter bar, but now it is possible to report lengths in terms of the atomic standard to a precision of 1 part in 100 000 000, almost a hundred times better than was previously possible.

The metric system has the advantage (not shared by the British system) that the various units of a physical quantity are related by factors of 10, thus considerably

UPI

It will require many years for the United States to change to the metric system, but evidence that the conversion is underway is beginning to appear.

---

* More accurately, we should say that all atoms of the isotope krypton-86 are exactly alike; we will learn about *isotopes* in the next chapter.

TABLE 1-1 METRIC UNITS OF LENGTH

10 mm = 1 cm
100 cm = 1 m
1000 m = 1 km

TABLE 1-2 LENGTH CONVERSION FACTORS

2.54 cm = 1 in.
39.37 in. = 1 m
1.609 km = 1 mi

TABLE 1-3 COMMONLY USED PREFIXES

SYMBOL	PREFIX	POWER OF 10	EXAMPLE
μ	micro-	10^{-6}	10^{-6} second = 1 μs
m	milli-	10^{-3}	10^{-3} meter = 1 mm
c	centi-	10^{-2}	10^{-2} meter = 1 cm
k	kilo-	10^{3}	10^{3} watts = 1 kW
M	mega-	10^{6}	10^{6} volts = 1 MV

simplifying any conversions that are necessary. For example,

$1 \text{ m} = 100$ centimeters (cm), or 10^2 cm

$1 \text{ cm} = 0.01$ m, or 10^{-2} m

$1 \text{ m} = 0.001$ kilometer (km), or 10^{-3} km

$1 \text{ km} = 1000$ m, or 10^3 m

The metric units of length are summarized in Table 1-1.

Conversion of Units

Occasionally, we will need to convert from the British system to the metric system or vice versa. For length conversions we use the correspondences shown in Table 1-2. Suppose that we wish to express 18 inches in terms of centimeters. Now, 1 inch (in.) is exactly equal to 2.54 cm. Therefore, we can form a *ratio* that is equal to unity:

$$\frac{2.54 \text{ cm}}{1 \text{ in.}} = 1$$

We can multiply (or divide) any quantity by unity without affecting its value. If we use the above ratio for unity, and multiply it by 18 in., we find

$$(18 \text{ in.}) \times \left(\frac{2.54 \text{ cm}}{1 \text{ in.}} \right) = 45.72 \text{ cm}$$

Powers of Ten—How We Use Them

One of the problems that we encounter when dealing with physical quantities is how to express very large and very small numbers in a convenient way. For example, to write that the distance from the Earth to the Sun is 150 000 000 000 meters or that the mass of a hydrogen atom is 0.000 000 000 000 000 000 000 000 001 673 kilograms is obviously quite awkward (and subject to errors unless the zeroes are counted carefully!). To overcome this difficulty in writing very large or very small numbers, we use a compact notation employing *powers of ten*. For example, notice that

$$10 \times 10 = 100 = 10^2$$
$$10 \times 10 \times 10 = 1000 = 10^3$$
$$10 \times 10 \times 10 \times 10 = 10\ 000 = 10^4$$

That is, the number of times that 10 is multiplied together appears in the result as the superscript of 10 (called the *exponent* of 10 or the *power* to which 10 is raised).

Any number can be expressed in powers-of-ten notation. For example,

$$147\ 000\ 000 = 1.47 \times 100\ 000\ 000 = 1.47 \times 10^8$$

Notice that in going from 147 000 000 to 1.47×10^8, we move the decimal *eight* places to the left; therefore, the exponent of 10 that appears in the result is 8. Similarly, in going from 1.47×10^8 to 147 000 000, we move the decimal eight places to the *right*.

Products of powers of 10 are expressed as

$$10^2 \times 10^3 = (10 \times 10) \times (10 \times 10 \times 10) = 10^5 = 10^{(2+3)}$$

That is, in general, the product of 10^n and 10^m is $10^{(n+m)}$:

$$10^n \times 10^m = 10^{(n+m)}$$

If the power of 10 appears in the denominator, the exponent is given a negative sign:

$$\frac{1}{10} = 0.1 = 10^{-1}$$

$$\frac{1}{100} = 0.01 = 10^{-2}$$

$$\frac{1}{1000} = 0.001 = 10^{-3}$$

In general,

$$\frac{1}{10^m} = 10^{-m}$$

Any decimal number can be expressed as a negative power of 10:

$$0.037 = \frac{37}{1000} = \frac{3.7}{100} = 3.7 \times 10^{-2}$$

Notice that in going from 0.037 to 3.7×10^{-2}, we move the decimal *two places* to the right; therefore, the exponent of 10 that appears in the result is -2. Similarly, in going from 3.7×10^{-2} to 0.037, we move the decimal two places to the *left*.

Calculations involving large or small numbers are made considerably easier by using the powers-of-ten notation:

$$400\ 000 \times 0.02 = (4 \times 10^5) \times (2 \times 10^{-2}) = (4 \times 2) \times 10^{(5-2)} = 8 \times 10^3$$

$$\frac{640\ 000}{4\ 000\ 000\ 000} = \frac{6.4 \times 10^5}{4 \times 10^9} = \frac{6.4}{4} \times 10^{(5-9)} = 1.6 \times 10^{-4}$$

Sometimes we use a prefix to a unit to express the appropriate unit. For example, *centi-* means $\frac{1}{100}$; therefore, *centimeter* means $\frac{1}{100}$ of a meter. The commonly used prefixes are listed in Table 1-3.

TABLE 1-4 THE RANGE OF DISTANCES IN THE UNIVERSE (ALL VALUES ARE APPROXIMATE.)

Radius of the Universe	100 000 000 000 000 000 000 000 000 m = 10^{26} m
Nearest galaxy	10 000 000 000 000 000 000 000 m = 10^{22} m
Nearest star	10 000 000 000 000 000 m = 10^{16} m
Earth–Sun	100 000 000 000 m = 10^{11} m
New York–Chicago	1 000 000 m = 10^{6} m
Length of a football field	100 m = 10^{2} m
Height of a child	1 m = 10^{0} m
Width of a finger	0.01 m = 10^{-2} m
Grain of salt	0.000 01 m = 10^{-5} m
Radius of an atom	0.000 000 000 1 m = 10^{-10} m
Nuclear radius	0.000 000 000 000 01 m = 10^{-14} m

TABLE 1-5 SOME USEFUL DISTANCES

1 light year, L.Y. (the distance light will travel in 1 year)	9.46×10^{15} m
Earth–Sun distance (called 1 *astronomical unit,* A.U.)	1.50×10^{11} m
Radius of Sun	6.96×10^{8} m
Earth–Moon distance	3.84×10^{8} m
Radius of Earth	6.38×10^{6} m
Radius of Moon	1.74×10^{6} m
Wavelength of yellow light	6.0×10^{-7} m
1 angstrom, Å	10^{-10} m
Radius of hydrogen atom	5.29×10^{-11} m
Radius of proton	1.2×10^{-15} m

TABLE 1-6 TIME UNITS

1 second = 9 192 631 770 vibrations of cesium atom	
1 minute = 60 s	
1 hour = 3600 s	
1 day = 86 400 s	
1 year = 3.156×10^{7} s	

Notice that *in.* occurs both in the numerator and the denominator of the left-hand side and therefore cancels, leaving the result expressed in cm. We can always use this technique to convert from one system of units to another.

The range of lengths and distances that we encounter in the Universe is truly enormous. Table 1-4 lists some representative values. Notice that the size of the Universe is about 10^{40} times the size of a nucleus! Some useful distances are given in Table 1-5.

Time

We all have a firm intuitive idea of the meaning of *length.* And although we have a similar feeling for *time,* it is more difficult to give expression to this concept in words. One possible definition: "*Time* is that which takes place between two events."

In order to *measure* time, we must have a series of regularly spaced *events,* such as the ticks of a clock. Ancient peoples used the apparent motion of the Sun as a crude clock. The interval between sunrise and sunset was reckoned to be *one day.* The Egyptians further divided the day and the night into 12 hours each, using shadow clocks (sun dials) to keep track of the daylight hours. But in this system the hours are not of equal duration because the length of the day changes with the seasons. Early attempts to reproduce constant fractions of a day included measuring the level of water in a large vat as water was allowed to trickle out through a small hole at the bottom.

Sun dials and water clocks eventually gave way to mechanical clocks. About 1300 A.D., the *escapement clock* was invented in which a toothed wheel, driven by a set of weights or a spring, engages a ratchet to regulate its turning. This device is basic to the operation of all mechanical clocks, even the modern variety. By the early 18th century, the great English clockmaker John Harrison had produced a clock for navigational purposes that maintained an accuracy of 15 seconds during a 5-month sea voyage—this was the first true *chronometer,* or precision clock.

The next important advance in timekeeping occurred in this century with the introduction of rapidly vibrating systems, such as tuning forks or quartz crystals, to regulate the motion of clock mechanisms. Miniaturized tuning-fork and quartz-crystal devices have recently been developed for use in wristwatches (see the photograph). Tuning-fork regulation can achieve an accuracy of about 1 second per day. Crystal-controlled clocks are capable of an accuracy of 1 part in 100 000 000 (10^8), which corresponds to 1 second in 3 years.

Even a precision as high as that possible with crystal control is not sufficient for many scientific purposes. Within the last few years methods that depend on *atomic* vibrations have been developed for controlling clocks. In fact, since 1967 the international standard of time has been based on the vibrations of cesium atoms. Thus, we now have atomic standards for two of the fundamental units of measure: the meter and the second. The various time units that we use are listed in Table 1-6 and the range of time intervals in the Universe is shown in Table 1-7.

BULOVA WATCH COMPANY, INC.

A quartz-crystal regulated wristwatch. The digital readout is by means of light emitting diodes.

TABLE 1-7 RANGE OF TIME INTERVALS IN THE UNIVERSE
(ALL TIMES ARE APPROXIMATE.)

Age of the Universe	1 000 000 000 000 000 000 s = 10^{18} s
Age of the Earth	100 000 000 000 000 000 s = 10^{17} s
Age of the Pyramids	100 000 000 000 s = 10^{11} s
Lifetime of a man	1 000 000 000 s = 10^9 s
4 months	10 000 000 s = 10^7 s
Light travels from Sun to Earth	1000 s = 10^3 s
Interval between heartbeats	1 s = 10^0 s
Period of a sound wave (typical)	0.001 s = 10^{-3} s
Period of a radio wave (typical)	0.000 001 s = 10^{-6} s
Light travels 1 foot	0.000 000 001 s = 10^{-9} s
Period of atomic vibration (typical)	0.000 000 000 000 001 s = 10^{-15} s
Period of nuclear vibration (typical)	0.000 000 000 000 000 000 001 s = 10^{-21} s

Mass

Unlike length and time, the third fundamental physical quantity — *mass* — is associated with and is an intrinsic property of *matter*. In fact, the mass of an object is a measure of the amount of matter in the object. We could specify the mass of a bar of gold, for example, in terms of the number of gold atoms in the bar. Because all gold atoms found in Nature are absolutely identical, any other gold bar that contains exactly the same number of gold atoms will have exactly the same mass. The mass of any amount of gold could be determined by counting the number of atoms in the sample. The counting operation is well defined and so we have a precise method of comparing the masses of different gold samples. The procedure could be extended to other materials by measuring the mass of every other type of atom in terms of the mass of the gold atom. All mass determinations could therefore be based on an atomic gold standard. There is, however, an obvious flaw in the argument: we know of no way to count precisely the number of atoms in any bulk sample of material because even 1 gram of gold contains about 3×10^{21} atoms! Although we can *compare* the masses of two different types of atoms (indeed, this can be done with high precision), we have no way to relate such atomic comparisons to comparisons of bulk samples of the materials. That is, we can measure, for example, the *relative* masses of atoms of gold and aluminum, but this knowledge does not assist us in determining the mass of an aluminum bar in terms of the mass of a gold atom or a gold bar. Therefore, we do not yet have a truly *atomic* standard for mass as we do have for length and time.

Since 1889 the international standard of mass has been a cylinder of platinum–iridium, housed in the International Bureau of Weights and Measures, and designated as *1 kilogram*. The United States standard is Kilogram No. 20 (see the photograph), which is located at the National Bureau of Standards, Gaithersburg, Maryland.

Although the *kilogram* is the standard unit of mass in the metric system, we will, for convenience in our discussions, sometimes use the smaller unit, the *gram* (1 kg = 1000 g). The relationship connecting the kilogram and the British mass unit is (approximately)

$$1 \text{ pound (lb)} = 0.454 \text{ kilogram (kg)}$$

Kilogram No. 20, the standard of mass for the United States. The cylinder of platinum–iridium is 39 mm in diameter and 39 mm high. This secondary standard was compared with the international standard in 1948 and was found to be accurate to within 1 part in 50 million (5×10^7).

For many purposes it is sufficient to use the approximate value, 1 kg = 2.2 lb.

The range of masses that we find in the Universe is even greater than those for length and time. The least massive object known is the electron, $m = 9.1 \times 10^{-31}$ kg, whereas the mass of the entire Universe is estimated to be about 10^{50} kg—a span of 80 factors of ten! The masses of some important objects are given in Table 1-8.

The Metric Units of Measure

In the metric system of units the fundamental physical quantities—length, time, and mass—are measured in the following units:

Length:	Meter (m) or centimeter (cm); 1 m = 100 cm
Time:	Second (s)
Mass:	Kilogram (kg) or gram (g); 1 kg = 1000 g

To convert a quantity from metric measure to British measure or vice versa, we need only two conversion factors:

$$1 \text{ in.} = 2.54 \text{ cm}$$
$$1 \text{ lb} = 0.454 \text{ kg} \tag{1-1}$$

A Derived Quantity: Density

The fundamental quantities, length, time, and mass, can be combined in various ways to provide units for different physical quantities. For example, as we will see, *speed* or *velocity* is measured in terms of *length per unit time* (miles per hour, meters per second, or some other combination). We will encounter many of these *derived* quantities as we proceed with our discussions. In addition to velocity, we will use acceleration, force, momentum, work, energy, power, and several others. Even though we will attach special names to the units for many of these quantities, it should be remembered that the *fundamental* definition of any physical quantity can always be made in terms of length, time, and mass.

As an example of a derived quantity, let us consider a simple but important case: *density*. If we cut a bar of

TABLE 1-8 SOME IMPORTANT MASSES

OBJECT	MASS (kg)
Sun	1.991×10^{30}
Earth	5.977×10^{24}
Moon	$7.35 \ \times 10^{22}$
Hydrogen atom	1.673×10^{-27}
Electron	9.108×10^{-31}

iron into a number of pieces with various sizes and shapes, the pieces will all have different masses. But each piece still consists of *iron* and it must have some property that is characteristic of iron. If one of the pieces is twice as large as another piece, the mass must also be twice as great. A piece three times as large would have three times the mass, and so forth. That is, the ratio of the mass to the volume is constant for a particular substance — this ratio is called the *density:*

$$\frac{\text{mass}}{\text{volume}} = \text{density}$$

or, in symbols,

$$\boxed{\frac{M}{V} = \rho} \tag{1-2}$$

Mass is measured in kilograms (or grams) and volume is measured in cubic meters (or cubic centimeters). Therefore, the units of density are kg/m^3 or g/cm^3. Some representative densities are listed in Table 1-9. Notice that the density of water is 1.00 g/cm^3. In fact, the kilogram was originally defined as the mass of 1000 cm^3 of water. It is probably easier to think of densities in terms of g/cm^3, instead of kg/m^3 because in these units the density of water is 1.

What do we know about the density of the Earth? If we pick up a rock and measure its mass and volume, we will find a density of 2 or 3 g/cm^3. A different kind of rock will have a different density. Furthermore, the interior of the Earth is believed to consist of molten iron with a very high density. Thus, various parts and pieces of the Earth have different densities. If we wish to find *the* density of the Earth, then we must be content with an *average* density. That is, we divide the *total* mass of the Earth by the *total* volume.

The radius of the Earth is 6.38×10^6 m. Therefore,

$$V = \tfrac{4}{3}\pi R^3 = \tfrac{4}{3}\pi (6.38 \times 10^6 \text{ m})^3$$
$$= 1.08 \times 10^{21} \text{ m}^3$$

And the mass of the Earth is

$$M = 5.98 \times 10^{24} \text{ kg}$$

Thus, the average density is

$$\rho = \frac{M}{V} = \frac{5.98 \times 10^{24} \text{ kg}}{1.08 \times 10^{21} \text{ m}^3}$$
$$= 5.5 \times 10^3 \text{ kg/m}^3$$

TABLE 1-9 DENSITIES OF SOME MATERIALS

MATERIAL	DENSITY (g/cm³)	(kg/m³)
Gold	19.3	1.93×10^4
Mercury	13.6	1.36×10^4
Lead	11.3	1.13×10^4
Iron	7.86	7.86×10^3
Aluminum	2.70	2.70×10^3
Water	1.00	1.00×10^3
Air	0.0013	1.3

or,

$$\rho = 5.5 \text{ g/cm}^3$$

Notice that this result confirms the indirect evidence that the interior of the Earth has a high density. The materials found on or near the surface have densities near 3 g/cm³. Therefore, the density of the interior must be quite high in order to make the *average* density equal to 5.5 g/cm³.

SUGGESTED READINGS

J. B. Conant, *Science and Common Sense.* (Yale Univ. Press, New Haven, Connecticut, 1951).

G. C. Gillespie, *The Edge of Objectivity.* (Princeton Univ. Press, Princeton, New Jersey, 1960).

Scientific American articles:

A. V. Astin, "Standards of Measurement," June 1968.

Lord Ritchie-Calder, "Conversion to the Metric System," July 1970.

QUESTIONS AND EXERCISES

1. Write down as many different units of length (modern or ancient) as you can remember or can find in a dictionary or encyclopedia. You should have no difficulty in finding 15 or 20.

2. An *acre* is defined to be 43 560 square feet (ft²). How many square meters are there in one acre?

3. How many feet are there in one kilometer?

4. A sprinter runs the 100-m dash in 10.0 s. What would be his time at 100 yd? (He runs each race at the same speed.)

5. Convert 3.5 miles to meters.

6. Express the age of the Earth (4.5 billion years) in seconds.

7. What is the mass of 1 cubic foot of water? Express the result in kilograms and in pounds.

8. Assume that the Sun consists entirely of hydrogen. (This is approximately correct.) Use the data in Table 1-8 and compute the number of hydrogen atoms in the Sun.

9. A 2 m $\times$ 3 m plate of aluminum has a mass of 324 kg. What is the thickness of the plate?

10. What is the mass of air in a room that measures 5 m $\times$ 8 m $\times$ 3 m?

11. Use the information in the caption of the photograph of Kilogram No. 20 and compute the density of the platinum–iridium material used to make the mass standard.

12. Use the data in Tables 1-5 and 1-8 and calculate the average density of the Moon. The surface rocks of the Moon brought back by the Apollo astronauts have densities near 3 g/cm³. Does it seem reasonable that the Moon has a high-density interior as does the Earth?

13. Two blocks, one of lead and one of aluminum, have the same mass. What is the ratio of their *volumes?*

2

THE MICROWORLD
OF PHYSICS

Matter in a variety of forms is all around us—the Earth, the seas, the air, and the materials from which our homes and cities are constructed. What makes up this matter—what is the *stuff* of which matter is composed? In the 19th century, chemists established the existence of the chemical elements, and many of the facts regarding chemical processes were explained on the basis of an *atomic* description of matter. Only during the last 50 years have we learned how atoms combine into molecules of various sorts and how protons and neutrons bind together to form the nuclear cores of atoms.

In this chapter we will discuss the fundamental building blocks of our material world—electrons, protons, and neutrons—and how they combine to form atoms, molecules, and nuclei. This is an introduction to the microworld of physics. In later chapters we will return to this subject and treat in more detail the inner workings of atoms and the structure of matter.

2-1 BASIC UNITS OF MATTER

Elements

Aristotle taught that all matter consists of varying proportions of four basic elements: earth, air, fire, and water. But even the ancient alchemists knew that there are certain substances other than Aristotle's four elements, substances that defy all attempts to break them down into simpler components. One of these substances is copper, a metal which was known to the Sumerians in about 3000 B.C. Some other materials have been known for almost as long and are mentioned in the Old Testament: silver, gold, and sulfur (called *brimstone* in the Bible). The metals tin, mercury, iron, and lead were also known to ancient Man.

These substances—copper, silver, gold, sulfur, tin, mercury, iron, and lead—which have been known and used for thousands of years, we now recognize as members of a class called *elements*, substances that

TABLE 2-1 THE CHEMICAL ELEMENTS[a]

ELEMENT	SYM-BOL	ATOMIC NUM-BER	ELEMENT	SYM-BOL	ATOMIC NUM-BER	ELEMENT	SYM-BOL	ATOMIC NUM-BER
Actinium	Ac	89	Hafnium	Hf	72	Promethium	Pm	61
Aluminum	Al	13	Helium	He	2	Protactinium	Pa	91
Americium	Am	95	Holmium	Ho	67	Radium	Ra	88
Antimony	Sb	51	Hydrogen	H	1	Radon	Rn	86
Argon	Ar	18	Indium	In	49	Rhenium	Re	75
Arsenic	As	33	Iodine	I	53	Rhodium	Rh	45
Astatine	At	85	Iridium	Ir	77	Rubidium	Rb	37
Barium	Ba	56	Iron	Fe	26	Ruthenium	Ru	44
Berkelium	Bk	97	Krypton	Kr	36	Samarium	Sm	62
Beryllium	Be	4	Lanthanum	La	57	Scandium	Sc	21
Bismuth	Bi	83	Lawrencium	Lw	103	Selenium	Se	34
Boron	B	5	Lead	Pb	82	Silicon	Si	14
Bromine	Br	35	Lithium	Li	3	Silver	Ag	47
Cadmium	Cd	48	Lutetium	Lu	71	Sodium	Na	11
Calcium	Ca	20	Magnesium	Mg	12	Strontium	Sr	38
Californium	Cf	98	Manganese	Mn	25	Sulfur	S	16
Carbon	C	6	Mendelevium	Md	101	Tantalum	Ta	73
Cerium	Ce	58	Mercury	Hg	80	Technetium	Tc	43
Cesium	Cs	55	Molybdenum	Mo	42	Tellurium	Te	52
Chlorine	Cl	17	Neodymium	Nd	60	Terbium	Tb	65
Chromium	Cr	24	Neon	Ne	10	Thallium	Tl	81
Cobalt	Co	27	Neptunium	Np	93	Thorium	Th	90
Copper	Cu	29	Nickel	Ni	28	Thulium	Tm	69
Curium	Cm	96	Niobium	Nb	41	Tin	Sn	50
Dysprosium	Dy	66	Nitrogen	N	7	Titanium	Ti	22
Einsteinium	Es	99	Nobelium	No	102	Tungsten	W	74
Erbium	Er	68	Osmium	Os	76	Uranium	U	92
Europium	Eu	63	Oxygen	O	8	Vanadium	V	23
Fermium	Fm	100	Palladium	Pd	46	Xenon	Xe	54
Fluorine	F	9	Phosphorus	P	15	Ytterbium	Yb	70
Francium	Fr	87	Platinum	Pt	78	Yttrium	Y	39
Gadolinium	Gd	64	Plutonium	Pu	94	Zinc	Zn	30
Gallium	Ga	31	Polonium	Po	84	Zirconium	Zr	40
Germanium	Ge	32	Potassium	K	19	(Unnamed)	?	104
Gold	Au	79	Praseodymium	Pr	59	(Unnamed)	?	105

[a] The atomic number of an element indicates the number of electrons possessed by an electrically neutral atom of the element. This method of specifying the ordering of the elements will prove useful when atomic structure is discussed.

cannot be reduced by any chemical means to simpler parts. We now know that there are 92 different natural elements, and that a dozen or so more can be produced artificially in the laboratory. (A list of the known elements, together with their chemical symbols and atomic numbers, is given in Table 2-1.) Although the number of *elements* is relatively small, these elements can be combined in various ways to produce the molecules of a truly enormous number of chemical compounds.

Atoms

An element cannot be separated into any simpler chemical constituents. What happens, then, if we divide a sample of an element into smaller and smaller pieces? Can we continue this process indefinitely and produce an arbitrarily small sample of the element? The answer to these questions was anticipated by the Greek philosopher Democritus (485–425 B.C.) who argued that all matter must be corpuscular in character. Democritus' reasoning was based on philosophical, not scientific grounds; he found it impossible to understand that *permanence* (that is, *matter*) and *change* can exist in the same world unless all matter consists of ultimate particles that can be rearranged as the result of change. These ultimate particles are *atoms,* the smallest bits of matter that retain the properties of an element.

The first scientific argument regarding the existence of atoms was made by John Dalton (1766–1844), an English chemist. Dalton's reasoning was based on a discovery that had been made by Antoine Lavoisier (1743–1794) concerning the way in which elements combine to form molecules. Lavoisier had found that when two elements combine to produce a distinct chemical compound, they always combine with a definite ratio of masses. For example, when hydrogen combines with oxygen to form water, each gram of hydrogen always combines with 8 grams of oxygen to produce 9 grams of water. All other chemical reactions conform to this same scheme. Dalton realized that this rule of Lavoisier (called the *law of definite proportions*) has a far-reaching consequence. How can it be that 1 g of hydrogen always requires 8 g of oxygen in order to be completely converted to water with no hydrogen or oxygen remaining? The answer must be that $\frac{1}{2}$ g of hydrogen combines with 4 g of oxygen; $\frac{1}{4}$ g of hydrogen combines with 2 g of oxygen; $\frac{1}{8}$ g of hydrogen combines with 1 g of oxygen; and so on. That is, there is some

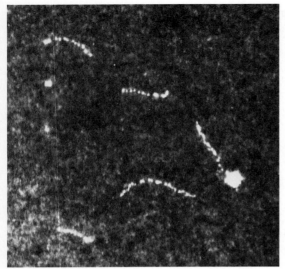

This photograph was made by Professor Albert V. Crewe with an electron microscope at the University of Chicago and shows several series of individual thorium atoms which are attached to long-chain molecules.

fundamental unit of hydrogen that always combines with some fundamental unit of oxygen in such a way that the hydrogen-to-oxygen mass ratio is always 1:8. Dalton therefore concluded that Lavoisier's law implies the existence of fundamental units of matter for all elements — these units are *atoms*.

Atoms are far too small to be visible even with the most powerful optical microscopes. The tiny one-celled organisms that we *can* see with a microscope contain about a billion billion (10^{18}) atoms! A typical atomic size is about 10^{-10} m, or about $\frac{1}{5000}$ of the wavelength of visible light. How, then, do we know that atoms really exist? Until recently, all our evidence was indirect (although still conclusive). But now the development of powerful new electron microscopes has enabled us, for the first time, actually to observe individual atoms. The technique is limited, however, and only the largest atoms can be clearly identified. Photographs have been taken, for example, of thorium atoms, as shown on this page. Electron microscopes are playing an increasingly important role in our efforts to understand the detailed structure of the matter which makes up our world.

Molecules

When two or more atoms join together, a *molecule* is formed. All *compounds* occur as molecules. Some *elements* also occur in molecular form. Chlorine, for example, does not occur naturally as separate atoms; chlorine gas is always in *molecular* form, two chlorine atoms bound together as a chlorine molecule. The smallest unit of matter identifiable as chlorine is the chlorine atom, but as found in Nature, chlorine invariably exists as *molecules* (Fig. 2-1b).

The *noble gases* (helium, neon, argon, krypton, xenon, and radon) generally exist as atoms; these elements are called *monatomic* (*one*-atom) gases (Fig. 2-1a). All other gaseous elements (for example, hydrogen, nitrogen, oxygen, and chlorine) occur as *diatomic* (*two*-atomic) molecules.

We can now make clear the distinction between atoms and molecules. An *atom* is the smallest unit of matter than can be identified as a certain chemical element. A *molecule* is the smallest unit of a given substance (an element or a compound) that exists in Nature. An atom is always an element; a molecule can be either an element or a compound.

In order to simplify the way in which we express the

Figure 2-1 (a) The gas helium occurs in *atomic* form. (b) The gas chlorine occurs as diatomic *molecules*.

composition of molecules, we use the following scheme. First, we use the element symbols in Table 2-1; for example, Cl stands for the element chlorine. Then, we use a subscript number to indicate the number of atoms of the element that occur in each molecule of the substance. Chlorine gas, for example, consists of molecules that contain two atoms of chlorine; therefore, the molecular symbol for chlorine is Cl_2.

A *water* molecule consists of two atoms of hydrogen (symbol, H) and one atom of oxygen (symbol, O), so the formula for water is H_2O. A molecule of *ammonia* consists of one nitrogen atom and three hydrogen atoms; the chemical formula is NH_3. Carbon and oxygen atoms combine in two different ways to form molecules. When two oxygen atoms combine with one carbon atom, *carbon dioxide* (CO_2) is formed. But when only one oxygen atom combines with a carbon atom, *carbon monoxide* (CO) is formed. Both CO and CO_2 are colorless, odorless gases. Carbon dioxide is used by plants in the growing process. Carbon monoxide, on the other hand, is a poisonous gas, often emitted in the exhaust fumes of automobile engines.

Carbon and oxygen combine as CO and CO_2 but not as C_2O or CO_4 or C_2O_5. After we have learned more of the atomic structure of matter we will see, in Chapter 19, the reason why atoms join together only in certain ways to produce molecules.

Electrons

The existence of atoms and molecules was deduced from chemical experiments that had been carried out with elements and compounds. Even after it was recognized that an atom represents the smallest bit of matter that can be identified as a particular chemical element, there was still no hint regarding the possible inner structure of atoms. No one knew whether atoms contained still smaller and more fundamental pieces of matter.

In 1897, J. J. Thomson (1856–1940) published a report of his experiments in which he had identified the *electron* as a basic constituent of matter. Indeed, the modern approach to the structure of matter begins with Thomson's discovery of the electron.

Thomson's experiments involved the study of rays (called *cathode rays*) that stream through the gas in a partially evacuated glass tube when the metal electrodes placed in the ends of the tube are given a high voltage with respect to one another. Thomson discovered that

Sir J. J. Thomson (1856–1940), discoverer of the electron and winner of the 1906 Nobel Prize in physics.

cathode rays can be deflected by electric and magnetic fields. He found that cathode rays are repelled by a negatively charged plate and are attracted toward a positively charged plate. Because electrical charges of the same sign repel one another and charges of different signs attract one another, Thomson concluded that cathode rays consist of *negative* electrical charges.

In his experiments, Thomson used different materials for the wires and plates which carried the high voltage, and he filled the tube with different gases. No matter what changes were made, the cathode rays always behaved in exactly the same way. Thomson concluded that cathode rays (which he called *electrons*) originate in matter, but they are not characteristic of the *type* of matter. Because electrons are all identical and are common to all types of atoms, they must be fundamental bits of matter. Thomson had been able to show:

(a) Cathode rays consist of *electrons* which are identical and are common to all types of matter.

(b) Electrons carry a *negative* charge.

(c) Electrons have a far smaller *mass* then even the lightest atom, hydrogen.

Further studies of the properties of electrons by the American physicist Robert A. Millikan (1868–1953) established the fact that all electrons carry exactly the *same* electrical charge. We denote the magnitude of the electron charge by the symbol e, and we measure the charge in terms of a unit called the *coulomb* (C):

$$\text{electron charge} = -e$$
$$e = 1.602\ 192 \times 10^{-19}\ \text{C}$$

with an uncertainty only in the last decimal place. For our purposes in this book we will use the approximate value,

$$e = 1.60 \times 10^{-19}\ \text{C}$$

We will discuss the electron charge and the unit of charge in more detail in Chapter 6 when we treat the subject of electrical forces.

Ions

Every normal atom carries equal amounts of positive and negative electrical charge; normal atoms therefore have *zero net charge* and are said to be electrically *neutral*. If an electron is removed from an atom, the atom, having lost a charge $-e$, then carries a net charge of $+e$.

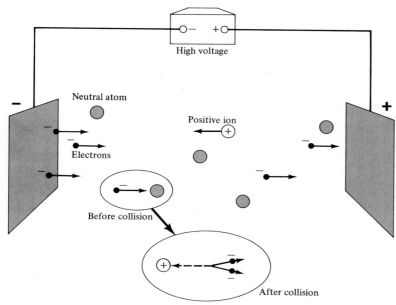

Figure 2-2 An electron in a cathode-ray beam strikes an electrically neutral gas atom and removes one of the atomic electrons. A positively charged ion results which adds to the positive-ray beam and the extra electron contributes to the cathode-ray beam. If the incident electron removes *two* electrons from the gas atom, we say that the atom is *twice ionized* and carries a charge of $+2e$.

Such atoms are called *ions*. (An ion that carries a charge $+e$ is said to be *singly ionized;* if two electrons are removed, the ion is *doubly ionized* and carries a charge $+2e$; and so forth.) Thomson also studied the positively charged ions produced when electrons strike gas atoms and knock out atomic electrons (Fig. 2-2). He found that these ions retain the properties of the original atomic species even though they have lost one or more electrons. Thus, there must be something else in atoms, in addition to electrons, that gives to atoms their particular characteristics. But this "something else" was not discovered until more than 15 years later.

2-2 THE NUCLEAR ATOM

Rutherford's Experiments with Alpha Particles

Soon after the discovery of *radioactivity* in 1896 (see Chapter 20), Ernest Rutherford (1871–1937) began a study of the radiations emitted in the process called α

Ernest Rutherford, leader of the early attempts to solve the mysteries of radio-activity. He identified the α particle as an ion of helium, established the nuclear model of the atom, and was the first to produce a nuclear reaction in the laboratory. For his work on radioactivity, Rutherford was awarded the 1908 Nobel Prize in chemistry. In 1931 he was made Lord Rutherford, Baron of Nelson.

Almost all of the milestone events in the early history of nuclear physics—the identification of radioactivity as a *nuclear* phenomenon, the proof that α particles are identical to helium ions, the nuclear model of the atom, and the first artificial nuclear disintegration—are associated with Lord Rutherford. When Rutherford's good friend and colleague A. S. Eve once playfully criticized him by charging that he only rode the crest of a wave, Rutherford responded, "Well, I made the wave, didn't I?"

radioactivity or α decay. This process is one in which α *rays* (now called α *particles*) are spontaneously ejected with high speeds from various radioactive elements such as uranium and radium. Rutherford soon proved that α particles are identical to doubly ionized helium atoms. Furthermore, he realized that these high-speed particles could be used as a tool to explore the interiors of atoms. Assisted by Hans Geiger and Ernest Marsden, Rutherford undertook a series of experiments to study the way in which rapidly moving α particles are deflected by thin sheets of various materials.

When the experiments were carried out, it was discovered that more of the α particles were deflected through large angles than had been expected. Rutherford was startled to learn that deflections of more than $90°$ had been observed and that one α particle in about 20 000 of those incident was turned completely around by gold foil only 4×10^{-7} m thick! Rutherford's reaction to this unexpected result is contained in a remark made during one of the last lectures he gave: "It was quite the most incredible event that has ever happened to me in my life. It was almost as incredible as if you had fired a 15-inch shell at a piece of tissue paper and it came back and hit you."

Rutherford was at a loss to explain this strange result. An α particle (which is about 7000 times as massive as an electron) cannot be turned around in a collision with an electron any more than a bowling ball can be turned around in a collision with a marble. Nor can the result be explained by the deflection of an α particle through small angles many times within the foil to add up to a large-angle deflection—there were simply too many α

particles that emerged from the side of the foil nearest the source. Clearly, there are extremely strong forces at work within the atom, forces that had not been foreseen in any theory of atoms.

Rutherford's Nuclear Model

The reason for the large-angle deflections of α particles was at first mystifying. But by 1911 Rutherford had found the solution. He concluded that there is only one way in which there could be a force within an atom of sufficient strength to turn a fast-moving α particle completely around—the entire positive charge and most of the mass of the atom must be concentrated in a tiny central core, the *nucleus* of the atom (Fig. 2-3). Surrounding the nucleus there are the atomic electrons, each element having its own particular number of these electrons.

According to Rutherford's model, if an α particle passes through the outer portion of an atom (see Fig. 2-4), the distance from the nucleus is so great that the repulsion is weak and only a small deflection results. But if the α particle happens to be on a path that takes it directly toward the nucleus, the repulsion of the positive core is so great that the α particle can be deflected through a large angle. By applying his theory to the experimental results, Rutherford was able to conclude that nuclei must be many times smaller than atoms: a typical atomic size is about 10^{-10} m, whereas the size of a nucleus such as that of gold is 10 000 times smaller, or about 10^{-14} m. Atoms are mostly empty space!

Probing an atom with an α particle is much the same as probing a peach with a long needle. By noting that the needle strikes something hard in the middle of the peach, it would be possible to deduce the existence and the size of the peach pit without ever having seen it.

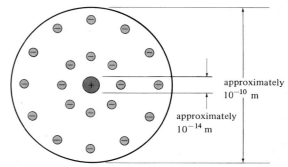

Figure 2-3 Rutherford's nuclear model of the atom. All of the positive charge and most of the mass of the atom are concentrated in a central nucleus whose size is only about 10^{-4} of that of the atom.

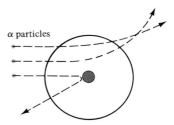

Figure 2-4 According to Rutherford's nuclear model of the atom, an α particle making a close collision with the nucleus can be deflected through a large angle. Usually, however, because of the small size of the nucleus, the α particle will not pass close to the nucleus and will suffer only a small deflection. Rutherford's analysis, based on a *nuclear* atom, is in complete agreement with the experimental results obtained by Geiger and Marsden.

2-3 THE COMPOSITION OF NUCLEI

Atomic Number

The question "Of what are atoms composed?" had been answered by Thomson and Rutherford: atoms consist of electrons and nuclei. And now there was an even more intriguing question to ask: "Of what are nuclei composed?"

Based on various chemical measurements, the elements had been placed in order according to increasing

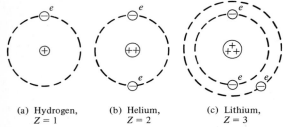

(a) Hydrogen, $Z = 1$ (b) Helium, $Z = 2$ (c) Lithium, $Z = 3$

Figure 2-5   The first three elements: hydrogen, helium, and lithium. Each atom has a number of atomic electrons equal to Z and each nucleus carries a positive charge of equal magnitude.

atomic mass. The lightest element, hydrogen, was given an *atomic number* of 1; helium was given atomic number 2; lithium was labeled with atomic number 3; and so on. By 1914 a series of experiments had established that the normal (un-ionized) atoms of each chemical element contain a particular number of electrons, with no two elements having the same number of atomic electrons. Each hydrogen atom contains just one electron; each helium atom contains two electrons; each lithium atom contains three electrons; and so on (see Fig. 2-5). Thus, the *atomic number* of an element is equal to the number of electrons in a normal atom of that element. The atomic number of an element is denoted by the letter Z. The first few chemical elements are

$$Z = 1 \quad \text{Hydrogen (H)}$$
$$Z = 2 \quad \text{Helium (He)}$$
$$Z = 3 \quad \text{Lithium (Li)}$$
$$Z = 4 \quad \text{Beryllium (Be)}$$
$$Z = 5 \quad \text{Boron (B)}$$
$$Z = 6 \quad \text{Carbon (C)}$$
$$Z = 7 \quad \text{Nitrogen (N)}$$
$$Z = 8 \quad \text{Oxygen (O)}$$

A complete list of atomic numbers (arranged alphabetically by element) is given in Table 2-1.

In every normal atom the negative charge of the atomic electrons must be exactly counterbalanced by the positive charge of the nucleus. Thus, the atomic number of an element not only specifies the number of electrons in an atom but also the amount of positive charge in the nucleus. An element with atomic number Z has Z electrons with a total charge of $-Ze$ and a total nuclear charge of $+Ze$ (where e is the magnitude of the electronic charge).

Protons

The hydrogen atom is the simplest of all atoms, and the nucleus of the hydrogen atom (called a *proton*) is the simplest of all nuclei. A proton is a particle that carries an electrical charge of magnitude exactly equal to that of an electron but of opposite sign; the mass of a proton is 1836 times the mass of an electron:

$$\text{electron charge} = -e = -1.60 \times 10^{-19} \text{ C}$$
$$\text{proton charge} = +e = +1.60 \times 10^{-19} \text{ C}$$
$$\text{electron mass} = m_e = 9.11 \times 10^{-31} \text{ kg}$$
$$\text{proton mass} = m_p = 1836 m_e = 1.67 \times 10^{-27} \text{ kg}$$

Just as the electron carries the basic unit of negative charge, the proton carries the basic unit of positive charge.

Neutrons

The nucleus of an atom with atomic number Z must contain exactly Z protons. But on the basis of chemical mass determinations we know, for example, that the mass of an oxygen atom ($Z = 8$) is 16 times greater than the mass of a hydrogen atom. Thus, the 8 protons in an oxygen nucleus that are required to balance the charge of the atomic electrons contribute only *half* of the mass of that nucleus. What contributes the other half? The answer to the puzzle was provided in 1932 by the English physicist James Chadwick (1891–1974), who found that nuclei contain, in addition to protons, uncharged particles with a mass approximately equal to the proton mass. These neutral particles are called *neutrons*.

With Chadwick's discovery of the neutron, the facts concerning atomic masses finally fell into place. The oxygen nucleus, for example, does contain exactly $Z = 8$ protons and the remaining mass is contributed by 8 neutrons. And the helium nucleus contains $Z = 2$ protons and 2 neutrons, giving an atomic mass equal to 4 times that of hydrogen. *All* nuclei (except hydrogen) contain neutrons as well as protons.

The total number of particles in a nucleus (protons and neutrons) is called the *mass number* of the nucleus and is denoted by the letter A. The number of neutrons in a nucleus is $N = A - Z$.

Isotopes

Every nucleus of a given element must contain the same number (Z) of protons, but these nuclei can contain different numbers of neutrons. Most hydrogen atoms have nuclei that consist of a single proton, but a small fraction (about 0.015 percent) of the hydrogen atoms that occur in Nature have one neutron in addition to the proton in their nuclei. This "heavy hydrogen" is called *deuterium*

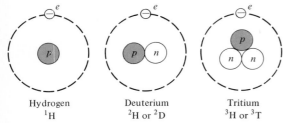

Hydrogen	Deuterium	Tritium
^{1}H	^{2}H or ^{2}D	^{3}H or ^{3}T

Figure 2-6   The three isotopes of hydrogen. Natural hydrogen consists primarily of the isotope with $A = 1$ and is labeled ^{1}H. The heavier isotopes, deuterium (^{2}H or ^{2}D) and tritium (^{3}H or ^{3}T) contain one neutron and two neutrons, respectively, in addition to the single proton. These schematic pictures of nuclei are actually not realistic. Quantum theory shows that nuclei (and atoms) are really "fuzzy" objects.

and is sometimes given its own chemical symbol (D). A third form of hydrogen atoms have nuclei containing *two* neutrons; hydrogen with $A = 3$ is called *tritium* (T), a radioactive species. The series of nuclei with the same value of Z but different values of A are called *isotopes* (Fig. 2-6).

The isotopes of a given element are distinguished by using a superscript to the element symbol to denote the mass number. For example, the oxygen isotopes found in Nature are the following:

^{16}O: 8 protons ($Z = 8$), 8 neutrons ($N = 8$), $A = 16$

^{17}O: 8 protons ($Z = 8$), 9 neutrons ($N = 9$), $A = 17$

^{18}O: 8 protons ($Z = 8$), 10 neutrons ($N = 10$), $A = 18$

(Older practice places the A number on the right instead of the left of the symbol: O^{16}, O^{17}, O^{18}.)

The isotopes of some of the light elements are listed in Table 2-2.

TABLE 2-2   THE ISOTOPES OF SOME LIGHT ELEMENTS

ELEMENT	Z	N	A	SYMBOL	REMARKS[a]
Hydrogen	1	0	1	^{1}H	Stable (99.985%)
	1	1	2	^{2}H or ^{2}D (deuterium)	Stable (0.015%)
	1	2	3	^{3}H or ^{3}T (tritium)	β Radioactive
Helium	2	1	3	^{3}He	Stable (0.00015%)
	2	2	4	^{4}He	Stable (99.99985%)
	2	4	6	^{6}He	β Radioactive
Lithium	3	3	6	^{6}Li	Stable (7.52%)
	3	4	7	^{7}Li	Stable (92.48%)
	3	5	8	^{8}Li	β Radioactive
Beryllium	4	3	7	^{7}Be	Radioactive
	4	4	8	^{8}Be	α Radioactive
	4	5	9	^{9}Be	Stable (100%)
	4	6	10	^{10}Be	β Radioactive
Boron	5	5	10	^{10}B	Stable (18.7%)
	5	6	11	^{11}B	Stable (81.3%)
	5	7	12	^{12}B	β Radioactive
Oxygen	8	8	16	^{16}O	Stable (99.76%)
	8	9	17	^{17}O	Stable (0.04%)
	8	10	18	^{18}O	Stable (0.20%)
	8	11	19	^{19}O	β Radioactive

[a] The numbers in parentheses are the relative natural abundances of the isotopes. The α- and β-radioactive isotopes do not occur in Nature and must be produced artificially. The isotope ^{7}Be (which must also be produced artificially) decays by still another mechanism; ^{7}Be captures into the nucleus one of the atomic electrons and becomes ^{7}Li. Radioactivity is discussed in Section 20-1.

The Atomic Mass Unit

The masses of nuclei are measured on a scale in which the mass of the atom of the most abundant carbon isotope (^{12}C: 6 protons and 6 neutrons) is exactly 12. Thus, we say that the mass of ^{12}C is 12 *atomic mass units* or 12 AMU. In metric units,

$$1 \text{ AMU} = 1.66 \times 10^{-27} \text{ kg} \qquad (2\text{-}1)$$

The proton and the neutron each have a mass of approximately 1 atomic mass unit:

$$\left.\begin{array}{l} \text{proton mass} = m_p = 1.0073 \text{ AMU} \\ \text{neutron mass} = m_n = 1.0087 \text{ AMU} \end{array}\right\} \qquad (2\text{-}2)$$

The mass of a nucleus is approximately (but not exactly) equal to the masses of the constituent protons and neutrons. Thus, the mass of ^{18}O is approximately 18 AMU and the mass of ^{235}U is approximately 235 AMU. In Chapter 20 we will learn how processes such as fission and fusion depend upon the fact that the nuclear mass is always slightly *less* than the combined mass of the constituent protons and neutrons.

2-4 ELEMENTARY PARTICLES

The Basic Ingredients of Matter

The increasingly detailed study of the nature of *matter* has revealed that

(a) All matter consists of *atoms*.
(b) All atoms consist of *electrons* and *nuclei*.
(c) All nuclei consist of *protons* and *neutrons*.

Moreover, similar studies of the nature of *light* (which we will discuss further in Chapters 15 and 17) have shown that

(a) Light, ultraviolet, infrared, X rays, and γ rays all consist of *waves,* and these waves are just different forms of *electromagnetic radiation.*
(b) When an atom emits light or an X ray, or when a nucleus emits a γ ray, the radiation emerges as a tiny bundle, called a *photon.*

We are therefore led to the conclusion that the composition and the behavior of all ordinary matter depends on only four basic units: *electrons, protons, neutrons,* and *photons*. These are the modern equivalent of Aristotle's four elements, earth, air, fire, and water.

Particles and Antiparticles

It is natural to inquire whether there exist any additional "elementary particles" and whether there is still another layer of "fundamentalism" that underlies the four basic units of ordinary matter. These questions began to be investigated during the 1930's and 1940's, and a series of startling discoveries were made. The first event of importance was the discovery in 1932 of a particle that appears to be the same as an electron except that it carries a *positive* charge. This new particle is called a *positron* and is usually indicated by the symbol e^+ (in order to distinguish it from the ordinary electron which is indicated by e or by e^-). Positrons do not exist in Nature but they can be produced in the laboratory.

Subsequent studies of the positron verified that this particle is indeed identical to the electron except for the opposite electrical charge. The positron is the electron's *antiparticle*. Further investigations of the concept of *antimatter* resulted in the discovery in the mid-1950's that both protons and neutrons also have corresponding antiparticles, the *antiproton* (symbol, $\bar{p}$) and the *antineutron* (symbol, $\bar{n}$). The antiproton carries a *negative* charge (that is, a charge *opposite* to that of the proton). The neutron and antineutron, although they have no electrical charge, nevertheless have opposite *magnetic* properties. In general, a particle is distinguished from its antiparticle partner by opposite *electromagnetic* properties.

Antiparticles are just as "elementary" as ordinary particles. Our world happens to be composed of ordinary matter (that is why it is "ordinary"!), but we can imagine a world that is composed of *antimatter*. For example, an atom of antihydrogen would consist of an antiproton and a positron. And a nucleus of antideuterium would consist of an antiproton and an antineutron. Nuclei of antideuterium and antihelium have actually been produced and identified in the laboratory. An antiworld would obey the same physical laws that we know and would operate in exactly the same way as our world. Although most of the evidence points to the conclusion that all of the galaxies of stars that we see in space are composed of ordinary matter, we cannot be certain that there are not some that consist of antimatter.

Neutrinos, Pions, and Muons

Early studies of radioactivity revealed that certain isotopes spontaneously emit particles later identified as electrons. This type of process is called β radioactivity or β decay. In the 1930's it was discovered that electrons (and positrons) are not the only particles emitted in β decay processes. In every β decay, accompanying the electron is a curious particle called a *neutrino*. The neutrino was not detected in the early studies of β radioactivity. In fact, it was not until 1953 that an extremely complex apparatus was constructed that permitted for the first time the direct detection of the neutrino. It is not surprising that the detection of the neutrino is difficult when it is realized that the neutrino not only interacts extremely weakly with matter but also has no mass and carries no electrical charge! In spite of the seeming "nothingness" of the neutrino, this elusive particle plays an important role in the type of nuclear interaction that leads to β decay.

Cosmic rays are produced in violent stellar explosions, travel through space, and enter the Earth's atmosphere at high speeds. These particles are primarily nuclei of hydrogen atoms (*protons*) but some heavier nuclei are also present. Studies of the collisions of cosmic rays with the nuclei of atmospheric atoms in the 1930's and 1940's led to the discovery of two new types of particles, quite different from any particle previously known. These particles are formed when a fast cosmic-ray particle collides with a nucleus. They exist for only a small fraction of a second and then decay, eventually producing ordinary electrons, neutrinos, and photons. In a high-speed collision between a cosmic-ray proton and a nucleus, the most likely event is the production of a *pi* (π) *meson* or *pion*. These particles occur in three forms with different electrical charge: positive (π^+), negative (π^-), or neutral (π^0). The charged variety of pions exist for only about 10^{-8} s before they undergo decay and form the second type of new particle, the *muon* (μ^+ or μ^-). (Neutral pions decay directly into two photons.) Muons, in turn, decay after about 10^{-6} s into electrons and neutrinos. Although pions live, in the free state, for only a hundred-millionth of a second, their properties have been closely studied. We now believe that pions are the principal agents responsible for the extremely strong force that exists between protons and neutrons and binds these particles together to form nuclei. Muons, on the other hand, are still mystery par-

This is the first photograph ever taken of an event initiated by a neutrino. The neutrino is incident from the left and interacts with a proton, producing a three-pronged event consisting of a muon, a pion, and a proton. The reaction is $\nu_\mu + p \rightarrow p + \mu^- + \pi^+$. The event takes place in a *liquid-hydrogen bubble chamber* in which the path of a charged particle is rendered visible by virtue of the tiny bubbles that form in its wake.

ticles—we have not yet discovered their fundamental role in Nature's scheme.

The investigation of pion and muon decays has shown that the neutrinos emitted in these events are different from the neutrinos emitted in nuclear β decay. Thus, there are two different types of neutrinos, electron neutrinos (ν_e) and muon neutrinos (ν_μ). Each of these neutrinos has an antiparticle partner ($\bar{\nu}_e$ and $\bar{\nu}_\mu$).

The Elementary Particle Zoo

Our list of elementary particles has now grown considerably from the original four particles. We now have two electrons (e^- and e^+), two protons (p and $\bar{p}$), two neutrons (n and $\bar{n}$), three pions (π^+, π^-, π^0), two muons (μ^+ and μ^-), four neutrinos (ν_e, $\bar{\nu}_e$, ν_μ, and $\bar{\nu}_\mu$), and the photon. This list is by no means complete. The detailed investigation of high-speed collisions of protons and

TABLE 2-3   ELEMENTARY PARTICLES AND THEIR ANTIPARTICLES[a]

NAME	PARTICLE	ANTI-ARTICLE	MASS (IN UNITS OF ELECTRON MASS)	DECAY OF PARTICLE (PRINCIPAL MODE)	HALF-LIFE[b] (s)
Photon	γ	(Same)	0	Stable	
Neutrino	ν_e	$\bar{\nu}_e$	0	Stable	
	ν_μ	$\bar{\nu}_\mu$	0	Stable	
Electron	e^-	e^+	1	Stable	
Muon	μ^-	μ^+	207	$e^- + \nu_\mu + \bar{\nu}_e$	2×10^{-6}
Pion	π^+	π^-	273	$\mu^+ + \nu_\mu$	2×10^{-8}
	π^0	(Same)	264	$\gamma + \gamma$	1.4×10^{-16}
Kaon	K^+	K^-	966	$\mu^+ + \nu_\mu$	0.8×10^{-8}
	K_1^0	$\overline{K_1^0}$	974	$\pi^+ + \pi^-$	6×10^{-11}
	K_2^0	$\overline{K_2^0}$	974	$\pi^+ + e^- + \bar{\nu}_e$	4×10^{-8}
Eta meson	η^0	(Same)	1073	$\pi^+ + \pi^- + \pi^0$	10^{-18}
Proton	p	$\bar{p}$	1836	Stable	
Neutron	n	$\bar{n}$	1839	$p + e^- + \bar{\nu}_e$	760
Lambda hyperon	Λ^0	$\overline{\Lambda^0}$	2182	$p + \pi^-$	2×10^{-10}
Sigma hyperons	Σ^+	$\overline{\Sigma^+}$	2328	$p + \pi^0$	6×10^{-11}
	Σ^0	$\overline{\Sigma^0}$	2332	$\Lambda^0 + \gamma$	$\sim 10^{-14}$
	Σ^-	$\overline{\Sigma^-}$	2341	$n + \pi^-$	10^{-10}
Xi hyperons	Ξ^0	$\overline{\Xi^0}$	2571	$\Lambda^0 + \pi^0$	2×10^{-10}
	Ξ^-	$\overline{\Xi^-}$	2583	$\Lambda^0 + \pi^-$	10^{-10}
Omega hyperon	Ω^-	$\overline{\Omega^-}$	3290	$\Lambda^0 + K^-$	8×10^{-11}

[a] The photon γ, the neutral pion π^0, and the eta meson η^0 are their *own* antiparticles.
[b] See Section 20-1 for a discussion of *half-life*.

nuclei has revealed more than a hundred additional types of short-lived particles. These particles have half-lives that range from 10^{-10} s down to about 10^{-23} s and masses up to 3 times the proton mass. Table 2-3 lists 37 of these elementary particles that have the longest lifetimes and are therefore the easiest to study. Many of the unstable particles have several possible decay modes; only the principal mode is shown in the table.

What is the significance of the large number of elementary particles that have been discovered? We now are confident that neutrinos are intimately connected with β, π, and μ decay processes and that pions and kaons (and perhaps heavier particles) are responsible for the force that binds nuclei together. But the reason for the existence of the multitude of other particles still eludes us. We continue to maintain our faith that there is some Grand Scheme that simply and clearly specifies the relationship of the elementary particles one to another. We have been permitted a glimpse, here and there, of powerful fundamental principles at work, but we have not yet succeeded in locating the key to the riddle of elementary particles.

2-5 MATTER IN BULK

In the latter part of this book we will concentrate on the atomic and nuclear aspects of matter. As we build up the basic physical ideas and principles that will permit us to pursue these discussions, we will draw our illustrations from the properties of matter in bulk form. That is, we will first study the way that aggregates of atoms and molecules behave; then, we will return to the details of the microworld of physics.

Whenever we discuss the behavior of a piece of bulk matter, we must remember that we are treating a collection of a very large number of atoms. The properties of any sample of bulk matter can be traced directly to the atomic constituents of the material and the way in which they are bound together. The magnetism of a piece of iron, the electrical conductivity of a copper wire, or the passage of light through a diamond all depend on the atomic character of the particular material.

The States of Matter

Ordinary matter can exist in three distinct states: solid, liquid, and gas. We are all familiar with the fact that

A high energy cosmic-ray particle is incident from the top on a nucleus in the photographic emulsion. A dense jet of pions is ejected in the direction of the incident particle. This photomicrograph spans a distance of 0.3 mm in the emulsion.

water can be found as liquid water, solid ice, or gaseous steam, depending on the temperature. Even a substance such as iron exhibits the same three states. At normal temperatures, iron is solid, but at a temperature of 1535 °C, solid iron melts and becomes a liquid; the boiling point of iron is 3000 °C, and at this temperature molten iron becomes a vapor of iron atoms.

Whether a substance exists in the solid, liquid, or gaseous form depends on the strength of the forces that exist between atoms of the material. In solids these forces are strong and hold the substance in a rigid structure. In gases the interatomic forces are almost nonexistent, and consequently, the atoms are free to move independently of one another. In liquids the forces are intermediate in strength; the substance is in a condensed state, but the atoms slide easily through the surrounding atoms.

When heat is added to a solid and the temperature rises, the atoms vibrate more rapidly but they maintain their same average positions. If the temperature is increased still further, there will come a point at which the atoms become sufficiently agitated that the interatomic bonds are broken and the substance becomes a liquid. The temperature at which a substance changes from the solid to the liquid state is called the *melting point.* For water this temperature is well defined (0 °C). Indeed, for all substances that exist in solid form as *crystals,* the melting point represents a temperature of sharp change in the state of the material. (Crystalline matter is matter that consists of atoms organized into regular and repeating patterns, as in salt crystals and diamonds; see Chapter 19.) Many of the substances that we see and use are crystals, although these crystals are often so small that we do not ordinarily think of the material as crystalline. A bar of iron, for example, does not appear to be a crystal; yet, if you examine a piece of iron with a microscope you will see that it consists of an enormous number of tiny crystalline pieces (microcrystals) that are welded tightly together. On the other hand, materials such as glass, plastics, or tar do not exist in crystalline form. Such materials do not exhibit sharply defined melting points. If you heat a piece of glass, you will find that it softens gradually instead of changing suddenly from solid to liquid. (In fact, glass is not really a solid at all—it is a supercooled liquid and tends to flow, very slowly and over long periods of time, even at normal temperatures.)

When a substance is melted, the bonds that hold the

atoms in the crystalline lattice must be broken. The breaking of these bonds requires the expenditure of an amount of energy which is different for different materials. Similarly, when a liquid boils (that is, changes to a gas), energy is required to overcome the bonds that hold together the atoms or molecules in the liquid. Every change of state involves changes in the pattern of interatomic forces and the expenditure (or release) of energy.

All of the ordinary substances with which we are familiar occur as solids, liquids, or gases. However, there exists an extremely important fourth state of matter called *plasma*. As a substance is heated, it passes first from the solid state to the liquid state and eventually into the gaseous state. Further heating causes the gas atoms to move more rapidly, and the collisions between atoms become more violent. When the speeds of the atoms have increased sufficiently, the collisions can cause electrons to be ripped off the atoms. At extremely high temperatures (millions of degrees), no electrons remain attached to the atomic nuclei. At this point we have a plasma—completely bare nuclei moving in a sea of free electrons. Plasma conditions can be achieved on Earth only in special laboratory equipment. But in stars, plasma is the normal, not the exceptional form of matter. The *fourth state of matter* is therefore the most common state throughout the Universe! We will have occasion to mention plasmas again in connection with the generation of power by fusion processes in reactors (Chapter 20).

Avogadro's Number

Even a small sample of bulk matter contains an enormous number of atoms. One cubic centimeter (1 cm^3) of iron, for example, contains more than 8×10^{22} atoms. The rule for finding the number of atoms (or molecules) in a sample is the following. First, we must remember that the smallest unit of a substance that occurs in Nature is the molecule. Our rule will therefore be stated in terms of the number of *molecules* in a sample. For a monatomic substance such as helium, the number of atoms in a sample is the same as the number of molecules. But for hydrogen gas (H_2), the number of atoms in a sample is twice the number of molecules.

Next, we define a quantity of a substance called a *mole*. To do so we need to know the *molecular mass* of the particular material. Table 2-4 gives some examples. Notice that the molecular mass of helium is the same as

TABLE 2-4 SOME MOLECULAR MASSES

SUBSTANCE	FORMULA	MOLECULAR MASS (AMU)
Hydrogen	H_2	2
Helium[a]	He	4
Carbon[a]	C	12
Oxygen	O_2	32
Aluminum[a]	Al	27
Iron[a]	Fe	56
Methane	CH_4	16
Water	H_2O	18
Ethyl alcohol	C_2H_5OH	46
Aluminum oxide	Al_2O_3	102

[a] In these cases the molecular mass is the same as the atomic mass.

BETTMANN ARCHIVE

Amedo Avogadro, Conte di Quaregna

the atomic mass, namely, 4 AMU. For hydrogen gas (H_2), the molecular mass is twice the atomic mass. In general, the molecular mass of any element or compound is the sum of the individual atomic masses. For example,

Molecular mass of water (H_2O)
$$= 2 \times \text{(atomic mass of hydrogen)}$$
$$+ 1 \times \text{(atomic mass of oxygen)}$$
$$= (2 \times 1 \text{ AMU}) + (1 \times 16 \text{ AMU})$$
$$= 18 \text{ AMU}$$

An amount of a substance which has a mass in grams equal to the molecular mass in AMU is called a *mole* of the substance. Thus, a mole of iron has a mass of 56 g and a mole of water has a mass of 18 g. One mole of a substance always has the *same number* of molecules. That is, 2 g of hydrogen gas contains the same number of molecules as 56 g of iron (*atoms* in this case) or 46 g of ethyl alcohol. The number of molecules in a mole is called *Avogadro's number* after Count Amedo Avogadro (1776–1856) who first hypothesized this important principle. The value of Avogadro's number is

$$N_0 = 6.022 \times 10^{23} \text{ molecules/mole}$$

Suppose that we wish to determine the number of molecules in 1 kg of water. The molecular mass of H_2O is 18 AMU; therefore, 1 kg of water amounts to $(1000 \text{ g})/(18 \text{ g/mole}) = 55.6$ moles. The number of molecules in the 1-kg sample is

$$\text{Number of molecules} = 55.6 \, N_0 = 55.6 \times (6.022 \times 10^{23})$$
$$= 3.35 \times 10^{25}$$

Atomic and Molecular Sizes

We are now in a position to perform an interesting calculation to estimate molecular sizes. Moreover, the method illustrates the point that an *approximate* calculation based on a simplified model can often be used to obtain a rough idea of the magnitude of a quantity without the necessity of performing a laborious computation. Approximate calculations and estimates are not to be scoffed at—indeed, they are of great value to the scientist in guiding his thinking and showing the way to approach a problem on a more fundamental basis.

From a knowledge of Avogadro's number, we can estimate the size, for example, of a water molecule. The mass of 1 mole of water is 18 g, and since the density of liquid water is 1 g/cm³, a mole of water will occupy 18 cm³. Therefore, a 1-cm³ sample of water will contain $\frac{1}{18}N_0$ molecules:

$$\text{Number of molecules of water in 1 cm}^3 = \frac{N_0}{18}$$

$$= \frac{6.022 \times 10^{23} \text{ molecules/mole}}{18 \text{ cm}^3/\text{mole}}$$

$$\cong 3 \times 10^{22} \text{ molecules/cm}^3$$

Imagine that 1 cm³ is divided into a large number of tiny cubes such that each cube contains just one water molecule. Clearly, each cube must have extremely small dimensions in order to fit 3×10^{22} cubes into 1 cm³. Each cube must have a side of length approximately 3.2×10^{-8} cm (or 3.2×10^{-10} m) for the total volume of 3×10^{22} such cubes to be

$$V = (3 \times 10^{22}) \times (3.2 \times 10^{-8})^3$$
$$= (3 \times 10^{22}) \times (32.8 \times 10^{-24})$$
$$= 3 \times 32.8 \times 10^{-2}$$
$$\cong 1 \text{ cm}^3$$

This crude calculation yields only an *estimate* of the size of a water molecule because each tiny cube could be *larger* than the volume actually occupied by a water molecule. Nevertheless, this estimate of about 3×10^{-10} m for the size of a water molecule is close to that measured by various precise techniques. A useful rule-of-thumb is that the size of an atom (and all atoms are *roughly* the same size) is about 10^{-8} cm $= 10^{-10}$ m. Molecules are larger and their sizes vary greatly—some molecules contain only two atoms but others (for example, protein molecules) can contain millions of atoms.

SUGGESTED READINGS

I. Asimov, *The Search for the Elements* (Basic Books, New York, 1962).

E. N. daC. Andrade, *Rutherford and the Nature of the Atom* (Doubleday, Garden City, New York, 1964).

Scientific American articles:

E. N. daC. Andrade, "The Birth of the Nuclear Atom," November 1956.

L. Badash, "How the 'Newer Alchemy' Was Received," August 1966.

QUESTIONS AND EXERCISES

1. Explain how Thomson was able to conclude that the cathode rays which he studied are *fundamental* bits of matter.

2. If two electrons are removed from an electrically neutral atom of ^{11}B, what will be the net charge on the resulting ion?

3. A certain isotope contains 7 protons and 8 neutrons. What is the isotope?

4. The atomic number of the metal *magnesium* is 12. There are three isotopes of magnesium that occur in Nature: ^{24}Mg, ^{25}Mg, and ^{26}Mg. How many protons and how many neutrons are there in each of these isotopes?

5. Naturally occurring hydrogen consists of two isotopes, and naturally occurring oxygen consists of three isotopes. The most common form of the water molecule is represented by the formula $^1H_2^{16}O$ and has a molecular mass of 18 AMU. There are 8 additional isotopic possibilities for the water molecule. Write down the formulas and the molecular masses for the other forms. Which isotopic species do you expect to be the *rarest* form of water? (Refer to Table 2-2.)

6. If two protons and two neutrons are removed from a nucleus of ^{16}O, what nucleus remains?

7. An α particle is the nucleus of a helium atom and therefore consists of two protons and two neutrons. When ^{212}Po undergoes α decay and emits an α particle, what new isotope is formed? (Use Table 2-1 in order to find the appropriate atomic numbers.)

8. ^{10}Be exhibits β radioactivity. When a nucleus of ^{10}Be emits an electron, what new element is formed?

9. What is the molecular mass of methanol (methyl alcohol), CH_3OH?

10. How many atoms are there in 1 cm³ of aluminum? (You will need the density of aluminum; see Table 1-9.)

11. What is the volume of 1 mole of water?

12. The mass of a mole of hydrogen sulfide (H_2S) is 34 g. What is the atomic mass of sulfur in AMU?

13. The English physicist, Lord Rayleigh (1842–1919), performed the following experiment in order to obtain an estimate of the size of an oil molecule. He allowed a droplet of oil (mass = 8×10^{-7} kg, density = 0.9 g/cm³) to spread out over a water surface. The area of the oil film was found to be 0.55 m². The oil would cover no greater area; any attempt to spread the film further resulted in tearing the film and leaving holes. The conclusion is that the film consists of a single molecular layer. Oil molecules are chainlike and have the property that only one end has an affinity for water. Therefore, when the oil is spread on water, the oil molecules "stand on their heads." The film thickness then corresponds to the *length* of an oil molecule. Use Lord Rayleigh's data and calculate the length of a molecule of the oil used in the experiment.

14. The density of air under normal conditions of pressure and temperature is 1.3 kg/m³. Air consists primarily of nitrogen gas (N_2, molecular mass = 28 AMU). What is the average distance between the molecules in air under normal conditions? (Assume that each molecule occupies an identical cubical volume and calculate the dimensions of the cube.)

3

MOTION

We live in a restless Universe. Everything around us—from the atoms that make up all matter to the distant galaxies of stars in space—is in ceaseless motion. Every physical process involves motion of some sort. Because *motion* is such an important feature of every physical process, it is the logical subject with which to begin our detailed study of physical phenomena. The ideas that are developed here will be used throughout this survey—in describing planetary motion, in discussing electrical current, and in studying the behavior of atoms and nuclear particles. *Motion* is at the heart of every physical process.

3-1 AVERAGE SPEED

Distance and Time

If an object is in one position at a certain time and is in a different position at a later time, we know that *movement* has occurred. How can we describe the details of movement in a meaningful way? When we take a trip by automobile and note the behavior of the speedometer, we see that we rarely travel very long at *constant* speed. For one reason or another, it is frequently necessary to slow down or speed up. By the time the trip is completed, we have traveled at many different speeds. But there is still one speed—the *average* speed—that can be applied to the entire trip. The idea of *average speed* draws upon two familiar concepts: distance and time. If our trip covered 30 miles and required 1 hour, we say that the average speed was 30 miles per hour, or 30 mi/h. That is,

$$\text{average speed} = \frac{\text{distance traveled}}{\text{time interval for the motion}} \qquad (3\text{-}1)$$

We can simplify this equation by substituting *symbols* for the words. We use

$$\bar{v} = \text{average speed}$$
$$x = \text{distance traveled}$$
$$t = \text{time interval for the motion}$$

Then, we can write

$$\bar{v} = \frac{x}{t} \qquad (3\text{-}2)$$

We use the symbol v for speed (instead of s) because we will shortly introduce the closely related concept of *velocity* which is customarily represented by v. The bar over the v indicates *average* speed.

The numerical value of average speed is obtained by dividing the number that represents the distance traveled by the number that represents the time interval. But average speed also has units or dimensions. The *dimensions* of average speed are those of distance divided by time. If distance is measured in miles and the time in hours, then the average speed is in miles/hour (miles per hour) or mi/h. Other possible dimensions for average speed are meters per second (m/s) or feet per minute (ft/min).

An automobile that travels 240 miles in 4 hours moves with an average speed of

$$\bar{v} = \frac{240 \text{ miles}}{4 \text{ hours}} = 60 \text{ mi/h}$$

And a sprinter who runs the 100-meter dash in 10 seconds moves with an average speed of

$$\bar{v} = \frac{100 \text{ meters}}{10 \text{ seconds}} = 10 \text{ m/s}$$

Notice the difference in the two ways that are used to express the average speeds in these examples. In the first case the result is given in mi/h and in the second case m/s is used. Does this matter? Yes and no. We cannot directly compare two speeds that are expressed

TABLE 3-1 SOME TYPICAL SPEEDS

Speed (m/s)   Speed (mi/h)

3×10^8 — Light (in vacuum)
10^8 — Distant galaxy (relative to Earth) 10^8
       Electron in hydrogen atom
10^6 10^6
       Earth in orbit
10^4 — X–15 rocket plane 10^4
       Sound (in air)
10^2 — Racing car 10^2
       Sprinter
1   Walking man   1

10^{-2}
       Snail
10^{-4}
       Rapidly moving glacier
10^{-6}
10^{-8}
       Growth of hair (human head)
10^{-10} Drifting apart of Africa and South America

in different units: 60 mi/h is *not* 6 times greater than 10 m/s. But, unless comparisons are to be made, there is nothing "wrong" with using different units. In fact, in this chapter we will express speeds in various British units (mi/h, ft/s) and also in metric units (m/s, km/s). In succeeding chapters we will gradually change over to the exclusive use of metric units.

Notice also that no mention was made in the examples of the *direction* of motion. The automobile in the first example could have made a more-or-less straight-line highway trip from Washington to New York or it could have made 120 circuits of a 2-mile racetrack. Later, we shall be concerned with the *direction* of motion, but for the purposes of discussing average speed, we need to consider only the *total distance traveled*, regardless of the direction of motion or whether the direction changed during the motion.

3-2 GRAPHICAL REPRESENTATION OF SPEED

Distance versus Time

One way in which we can represent the motion of an object is by plotting a graph of the distance versus the time. Figure 3-1 shows such a graph. In the time interval between $t = 0$ and $t = 2$ s, the object moves 6 m. Furthermore, between $t = 3$ s and $t = 5$ s, the object again moves 6 m. Therefore, the average speed for each of these time intervals is

$$\bar{v} = \frac{6 \text{ m}}{2 \text{ s}} = 3 \text{ m/s}$$

In this case it does not matter what time interval we choose; we always obtain $\bar{v} = 3$ m/s. Even if we choose the interval from $t = 0$ to $t = 4$ s, we find $\bar{v} = 12$ m/4 s $= 3$ m/s. We therefore conclude that the average speed is *always* 3 m/s; that is, the motion takes place with *constant speed*. Whenever we have a case in which the motion can be represented by a *straight line* in a distance–time graph, the motion takes place with *constant speed*.

Instantaneous Speed

Although straight-line graphs are important and occur often, we also find cases in which the plotted points define a *curve*. If a distance–time graph is not a straight

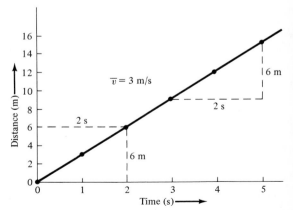

Figure 3-1 A straight-line distance–time graph represents motion at *constant speed*. In each 2-s interval, the object moves 6 m. Therefore, the average speed is $\bar{v} = 6$ m/2 s $= 3$ m/s.

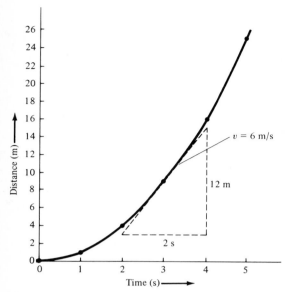

Figure 3-2 The *instantaneous speed* at a given instant of time is obtained by measuring the slope of the line that is tangent to the distance–time graph at that point. Here, $v = 12$ m/2 s $= 6$ m/s.

TIME (s)	DISTANCE (m)	INSTANTANEOUS SPEED (m/s)
0	0	0
1	1	2
2	4	4
3	9	6
4	16	8
5	25	10

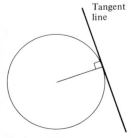

Figure 3-3

line, how do we determine the speed? Look at Fig. 3-2 which shows such a case. What does the upward curvature mean? As time goes on, the object travels a greater and greater distance in each second of motion – that is, the object is picking up speed. It moves a greater distance between $t = 1$ s and $t = 2$ s than it does between $t = 0$ and $t = 1$ s. If we take, for example, the first 2 seconds of motion, we find for the average speed, $\bar{v} = 4$ m/2 s $= 2$ m/s. But if we choose the first four seconds, we find $\bar{v} = 16$ m/4 s $= 4$ m/s. In this case the speed is continually changing and by considering different time intervals, we obtain different values for the average speed. In order to determine the speed precisely at $t = 3$ s (*not* the average speed between $t = 2$ s and $t = 4$ s), we must consider a very tiny time interval around $t = 3$ s. There is no restriction on how small an interval we may choose: we can imagine an interval of 10^{-2} s or 10^{-6} s or 10^{-15} s. By making the interval smaller and smaller, the average speed for that interval more and more nearly approaches the speed *exactly at* $t = 3$ s. This is the *instantaneous speed*.

If we choose a very small time interval around $t = 3$ s, we will not be able to see the straight line connecting the points. Therefore, we extend the line in both directions and obtain the sloping dashed line in Fig. 3-2. We now use this straight line as before to determine the speed – but now it is the *instantaneous speed at* $t = 3$ s. As shown on the graph, a 2-s time interval corresponds to a distance change of 12 m, so the instantaneous speed is $v = 12$ m/2 s $= 6$ m/s. (We use the symbol v for the *instantaneous* speed and we reserve the symbol $\bar{v}$ to indicate the *average* speed.) By following this procedure and obtaining the *slope* of the curve at various points, we can determine the instantaneous speed at any instant of time. The table at the left lists the distances and the instantaneous speeds at 1-s intervals for the distance–time curve in Fig. 3-2. (Verify as many of these values as you can. Be careful in drawing the slopes – it's not easy!)

The sloping dashed line in Fig. 3-2 is called the *tangent line*. We do not need to inquire as to the precise definition of the tangent line, but it is the line that just touches the curve at a given point. For the case of a circle, it is the line at right angles to the radius line drawn to the same point (see Fig. 3-3). For other types of curves, the tangent line may not be easy to draw because several different lines may appear to be just touching the curve. Try this for a few points on the curve in Fig. 3-2.

3-3 ACCELERATION

Rate of Change of Speed

In order to make an automobile go faster, you "step on the gas" and the speed of the car increases — you *accelerate*. (An automobile gas pedal is appropriately called an "accelerator.") Applying the brakes also causes an automobile to accelerate (in a negative sense) — braking involves slowing down. In each case there is a *change* in the speed, that is, an *acceleration*. Before we define acceleration, let us review the definition of speed.

The definition of *speed*, as we have seen, is the change in *position* per unit time:

$$\text{speed} = \frac{\text{position (or distance) change}}{\text{time required for change}}$$

If the object is at the position $x = x_0$ at time $t = t_0$ and is at $x = x_1$ at the later time $t = t_1$ (see Fig. 3-4), the average speed is

$$\bar{v} = \frac{x_1 - x_0}{t_1 - t_0} \qquad (3\text{-}3)$$

This is just Eq. 3-2 in more precise form. We can simplify this still further by writing $x_1 - x_0 = \Delta x$ (see Fig. 3-4) and $t_1 - t_0 = \Delta t$. The symbol Δx ("delta x") means "the change in x" (it does *not* mean "Δ multiplied by x"). Similarly, Δt means "the change in t." Then, we have

$$\boxed{\bar{v} = \frac{\Delta x}{\Delta t}} \qquad (3\text{-}4)$$

We can follow this same reasoning for the case of *acceleration*. The definition of acceleration is the change in *speed* per unit time:

$$\text{acceleration} = \frac{\text{change in speed}}{\text{time required for change}} \qquad (3\text{-}5)$$

If the speed is v_0 at $t = t_0$ and is v_1 at $t = t_1$, then the average acceleration is

$$\boxed{\bar{a} = \frac{v_1 - v_0}{t_1 - t_0} = \frac{\Delta v}{\Delta t}} \qquad (3\text{-}6)$$

We can see the similarity in the definitions of speed and acceleration by referring to Fig. 3-5. In Fig. 3-5a we have a *distance–time* graph, and in Fig. 3-5b we have a

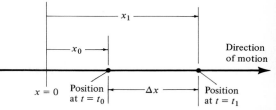

Figure 3-4   An object moves in the *x*-direction from x_0 to x_1; the change in position is expressed as $x_1 - x_0 = \Delta x$.

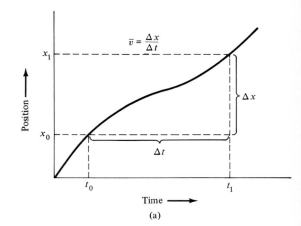

(a)

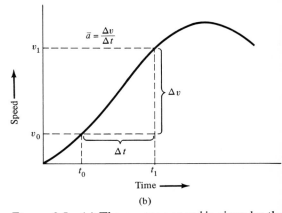

(b)

Figure 3-5   (a) The *average speed* is given by the change in *position* divided by the time. (b) The *average acceleration* is given by the change in *speed* divided by the time. Where the speed-time graph has an upward slope, the object is speeding up; where the slope is downward (or negative), the object is slowing down. Notice that in each case we refer to the *average* speed or *average* acceleration during the interval Δt because neither graph is a straight line.

speed–time graph. We obtain the average speed (or average acceleration) by dividing the change in position Δx (or the change in speed Δv) by the corresponding time interval Δt. Notice in Fig. 3-5a that the distance–time graph is not a straight line; this means that the instantaneous speed changes with time. Similarly, Fig. 3-5b shows that the instantaneous acceleration changes with time. During the early part of the speed–time graph, the slope is upward, indicating an increase in speed or positive acceleration. Near the end of the time interval covered by the graph, the slope is downward, indicating a decrease in speed or negative acceleration.

The Calculation of Acceleration

Let us apply Eq. 3-6 to the motion represented in the distance–time graph of Fig. 3-2 and in the accompanying table of data. The distance–time graph curves *upward,* indicating that the speed *increases* with time. We plot the instantaneous speeds shown in the table on page 38 to obtain the speed–time graph in Fig. 3-6. The fact that this graph is a straight line means that the motion takes place with *constant* acceleration. Consider the speeds at $t = 1$ s and at $t = 3$ s and apply Eq. 3-6:

$$a = \frac{6 \text{ m/s} - 2 \text{ m/s}}{3 \text{ s} - 1 \text{ s}} = \frac{4 \text{ m/s}}{2 \text{ s}} = 2 \text{ m/s}^2$$

Notice that the dimensions of acceleration are the dimensions of speed (m/s) divided by the dimensions of time (s); that is,

$$a = \frac{\Delta v}{\Delta t}$$

and the units are

$$\frac{\text{m/s}}{\text{s}} = \frac{\text{m}}{\text{s}^2}$$

so that acceleration is expressed in units of m/s². The result we have obtained means that the speed increases uniformly at a rate of 2 m/s *each* second. (In words, we say that the acceleration is "2 meters per second per second" or "2 meters per second squared.")

Acceleration can be expressed in other units — for example, cm/s², mi/h², or even (mi/h)/s — by multiplying the original result by the unity factors that convert meters to centimeters, meters to miles, seconds to hours, and so forth. For example,

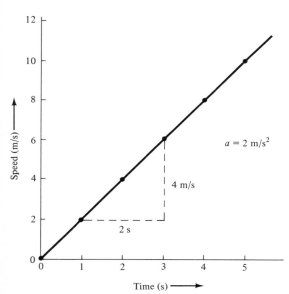

Figure 3-6   Speed–time graph for the motion represented in the distance–time graph in Fig. 3-2. The straight line indicates that the motion takes place with *constant* acceleration.

$$2 \text{ m/s}^2 = 2 \frac{\text{m}}{\text{s}^2} \times \frac{3600 \text{ s}}{1 \text{ h}} \times \frac{3600 \text{ s}}{1 \text{ h}} \times \frac{1 \text{ mi}}{1609 \text{ m}}$$

$$= 1.61 \times 10^4 \text{ mi/h}^2$$

Average Speed — Another Formula

When an object is uniformly accelerated, the speed increases at a constant rate (Fig. 3-6). In some cases we need to know the average speed during a certain interval of accelerated motion. How do we do this? We add the initial and final values and divide by 2. If an object has a speed v_0 at time zero and has a speed v at a time t, the average speed for the time interval ending at t is (see Fig. 3-7)

$$\bar{v} = \tfrac{1}{2}(v_0 + v) \tag{3-7}$$

We will find this expression for $\bar{v}$ useful in the calculation in the next section.

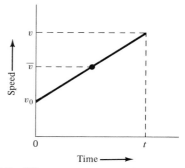

Figure 3-7 The average speed during the time t is $\bar{v} = \tfrac{1}{2}(v_0 + v)$.

3-4 ACCELERATED MOTION

Equations for the Analysis of Motion

Equations 3-4 and 3-6 are the basic definitions of speed and acceleration. But these equations alone do not permit us to analyze all types of motion. How can we use the defining equations to obtain expressions that can be applied to various cases of motion?

Suppose that we consider a case in which an object moves with constant acceleration. (This is the type of motion we will usually discuss in this book; when the acceleration is constant, we have $a = \bar{a}$.) If the acceleration is given and we are asked to find the speed, we start with the equation that defines acceleration. We begin describing the motion at an instant which, for convenience, we call time *zero*; that is, we set $t_0 = 0$. Then, we wish to find the speed v at some later time t. If the initial speed is v_0 (that is, $v = v_0$ at $t = 0$), the *change* in speed between time zero and t is $v - v_0$. Therefore,

$$a = \frac{\Delta v}{\Delta t} = \frac{v - v_0}{t - 0} = \frac{v - v_0}{t} \tag{3-8}$$

We can rewrite this expression for a by solving for v:

$$\boxed{v = v_0 + at} \tag{3-9}$$

This equation states that the instantaneous speed v at any time t is equal to the initial speed v_0 plus the additional speed at that is acquired by virtue of the constant acceleration. Equation 3-9 is not a new equation; it is the defining equation for the acceleration rewritten to express the way the speed changes with time.

Suppose that an object is started into motion with an initial speed of 40 ft/s and is subject to an acceleration of 20 ft/s². What will be the speed after 4 s? Using Eq. 3-9,

$$v = 40 \text{ ft/s} + (20 \text{ ft/s}^2) \times (4 \text{ s})$$
$$= 40 \text{ ft/s} + 80 \text{ ft/s} = 120 \text{ ft/s}$$

Equation 3-9 expresses the speed in convenient form. But this is not an essential equation and need not be memorized. The same results can always be obtained by using the basic equation, Eq. 3-8. How would the solution to this example look if Eq. 3-8 were used instead of Eq. 3-9?

Next, we consider a different question, but also for the case of constant acceleration. How far does an object travel in a certain time? Using our previous results for speed and acceleration, we can easily obtain an expression for the distance traveled by an object undergoing accelerated motion. If the object starts from the origin ($x_0 = 0$) at $t_0 = 0$, and moves to a position x at time t, then according to Eq. 3-4 the average speed is

$$\bar{v} = \frac{\Delta x}{\Delta t} = \frac{x}{t}$$

or, solving for the position x,

$$\boxed{x = \bar{v}t} \tag{3-10}$$

This equation is valid even for the case in which the object is accelerated if we are careful to use the *average* speed during the time t. For $\bar{v}$ we use the result given in Eq. 3-7; then,

$$x = \tfrac{1}{2}(v_0 + v)t$$
$$= \tfrac{1}{2}v_0 t + \tfrac{1}{2}vt$$

Using Eq. 3-9 for v, we have

$$x = \tfrac{1}{2}v_0 t + \tfrac{1}{2}(v_0 + at)t$$
$$= \tfrac{1}{2}v_0 t + \tfrac{1}{2}v_0 t + \tfrac{1}{2}at^2$$

or, finally,

$$\boxed{x = v_0 t + \tfrac{1}{2}at^2 \quad \text{(for constant acceleration)}} \tag{3-11}$$

This equation states that the distance traveled is equal

to $v_0 t$ (the distance that would be traveled *without* acceleration) plus a term that depends on the acceleration and is proportional to the *square* of the time. Figure 3-2 shows a distance–time graph for a case of accelerated motion. The distance *increases* with the *square* of the time and the curve is a *parabola*.

In Eq. 3-11 we see symbols on the right-hand side that stand for speed, time, and acceleration. Check that the combined units of the terms $v_0 t$ and $\frac{1}{2} a t^2$ are both the same as the units of x (namely, length).

A case that we encounter frequently is the one in which an object begins its motion from a condition of rest. That is, the initial speed v_0 is zero, and Eq. 3-11 simplifies to

$$x = \tfrac{1}{2} a t^2 \qquad \text{(for constant acceleration, starting from rest)} \qquad (3\text{-}12)$$

We will find this equation useful in many situations.

Suppose that an automobile, moving at a constant speed, travels 200 m in 5 s. At the instant we call $t = 0$, the driver "steps on the gas," and the automobile accelerates at a constant rate of 4 m/s². How far has the automobile moved after 5 s of accelerated motion?

First, we need the initial speed:

$$v_0 = \frac{200 \text{ m}}{5 \text{ s}} = 40 \text{ m/s}$$

Then, using Eq. 3-11,

$$x = (40 \text{ m/s}) \times (5 \text{ s}) + \tfrac{1}{2}(4 \text{ m/s}^2) \times (5 \text{ s})^2$$
$$= 200 \text{ m} + 50 \text{ m}$$
$$= 250 \text{ m}$$

Notice that during the period of acceleration, the automobile moves the same 200 m it would have traveled without acceleration *plus* 50 m due to the acceleration.

Describing Motion with Areas

We can make an interesting geometrical interpretation of the equations we have just derived. Figure 3-8a shows a speed–time graph for a case of motion with constant speed v. We know from Eq. 3-10 that the distance traveled in this situation is $x = vt$. Now look at the shaded rectangle in Fig. 3-8a. This rectangle has a length (along the time axis) of t and a height (along the speed axis) of v. Therefore, the area of the shaded rectangle is $v \times t$. That is, the area beneath the line that

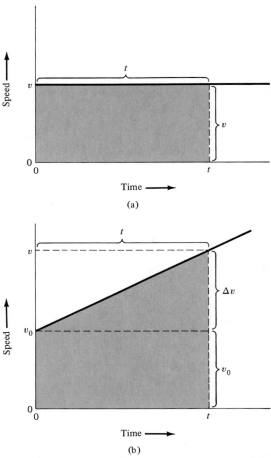

Figure 3-8 The area beneath the line that gives the speed in a speed–time graph is equal to the distance traveled.

gives the speed in a speed–time graph is equal to the distance traveled:

$$\text{area} = v \times t = x = \text{distance traveled}$$

This result is not limited to cases in which the speed is constant. Figure 3-8b shows a speed–time graph for motion with constant acceleration, starting with an initial speed v_0 (compare Figs. 3-6 and 3-7). The area beneath the speed line can be considered to consist of two parts. The lower part is rectangular and has an area $v_0 \times t$. The upper part is triangular and has an area $\frac{1}{2} \Delta v \times t$. Therefore, the total area of the shaded region is

$$\text{area} = v_0 \times t + \tfrac{1}{2} \Delta v \times t$$

If we divide and multiply the second term by t, we have

$$\tfrac{1}{2} \Delta v \times t = \tfrac{1}{2} \frac{\Delta v}{t} \times t^2$$

But the change in speed Δv divided by the time t is the acceleration a (Eq. 3-6). Thus,

$$\tfrac{1}{2} \Delta v \times t = \tfrac{1}{2}at^2$$

and the expression for the area becomes

$$\text{area} = v_0 t + \tfrac{1}{2}at^2 = x = \text{distance traveled}$$

where we have used Eq. 3-11 to identify the distance traveled.

We can extend this idea still further to cases that involve changing accelerations. Figure 3-9 shows such a case. If we wish to know the distance traveled between time zero and time t, we must find the area beneath the curve extending to the dashed line. We can do this by dividing the region into a number of small time intervals, t_1 to t_2, t_2 to t_3, and so forth. The area beneath the speed curve between t_1 and t_2, for example, is equal to the speed at $t = t_1$ multiplied by $\Delta t = t_2 - t_1$ *plus* the small triangular area at the top of the shaded rectangle (see Fig. 3-9). Even though the speed graph is a curve instead of a straight line, each small area above a rectangle is approximately triangular, and we can calculate the total area beneath the curve by summing the areas of all the rectangles and triangles.

If, in Fig. 3-9, we make each time interval Δt very small, we of course have many more rectangular areas to sum. But, at the same time, we reduce the importance of the triangular areas. By making Δt exceedingly small, we can minimize the triangular areas to any desired

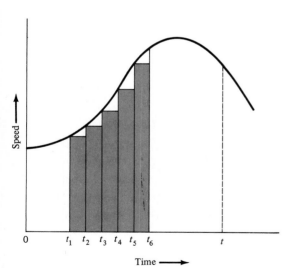

Figure 3-9 The distance traveled between time zero and time t is equal to the area beneath the speed–time curve and can be computed by summing a large number of rectangular areas.

degree and can therefore find the total area to any desired accuracy by summing the rectangular areas alone. This way of determining the area beneath a curve leads to the methods of *calculus,* but these techniques need not concern us here.

We have seen how it is possible to find the *distance* that an object travels by determining the area beneath its *speed–time* curve. It is also possible to find the *speed* of an object by determining the area beneath the *acceleration–time* curve. Can you see how this is done?

3-5 FREE FALL

Motion under the Influence of Gravity

An important case of accelerated motion is one that we see every day: the motion of a falling object. When an object is dropped, the gravitational attraction of the Earth causes the object to fall with continually increasing speed—the object is accelerated by gravity. The first systematic investigation of the behavior of falling objects was carried out by Galileo Galilei (1564–1642). In Galileo's time, a new attitude toward scientific thought was emerging. Instead of following the philosophy of Aristotle, who reached conclusions based on reasoning alone, science began to be guided by conclusions based on experiment and observation, coupled with logic and reasoning. This approach we call the *scientific method.* Galileo recognized this attitude as the only proper way to advance our understanding of Nature, and all of his writings reveal a truly modern approach to science. In his study of falling objects, Galileo performed careful experiments and made precise measurements of distances and times. He was able to show that the distance through which an object falls (starting from rest) is proportional to the square of the time of fall. This important conclusion is the same as we have derived (Eq. 3-11) from the equations of motion for constant acceleration.

The acceleration experienced by all objects that fall near the surface of the Earth is the same (in the absence of air resistance). The stroboscopic photograph on page 48 shows the correctness of this statement. Two balls of unequal mass are released from the same height at the same time (by means of an electrical release). The camera shutter remains open and an intense stroboscopic

Acceleration of a Drag Racer

In November 1973, Don ("Big Daddy") Garlits of Tampa, Florida, established a drag-racing record in the top-fuel class of the National Hot Rod Association by accelerating from rest to a speed of almost 250 mi/h in a distance of $\frac{1}{4}$ mi. In order to compute Don Garlits' acceleration (which we assume to be constant), we first require the average speed. The start was from rest ($v_0 = 0$). So the average speed was

$$\bar{v} = \tfrac{1}{2}(v_0 + v) = \tfrac{1}{2}(0 + 250 \text{ mi/h}) = 125 \text{ mi/h}$$

Solving Eq. 3-10 for the time t, we find

$$t = \frac{x}{\bar{v}} = \frac{\frac{1}{4} \text{ mi}}{125 \text{ mi/h}} = \frac{1}{500} \text{ h} = \frac{1}{500} \text{ h} \times \frac{3600 \text{ s}}{1 \text{ h}} = 7.2 \text{ s}$$

for the time to travel $\frac{1}{4}$ mi. Then, since the initial speed was zero ($v_0 = 0$), Eq. 3-9 gives

$$a = \frac{v}{t} = \frac{250 \text{ mi/h}}{7.2 \text{ s}} = 34.7 \text{ (mi/h)/s}$$

Notice that the result is given in *mixed* units, (mi/h)/s, instead of in mi/h² or mi/s². In this case it seems easier to appreciate the magnitude of the acceleration by using unconventional units: the speed increased by 34.7 mi/h during each second of travel.

Actually, Don Garlits attained his final speed of 247 mi/h with a run of only 5.78 s, not 7.2 s as we have calculated here. This means that Garlits' acceleration was not constant. The acceleration was greater than 34.7 (mi/h)/s at the start and decreased to less than 34.7 (mi/h)/s as the run progressed. (Can you see why?) The average acceleration was (247 mi/h)/(5.78 s) = 42.7 (mi/h)/s.

Don Garlits

NATIONAL HOT ROD ASSOCIATION

lamp flashes at intervals of $\frac{1}{40}$ s. The resulting picture demonstrates that the two balls do indeed fall at the same rate. By using such techniques, it is also possible to determine the value of the acceleration of falling objects. Near the surface of the Earth, the acceleration due to gravity (which we denote by the symbol g) is

$$g = 9.8 \text{ m/s}^2 = 32 \text{ ft/s}^2 \qquad (3\text{-}13)$$

These values are only approximate, but they will suffice for all of our purposes here.

The gravitational attraction that the Earth exerts on an object decreases as the distance between the object and the center of the Earth increases. Consequently, the acceleration experienced by an object falling toward the Earth from a great height will be significantly less than g. At a height of 4000 mi above the surface of the Earth (corresponding to a distance of one Earth radius above the surface), the acceleration of a falling object is $\frac{1}{4}g$.

Motion near the Surface of the Earth

Using the equations derived in the preceding section together with the value of g given in Eq. 3-13, we can now quantitatively discuss the vertical motion of an object moving freely near the surface of the Earth. Sup-

---

Galileo Galilei (1564–1642). Galileo was the son of a nobleman of Florence. He was educated at the University of Pisa and held posts of professor of mathematics at Pisa, Padua, and Florence. Galileo was one of the first systematic practitioners of the modern scientific method. His careful experiments on falling bodies and his well-constructed logical arguments established mechanics as a science and paved the way for Newton to formulate a complete set of laws of motion. Although Galileo did not invent the telescope, he made the first practical instrument and with it discovered the mountains of the Moon, Jupiter's satellites, Saturn's rings, sunspots (from the movement of which he inferred the rotation of the Sun), and the phases of Venus. Because of his support of the Copernican theory that the Sun, not the Earth, is the center of the solar system, he incurred the wrath of Church authorities and was removed from his academic posts. He remained active until his death, although in his later years he was plagued by near blindness, disease, and domestic troubles. Galileo remained a staunch supporter of the heliocentric theory even though the restrictions placed upon him by the Church prevented him from speaking publicly about his views. He was also prohibited from publishing his scientific conclusions, but his last (and greatest) book was smuggled to Holland where it was published four years before his death.

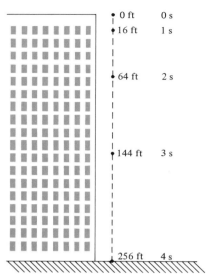

0 ft	0 s
16 ft	1 s
64 ft	2 s
144 ft	3 s
256 ft	4 s

Figure 3-10   An object dropped from the top of a building accelerates uniformly downward at a rate of 32 ft/s².

pose that we drop an object from a high building (see Fig. 3-10). How far will it have fallen and what will be its speed after 4 s? Because the initial velocity is zero ($v_0 = 0$), we use Eq. 3-12 for x and Eq. 3-9 for v:

$$x = \tfrac{1}{2}gt^2 \quad \text{and} \quad v = gt$$

where g has been inserted for the acceleration. Substituting $g = 32$ ft/s² and $t = 4$ s, we find

$$x = \tfrac{1}{2} \times (32 \text{ ft/s}^2) \times (4 \text{ s})^2 = 256 \text{ ft}$$
$$v = (32 \text{ ft/s}) \times (4 \text{ s}) = 128 \text{ ft/s}$$

Conversely, if we drop a stone from the top of a cliff and note that it requires 4 s to strike the water at the base, we can conclude that the height of the cliff is 256 ft.

The acceleration of gravity always acts downward, toward the center of the Earth. An object that is thrown *upward* will experience a *downward* acceleration. Thus, the upward velocity will be gradually reduced to zero, the object will cease to rise, and finally it will begin to fall downward. During this entire process the acceleration is *constant* (in *magnitude* and in *direction*), but the direction of motion changes from upward to downward. Note the following point: if an object is thrown upward, after it reaches its highest point the downward part of

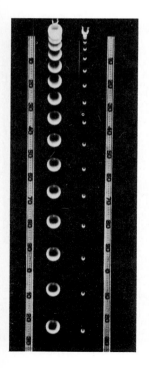

Aristotle argued that a large stone is more strongly attracted toward the Earth than a small stone and therefore a large stone should fall more rapidly than a small stone. (There is an element of truth in this statement since a small stone will suffer a relatively larger retardation effect due to air resistance than will a large stone. The large stone will therefore fall slightly faster than the small stone. The effect is, however, rather small for any short fall.) Aristotle did not perform any experiments to test his conclusion. It remained for Simon Stevinus (1548–1620), a Dutch mathematician and scientist, to drop two balls of different mass from a high building and to demonstrate that they reached the ground at the same time. The origination of this experiment has been incorrectly attributed to Galileo, who is said to have dropped a cannon ball and a musket ball from the Leaning Tower of Pisa to show that each experienced the same acceleration, but he probably never actually performed this experiment.

Two balls of unequal mass are shown in this stroboscopic photograph to fall at the same rate. (Can you use this photograph to determine the value of g? The successive positions of the balls are shown at intervals of $\frac{1}{40}$ s and the markings on the meter sticks signify intervals of 10 cm.)

the motion is exactly the same as if it had been *dropped* from the point of maximum height.

Effects of Friction on Free Fall

In a real case, a falling object will not continue to accelerate indefinitely. For any fall from a great height, the frictional effects of air resistance are important and eventually the acceleration is reduced to zero; the object then falls at constant speed (called the *terminal* speed). A sky-diver, for example, reaches a terminal speed of about 125 mi/h. Air resistance is not a factor, of course, in a vacuum. Therefore, all objects, regardless of the way in which they are influenced by air resistance, will fall at the same rate in vacuum.

A sky-diver falls at a constant speed because the effects of air friction prevent any further acceleration.

3-6 VECTORS

Magnitude and Direction

Speed is a quantity that can be specified by means of a single number (plus the appropriate units). Thus, we can say that the speed of an automobile is 40 mi/h. If you want to travel the 200 miles from *A* to *B* in 5 hours, you will be able to do so by maintaining an average speed of 40 mi/h. But you will never arrive at *B* unless you proceed in the right direction! In order to describe motion completely, it is necessary to give both the *speed* and the *direction* of the motion. We know precisely the motion of an automobile if we say that it is traveling *northeast* at 40 mi/h.

A quantity that requires for its complete specification a *size* (or *magnitude*) and a *direction* is called a *vector*. The quantity that combines speed and direction of motion is a vector, called the *velocity* vector. We will encounter many other physical quantities that are also vectors: force, momentum, electric field, and so forth. Quantities such as mass, time, and temperature have only magnitudes, not directions; these quantities are called *scalars*.

We will use boldface characters for vectors and light-face italic characters for scalars. Thus, we will use **v** to represent velocity and we will continue to use *v* to represent *speed*. (*Speed* is the magnitude of *velocity,* and *v* is the magnitude of **v**.) We will use the term "velocity" when we wish to convey the importance of both magnitude and direction; the term "speed" will be used when

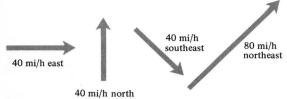

40 mi/h east

40 mi/h north

40 mi/h southeast

80 mi/h northeast

Figure 3-11   Four different velocity vectors.

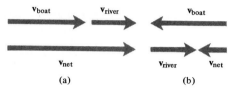

$\mathbf{v}_{boat}$ $\mathbf{v}_{river}$ $\mathbf{v}_{boat}$

$\mathbf{v}_{net}$ $\mathbf{v}_{river}$ $\mathbf{v}_{net}$

(a)   (b)

Figure 3-12   Two simple cases of vector addition. In each case, $\mathbf{v}_{net} = \mathbf{v}_{boat} + \mathbf{v}_{river}$, but in (a), $\mathbf{v}_{boat}$ and $\mathbf{v}_{river}$ are in the *same* direction, whereas in (b), they are in *opposite* directions.

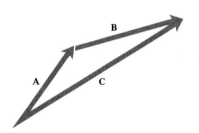

B

A   C

Figure 3-13   The vector addition, $\mathbf{C} = \mathbf{A} + \mathbf{B}$.

A   −A

Figure 3-14   The vector A and its negative.

we are interested only in the rate at which an object moves.

In diagrams, vectors will be represented by arrows. The length of an arrow indicates the magnitude of the vector and the direction of the arrow indicates the direction of the vector (see Fig. 3-11).

Vector Addition and Subtraction

The manipulation of *numbers* requires only the basic operations of arithmetic. But how do we handle *vectors?* Vectors are more complicated than numbers because they combine the essential property of numbers (namely, *magnitude*) with the additional property of *direction.* Even so, we can define in a simple way the addition and subtraction of vector quantities. We need only a few basic rules:

(a) A certain boat is capable of moving with a speed of 4 mi/h in still water. If this boat travels downstream in a river, running with a current of 3 mi/h, the net speed of the boat relative to the land will be 4 mi/h + 3 mi/h = 7 mi/h. Figure 3-12a shows the velocity vector diagram for this situation. The vector $\mathbf{v}_{net}$ is the *vector sum of* $\mathbf{v}_{boat}$ and $\mathbf{v}_{river}$; that is, $\mathbf{v}_{boat} + \mathbf{v}_{river} = \mathbf{v}_{net}$.

If the same boat travels upstream, running against the current, the net speed of the boat will be reduced to 1 mi/h. Figure 3-12b, shows the way in which the velocity vectors are combined in this case. The vector $\mathbf{v}_{net}$ is again the sum of $\mathbf{v}_{boat}$ and $\mathbf{v}_{river}$, but now these two vectors have *opposite* directions so that the magnitude of the sum $\mathbf{v}_{river} + \mathbf{v}_{boat}$ is only 1 mi/h.

Notice how we obtained the sum vector in these two diagrams. In each case we started with the vector $\mathbf{v}_{boat}$; then, we placed the origin of the vector $\mathbf{v}_{river}$ at the head (the arrow end) of $\mathbf{v}_{boat}$. The sum vector $\mathbf{v}_{net}$ was obtained by connecting the origin of $\mathbf{v}_{boat}$ with the head of $\mathbf{v}_{river}$. We follow exactly this same procedure if we wish to find the vector sum, $\mathbf{C} = \mathbf{A} + \mathbf{B}$, of two vectors, $\mathbf{A}$ and $\mathbf{B}$, that do not lie along the same straight line. Figure 3-13 shows this general case of vector addition.

(b) The *negative* of a vector $\mathbf{A}$ is another vector, $-\mathbf{A}$, which has the same magnitude as $\mathbf{A}$ but has the opposite direction (Fig. 3-14). If $\mathbf{A} = 30$ mi/h *northeast,* then $-\mathbf{A} = 30$ mi/h *southwest.*

(c) How do we *subtract* one vector from another? If we are given the vectors $\mathbf{A}$ and $\mathbf{B}$, how do we calculate

$C = A - B$? This operation is carried out by using the procedures in (a) and (b). First, knowing B, we can find the vector $-B$; then we write (just as we can with *numbers*),

$$C = A - B = A + (-B)$$

That is, to subtract B from A, we *add* $-B$ to A (Fig. 3-15).

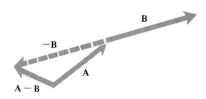

Figure 3-15 The subtraction, $A - B$, is the same as adding $-B$ to A.

An important point regarding vector addition (or subtraction) can be seen in Fig. 3-13. The magnitude of the vector A (that is, A) plus the magnitude of the vector B (that is, B) is *greater* than the magnitude of the vector C (that is, C). Thus, even though $C = A + B$, the magnitudes are *not* equal: $C \neq A + B$.

Vector Components

There are many different shapes of the vector diagrams that represent the addition of two vectors, A and B, to form a sum vector C (Fig. 3-16). A particularly interesting vector triangle is that in which the vectors A and B are *perpendicular* (Fig. 3-16c). In this case, we can calculate the magnitude of the sum vector C by using the Pythagorean theorem of plane geometry:

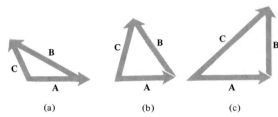

Figure 3-16 The vector A is added to different vectors B (all of which have the same magnitude) to produce different sum vectors C. In (c) the vectors A and B are at right angles.

$$C^2 = A^2 + B^2 \qquad (3\text{-}14)$$

or

$$C = \sqrt{A^2 + B^2} \qquad (3\text{-}15)$$

For example, suppose that a boat which can move with a speed of 4 mi/h in still water attempts to proceed directly *across* a river in which the current flows with a speed of 3 mi/h. What will be the boat's speed relative to the land? Figure 3-17 shows the vector triangle for this case. The sum of v_{boat} and v_{river} (which are perpendicular vectors) produces the diagonal vector v_{net} which represents the boat's true velocity relative to the land. The speed of the boat (that is, the magnitude of v_{net}) is

$$\begin{aligned} v_{net} &= \sqrt{(v_{boat})^2 + (v_{river})^2} \\ &= \sqrt{(4 \text{ mi/h})^2 + (3 \text{ mi/h})^2} \\ &= \sqrt{16 + 9} \text{ mi/h} = \sqrt{25} \text{ mi/h} \\ &= 5 \text{ mi/h} \end{aligned}$$

An even more generally useful property of right-angled vector triangles is the following. Suppose that

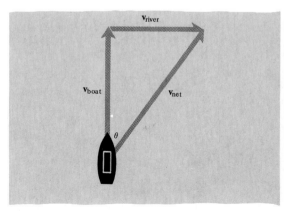

Figure 3-17 A boat that attempts to move directly across a river is carried on a diagonal path because of the velocity of the river current.

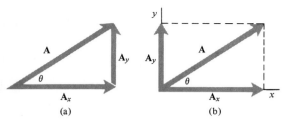

Figure 3-18   Because $\mathbf{A} = \mathbf{A}_x + \mathbf{A}_y$, we can decompose the vector $\mathbf{A}$ into the component vectors, $\mathbf{A}_x$ and $\mathbf{A}_y$.

two vectors, $\mathbf{A}_x$ and $\mathbf{A}_y$ are added to give the vector $\mathbf{A}$: $\mathbf{A} = \mathbf{A}_x + \mathbf{A}_y$, as in Fig. 3-18a. If $\mathbf{A}_x$ and $\mathbf{A}_y$ represent some physical quantity—for example, the forces exerted on an object by two persons pulling in different directions—then there is absolutely no difference between the effects produced by $\mathbf{A}_x$ and $\mathbf{A}_y$ taken together and the effect produced by the sum vector $\mathbf{A}$. That is, $\mathbf{A}$ and $\mathbf{A}_x + \mathbf{A}_y$ are entirely equivalent. This being the case, we can turn the situation around and *decompose* any vector $\mathbf{A}$ into two other vectors whose sum is equal to $\mathbf{A}$. If these vectors are at right angles (for example, if they lie along the axes of an *x-y* coordinate system as shown in Fig. 3-18b), they are called the *component* vectors of $\mathbf{A}$.

We can write expressions for A_x and A_y in terms of A and the angle θ (see Fig. 3-18) by using simple trigonometry. (A review of the essential definitions of trigonometry are given in the Appendix.) For example, from the definitions of the sine and the cosine of an angle, we can write

$$\sin \theta = \frac{A_y}{A}; \qquad \cos \theta = \frac{A_x}{A}$$

Therefore, the components A_x and A_y are

$$\boxed{\begin{aligned} A_x &= A \cos \theta \\ A_y &= A \sin \theta \end{aligned}}$$

(3-16a)

(3-16b)

In Fig. 3-19a, a block is acted upon by a force $\mathbf{F}$ which has a magnitude of 30 units and is directed at an angle of 30° above the horizontal. What are the *x* (horizontal) and *y* (vertical) components of $\mathbf{F}$? The component diagram is shown in Fig. 3-19b. Using Eqs. 3-16a and 3-16b, we can write

$$F_x = F \cos \theta = 30 \times \cos 30°$$
$$F_y = F \sin \theta = 30 \times \sin 30°$$

From the table on page 543 we find $\cos 30° = 0.866$ and $\sin 30° = 0.500$. Therefore,

$$F_x = 30 \times 0.866 = 26.0 \text{ units}$$
$$F_y = 30 \times 0.500 = 15.0 \text{ units}$$

In the example illustrated in Fig. 3-17, suppose that we wish to know the angle between $\mathbf{v}_{\text{boat}}$ and $\mathbf{v}_{\text{net}}$. Denoting this angle by θ and identifying $\mathbf{v}_{\text{boat}}$ as equivalent to $\mathbf{A}_x$ and $\mathbf{v}_{\text{net}}$ as equivalent to $\mathbf{A}$, we have

$$\cos \theta = \frac{v_{\text{boat}}}{v_{\text{net}}} = \frac{4 \text{ mi/h}}{5 \text{ mi/h}} = 0.8$$

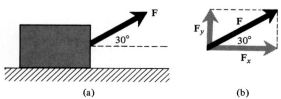

Figure 3-19   Resolving the force $\mathbf{F}$ into components $\mathbf{F}_x$ and $\mathbf{F}_y$.

Looking down the cosine column of the table in the Appendix, we find that cos $\theta = 0.8$ corresponds to an angle θ of approximately $37°$.

3-7 MOTION IN TWO DIMENSIONS

Parabolic Motion

If we drop an object from a certain height, we know from experience that the object will move straight downward with increasing speed. What will happen if, instead of *dropping* the object, we *throw* it parallel to the ground (that is, we give the object an initial horizontal velocity)? Again, we know from experience that the object will follow a path that curves toward the ground. The photograph at the right shows both of these situations. The ball on the left was dropped straight downward, whereas the ball on the right (which was released at the same instant) was given an initial velocity in the horizontal direction. The right-hand ball is seen to follow a curved path (actually, a *parabola*).

We can use the idea of vector components to analyze the motion of the two objects. At any instant of time, the motion of each object can be described by a velocity vector. For the object on the left, the velocity vector always points downward. But for the object on the right, the velocity vector at first points in the horizontal direction and then gradually turns downward as the object curves toward the ground. This velocity vector can be resolved into two components: a horizontal component $\mathbf{v}_x$ and a vertical component $\mathbf{v}_y$ (see Fig. 3-20).

How do the two velocity components change with time? A change in velocity requires an acceleration, and in this case there is only one cause of acceleration: the downward pull of gravity. Up to this point we have ignored an important fact: *acceleration has direction and is therefore a vector quantity*. The direction of gravitational acceleration is *downward*. That is, there is a vertical acceleration but *no horizontal acceleration*. With no acceleration to affect its value, the horizontal velocity must remain constant. That is, $v_x = $ constant. Because of the downward acceleration due to gravity, the vertical velocity, at a time t after release, is $v_y = gt$. This is just the familiar equation, $v = v_0 + at$, with $v_0 = 0$ and with g substituted for the acceleration. Therefore, the motion of the object is summarized by the equations,

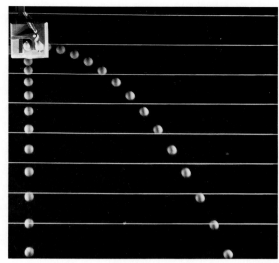

Stroboscopic photograph of two balls released simultaneously, one with an initial velocity in the horizontal direction. The vertical motions are exactly the same, and the horizontal component of the velocity of the right-hand ball remains constant.

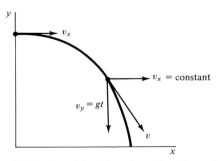

Figure 3-20 An object is released with an initial horizontal velocity v_{0x}. The horizontal velocity remains constant whereas the vertical velocity increases with time.

$$v_x = \text{constant} = \text{original projection speed}$$
$$v_y = gt$$

(3-17)

This analysis indicates that the downward motion of the object, under the influence of gravity, is independent of the horizontal motion. That is, the vertical velocity, $v_y = gt$, does not depend in any way on the speed with which the object was thrown horizontally—*the vertical and horizontal motions do not affect one another.* Can this really be true? The stroboscopic photograph shows that it is. The ball on the right falls downward with exactly the same acceleration as the ball on the left. The acceleration due to gravity acts only on the vertical component of the velocity, and it acts in exactly the same way on each ball. Therefore, there is no difference in the vertical motions of the two balls. There is a difference only in the horizontal motions due to the fact that the right-hand ball has an initial horizontal velocity. Notice that this analysis proves the following point which is at first rather startling. If a bullet is dropped from the same height and at the same instant that a bullet is fired horizontally over a flat surface, the two bullets will strike the ground simultaneously.

Suppose that a ball is thrown straight outward from a cliff which is 64 feet above the sea; the initial horizontal

Parabolic or Vertical Motion?

The photograph shows a "stick" of bombs being dropped from a B-17 over Europe during World War II. The bombs appear to be falling straight down from the aircraft. But the bomber is in motion as the bombs are released in succession. The bombs remain in a vertical column because each has the same horizontal velocity as the bomber. The diagram shows that each bomb follows a parabolic path. Will all of the bombs strike the ground at the same point? What would the "stick" look like if the bomber were traveling *half* as fast?

U.S. AIR FORCE

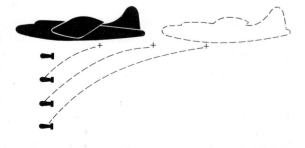

velocity of the ball is 25 ft/s. How far from the point directly under the initial position will the ball strike the water? Figure 3-21 shows the situation.

Because the initial vertical velocity is zero, we use Eq. 3-12 and write

$$y = \tfrac{1}{2}gt^2$$

where y has been used to denote the vertical direction and where g has been substituted for the acceleration. The motion starts at $y = 0$, so all subsequent values of y are negative. Notice also that the acceleration due to gravity is *downward*, so that g has a negative value. The final value of y is -64 ft, and substituting $g = -32$ ft/s², we have

$$-64 = -16t^2$$

Solving for t,

$$t = \sqrt{\frac{64}{16}} = \sqrt{4} = 2 \text{ s}$$

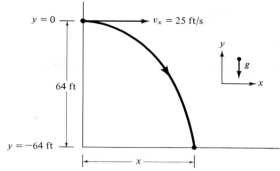

Figure 3-21   A ball is released from a height of 64 ft with an initial horizontal velocity of 25 ft/s.

The Parabolic Motion of a Ball

The fact that a ball projected into the air will follow a curving path is familiar to everyone. If someone throws or bats a ball to you, you know almost instinctively where you must position yourself to catch the ball. With a bit of experience, only a small part of the initial flight of the ball needs to be viewed in order to "compute" where the point of catch will be (see Fig. 3-22). Unconsciously, you judge the vertical and horizontal components of the motion during the first part of the motion, and your brain then "computes" where the ball will fall. Because of air resistance, the path is not exactly parabolic, but the brain takes this into account in its computation. And all this goes on even if you have never analyzed mathematically the motion of a ball! Provided with a minimum of information, the human brain is a marvelous computer. With more experience, the computation time can be significantly reduced. Have you ever noticed how a good centerfielder will be "off with the crack of the bat" to make a catch?

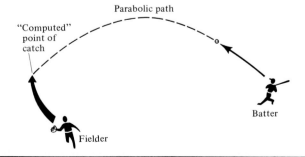

Figure 3-22

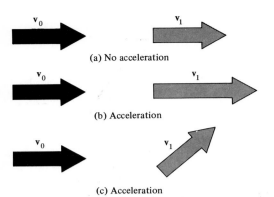

(a) No acceleration

(b) Acceleration

(c) Acceleration

Figure 3-23 Acceleration involves a change in the velocity vector. If there is no change (a), there is no acceleration. If there is a change either in the magnitude (b) or in the direction (c) of the velocity vector, then there has been an acceleration.

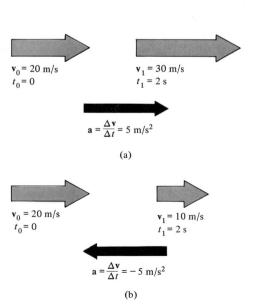

$v_0 = 20$ m/s
$t_0 = 0$

$v_1 = 30$ m/s
$t_1 = 2$ s

$a = \dfrac{\Delta v}{\Delta t} = 5$ m/s²

(a)

$v_0 = 20$ m/s
$t_0 = 0$

$v_1 = 10$ m/s
$t_1 = 2$ s

$a = \dfrac{\Delta v}{\Delta t} = -5$ m/s²

(b)

Figure 3-24 Two cases of acceleration. Although the *magnitude* of the acceleration vector is the same in each case, the *directions* are opposite.

The distance x from the cliff to the point where the ball strikes the water is equal to the original horizontal velocity multiplied by the time of fall:

$$x = v_x t = (25 \text{ ft/s}) \times (2 \text{ s}) = 50 \text{ ft}$$

Notice that the problem is solved in two steps: first, we find the time of fall, and then we use this information to compute the horizontal distance traveled.

Acceleration — The General Definition

Acceleration means *change in velocity*. But, as we have seen, velocity is a *vector* quantity. There are two ways that a vector can change: in *magnitude* and in *direction*. A change in either (or both) of these characteristics of the velocity vector implies an acceleration. An automobile that is moving north at 40 mi/h at time t_0 and is moving north at 60 mi/h at time t_1 has undergone acceleration. But the automobile has also undergone acceleration if, at time t_1, the velocity is 40 mi/h *eastward*. In the latter case, the *speed* is the same, but the *direction* is different — this can happen only if there is an acceleration.

Figure 3-23 illustrates these points. In each of the three cases shown, v_0 is the velocity vector at time $t = t_0$ and v_1 is the velocity vector at a later time $t = t_1$. In Fig. 3-23a, $v_0 = v_1$ and there is no acceleration. In Fig. 3-23b, the magnitude, but not the direction, of the velocity vector changes; and in Fig. 3-23c, the direction, but not the magnitude, of the velocity vector changes. Both of these latter cases involve acceleration.

Acceleration is a vector quantity. We know, for example, that the direction of the acceleration due to gravity is always *downward*. Whenever there is a change in an object's velocity vector, we can identify the direction of the acceleration as the direction of the velocity *change*, $v_1 - v_0$ or Δv. This becomes clear if we write the defining equation for acceleration (Eq. 3-6) as a *vector* equation:

$$a = \frac{v_1 - v_0}{t_1 - t_0} = \frac{\Delta v}{\Delta t} \qquad (3\text{-}18)$$

The vectorial property of acceleration is illustrated in Figs. 3-24 and 3-25. In Fig. 3-24a, we have an initial velocity of 20 m/s to the right; 2 s later the velocity is 30 m/s in the same direction. The velocity change Δv is 30 m/s − 20 m/s = 10 m/s and is also directed to the right. Therefore, the acceleration is $\Delta v/\Delta t = 5$ m/s², to the right. In Fig. 3-24b, we have the same initial velocity, but the final velocity is *smaller* in this case. Using $\mathbf{a} = \Delta\mathbf{v}/\Delta t$, we find the same magnitude for the acceleration, 5 m/s², but now the velocity change $\Delta\mathbf{v}$, and therefore the acceleration $\mathbf{a}$, is directed to the *left*. In case (a), we say that the acceleration is *positive* (speeding up), and in case (b), we say that the acceleration is *negative* (slowing down).

Figure 3-25 shows a case in which the velocity changes both in magnitude and in direction. The initial velocity is $\mathbf{v}_0 = 3$ m/s toward the east and the final velocity is $\mathbf{v}_1 = 5$ m/s directed 37° east of north. What is the velocity change vector $\Delta\mathbf{v}$? We note that $\Delta\mathbf{v}$ is the vector that must be added to $\mathbf{v}_0$ in order to produce $\mathbf{v}_1$. That is,

$$\mathbf{v}_1 = \mathbf{v}_0 + \Delta\mathbf{v}$$

The diagram shows this addition. For the particular values and directions of $\mathbf{v}_0$ and $\mathbf{v}_1$, $\Delta\mathbf{v} = 4$ m/s toward the north. (Compare this triangle with that in Fig. 3-17.) If the velocity change required 2 s, then we would have $\mathbf{a} = 2$ m/s² (north).

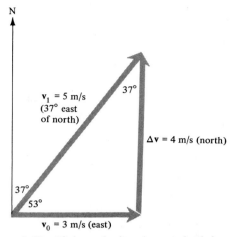

Figure 3-25 If the velocity changes both in magnitude and direction, we find $\Delta\mathbf{v}$ from the relation $\mathbf{v}_1 = \mathbf{v}_0 + \Delta\mathbf{v}$.

Circular Motion

Along with projectile motion, the most frequently encountered case of two-dimensional motion is motion in a circle. If an object moves in a circular path with constant speed, we say that the object undergoes *uniform circular motion*. The blades of a fan and the hands of a clock undergo uniform circular motion, and the planets moving around the Sun undergo (almost) uniform circular motion.

The *period* τ of circular motion is the time required to complete one revolution or cycle of the motion. For example, the period of the Earth's rotation around the Sun is 1 year: $\tau = 1$ y $= 3.16 \times 10^7$ s. The radius of the motion is 1 A.U. or 1.50×10^{11} m. From this information we can compute the speed of the Earth's motion:

$$v = \frac{\text{distance}}{\text{time}} = \frac{\text{circumference of orbit}}{\text{period}}$$

$$= \frac{2\pi r}{\tau}$$

$$= \frac{2\pi \times (1.50 \times 10^{11} \text{ m})}{3.16 \times 10^7 \text{ s}}$$

$$= 2.99 \times 10^4 \text{ m/s} \cong 30 \text{ km/s}$$

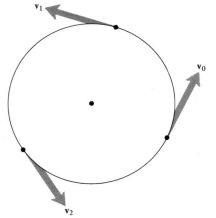

Figure 3-26 An object undergoing uniform circular motion moves with constant *speed,* but the velocity vector continually changes direction.

If we examine the velocity vector of an object moving uniformly in a circle, we find that although the *magnitude* (that is, the speed) is constant, the *direction* of the velocity is continually changing. An object moving uniformly in a circle is therefore continually accelerated. Figure 3-26 shows an object undergoing uniform circular motion. Notice that the arrows representing the velocity **v** are all the same length but that the directions change as the motion proceeds. The object is being accelerated.

In order to move in a circle, an object must be pulled toward the center of the circle. This pull is called the *centripetal* (or center-seeking) *force.* (If the pull is exerted by a string, it is sometimes said that the object exerts a *centrifugal force* on the string.) The pull alters the direction of **v** and guides the object in its circular path. The acceleration imparted to the object is called the *centripetal* acceleration and depends on the speed of the object and the radius of the circular path in the manner described in the next section.

Centripetal Acceleration

We can derive an expression for the centripetal acceleration by referring to Fig. 3-27. At the time t_0 the velocity vector is $\mathbf{v}_0$ and at the later time t_1 the velocity vector is $\mathbf{v}_1$. During the time interval $t_1 - t_0 = \Delta t$, the motion progresses through the angle $\Delta\theta$. We consider uniform circular motion so that the magnitudes of $\mathbf{v}_0$ and $\mathbf{v}_1$ are equal: $v_0 = v_1$. As the motion proceeds from the position at t_0 to the position at t_1, the object describes an arc with a length Δs; the length of the corresponding chord (dashed line) is Δx. Above the circular diagram is the vector triangle which indicates how the velocity change vector $\Delta\mathbf{v}$ is obtained from $\mathbf{v}_0$ and $\mathbf{v}_1$. Notice that the angle $\Delta\theta$ between the two velocity vectors is the same as the angle between the two radius lines.

Figure 3-28 shows two triangles. On the left is the triangle formed by the chord Δx and the two radii. On the right is the triangle formed by the two velocity vectors and $\Delta\mathbf{v}$; because $v_0 = v_1$, we abbreviate both as v. Notice that these two triangles are *similar;* that is, they are both isosceles triangles (two sides equal) with the same angle $\Delta\theta$ between the equal sides. Because the triangles are similar, the ratios of the lengths of the sides are equal:

$$\frac{\Delta v}{v} = \frac{\Delta x}{r}$$

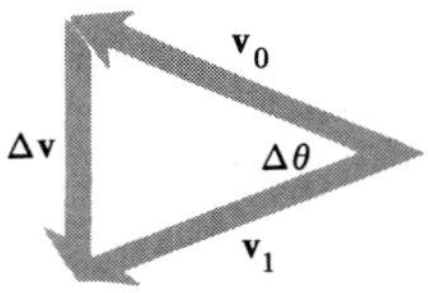

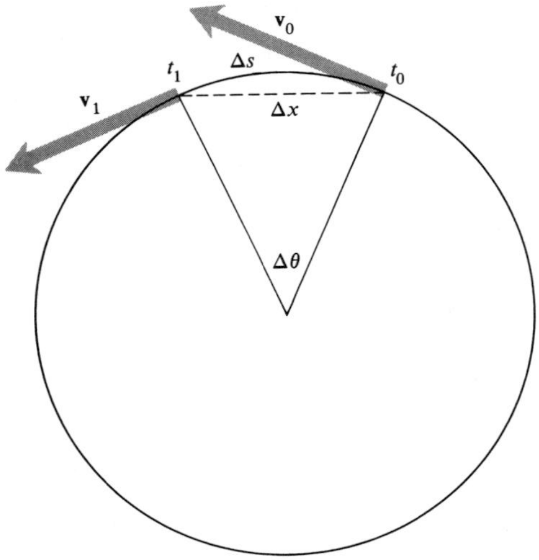

Figure 3-27 Velocity diagram for uniform circular motion.

Therefore, the magnitude of $\Delta\mathbf{v}$ is

$$\Delta v = \frac{v \times \Delta x}{r}$$

The average centripetal acceleration $\bar{a}_c$ is obtained by dividing Δv by Δt:

$$\bar{a}_c = \frac{\Delta v}{\Delta t} = \frac{v \times \Delta x / r}{\Delta t}$$

Rearranging, we can write

$$\bar{a}_c = \frac{v}{r} \times \frac{\Delta x}{\Delta t}$$

Now, v is the constant value of the instantaneous speed. The quantity $\Delta x / \Delta t$ is the *average* speed between t_0 and t_1. Notice, however, that the length of Δx is slightly smaller than the length of the chord Δs which describes the actual motion of the object. Therefore, $\Delta x / \Delta t$ is not really equal to v. But, if we imagine that Δt (and, hence, $\Delta\theta$) becomes very small, then Δx is nearly equal to Δs. Indeed, by making Δt sufficiently small we can cause Δx to approach Δs with any desired accuracy. Also, as we make Δt smaller, the *average* centripetal acceleration approaches the *instantaneous* centripetal acceleration. Therefore, we can replace $\Delta x / \Delta t$ by $\Delta s / \Delta t = v$ and $\bar{a}_c$ by a_c. The result is

$$\boxed{a_c = \frac{v^2}{r}} \qquad (3\text{-}19)$$

This important equation gives the magnitude of $\mathbf{a}_c$; but what is the *direction* of the centripetal acceleration? Look at the vector diagram in the upper part of Fig. 3-27. As $\Delta\theta$ becomes very small $\mathbf{v}_0$ and $\mathbf{v}_1$ nearly coincide. Then, $\Delta\mathbf{v}$ is a vector essentially perpendicular to both $\mathbf{v}_0$ and $\mathbf{v}_1$; that is, $\Delta\mathbf{v}$ points toward the *center* of the circular path (Fig. 3-29).

We can use Eq. 3-19 to calculate the speed (and from this, the period) of an artificial satellite in a circular orbit close to the Earth. The centripetal acceleration of the satellite is the acceleration due to gravity. At a height of $150 \text{ km} = 0.15 \times 10^6$ m, which is typical for such satellites, the acceleration due to gravity is essentially the same as the surface value, namely, $g = 9.8$ m/s². Solving Eq. 3-19 for v, we have

$$v = \sqrt{gr}$$

The value of r is the radius of the Earth (6.38×10^6 m) plus the height of the satellite orbit:

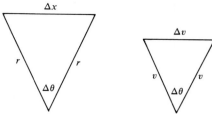

Figure 3-28   The two triangles obtained from the construction in Fig. 3-27 are *similar*. Here, v stands for v_1 and v_2 which are equal.

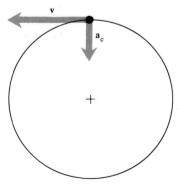

Figure 3-29   The directions of the velocity $\mathbf{v}$ and the centripetal acceleration $\mathbf{a}_c$ for circular motion. The vector $\mathbf{a}_c$ always points toward the center of the circular path.

$$v = \sqrt{(9.8 \text{ m/s})^2 \times (6.38 \times 10^6 \text{ m} + 0.15 \times 10^6 \text{ m})}$$
$$= \sqrt{(9.8 \text{ m/s}^2) \times (6.53 \times 10^6 \text{ m})}$$
$$= 8 \times 10^3 \text{ m/s} = 8 \text{ km/s}$$

The period of the motion is

$$\tau = \frac{2\pi r}{v}$$
$$= \frac{2\pi \times (6.53 \times 10^6 \text{ m})}{8 \times 10^3 \text{ m/s}}$$
$$= 5130 \text{ s} = 85 \text{ min}$$

That is, a satellite in a low circular orbit around the Earth will have a speed of approximately 8 km/s and will require about an hour and a half to circle the Earth. Many artificial satellites, such as Skylab, have orbits with similar speeds and periods.

SUGGESTED READINGS

S. Drake, *Discoveries and Opinions of Galileo* (Doubleday, Garden City, New York, 1957).

L. Fermi and G. Bernardini, *Galileo and the Scientific Revolution* (Basic Books, New York, 1951).

Scientific American articles:

I. B. Cohen, "Galileo," August 1949.

W. A. Heiskanen, "The Earth's Gravity," September 1955.

QUESTIONS AND EXERCISES

1. An ocean liner makes a 3600-km voyage in 8 days, 8 h. What was the average speed during the trip?

2. A runner completes a 400-m race in 44 s. What was his average speed?

3. Make a distance–time graph that represents the motion of an object which travels at a constant speed of 8 m/s for 1 min.

4. At $t = 0$ an object is at the position $x = 0$. At $t = 3$ s the object is at $x = 15$ m and at $t = 8$ s the object is at $x = 95$ m. What is the average speed for the first 3 seconds and for the interval $t = 3$ s to $t = 8$ s? What is the average speed for the entire 8-second period?

5. An object moving with a speed of 25 m/s is uniformly accelerated at a rate of 2 m/s². What is the speed after 12 s of acceleration?

6. An object leaves the point $x = 0$ at time $t = 0$ with a speed of 20 m/s and an acceleration of 3 m/s². Where is the object located at $t = 5$ s? What is the speed of the object at $t = 8$ s?

7. An automobile is traveling with a speed of 60 mi/h. At $t = 0$ the driver applies the

brakes and decelerates uniformly to a speed of 20 mi/h in 8 s. What was the value of the acceleration? Express the result in (mi/h)/s and in ft/s².

8. A rock is dropped from the top of a cliff and 5 s later the rock strikes the ground. How high is the cliff?

9. An object is dropped from the top of a 256-ft building. How long does it take for the object to reach the ground?

10. An object is thrown vertically upward with an initial speed of 128 ft/s. Compute the height and the speed of the object at 1-s intervals until the object strikes the ground. Draw a graph of speed versus time.

11. Draw a distance–time graph that exhibits (in different parts of the curve) motion with positive acceleration, negative acceleration, and zero instantaneous speed. Identify the various parts of the curve that show these features.

12. A parachutist is falling with a terminal speed of 120 mi/h. It requires 2 s for his parachute to deploy and decrease his vertical speed to 20 mi/h. What acceleration does the parachutist experience? Express the result in (mi/h)/s and in units of g.

13. List several familiar situations in which the effects of air resistance are important. What would happen in each case if air resistance were not present?

14. The vector **A** points north and has a magnitude of 2 units. The vector **B** points west and has a magnitude of 3 units. Use a graphical construction and find the sum **A** + **B**. What is the *magnitude* of **A** + **B**? What is the *direction* of **A** + **B**? First, use a ruler and a protractor to obtain the result; then, perform a calculation to check the original result.

15. The vector **C** has a magnitude of 2 units and is directed along the $+x$ axis. The vector **D** has a magnitude of 2 units and is directed along the $-y$ axis. What is the magnitude and direction of the vector **C** − **D**?

16. A certain moving object has velocity components of 9 m/s *north* and 12 m/s *east*. What is the *speed* of the object?

17. A vector **A** has a magnitude of 40 units and is directed at an angle of 25° with respect to the x axis. What are the magnitudes of the component vectors, A_x and A_y?

18. Two forces, $F_x = 15$ units and $F_y = 40$ units, act on an object. What is the resultant force (magnitude and direction)?

19. Consider the two vectors: **A** has a magnitude of 6 units and is directed at an angle of 30° with respect to the x axis; **B** has a magnitude of 10 units and is directed at an angle of 45° with respect to the x axis. Find the resultant vector, **C** = **A** + **B**. Proceed in the following way: First, compute the components of **A** and **B**. Next, express the components of **C** as $C_x = A_x + B_x$ and $C_y = A_y + B_y$. Finally, knowing the components of **C**, compute the magnitude and direction of **C**. Make a sketch of the procedure.

20. A ball is thrown with a horizontal velocity of 30 ft/s from the top of a building. A short time later, the ball lands in the street, 75 ft from the building. What is the height of the building?

21. A projectile is launched with an initial velocity of 1000 ft/s at an angle of 30° above the horizontal. The ground over which the projectile travels is flat. How far away from the launch site will the projectile land? [Neglect the effects of air resistance and proceed in the following way. First, compute the vertical and horizontal components of the initial velocity. Next, consider the vertical component and find the time t at which maximum height is reached; that is, find the time at which $v_y = 0$. The projectile will therefore be in motion for a total time $2t$. (Why?) Finally, calculate the horizontal distance traveled during the time $2t$.]

22. The shaft of an electric motor makes 7.2×10^4 revolutions in 1 h. Express the angu-

lar speed in revolutions per minute (rpm) and in degrees per second. What is the period of the motion?

23. What is the speed of a point on the Earth's surface at the Equator due to the rotation of the Earth on its own axis?

24. The stars in the Milky Way all undergo a general circular rotation around the center of the Galaxy. The Sun is at a distance of about 3×10^4 L.Y. from the center of the Galaxy and is moving with a speed of about 300 km/s. What period of time is required for the Sun to make one revolution?

25. An object is moving in a circular path of radius 9 ft and is experiencing an acceleration of $2g$. What is the speed of the object? What is the period of the motion?

26. The Moon moves in a nearly circular orbit around the Earth. The period of the motion is 27.3 days, and the Earth–Moon distance is 3.84×10^8 m. What is the centripetal acceleration of the Moon in m/s^2? What is the origin of this acceleration?

4

FORCE

In the preceding chapter we developed methods for describing and analyzing motion. But what *causes* motion? We know that if we push or pull (sufficiently hard) on an object, the object can be set into motion. Or, if we apply a restraining push or pull to an object already in motion, the object can be slowed down and brought to rest. In every such case, some *force*—represented by a push or pull—must be applied to an object in order to change its state of motion. In this chapter we will examine the concept of force and its relation to the behavior of objects at rest and in motion.

4-1 FORCE AND INERTIA

Intuitive Ideas

The intuitive notion that a force is a push or a pull is entirely consistent with the precise physical definition of this important quantity. We have other, equally correct intuitive ideas about force. For example, if we push in a certain direction on an object at rest, the object tends to move in that direction. Or, if we wish to stop a moving object, we must exert a push in the direction opposite to that of the object's motion. That is, force has *direction* as well as *magnitude*—force is a *vector* quantity. We also appreciate the fact that the *mass* of an object is important in any effort to change its state of motion. A kick applied to a soccer ball will send the ball flying; but a kick applied to a bowling ball will result in only a

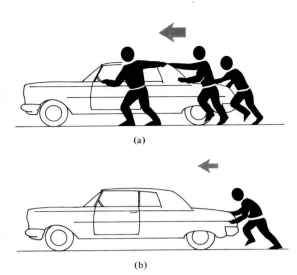

(a)

(b)

Figure 4-1 (a) Because of an automobile's large inertia, it is difficult to set into motion. (b) Once in motion, much less effort is required to maintain the motion.

slight motion of the ball (and a bruised toe). The property of an object that tends to resist any change in its state of motion is called *inertia* — the measure of an object's inertia is its *mass*.

An automobile clearly has a large inertia: it requires a considerable effort to start an automobile into motion by pushing. But, once in motion along a flat surface, considerably less effort is required to maintain the motion (Fig. 4-1). The inertia of the automobile resists any *change* in the state of motion. In fact, if there were no friction to slow the automobile, it would continue in motion with constant velocity indefinitely, even with *no* applied force. This, of course, is an idealized case because friction can never be completely eliminated in any real system, and a small push of some sort is always required to maintain the motion. Nevertheless, in many situations friction can be reduced to the point that it is a negligible contribution to the effect under study and so it is still useful to consider these idealized "frictionless" cases. For example, a flat disk (such as a hockey puck) will coast at nearly constant velocity for a great distance over smooth, flat ice. In many of the following discussions we will, for simplicity, assume that frictional effects are unimportant. However, at the end of this chapter we will give some details concerning the effects of friction.

Newton's First Law — The Law of Inertia

The tendency for the inertia of an object to resist a change in its state of motion or rest is embodied in *Newton's first law of motion,* sometimes called the *law of inertia*. If there is no force applied to an object at rest, it will remain at rest. But it is also true that if there is no force acting on an object in motion, it will continue to move in a straight line with constant velocity. These two facts can be combined in the statement,

> If the net force acting on an object is zero, then the acceleration of the object is zero and it moves with constant velocity.

In symbols, Newton's first law becomes

$$\mathbf{F} = 0 \quad \text{implies} \quad \mathbf{a} = 0 \quad \text{or} \quad \mathbf{v} = \text{constant} \tag{4-1}$$

It is important to realize that Eq. 4-1 is a *vector* equation. That is, $F_x = 0$ implies v_x is constant, regardless of the force and the motion in the y-direction. This is exactly the case of motion under the influence of gravity discussed in Section 3-7. The Earth's gravity exerts a force on an object in the downward direction (that is, the y-direction) but not in the x-direction. Therefore, the vertical motion is accelerated, but the horizontal motion takes place with constant velocity ($F_x = 0$ and v_x = constant).

Aristotle taught that the normal state of every object is to be at rest and that all motion is due to external influences (forces). Newton's reasoning showed Aristotle's logic to be incorrect. The "natural condition" of an object (that is, the state of the object in the absence of an external force) is one of zero acceleration. Thus, motion with constant velocity is as natural a condition as one of rest.

Sir Isaac Newton (1642–1727). In 1661 Newton entered Trinity College, Cambridge, but the university was closed in 1665 when the bubonic plague crept beyond the confines of London, and he returned to his home in Woolsthorpe. Here Newton spent two years engaged in a series of experiments in optics and in the initial development of his gravitation theory. In 1667 he returned to Trinity College as a fellow. While still a young man, Newton made enormous contributions to physics and mathematics. He formulated the theory of dynamics based on his famous three laws; he established his theory of universal gravitation by applying the description to planetary and cometary motions; he significantly advanced the science of optics and invented the reflecting telescope, the astronomical instrument most commonly used today; and in mathematics, he introduced the binomial theorem and invented (independently of Leibniz) the calculus. In his later years, Newton turned to theology and mysticism—almost half of his writings (a total of some 5 million words) are on these subjects. He became quarrelsome and suspicious, developing a rather unpleasant personality. In 1699 Newton became Warden of the Mint, and later he was advanced to Master of the Mint. Although some biographers have downgraded Newton's administrative posts, he actually performed an extremely valuable service by reorganizing the Mint. Because it was common practice to clip tiny amounts of silver from the hammered coins of the day, the value of British coinage was always in doubt and, consequently, trade and commerce could only creep along. Newton not only introduced coins with milled edges so that clipping could be readily detected, he increased the output of silver coins at the Mint by almost a factor of ten. Thus, Newton, who had such a profound influence on science, also made a significant contribution to British economics.

NIELS BOHR LIBRARY, AIP

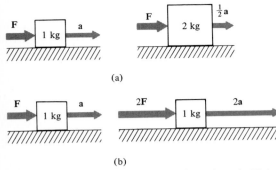

(a)

(b)

Figure 4-2 (a) A given force will produce half the acceleration if the mass is doubled. (b) If the force applied to a given object is doubled, the acceleration is also doubled.

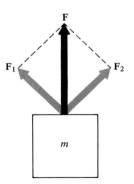

Figure 4-3 The application of F_1 and F_2 to the object produces the net force F.

4-2 DYNAMICS

Newton's Second Law

Newton's first law provides us with a description of the motion of an object in the event that there is zero force applied. But what happens when there *is* an applied force? The *absence* of force results in motion with constant velocity. Therefore, *accelerated motion* must be due to the presence of a force. Indeed, we can define force in the following way:

A force is any influence that can cause a change in the state of motion of an object.

If an object is at rest or moving with constant velocity, the *net* force on the object must be zero. (Forces may be acting on such an object, but they all cancel and sum to zero.) An object that is accelerating is necessarily being acted upon by a net force.

It is easy to appreciate that a given force will produce a greater effect (that is, a greater acceleration) if applied to an object with a mass of 1 kg than if applied to a 2-kg object. Similarly, when applied to a given object, the greater of two forces will produce the greater acceleration (Fig. 4-2). Newton expressed these facts in his *second law of motion*:

$$\mathbf{F} = m\mathbf{a} \qquad (4\text{-}2)$$

This equation not only expresses the direct proportionality between force and mass and between force and acceleration, but because it is a *vector* equation, there is the additional statement that the *direction* of the force is the same as the *direction* of the resulting acceleration.

The force $\mathbf{F}$ that appears in Eq. 4-2 is the *net* force acting on the object with mass m. That is, $\mathbf{F}$ is the *vector sum* of all the individual forces (including any frictional force) acting on m. If, as shown in Fig. 4-3, a force $\mathbf{F}_1$ acts on an object in the direction northwest and another force $\mathbf{F}_2$, of the same magnitude, acts in the direction northeast, the *net* force $\mathbf{F}$ on the object is due North. There is no difference between the application to the object of the two separate forces, $\mathbf{F}_1$ and $\mathbf{F}_2$, and the application of the force $\mathbf{F}$ alone.

Every acceleration—whether that of an automobile on the street, of a rock that is dropped from the top of a building, or of the Earth in its orbit around the Sun—is the direct result of a force. But an acceleration is not produced with every application of a force. For example, as shown in Fig. 4-4, two forces of equal magnitude applied in opposite directions to a block will produce no acceleration because the *net* force on the block is zero.

Every object moves in accordance with the forces acting on the object and the acceleration is given by $\mathbf{a} = \mathbf{F}/m$. An airplane wing, for example, has several forces acting on it (Fig. 4-5). The aircraft will be accelerated forward as long as the thrust exceeds the drag, and it will rise at an increasing rate if the lift is greater than the downward force due to gravity.

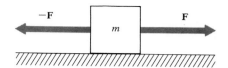

Figure 4-4 The *net* force on the block is zero, $\mathbf{F} + (-\mathbf{F}) = 0$, and so there will be no acceleration of the block.

The Unit of Force

According to Newton's second law, Eq. 4-2, the force applied to an object is equal to the product of the object's mass and acceleration. Thus, if a 1-kg object is accelerating at a rate of 1 m/s², the net applied force must be 1 kg-m/s². To this force we give the special name, 1 newton (N). That is,

$$1 \text{ N} = 1 \text{ kg-m/s}^2 \qquad (4\text{-}3)$$

Summarizing Newton's second law,

$$\mathbf{F} = m\mathbf{a}$$

where

$\mathbf{F} = force$ in newtons (N)

$m = mass$ in kilograms (kg)

$\mathbf{a} = acceleration$ in meters per second per second (m/s²)

We must remember that the direction of the acceleration is the same as the direction of the *net* force acting on the object.

Accelerated Motion

Let us now consider some specific cases of accelerated motion due to the application of external forces to various objects.

Suppose that a 1000-kg automobile is moving with a velocity of 20 m/s. What (constant) braking force is necessary to slow the automobile uniformly to rest in 5 s?

The acceleration of the automobile is

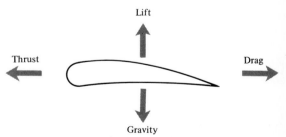

Figure 4-5 The forces acting on an airplane wing. The *thrust* is provided by the engine and the *drag* is due to air resistance (friction). The upward force (*lift*) is generated by the air flow over the wing structure, and *gravity* acts downward. The aircraft accelerates in the direction which is the vector sum of these four forces.

$$a = \frac{v_1 - v_0}{t_1 - t_0} = \frac{0 - 20 \text{ m/s}}{5 \text{ s} - 0}$$
$$= -4 \text{ m/s}^2$$

where the negative sign means that the acceleration is in the direction *opposite* to the motion. The required braking force is

$$F = ma = (1000 \text{ kg}) \times (-4 \text{ m/s}^2)$$
$$= -4000 \text{ N}$$

where the negative sign means a *retarding* force.

When a missile is fired, the rocket engine burns for a short period and causes a (nearly) constant force to be exerted on the missile. If a 2000-kg missile is subjected to a force of 80 000 N for 10 s, what will be the final velocity?

We know from the discussion in the preceding chapter that the final velocity will be equal to the product of the acceleration and the time. And we know that the acceleration is equal to the force divided by the mass. Thus,

$$v = at = \frac{F}{m} \times t$$
$$= \frac{80\,000 \text{ N}}{2000 \text{ kg}} \times 10 \text{ s}$$
$$= 400 \text{ m/s or } 1440 \text{ km/h}$$

Acceleration is sometimes expressed in terms of the normal acceleration due to gravity g. Thus, an acceleration of $3g$ is $3 \times 9.8 \text{ m/s}^2 = 29.4 \text{ m/s}^2$. In the example above, the missile acceleration is 80 000 N/2000 kg = 40 m/s² or approximately 4 g. Many persons will "black out" when subjected to an acceleration in excess of about $5g$ if the direction of acceleration is from foot to head because the blood then tends to drain from the head and dulls the brain. If the acceleration is directed perpendicular to the foot–head direction, considerably higher accelerations can be tolerated. Astronauts are always seated perpendicular to the acceleration at blast-off.

Finally, let us consider a more complicated situation. Figure 4-6a shows a block acted upon by three forces. What is the net effect of these forces? We know that the individual forces must be combined vectorially to produce $\mathbf{F}_{\text{net}}$:

$$\mathbf{F}_{\text{net}} = \mathbf{F}_1 + \mathbf{F}_2 + \mathbf{F}_3$$

A vector equation such as this can also be expressed in terms of components. That is, we can write *two equa-*

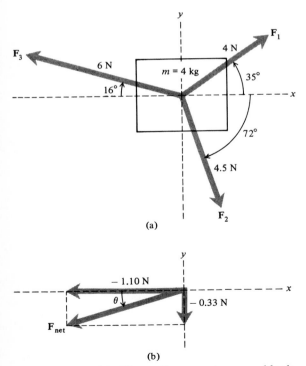

(a)

(b)

Figure 4-6 (a) Three forces act on a block. (b) The net force is the vector sum of the three individual forces.

tions, one for the *x*-components and one for the *y*-components:

$$F_{\text{net},x} = F_{1,x} + F_{2,x} + F_{3,x}$$
$$F_{\text{net},y} = F_{1,y} + F_{2,y} + F_{3,y}$$

Using the values shown in Fig. 4-6a, we find

F₁: $F_{1,x} = 4 \cos 35° = 3.28$
F₂: $F_{2,x} = 4.5 \cos 72° = 1.39$
F₃: $F_{3,x} = -6 \cos 16° = -5.77$
$\phantom{\textbf{F}_3:\ } F_{1,x} + F_{2,x} + F_{3,x} = \overline{-1.10}\ \text{N}$

$$F_{1,y} = 4 \sin 35° = 2.29$$
$$F_{2,y} = -4.5 \sin 72° = -4.28$$
$$F_{3,y} = 6 \sin 16° = 1.65$$
$$F_{1,y} + F_{2,y} + F_{3,y} = \overline{-0.33}\ \text{N}$$

The three numbers listed here actually sum to −0.34 N. However, this result is subject to round-off error in the individual numbers. The correct sum is −0.33 N.

Thus, the net force has the following components

$$F_{\text{net},x} = -1.10\ \text{N}$$
$$F_{\text{net},y} = -0.33\ \text{N}$$

These force components are shown (to a different scale) in Fig. 4-6b. The value of the angle θ is determined by finding the angle whose tangent is $0.33/1.10 = 0.30$; this angle is 16.7°. The magnitude of the net force can be found by using the Pythagorean theorem:

$$F_{\text{net}} = \sqrt{(-1.10\ \text{N})^2 + (-0.33\ \text{N})^2} = 1.15\ \text{N}$$

Therefore, the acceleration of the block is

$$a = \frac{F_{\text{net}}}{m} = \frac{1.15\ \text{N}}{4\ \text{kg}} = 0.29\ \text{m/s}^2$$

and the direction is to the left, 16.7° below the *y*-axis.

All problems involving the addition of several forces (or any other vectors) can be solved by using this method. First, decompose each vector into *x*- and *y*-components. Then, add the *x*-components and the *y*-components separately. Finally, construct the resultant vector from these sums, obtaining the magnitude and the direction.

Reference Frames

Newton's first and second laws involve the concepts of "rest," "motion with constant velocity," and "acceleration." In order to determine each of these quantities, a *reference frame* or *coordinate system* must be chosen with respect to which measurements can be made. One must be careful in choosing reference frames because not all frames are equally useful. Suppose that you are riding in an automobile that is accelerating. The observations you make with respect to your reference frame (the automobile) indicate that the rest of the world is accelerating past you. Newton's law, $F = ma$, states that a force must be associated with any acceleration. But you know that the world accelerating past you is not being subjected to a force. What is wrong with the analysis? The problem is that you are using the law, $F = ma$, incorrectly. *Newton's laws are not valid in an accelerating frame.* Newton understood this problem and realized that in order for his laws to describe motion correctly it is necessary to use a nonaccelerating reference frame.

How does one find a nonaccelerating reference frame? An *Earth-frame* (that is, a frame fixed with respect to the Earth) is not really an acceptable frame because the Earth undergoes a complicated accelerated motion in moving around the Sun and rotating on its axis. The magnitude of the Earth's acceleration is rather small, however, and for describing small-scale motions (such as laboratory experiments), an Earth-frame proves quite adequate. If we wish to use Newton's laws to describe the motion of satellites or planets, then an Earth-frame is not satisfactory and we must use a frame that is fixed with respect to the Sun, or even better, with respect to the distant stars.

A reference frame fixed with respect to the distant stars is not the only frame in which Newton's laws provide a correct description of the motion of objects. *Any* nonaccelerating frame is equally valid. If Newton's laws are true in some reference frame (for example, the distant-star frame), then the laws are also true *in any other frame that moves with constant velocity with respect to the first frame* (Fig. 4-7). All frames in which Newton's laws are valid are called *inertial reference frames*.

We now see the reason why Newton's first law implies that the states of "rest" and "motion with constant velocity" are equally natural states of motion. An object which moves with constant velocity in one inertial reference frame can be at rest in another inertial frame (namely, the frame that moves with the object), and Newton's laws are valid in both frames. There is no *one* reference frame that is preferred over all others — there is no reference frame at "absolute rest" and even the concept of "absolute rest" is meaningless.

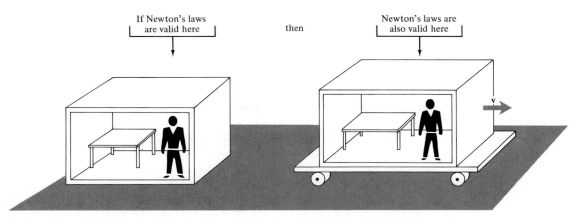

Figure 4-7 An *inertial reference frame* is one in which Newton's laws of motion are valid. Any reference frame which moves with constant velocity with respect to an inertial reference frame is also an inertial reference frame.

Mass and Weight

Mass and *weight* are terms that are frequently used interchangeably. However, mass and weight are *not* the same (even the physical units are different) and the terms refer to separate physical concepts. *Mass* is an intrinsic property of matter. An object contains the same amount of material (that is, the same number of atoms) regardless of its location, whether on the surface of the Earth, on the Moon, or in space. Mass is a measure of the inertia possessed by an object, that is, the tendency of the object to resist changes in its state of motion. The same force (1 N) is required to impart an acceleration of 1 m/s^2 to an object with a mass of 1 kg *no matter where that object is located.*

Weight is the gravitational force acting on an object and this force *does* depend on the location of the object. The gravitational force on an object near the surface of the Earth is greater (by about a factor of 6) than the gravitational force on the same object if it is on the surface of the Moon. And if the object is in distant space, far from any body that will gravitationally attract it, the gravitational force on the object, and hence its weight, will be zero.

The expression for computing the *weight* of an object is just Newton's equation $F = ma$. Since the weight w of an object with a mass m is the gravitational force on that object, we must replace F with w and the acceleration a with the acceleration due to gravity g. Thus, $F = ma$ becomes

Acceleration and Force

When an automobile accelerates, you feel yourself forced backward into the seat; that is, the automobile (through the seat) exerts on you a force which causes you to accelerate. Conversely, if the automobile decelerates rapidly from a high speed, you feel the pull of the seat belt and shoulder harness. If the deceleration is sufficiently high (as would be the case in a collision), the restraining force exerted by the belt and harness can do damage to the rider. Experiments with baboons have shown that a deceleration of 32g (that is, 32 times 32 ft/s² or 1024 ft/s², which corresponds to coming to a stop from 35 mi/h in 2 ft) will result in fatal injuries in about half of the cases if the subjects are restrained only with lap belts. On the other hand, with both a lap belt and a shoulder harness, the deceleration for 50 percent fatality increases to about 100g. (At a deceleration of 100g the total force on an individual would be 100 times his own weight!)

In order to decrease the probability of fatal injury in a collision, the *air-bag* system has been developed for use in automobiles. A "crash sensor" triggers the inflation of a balloonlike bag in front of the rider. The time required for inflating the bag is only a small fraction of a second, quick enough to cushion the rider and to prevent collision with the interior of the vehicle. The bag then rapidly deflates so that the rider will not "rebound." The entire cycle requires less than one second.

Tests with the air-bag system, again using baboons as stand-ins for humans, have shown only minor injuries for decelerations as high as 120g. Other tests with human volunteers protected with air bags showed no injuries when the automobile was driven into a fixed barrier at 30 mi/h or into a parked car at 60 mi/h (equivalent to 22g).

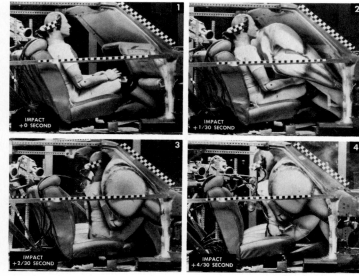

GENERAL MOTORS CORPORATION

Photographs of an air-bag system for collision protection taken at intervals of $\frac{1}{30}$ of a second as an impact sled is driven into a fixed barrier. After only $\frac{1}{15}$ s, the bag is fully inflated and cushions the dummy rider. In the fourth photograph, the bag is beginning to deflate.

$$\boxed{w = mg} \qquad\qquad (4\text{-}4)$$

The dimensions of weight are the same as those of force, namely, *newtons*. The weight of a 100-kg man on the surface of the Earth, where $g = 9.8$ m/s², is

$$w_{\text{on Earth}} = (100 \text{ kg}) \times (9.8 \text{ m/s}^2) = 980 \text{ N}$$

and on the surface of the Moon, where $g = 1.62$ m/s², the weight of the same man would be

$$w_{\text{on Moon}} = (100 \text{ kg}) \times (1.62 \text{ m/s}^2) = 162 \text{ N}$$

Summarizing,

> *Mass:* a measure of inertia; the amount of matter in an object.
> *Weight:* the gravitational force acting on a body.

In the British system of units, the term "pound" once referred to *weight* and was the standard unit of *force*. But now the pound is legally defined as a unit of *mass:* 1 lb = 0.45359 kg. Thus, it is not correct to say that a man "weighs" 200 lb. Instead, we should say that he has a *mass* of 200 lb; the equivalent *weight* is

$$w = (200 \text{ lb}) \times \left(\frac{0.45359 \text{ kg}}{1 \text{ lb}}\right) \times (9.8 \text{ m/s}^2) \cong 890 \text{ N}$$

We frequently determine mass by the process of "weighing." Two such methods are illustrated in Fig. 4-8. In each case we are actually comparing the gravitational force mg on the object with unknown mass to that on an object with a standard mass. If the value of g is the same in each measurement, then the method amounts to the comparison of masses. (But would a spring balance that had been calibrated in mass units on Earth read masses properly on the Moon?)

It is important to note, however, that the basic concept of *mass* does not involve the influence of gravity nor does it even require the existence of the gravitational force. A net force of 1 N applied to a mass of 1 kg will always produce an acceleration of 1 m/s², regardless of the nature of the particular force. If the force is due to the muscular action of an astronaut in deep space, far from the gravitational effect of any star or planet, the result will still obey the relation $F = ma$. Weight, on the other hand, *is* a force. The effect of the gravitational force on a particular object is exactly the same as any other force (with the same magnitude and

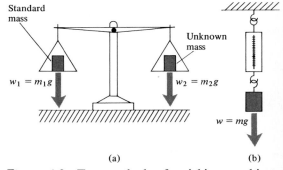

Figure 4-8 Two methods of weighing an object. (a) The gravitational force $m_2 g$ on an unknown mass is compared with the gravitational force $m_1 g$ on a standard mass. If the pans are in balance, then the two weights are the same, $w_1 = w_2$, and, since the value of g is the same at the positions of the two objects, the masses are then also equal, $m_1 = m_2$. (b) The weight of the object is measured by the amount of stretching of the spring in the spring balance. (The spring has been calibrated by using a series of standard masses.)

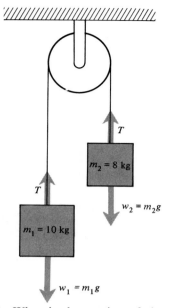

Figure 4-9   What is the motion of the system?

direction). Thus, a gravitational force of 1 N acting on a mass of 1 kg will produce an acceleration of 1 m/s². In this case we say that the object *weighs* 1 N.

If gravity were everywhere the same, an object would always have the same weight. But, of course, gravity does vary from place to place. (Even over the surface of the Earth, the value of the gravitational acceleration g varies by about 0.5 percent.) It is this variation that causes an object to have different *weights* even though its *mass* is constant.

Weight as a Force

Weight is simply one of the forces which we include in our list of pushes and pulls that can cause changes in motion. In Fig. 4-9 we consider the effect of gravity on two blocks connected by a string that passes over a pulley. The weight of the larger block m_1 is $(10 \text{ kg}) \times (9.8 \text{ m/s}^2) = 98$ N, and that of the smaller block m_2 is $(8 \text{ kg}) \times (9.8 \text{ m/s}^2) = 78.4$ N. Each weight (that is, each gravitational force) is directed downward. What is the effect of these forces on the motion of the system?

We can analyze the problem by writing $F = ma$ for each block. The *net* force on m_1 is the sum of its weight m_1g (downward) and the pull T of the string (upward). Similarly, the *net* force on m_2 is the sum of its weight m_2g and the pull of the string. Notice that the pull of the string has the same magnitude and direction in each case; we neglect any frictional effects in the pulley. Notice also that the accelerations of the two blocks have the same magnitude but opposite directions; that is, $a_2 = -a_1$. If we choose the downward direction as positive, we can write

$$m_1g - T = m_1a_1$$
$$m_2g - T = m_2a_2 = -m_2a_1 \qquad (4\text{-}5)$$

Solving each equation for T and equating the results, we find

$$m_1g - m_1a_1 = m_2g + m_2a_1$$

Finally, solving for a_1,

$$a_1 = \frac{m_1 - m_2}{m_1 + m_2} g \qquad (4\text{-}6)$$

Substituting numerical values,

$$a_1 = \frac{10 \text{ kg} - 8 \text{ kg}}{10 \text{ kg} + 8 \text{ kg}} \times (9.8 \text{ m/s}^2)$$

$$= 1.09 \text{ m/s}^2$$

Thus, m_1 accelerates *downward* at a rate 1.09 m/s² and m_2 accelerates *upward* at the same rate.

In 1784 the English scientist George Atwood (1746–1807) set out to make an accurate measurement of the acceleration due to gravity g. Such a determination can be made by timing the free fall of an object through various distances. But this technique involves the accurate measurement of very short time intervals, and the clocks available to Atwood were not sufficiently precise. In order to overcome this difficulty, Atwood used a pulley system similar to that shown in Fig. 4-9 (called an *Atwood machine*) to "dilute." the effect of gravity. As we have seen in the example, if m_1 and m_2 have nearly the same mass, the acceleration will be much smaller than g. The motion will therefore be much slower than that of a freely falling object and the time measurements will be much easier to make.

Next, look at Fig. 4-10. Here, we have a block on a plane which is inclined at an angle of 30° with respect to the horizontal. We assume that the friction between the block and the plane is negligible. (Later in this chapter, we will consider cases in which friction is important and must be included in the calculation.) We know that the block will slide down the plane due to the influence of gravity. What is the acceleration in this case?

First, we note that the acceleration of the block down the plane must be due to a force that acts in this direction. The only force that can cause this motion is the gravitational force, the weight of the block. But the weight mg is directed *downward*. However, this downward force does have a component *along* the plane (and also a component *perpendicular* to the plane). The magnitude of this force component is $mg \sin 30°$. This is the net force acting on the block and it moves with an acceleration

$$a = \frac{F_{\text{net}}}{m} = \frac{mg \sin 30°}{m} = g \sin 30°$$

$$= (9.8 \text{ m/s}^2) \times (0.500) = 4.9 \text{ m/s}^2$$

Notice that the mass of the block does not enter the final result; this is the same as the case of free fall. (Can you see why these two cases are similar?)

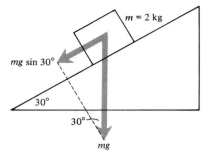

Figure 4-10   If friction is unimportant, what is the acceleration of the block down the plane?

"Weightlessness" in Space—What Does It Mean?

We have often heard the comment that an astronaut in a space vehicle orbiting the Earth or cruising to the Moon is "weightless." What does this term really mean? Suppose that an astronaut is in a vehicle that is undergoing unpowered flight at a distance of 6380 km above the Earth's surface. If we were to determine g at this position, we would find a value of approximately 2.45 m/s². Then the weight of a 90-kg astronaut would be, according to Eq. 4-4,

$$w = mg = (90 \text{ kg}) \times (2.45 \text{ m/s}^2) = 220.5 \text{ N}$$

This weight is only one-quarter of that which we would measure for the same astronaut at the surface of the Earth—but it is certainly not zero!

Where, then, does the idea of "weightlessness" come from? Suppose that the astronaut attempts to determine his weight in the conventional manner by standing on a bathroom-type scale. The scale will read zero. If the astronaut really has weight, why is it not registered by the scale? Stop and think how you use a bathroom-type scale. The scale rests on the floor and you stand on top of the scale. Gravity pulls you downward and you exert a force on the springs of the scale. The scale registers the force—this is your weight. But now suppose that the floor beneath the scale is suddenly removed and you and the scale fall freely downward. You and the scale are both accelerating downward at the same rate. It is no longer possible for you to exert a force on the scale—gravity pulls the scale toward the Earth just as rapidly as gravity pulls you toward the scale. You are "weightless."

It is in this same sense that an astronaut in space is "weightless." In unpowered flight, the space vehicle is falling freely—either around the Earth in orbit or perhaps around the Sun. The astronaut in the space vehicle is also falling freely and he exerts no force on the sides of the vehicle or on any scale within the vehicle. The astronaut, or any object, floats inside the vehicle in a state of "weightlessness."

As long as the value of g is not zero at the particular location, any object will have a weight given by $w = mg$, even if the object is in free fall. The condition of "weightlessness" means only that gravity cannot press a freely falling object against another freely falling object.

4-3 ACTION AND REACTION

Newton's Third Law

Whenever we push on an object we always experience a reluctance of the object to move: the object resists the applied force and *pushes back*. Even if we push very gently, we can always *feel* the object; this sensation of feeling is due to the force that the object exerts on the hand or finger. Every object reacts in this way to the application of a force. The reaction force is in the direction opposite to that of the applied force and the magnitudes of the two forces are exactly equal. This is *Newton's third law,* the law of action and reaction:

> If object 1 exerts a force on object 2, then object 2 exerts an equal force, oppositely directed, on object 1.

According to Newton's third law, *every* force is accompanied by an equal and opposite reaction force. Thus, no force can occur alone—all forces occur in pairs. A book resting on a table exerts a downward force on the table (the weight of the book), and the table exerts an upward force on the book (Fig. 4-11). Similarly, the table exerts a downward force on the floor and the floor exerts a force on the foundation and the Earth. In each case there is a corresponding reaction force exerted upward from the Earth through the foundation, the floor, and the table to the book.

If all forces occur in pairs, how can movement ever take place? Should not all forces just cancel? The important point to remember is that Newton's second law states that motion (acceleration) is due to the net force *on* an object. It does not matter whether that object is exerting reaction forces on other objects; if there is a net force acting *on* an object, that force governs the motion of the object. Suppose that a man braces himself against a wall and pushes on a box (Fig. 4-12). The man pushes on the wall, and the wall pushes back. Similarly, the box pushes back on the man. The net force on the man is zero and so he does not move. But what forces act on the box? Only the force exerted by the man. Therefore, the net force on the box is not zero, and the box will move away from the man. (The wall does not move. Are there forces on the wall not shown in the diagram?)

Another way to see that the occurrence of forces in pairs does not prevent objects from moving is to note that the two forces in a pair always act on *different* objects. In Fig. 4-12, the man exerts a force on the block and the block exerts a force on the man. These two forces constitute an action–reaction pair. But one force acts on the block and the other acts on the man. To solve a problem of motion, always proceed by identifying all of the forces that act *on* the object. Only these forces determine the object's acceleration.

As we will see in the next chapter, Newton's law of action and reaction leads to a powerful physical principle—the law of momentum conservation.

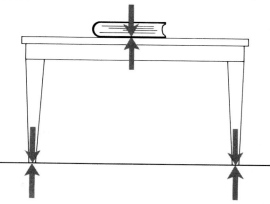

Figure 4-11 Forces and reaction forces existing among a book, a table, and the floor.

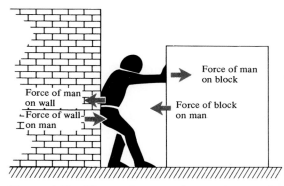

Figure 4-12 The net force on the man is zero; he does not move. The net force on the box is *not* zero and the box *will* move.

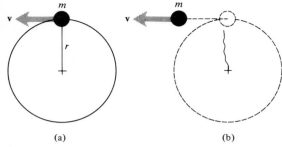

(a)   (b)

Figure 4-13   (a) The ball moves in a circular path because of the centripetal force exerted on the ball through the string. (b) If the string breaks, the force on the ball becomes zero and it moves off in a straight line.

Centripetal Reaction

An interesting case involving a reaction force occurs in circular motion. Suppose that you have a ball attached to the end of a string and that you whirl it in a circle, as in Fig. 4-13a. What forces are at work here? Because the ball does not move in a straight line, we know that it is being accelerated. (Remember, a change in direction of the velocity vector implies acceleration; see Section 3-7). The acceleration of the ball (the centripetal acceleration) is caused by the force you exert by pulling on the string. We call this the *centripetal force,* and it is equal to the mass of the ball multiplied by the acceleration:

$$F_c = ma_c = \frac{mv^2}{r} \qquad (4\text{-}7)$$

where we have used Eq. 3-19 for the centripetal acceleration.

According to Newton's law of action and reaction, there must be a reaction force that is equal in magnitude and opposite in direction to the centripetal force. The centripetal force is the inward pull of the string on the ball. The reaction force is the outward pull of the ball on the string; this we call the centripetal reaction or the *centrifugal force*.

One often hears a totally incorrect statement concerning centrifugal force. It is said that the ball moves with the restraining string at its maximum extension because there is an outward centrifugal force acting on the ball. There is actually only one force that acts on the ball—the inward centripetal force. The only outward force is the centrifugal force exerted *by* the ball on the string.

What happens if the string breaks? At one instant the ball is moving with a certain velocity **v** and is acted upon by the centripetal force $\mathbf{F}_c$, as in Fig. 4-13a. At the next instant, the centripetal force disappears (because it had been transmitted through the string). Suddenly, the ball finds itself subject to no external force. According to Newton's law of inertia (the first law), the ball must move with constant velocity. This constant velocity is the velocity **v** that existed the instant the string broke (Fig. 4-13b). That is, the ball moves away on a path that is tangent to the circle of previous motion. The ball does *not* move directly outward in line with the string direction. (This is another false statement that one sometimes hears.)

Forces in Equilibrium

Newton's laws provide us with a method for describing the behavior of objects in motion. The same laws also allow us to analyze the forces acting on stationary objects. If an object is at rest, we know that both the acceleration and the velocity are zero. Newton's first law (Eq. 4-1) then tells us that the force acting on the object is also zero. But we must remember that the equation $\mathbf{F} = 0$ refers to the *net* force on the object. Generally, there will be several individual forces acting on an object. In some cases we may not know all of the various forces that are acting; but if the object is at rest, we are assured that the vector sum of all the individual forces is precisely *zero:*

> For an object at rest, the vector sum of all of the forces acting on the object is zero.

When the forces on an object sum to zero, we say that the object is in *equilibrium* (or that the *forces* are in equilibrium). A book at rest on a table (Fig. 4-11) is in equilibrium because the downward force of gravity is exactly equal to the upward reaction force of the table.

If more than two forces act on an object, we usually compute the net force by using the method of components, discussed in Section 4-2. For an object to be in equilibrium, the net force must be zero. This means that *both* $F_{\text{net},x}$ and $F_{\text{net},y}$ must be zero.

Examples of Equilibrium

Consider the case shown in Fig. 4-14. Here a ball with mass $M = 6$ kg is attached to a fixed support by means of a string. A force $\mathbf{F}$ pulls in a horizontal direction on another string attached to the ball until the first string makes an angle of 30° with the vertical. In this situation the ball is at rest. There are three forces acting on the ball: the horizontal force $\mathbf{F}$, the downward force due to gravity $\mathbf{F}_G$, and the force $\mathbf{T}$ due to the tension in the support string. Because the ball is at rest and therefore in equilibrium, we know that both the x- and y-components of the net force are zero. In this case, $\mathbf{F}$ has no y-component and $\mathbf{F}_G$ has no x-component. Therefore, the x-component of $\mathbf{T}$ is equal to $\mathbf{F}$ and the y-component of $\mathbf{T}$

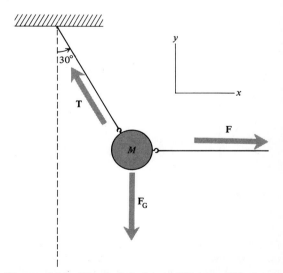

Figure 4-14   The ball is in equilibrium. What are the magnitudes of the three force vectors?

is equal to F_G. First, we compute F_G, the *weight* of the ball:

$$F_G = Mg = (6 \text{ kg}) \times (9.8 \text{ m/s}^2) = 58.8 \text{ N}$$

Next, equating the y-components of the forces,

$$T_y = T \cos 30° = F_G$$

from which,

$$T = \frac{F_G}{\cos 30°} = \frac{58.8 \text{ N}}{0.866} = 67.9 \text{ N}$$

Finally, equating the x-components of the forces,

$$T_x = T \sin 30° = F$$

from which

$$F = (67.9 \text{ N}) \times (0.500) = 33.95 \text{ N}$$

A slightly different type of equilibrium situation is shown in Fig. 4-15. Here, a 50-kg block rests on a perfectly smooth inclined plane. (We ignore the effects of friction for the moment; see the next section.) The block is held in position and prevented from sliding down the plane by the application of the force **F** which acts parallel to the plane. Two forces that obviously act on the block are the applied force **F** and the downward gravitational force **F**$_G$. But there is a third force acting as well—the reaction force of the plane. Notice that this force acts in the direction *perpendicular* to the plane. (In mathematical language, "normal" means "perpendicular"; for this reason, the force is usually called the *normal* force and represented by the symbol **N**.) Can you see why the reaction force acts in the perpendicular direction? Suppose, instead, that this force were to act directly upward as it does in the case of the book resting on the table (Fig. 4-11). Then, if the restraining force **F** is removed, the upward and downward forces would exactly cancel and the block would remain stationary. But we know this will not happen; the block will slide down the plane. The reason is that the gravitational force has a component along the plane but the normal force does not. This results in a net force along the plane when **F** is removed.

In order to analyze the situation, we select a set of xy-axes such that the x-axis is parallel to the inclined plane and the y-axis is perpendicular to the plane, as in Fig. 4-15b. The vector **F** therefore lies along the x-axis and the vector **N** lies along the y-axis. The vector **F**$_G$ lies along neither axis, but we can resolve **F**$_G$ into compo-

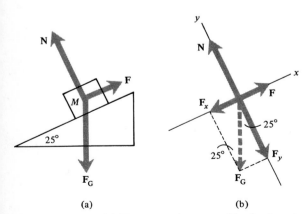

(a)   (b)

Figure 4-15   (a) Forces acting on a block at rest on an inclined plane. (b) Force diagram resulting from resolving **F**$_G$ into components acting parallel to and perpendicular to the plane.

nents that do lie along the axes; this is shown in Fig. 4-15b. Now we have four vectors that must represent an equilibrium situation. That is, $\mathbf{F}_y$ and $\mathbf{N}$ must exactly cancel, and likewise for $\mathbf{F}_x$ and $\mathbf{F}$. We can solve for the magnitudes of the forces as follows:

$$F_G = Mg = (50 \text{ kg}) \times (9.8 \text{ m/s}^2) = 490 \text{ N}$$
$$F = F_x = F_G \sin 25° = (490 \text{ N}) \times (0.423) = 207 \text{ N}$$
$$N = F_y = F_G \cos 25° = (490 \text{ N}) \times (0.906) = 444 \text{ N}$$

In this case we could have chosen a horizontal x-axis and a vertical y-axis. We could then have resolved $\mathbf{N}$ and $\mathbf{F}$ into x- and y-components. The results for F_G, F, and N would have been the same.

4-5 FRICTION

Kinetic Friction

Suppose that a block rests on a level surface and that you set the block into motion with a brief push. When you remove your hand from the block, the applied force drops to zero and, according to Newton's first law, the block should continue to move with constant velocity. But you know this does not happen; instead, the velocity of the block decreases until it stops. The reason is that a force *does* act on the sliding block; this force is the frictional force between the block and the surface over which it moves. Friction exists in any physical situation in which an object moves over or through matter. Friction exists between a rolling automobile tire and the highway, between a swimming fish and the water, and between a rifle bullet and the air. Only when an object moves through a perfect vacuum (an unrealizable situation) will friction disappear completely. By polishing two surfaces which slide or roll over one another or by supplying a lubricating fluid between the surfaces, frictional effects can be reduced. In some cases it is possible to reduce friction to the point that it can be ignored in analyzing a problem. But generally we must include in our computations an allowance for the effects of friction.

The type of friction that exists between an object and the surface over which it moves is called *kinetic* (or *sliding*) friction. The vector $\mathbf{f}_k$ that represents the kinetic friction acting on an object is in the direction opposite to the motion of the object (Fig. 4-16).

How can we learn about the magnitudes of frictional forces? Consider a block that is sliding over a level sur-

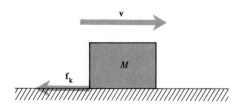

Figure 4-16 The frictional force $\mathbf{f}_k$ acting on a sliding block is in the direction opposite to the motion of the block described by the velocity vector $\mathbf{v}$).

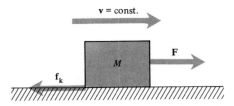

Figure 4-17 If an applied force **F** is required to maintain the block sliding with constant velocity, then the frictional force $\mathbf{f}_k$ is equal in magnitude to the applied force.

face. A certain force **F** is required to maintain the motion with constant velocity (Fig. 4-17). Since **v** = constant, we know that the net force on the block must be zero. Therefore, we conclude that in this situation the frictional force $\mathbf{f}_k$ exactly cancels the applied force **F**; that is, the magnitudes of the two forces are equal, $f_k = F$. Thus, a measurement of an applied force which causes motion with constant velocity is a direct determination of the force of kinetic friction.

If we repeat the force measurement with a different area of contact between the block and the surface (for example, by standing the block on end), we find that the frictional force does not change. Or if we slide the block with a different constant velocity, the frictional force also remains the same. That is, for a given mass of material sliding over a particular surface, the force of kinetic friction is independent (or nearly so) of the area of contact and the velocity. However, suppose that we increase the mass of the material that is sliding by placing an additional block on top of the original block (Fig. 4-18). In this case, we find that a greater applied force is necessary to maintain the motion with constant velocity. The frictional force is therefore greater than before. In fact, the frictional force is directly proportional to the mass of the block. We can state this result in a more general way that is applicable also to situations in which the surface is not level: The force of kinetic friction is directly proportional to the normal force exerted by the surface on the sliding object. If we write the proportionality constant as μ_k, we can express this conclusion in equation form:

$$f_k = \mu_k N \qquad (4\text{-}8)$$

The quantity μ_k is called the *coefficient of kinetic friction*. (Notice that in Figs. 4-16–4-18 we have not shown the downward gravitational force $\mathbf{F}_G$ exerted by the block on the surface nor the normal force **N** exerted by the surface on the block. In each case, however, the frictional force is proportional to this normal force.)

Values of the coefficient μ_k vary widely, depending on the textures of the materials in contact. Clearly, the friction between a rough-sawn piece of wood sliding over concrete is greater than that between the same mass of polished steel sliding over a lubricated surface of smooth glass. This means that μ_k for the wood–concrete system is much greater than μ_k for the steel–glass system. In fact, μ_k (wood–concrete) is about 0.7, whereas μ_k (steel–glass, lubricated) can be as small as

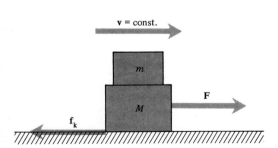

Figure 4-18 Increasing the mass of the sliding block increases the force necessary to maintain a constant velocity and therefore increases the frictional force.

0.05. (These values are, of course, only approximate because the coefficient of friction depends strongly on the precise nature of the surfaces.)

What force would be necessary to push a 10-kg block of wood with constant velocity over a horizontal concrete surface? The force will be equal to the frictional force given by Eq. 4-8, and the normal force N is equal to the weight of the block. Therefore,

$$F = f_k = \mu_k N = \mu_k \times Mg$$
$$= (0.7) \times (10 \text{ kg}) \times (9.8 \text{ m/s}^2)$$
$$= 68.6 \text{ N}$$

For the same mass of steel sliding over a glass surface, a much smaller force is necessary:

$$F = (0.05) \times (10 \text{ kg}) \times (9.8 \text{ m/s}^2)$$
$$= 4.9 \text{ N}$$

Static Friction

Friction exists not only in sliding and rolling motions but in static situations as well. Suppose that you pull with a steady force on a block that rests on a flat surface. If the pull is only a small force, the block will not move. A slightly greater force may still not be sufficient to move the block (Fig. 4-19a,b,c). As the pulling force is further increased, there will eventually come a point at which the block is set into motion (Fig. 4-19d). What was happening before the block began to move? In each case the block was at rest and so the net force was zero. This means that there was another horizontal force which exactly cancelled the pulling force; this force is the force of *static* friction. The *maximum* force of static friction occurs just before the block moves. This force is expressed in equation form in the same way that we described kinetic friction:

$$f_{s \text{ max}} = \mu_s N \quad \text{or} \quad f_s \le \mu_s N \qquad (4\text{-}9)$$

where μ_s is the *coefficient of static friction*.

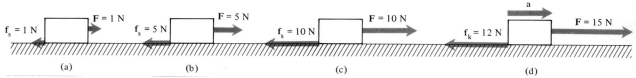

Figure 4-19 As the pulling force $\mathbf{F}$ is increased, (a), (b), (c), the force of static friction increases to maintain the block in static equilibrium. (d) The pulling force is finally sufficiently great to move the block; the net force on the block is 15 N − 12 N = 3 N. Notice that as in the previous diagrams, the forces $\mathbf{F}_G$ and N are not shown.

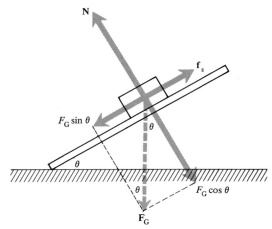

N

f_s

$F_G \sin \theta$

θ

θ

θ

$F_G \cos \theta$

F_G

Figure 4-20   The maximum angle θ at which the block remains at rest determines the coefficient of static friction between the two surfaces: $\mu_s = \tan \theta$.

The value of μ_s for any pair of surfaces is always greater than the corresponding value of μ_k. For example, for steel on dry wood, $\mu_k = 0.2$, whereas $\mu_s = 0.4$. That is, it requires twice the force to set a steel block into motion across a wood surface as it does to maintain the motion at constant velocity once the block is moving.

There is a simple way to measure μ_s by using gravity as the pulling force. Suppose that we place a block of some material on a plank of the desired surface material. Next, we raise one end slowly. As soon as the block begins to slide down the inclined surface, we measure the angle θ between the plane of the plank and the horizontal (Fig. 4-20). The force diagram that we draw applies to the situation immediately *before* motion begins because this is the case of maximum static friction. (After the block is in motion, we have the case of kinetic friction.) The force diagram (Fig. 4-20) is the same as that in Fig. 4-15. The frictional force f_s is equal to $F_G \sin \theta$ (which is the component of F_G along the plane), and the normal force is $F_G \cos \theta$ (which is the component of F_G perpendicular to the plane). The coefficient of static friction is the ratio of these two quantities:

$$\mu_s = \frac{f_s}{N} = \frac{F_G \sin \theta}{F_G \cos \theta} = \frac{\sin \theta}{\cos \theta} = \tan \theta \qquad (4\text{-}10)$$

Thus, if a certain block remains at rest on a particular plank until θ is increased to 28°, the value of μ_s for this combination of surfaces is $\tan 28° = 0.53$.

SUGGESTED READINGS

E. N. da C. Andrade, *Sir Isaac Newton* (Doubleday, Garden City, New York, 1954).

W. Bixby, *The Universe of Galileo and Newton* (American Heritage, New York, 1964).

Scientific American articles:

C. B. Boyer, "Aristotle's Physics," May 1950.

F. Palmer, "Friction," February 1951.

QUESTIONS AND EXERCISES

1. A large mass is suspended from a ceiling by a length of string. Hanging from the bottom of the mass is another length of the same string. Which string will break when a steady downward pull is exerted on the lower string? Which string will break if a sudden downward jerk is applied to the lower string?

2. What are the various controls in an automobile that can produce *acceleration?* (There are at least four.)

3. A 1000-kg automobile is moving with a velocity of 30 m/s. A braking force of 6000 N is applied for 4 s. What is the final velocity of the automobile?

4. Three forces, each of magnitude 10 N, act on a certain object. In what configuration will the net force on the object be as large as possible? Determine whether there is a configuration such that the net force on the object is *zero*.

5. A 50-kg object starts from rest and is subject to a constant force of 5 N. What is the acceleration of the object? What is its velocity after 5 s?

6. Two forces are applied to a 10-kg object: 5 N *north* and 5 N *east*. What is the magnitude and the direction of the acceleration?

7. Two students are attempting to stretch a spring between two posts. Which of the following methods is better? (Explain carefully.) (a) Each student grasps an end of the spring and each pulls his end toward one of the posts. (b) The students attach one end of the spring to one of the posts and then both pull the other end of the spring toward the other post.

8. Two forces act on a certain object: F_1, 3 N to the right and 55° above the *x*-axis, and F_2, 5 N to the right and 18° below the *x*-axis. What additional force must be applied to the object to make the net force zero?

9. What is the weight of a 120-kg man at the surface of the Earth? What would be his weight on the surface of the planet Mars where $g = 3.7$ m/s²?

10. A 5-kg object rests on the surface of the Earth. What is the gravitational force on the object? What is its acceleration? Explain your answer to the second question.

11. An object is thrown upward, rises to its maximum height, and then falls back to the starting point. What is the acceleration of the object during each phase of its motion?

12. An astronaut carries a calibrated spring scale to the planet Jupiter. There he weighs a 1-kg mass and finds that the spring scale reads 25.8 N. What is the value of g on Jupiter?

13. When an object is dropped it accelerates toward the Earth. This acceleration is due to the gravitational force exerted on the object by the Earth. According to Newton's third law, an equal and opposite force is exerted on the Earth by the object. Therefore, the Earth should also move toward the object. Does it?

14. A 1.5-kg ball is swung in a circle at the end of a 2-m string. The ball makes one revolution in 1.6 s. What is the tension in the string? (Neglect gravity here and assume that the ball and the string move in a horizontal plane.)

15. In the previous exercise gravity was neglected. Rework the problem taking into account the effect of gravity. (The string will now describe a cone with the apex at the holding point.) It will be helpful to draw a diagram showing the two forces acting on the ball. The upward component of the tension in the string must equal the downward force *mg*. The horizontal component of the tension is the centripetal force acting on the ball. Notice that the radius of the ball's orbit is now *smaller* than in Exercise 14 and so the speed is correspondingly smaller. Find the tension in the string and the angle that the string makes with the horizontal.

16. In Fig. 4-14, suppose that the force **F**, instead of being horizontal, is at an angle of 30° above the horizontal. Analyze the forces on the ball in this case.

17. A 12-kg object is supported by two strings. One string makes an angle of 60° with respect to the horizontal to the right of the object and the other string is at the same angle but to the left of the ball. What is the tension in each string?

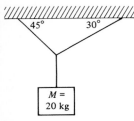

45°   30°

M =
20 kg

Exercise 18

18. What is the tension in each of the three strings used to support the block in the diagram? (Consider the *x*- and *y*-components of each force.)

19. The coefficient of kinetic friction between a 15-kg block and the surface over which it slides is $\mu_k = 0.55$. If a horizontal force of 100 N is applied to the block, what will be its acceleration?

20. Make a sketch of how you imagine the view through a microscope would appear if you examined the contact between surfaces of (a) wood and concrete, and (b) steel and glass. Account for the difference in frictional effects on the basis of this microscopic view.

21. What horizontal pulling force is necessary to maintain a wood block ($M = 8$ kg) moving with a velocity of 1.5 m/s over a flat surface of steel. The coefficient of kinetic friction is $\mu_k = 0.2$.

22. In the diagram below, find the force F and the tension in each of the strings.

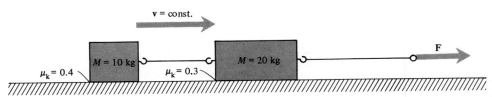

v = const.

M = 10 kg   M = 20 kg   F

$\mu_k = 0.4$ $\mu_k = 0.3$

23. The coefficient of static friction between rubber and concrete is $\mu_s = 0.9$. What is the maximum angle of a paved street on which it would be possible to park an automobile? What is the maximum angle of any street or road that you know of?

24. A block will slide down a certain plank when the angle between the plank and the horizontal is 33°, but the block will not slide if the angle is 29°. From this information how closely can you estimate the coefficient of static friction?

5

LINEAR MOMENTUM, TORQUE, AND ANGULAR MOMENTUM

In this chapter we continue the study of Newtonian dynamics. Although we can (at least, in principle) solve any problem in dynamics by analyzing the forces and accelerations according to Newton's second law, $\mathbf{F} = m\mathbf{a}$, there are many instances in which we do not know in detail all of the forces that are acting. For example, suppose that a bullet is fired into a block of wood that rests on a table. We know from experience that the bullet will bury itself in the wood and that the block will be set into motion. What will be the motion of the block? In order to solve this problem using Newton's second law, we must know the force exerted by the bullet on the block at each instant while the bullet is coming to rest in the block. The process by which this happens is quite complicated because it involves the detailed structure and elastic properties of the wood fibers. We are helpless to solve problems such as this if we have only $\mathbf{F} = m\mathbf{a}$ at our disposal.

Newton realized that there are other dynamical quantities, in addition to force and acceleration, that are important. By introducing a quantity that we now call *momentum* and by making use of the law of action and reaction, Newton was able to formulate a powerful principle—the principle of momentum conservation—which greatly simplifies the solving of many types of problems. Indeed, momentum conservation along with energy conservation (which we discuss in Chapter 7), are two of the most effective tools that we possess to solve problems of dynamics.

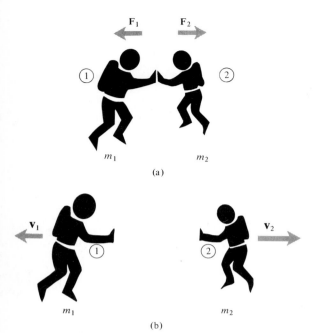

(a)

(b)

Figure 5-1 (a) Two astronauts in deep space push on each other: the two forces, $\mathbf{F}_1$ and $\mathbf{F}_2$, are exactly equal in magnitude. (b) The two astronauts are set into motion by the pushes. The momenta of the two astronauts are equal in magnitude and opposite in direction: $m_1\mathbf{v}_1 = -m_2\mathbf{v}_2$.

5-1 LINEAR MOMENTUM

(Mass) × (Velocity)

Consider two astronauts who are in deep space and therefore are isolated from influences by any other bodies. If the astronauts push on each other, then, according to Newton's third law, the two forces are equal in magnitude and opposite in direction (Fig. 5-1a). That is, astronaut ① pushes on astronaut ② with exactly the same force (oppositely directed) with which astronaut ② pushes on astronaut ①. Because each astronaut has a net force acting on him, each begins to move away from the other. Soon, they are out of each other's reach; then, the forces drop to zero and thenceforth they move with constant velocity (Fig. 5-1b).

The velocity with which astronaut ① moves is $\mathbf{v}_1 = \mathbf{a}_1 t$, where $\mathbf{a}_1$ is the acceleration of ① due to the force exerted on ① by ②, $\mathbf{a}_1 = \mathbf{F}_1/m_1$, and where t is the time during which the astronauts were in contact:

$$\mathbf{v}_1 = \mathbf{a}_1 t = \frac{\mathbf{F}_1}{m_1} t$$

Similarly, the velocity of astronaut ② is

$$\mathbf{v}_2 = \mathbf{a}_2 t = \frac{\mathbf{F}_2}{m_2} t$$

Rewriting these equations, we have

$$m_1\mathbf{v}_1 = \mathbf{F}_1 t \qquad \text{and} \qquad m_2\mathbf{v}_2 = \mathbf{F}_2 t$$

But t is the same time in each case (why?), and the two forces are equal and opposite, $\mathbf{F}_1 = -\mathbf{F}_2$. Therefore, we conclude that

$$m_1\mathbf{v}_1 = -m_2\mathbf{v}_2 \qquad (5\text{-}1)$$

If astronaut ① is *more massive,* he will move away from the site with the *smaller* velocity.

Notice that, in this derivation, the forces need not be constant in time. At every instant we have $\mathbf{F}_1 = -\mathbf{F}_2$, and so the total effect during the time that the astronauts are in contact is still equal in magnitude and opposite in direction.

The result expressed by Eq. 5-1, which is based on a combination of Newton's second and third laws, has extremely far-reaching consequences. Newton realized the great importance of the quantity that is the product of an object's mass and velocity. We call this product the *linear momentum* of the object (Newton called it "quantity of motion") and represent it by the symbol **p**, a vector:

$$\boxed{\text{linear momentum} = \mathbf{p} = m\mathbf{v}} \qquad (5\text{-}2)$$

Although **p** is frequently called the *momentum,* we emphasize here the full term, *linear momentum,* in order to distinguish it from *angular momentum,* which will be introduced later in this chapter.

The dimensions of linear momentum are those of (mass) × (velocity), namely kg-m/s. Thus, the linear momentum of a 3-kg mass moving with a velocity of 6 m/s has the magnitude $p = 18$ kg-m/s.

The Conservation of Linear Momentum

In the astronaut example, before the pushes were exerted neither astronaut was in motion, so the total linear momentum of the pair was zero. After the interaction, the momentum of astronaut ① is $\mathbf{p}_1 = m_1\mathbf{v}_1$ and the momentum of astronaut ② is $\mathbf{p}_2 = m_2\mathbf{v}_2$. Using Eq. 5-1, we find that $\mathbf{p}_1 = -\mathbf{p}_2$, so the total linear momentum is

$$\mathbf{p}_1 + \mathbf{p}_2 = m_1\mathbf{v}_1 + m_2\mathbf{v}_2 = 0$$

Therefore, the total linear momentum *before* the interaction (namely, *zero*) is equal to the total linear momentum *after* the interaction:

$$\boxed{\mathbf{p}_{\text{total (before)}} = \mathbf{p}_{\text{total (after)}}} \qquad (5\text{-}3)$$

Equation 5-3 actually expresses one of the important *conservation principles* of physics. This principle, or law, of linear momentum conservation can be stated as follows:

> For any system of objects that are not subject to any outside forces, the total linear momentum of the system remains constant.

Notice that the only restriction on the application of the principle is that the system must be *isolated*—not subject to forces that are exerted by any agency that lies outside the system. The objects *inside* the system can influence each other in any imaginable way and the principle is still valid. Even the most complicated forces can act within the system and the total linear momentum of the system still remains constant. No evidence has ever

Rocket Propulsion—How It Works

When a toy balloon is blown up and then released, the air rushing out of the filling hole propels the balloon forward and it shoots off on an erratic flight. Why does the balloon move? Does the escaping air push against the atmospheric air and transmit a force to the balloon? Emphatically, no! In fact, the balloon-rocket would work even better in *vacuum* (because there would be no air resistance). The toy balloon is driven by the same mechanism that propels a modern rocket into space: the driving force is a reaction force and the system moves in accordance with the principle of momentum conservation.

Consider the rocket device illustrated schematically in Fig. 5-2. The rocket is isolated in deep space and is at rest in some inertial reference frame. In the rear of this rocket are several tubes through which small pellets can be ejected, all in the same direction. Each pellet has a mass m and is ejected from the rocket with a velocity v. The mass of the pellets is small compared to the mass M of the rocket.

The initial linear momentum of the rocket is zero. When the first pellet is ejected, it carries a linear momentum mv (directed toward the rear). In order for the total linear momentum of the system (rocket plus pellet) to be conserved, the rocket must move *forward* with the same linear momentum, mv. Thus, the linear momentum of the rocket is $MV = mv$, and the rocket moves

Figure 5-2 The rocket is propelled forward by the reaction to the ejection of the pellets from the rear. (Conservation of momentum!)

The launch of the Saturn 1B rocket carrying Skylab 4 on November 16, 1973.

with a velocity $V = (m/M)v$. Each time a pellet is ejected, the rocket acquires another velocity increment $(m/M)v$.

If the pellets are ejected one by one, the rocket moves in a jerky fashion. But if the time between the ejection of successive pellets is made quite small, the acceleration of the rocket will be smooth. The rate of buildup of velocity can be increased if either m or v (or both) is made larger. There is no particular advantage in making m larger: we could achieve the same result simply by ejecting more small pellets. Therefore, the emphasis must be on making v larger. All present-day rockets are propelled by *gas molecules* which are produced by the combustion of solid or liquid fuels and ejected at very high speeds.

Because space rockets are required to operate outside the Earth's atmosphere, the oxidizer for the combustion process must be included as a part of the fuel supply. In liquid-fueled rockets, the oxidizer is usually liquid oxygen and the fuel may be kerosene or liquid hydrogen. The liquid-oxygen–liquid-hydrogen system is one of the most efficient chemical fuels known, providing almost 40 percent more thrust per unit mass of fuel than the liquid-oxygen–kerosene system.

---

been found that any phenomenon in Nature is in disagreement with the principle of momentum conservation. This is indeed a powerful physical principle!

Suppose that a 10-g rifle bullet is fired at a speed of 200 m/s into a block with a mass of 10 kg (Fig. 5-3). In coming to rest in the block, the bullet interacts with the block through forces that are difficult (if not impossible) to analyze in detail. Nevertheless, by using the principle of momentum conservation, we can compute the recoil velocity of the block after the bullet has become embedded. The initial linear momentum of the bullet–block combination is the linear momentum of the bullet alone (because the block is at rest):

$$
\begin{aligned}
p_{\text{total (before)}} &= p_{\text{bullet}} \\
&= (10 \text{ g}) \times (200 \text{ m/s}) \\
&= 2 \text{ kg-m/s}
\end{aligned}
$$

After the bullet has come to rest in the block, the total linear momentum is the mass of the bullet–block combination multiplied by the recoil velocity v. Because the bullet mass is small compared to the mass of the block, the combination has a mass of approximately 10 kg and, hence, the final linear momentum is

$$
p_{\text{total (after)}} = (10 \text{ kg}) \times v
$$

Equating $p_{\text{total (before)}}$ and $p_{\text{total (after)}}$, we find

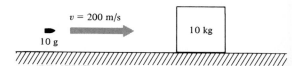

Figure 5-3 Momentum conservation can be used to compute the recoil velocity of the block after the bullet has become embedded.

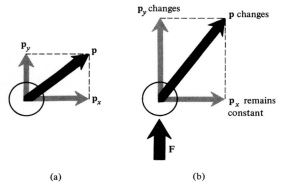

(a)   (b)

Figure 5-4 A force that is applied to an object changes only the momentum component along the direction of the force; the momentum component perpendicular to the direction of the force is not affected.

$$2 \text{ kg-m/s} = (10 \text{ kg}) \times v$$

from which

$$v = 0.2 \text{ m/s}$$

The Vector Character of Linear Momentum

Linear momentum is a *vector* quantity. The conservation of a vector quantity means that the components of the vector are *independently conserved*. If an object has a linear momentum vector $\mathbf{p}$ with components $\mathbf{p}_x$ and $\mathbf{p}_y$, then $\mathbf{p} = \text{const.}$ means that $\mathbf{p}_x = \text{const.}$ and $\mathbf{p}_y = \text{const.}$ If a force is applied to an object, the momentum will change. However, if the force is applied along the *y*-direction, only $\mathbf{p}_y$ will change; $\mathbf{p}_x$ will not change. Figure 5-4 shows such a case. A ball is moving originally with a linear momentum $\mathbf{p} = \mathbf{p}_x + \mathbf{p}_y$. A force $\mathbf{F}$ is applied to the ball in the *y*-direction. The result is that $\mathbf{p}_y$ increases which causes $\mathbf{p}$ to change, but $\mathbf{p}_x$ is unaffected. Notice that this is the same as the case we studied in Section 3-7. A ball that is projected outward will have an accelerated vertical motion which is the same as that of a dropped ball; the horizontal motion of the ball is not affected by gravity and this motion proceeds with constant velocity (that is, with constant linear momentum).

Next, consider the example illustrated in Fig. 5-5. A cart ($M = 50$ kg) is moving along a horizontal surface with a velocity of 4 m/s. At some point a mass of clay ($m = 30$ kg) falls into the cart with an impact velocity of 6 m/s; the clay sticks to the floor of the cart and moves along with it. What is the final velocity of the loaded cart? To solve this problem we use the fact that both $\mathbf{p}_x$ and $\mathbf{p}_y$ for the system (cart plus clay) are conserved. First, we notice that before impact the total *x*-component of the momentum is the momentum of the cart: $p_x(\text{before}) = (50 \text{ kg}) \times (4 \text{ m/s}) = 200$ kg-m/s. After impact, the *x*-component of the momentum is $p_x(\text{after}) = $

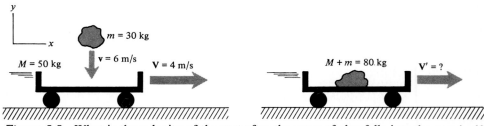

Figure 5-5 What is the velocity of the cart after the mass of clay falls into the cart bed?

$(50 \text{ kg} + 30 \text{ kg}) \times V'$. Equating these two quantities,

$$p_x(\text{before}) = p_x(\text{after})$$
$$200 \text{ kg-m/s} = 80 \text{ kg} \times V'$$

Solving for V', we find

$$V' = \frac{200}{80} = 2.5 \text{ m/s}$$

What about the vertical component of the momentum? Before impact, the system's y-component of momentum is the momentum of the clay: $p_y(\text{before}) = (30 \text{ kg}) \times (6 \text{ m/s}) = 180 \text{ kg-m/s}$. But after impact, there is no vertical motion so that $p_y(\text{after}) = 0$. Is this a violation of the law of momentum conservation? No, it is not. We must remember that the cart is in contact with the Earth so that the vertical momentum of the falling clay is transferred through the cart to the Earth. The mass of the Earth is so great in comparison with the mass of the clay that there is no detectable change in the motion of the Earth.

Finally, let us consider a more complicated example. Suppose that two boys are sliding in "saucers" across smooth ice (that is, friction is negligible). They move on parallel paths, each with a velocity of 3 m/s, as shown in Fig. 5-6a. The mass of each boy–saucer combination is 60 kg. In addition, each boy has with him a (huge!) 10-kg snowball. At a certain instant, each boy throws his snowball to the other boy who catches it. Both snowballs are thrown perpendicular to the motion of the saucers with velocities of 4 m/s. What is the motion of each boy after catching the monster snowball?

First, we note that the situation is symmetric; that is, all of the masses and velocities that apply for one boy also apply for the other. Therefore, we need to consider the effect of the exchange of snowballs only on the boy on the right-hand side. Whatever deflection he experiences (to the right) will also be experienced by the boy on the left-hand side, except that his deflection will be to the left. We also note that the momentum exchange takes place perpendicular to the original direction of motion. Therefore, the velocity $\mathbf{V}$ in the y-direction will be unaffected by the exchange.

We divide the problem into two parts—the throwing of one snowball and the catching of the other snowball. Momentum conservation applies to each part. When the boy on the right throws his snowball, the x-momentum must be conserved:

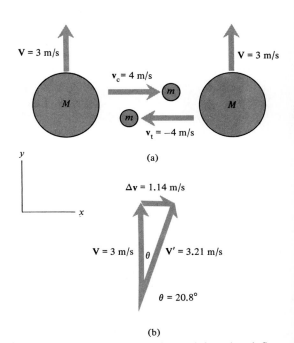

Figure 5-6 Diagram for determining the deflection caused by exchanging the masses m. $M = 60$ kg and $m = 10$ kg.

$$p_x(\text{before}) = p_x(\text{after})$$
$$0 = mv_t + M\ \Delta v_t$$
$$0 = (10\ \text{kg}) \times (-4\ \text{m/s}) + (60\ \text{kg}) \times \Delta v_t$$

where v_t is the velocity of the thrown snowball and where Δv_t is the velocity change experienced because of the throwing action. Solving for Δv_t, we find

$$\Delta v_t = \frac{40\ \text{kg-m/s}}{60\ \text{kg}} = \frac{2}{3}\ \text{m/s}$$

Next, the boy on the right catches the snowball thrown to him. When this takes place, the initial momentum is that of the incoming snowball plus the x-momentum of the saucer which resulted from throwing the first snowball. Then,

$$p_x(\text{before}) = p_x(\text{after})$$
$$mv_c + M\ \Delta v_t = (M + m)\ \Delta v$$

$(10\ \text{kg}) \times (4\ \text{m/s})$
$\quad + (60\ \text{kg}) \times (\tfrac{2}{3}\ \text{m/s}) = (60\ \text{kg} + 10\ \text{kg}) \times \Delta v$
$40\ \text{kg-m/s} + 40\ \text{kg-m/s} = (70\ \text{kg}) \times \Delta v$

where v_c is the velocity of the snowball that is caught and where Δv is the total velocity change in the x-direction. Solving for Δv, we have

$$\Delta v = \frac{80\ \text{kg-m/s}}{70\ \text{kg}} = 1.14\ \text{m/s}$$

The velocity diagram is shown in Fig. 5-6b. The angle of deflection θ is that angle whose tangent is $1.14/3 = 0.38$; this angle is $20.8°$. The value of the final velocity V' can be found by using the Pythagorean theorem:

$$V' = \sqrt{(1.14\ \text{m/s})^2 + (3\ \text{m/s})^2} = 3.21\ \text{m/s}$$

5-2 TORQUE

Rotation Due to Forces

Thus far, we have discussed the action of forces on objects that could be considered to be particles. That is, we have not treated effects that depend on the *sizes* of the objects and *where* on the objects the forces act. Consider now the rod in Fig. 5-7 which is free to pivot around a stationary shaft. What is the effect of applying a force to this rod? The effect depends critically on the point of application and the direction of the force. It is easy to see in the diagram that the force $\mathbf{F}_1$ will cause

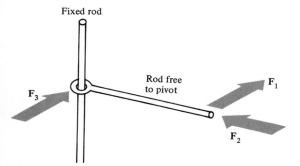

Figure 5-7 Three forces act on a movable rod. Only $\mathbf{F}_1$ exerts a torque on the rod and can therefore cause the rod to rotate around the pivot point.

the rod to pivot around the vertical shaft. The forces F_2 and F_3, however, will produce no rotation: F_2 acts directly *toward* the pivot point and F_3 acts directly *on* the pivot point. Neither of these forces will cause the rod to move. Any force that acts on an object in such a way that it tends to produce a rotation is said to exert a *torque* on the object. In Fig. 5-7, the force F_1 exerts a torque on the rod, but the forces F_2 and F_3 do not.

How do we measure torque? If we look at the force F_1 in Fig. 5-7, it is clear that the tendency of the rod to rotate will be increased if we increase the force. That is, the magnitude of the torque T produced by a force is proportional to the magnitude of the force. Also, if the rod in Fig. 5-7 is made longer and if we move the point of application of the force farther away from the pivot, we will have increased the length of the "lever arm" and will have increased the torque. That is, the magnitude of the torque is also proportional to the distance r from the pivot point to the point at which the force is applied. Therefore, in Fig. 5-8a, the torque exerted by the force F on the rod is equal to the product of F and the distance r: $T = Fr$. This is a special case, however, because F is perpendicular to the rod. Figure 5-8b shows the case in which F makes an angle θ with the rod. We can represent F in terms of two component vectors: $F_\perp$, which is perpendicular to the rod and $F_\parallel$, which is parallel to the rod. We know from the previous discussion that $F_\parallel$ exerts no torque on the rod and produces no rotation; the entire torque is due to $F_\perp$. Therefore, we can write the expression for the torque as

$$T = F_\perp r = Fr \sin \theta \qquad (5\text{-}4)$$

where we have used the fact that $F_\perp = F \sin \theta$.

The dependence of the torque on the angle θ is illustrated in Fig. 5-9. Here, a ball is attached to a thin rod which can pivot around the point P. In each of the three positions shown, the gravitational force, $F = mg$, is the same in magnitude and in direction. When the support rod is horizontal, the torque exerted by gravity is maximum: $T = mgr$. When the support rod is vertical, the torque is zero. At intermediate positions, the torque is given by $T = mgr \sin \theta$.

Couples

We have learned that if the net force on an object is zero, then the acceleration experienced by the object is

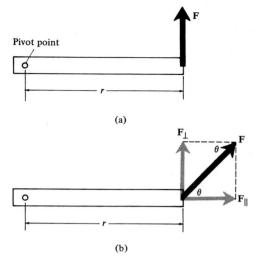

(a)

(b)

Figure 5-8   (a) The torque exerted on the rod by F is $T = Fr$. (b) The component $F_\parallel$ exerts no torque on the rod; therefore, $T = F_\perp r = Fr \sin \theta$.

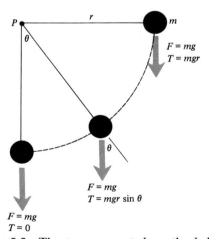

Figure 5-9   The torque exerted on the ball and rod by the gravitational force depends on the angle θ. Notice that in the horizontal position, $\theta = 90°$; because $\sin 90° = 1$, the torque is $T = mgr$.

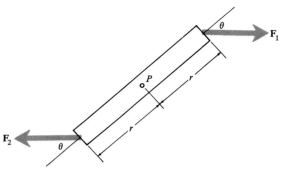

Figure 5-10   The force pair, $\mathbf{F}_1$ and $\mathbf{F}_2$, constitute a *couple*, which produces rotation but no net translational motion (because $\mathbf{F}_1 + \mathbf{F}_2 = 0$).

also zero. When discussing such situations we have always pictured the forces acting at the same point on the object (see, for example, Fig. 4-3). Suppose, however, that we allow two forces with equal magnitudes and opposite directions (so that $\mathbf{F}_{net} = 0$) to act at different points on an object. This situation is shown in Fig. 5-10. Each force acts at a distance r from the center of the rod (point P) and the direction of each force makes an angle θ with the direction of the rod. We can see that each force exerts a torque $T = Fr \sin \theta$ around point P. Because each torque produces a rotation in the same direction, the two effects add. But we still have $\mathbf{F}_{net} = 0$ so that there is no net translational motion of the rod; that is, the point P remains in the same position. Two forces that act together to produce rotation but no net translational motion are called a *couple*.

5-3 CENTER OF MASS

A Uniquely Useful Point

In Section 3-7 we discussed the motion of an object thrown into the air with some horizontal velocity component. We found that the motion takes place along a curved path (a *parabola*). This parabolic motion is easy to visualize and understand for a small object or particle. But what about the motion of an object with a complicated shape? The tip of a football, for example, undergoes an intricate rolling motion when the football is kicked end over end. A simpler case is shown in Fig. 5-11. Here, we have two balls with unequal masses attached by a thin, rigid rod. When this out-of-balance dumbbell is thrown into the air, the path of each ball is a looping curve; neither ball moves on a parabolic path (shown by the dashed curve in Fig. 5-11). There is one point on the dumbbell, however, that *does* move along the parabola. This point is located on the connecting rod and does not correspond to either the large or the small ball. If we were to focus on this point and ignore the rest of the dumbbell, we would see the exactly parabolic path that our simple analysis predicts (for the case of no air resistance). This unique point in the object which follows the predicted parabolic path is called the *center of mass* of the object.

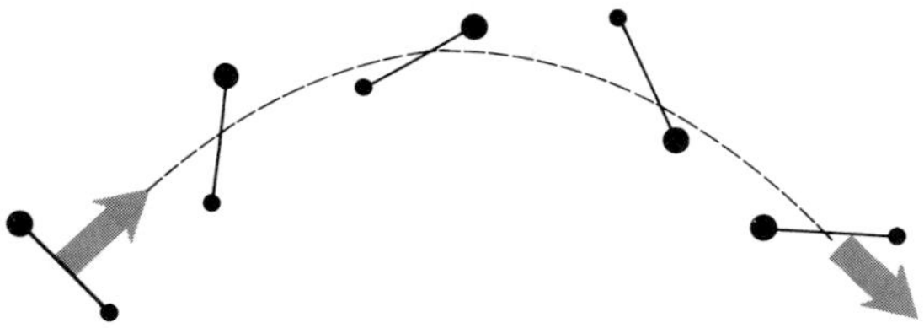

The center of mass of an object is the point at which we can, for the purposes of analyzing the motion, consider the net force on the object to be acting. We have

Figure 5-11   When an out-of-balance dumbbell is thrown into the air, neither ball follows a parabolic path (dashed curve). Only the *center of mass* of the dumbbell, located between the balls, moves always along the parabola.

already been making use of this idea without so stating. For example, we have represented the gravitational force on a block of material by a single vector, showing the force acting on the center of the block. In reality, the block is composed of a large number of individual particles and the gravitational force acts independently on each particle, as indicated in Fig. 5-12a. Instead of dealing with a large number of individual forces, it is more convenient to sum the various individual forces and to analyze the gravitational effect on the block in terms of a single force vector acting on the center of mass of the block (Fig. 5-12b).

If we consider an object that has a regular shape (for example, a sphere or a rectangular sheet) and which is *homogeneous* in composition (that is, the density is the same throughout the material), then it is easy to understand that the center of mass is simply the geometrical center of the object. Thus, we represent the force on a sphere as acting on its center and the force on a block as acting on *its* center (Fig. 5-12b). But how do we find the center of mass of an irregular object such as the out-of-balance dumbbell in Fig. 5-11? One way to do this is to make use of the idea of *torque,* as discussed in the preceding section.

Suppose that we adjust the dumbbell until it just balances on a sharp edge that acts as a pivot point (Fig. 5-13). If the dumbbell is moved in either direction from this point, one end or the other will rotate downward until it strikes the ground. When in balance there is no rotation and therefore no net torque on the dumbbell; the balance point is the center of mass. The absence of a net torque around the pivot point means that we can equate the clockwise torque around the pivot point to the counterclockwise torque around this same point. The clockwise torque is due to the gravitational force on the smaller ball and is equal to m_2gr_2. The counterclockwise torque is due to the gravitational force on the larger ball and is equal to m_1gr_1. Equating these torques, we have

$$m_1gr_1 = m_2gr_2$$

and cancelling the common factor g,

$$m_1r_1 = m_2r_2$$

Combining this result with a measurement of the distance between the two balls, $R = r_1 + r_2$, and the masses, m_1 and m_2, we can always find the position of the center of mass.

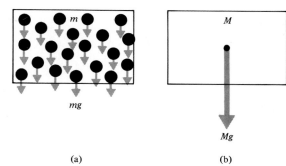

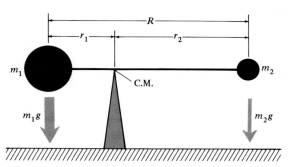

Figure 5-12 (a) The gravitational force acts on each of the particles comprising a block of material. (b) For purposes of analyzing the motion of the block, it is more convenient to add together all of the individual forces and to represent the gravitational force on the block by a single vector acting on the center of mass of the block.

Figure 5-13 To find the center of mass (C.M.) of the misshapen dumbbell, the dumbbell is balanced on a pivot point. Then, the torque around the pivot point due to the gravitational force on the large ball is equal in magnitude to the torque due to the gravitational force on the small ball. Equating these two torques determines the position of the center of mass.

Notice two points concerning this analysis. *First,* we used our previous conclusion about the center of mass of regular objects by considering the gravitational forces to act on the *centers* of the two balls. Only because of this were we able to specify easily the distances r_1, r_2, and R. *Second,* we used the idea of torque due to the gravitational force in order to write down the essential equation. But suppose that we wish to find the center of mass of the Earth–Moon system. There is no rod connecting these objects, and even if there were, there is no balance edge in space with which to locate the center of mass. However, we can still *imagine* that this type of measurement could be made, and we can write down the equations just as we did for the dumbbell. The geometry of the dumbbell and Earth–Moon cases is exactly the same and the procedure for calculating the position of the center of mass is likewise the same.

The method for calculating the position of the center of mass for an object with irregular shape and with density variations requires complicated mathematics. We will always consider only objects with simple shapes.

The Use of Torque in Statics Problems

In calculating the position of the center of mass of the dumbbell (Fig. 5-13), we made use of an important property of torque. For any system that is static (that is, in equilibrium), the net torque around any pivot produced by the forces acting on the system must be zero. Let us see how we can apply this idea in a more complicated situation. Figure 5-14 shows a loaded plank that is supported at each end. How much of the load is carried by each of the supports?

Because the loaded plank is in equilibrium, we can solve this problem by setting equal to zero the sum of the torques around the pivot point. But where is the pivot point in this case? It really does not matter which point we use as the pivot point. The sum of the torques will be zero around *any* point we choose. Let us select the left end of the plank as our pivot point. (Try working this example by choosing a different pivot, for example, the middle of the plank or even a point 2 m to the left of the left end of the plank.) The force vectors in Fig. 5-14 represent all of the forces acting on the plank. F_1 and F_5 are the reaction forces exerted by the supports; the sum of these forces is equal to the total weight of the plank

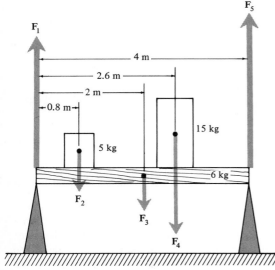

Figure 5-14 What is the load carried by each of the supports?

and the two masses. This gives us one of the equations we will need:

$$F_1 + F_5 = \text{(total mass)} \times g$$
$$= (5 \text{ kg} + 15 \text{ kg} + 6 \text{ kg}) \times (9.8 \text{ m/s}^2)$$
$$= (21 \text{ kg}) \times (9.8 \text{ m/s}^2)$$
$$= 205.8 \text{ N}$$

Next, we sum the torques around the left end of the plank. There are three clockwise torques, due to the two masses and the plank; we will denote these torques as *positive*. There is one counterclockwise torque, due to the reaction force F_5; we will denote this torque as *negative*. (The torque due to F_1 is zero because this force vector passes through the pivot point.) Notice that F_2, F_3, and F_4 have each been drawn acting on the center of mass of the particular object. Calculating the various torques, we find

$$T_2 = (mgr)_2 = (5 \text{ kg}) \times (9.8 \text{ m/s}^2) \times (0.8 \text{ m})$$
$$= 39.2 \text{ N-m}$$
$$T_3 = (mgr)_3 = (6 \text{ kg}) \times (9.8 \text{ m/s}^2) \times (2 \text{ m})$$
$$= 117.6 \text{ N-m}$$
$$T_4 = (mgr)_4 = (15 \text{ kg}) \times (9.8 \text{ m/s}^2) \times (2.6 \text{ m})$$
$$= 382.2 \text{ N-m}$$
$$T_5 = -F_5 \times (4 \text{ m})$$

Summing these torques and setting the total equal to zero, we have

$$539 \text{ N-m} - F_5 \times (4 \text{ m}) = 0$$

or

$$F_5 = \frac{539 \text{ N-m}}{4 \text{ m}} = 134.75 \text{ N}$$

Using the previous result for $F_1 + F_5$, we find

$$F_1 = 205.8 \text{ N} - F_5 = 71.05 \text{ N}$$

Thus, the right-hand support carries the greater load. (We expected this result because the larger mass is nearer the right-hand support.)

5-4 ANGULAR MOMENTUM

A New Conservation Principle

An object that slides frictionlessly over a smooth horizontal surface will move with constant velocity. We can

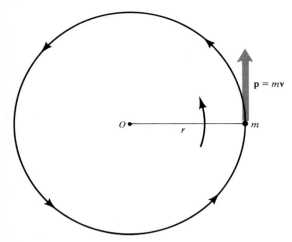

Figure 5-15   The angular momentum of the object is $L = pr = mvr$.

interpret such motion in terms of Newton's first law ($\mathbf{F} = 0$ implies $\mathbf{a}$ is zero), or we can state, with equal correctness, that the object is obeying the principle of momentum conservation ($\mathbf{F} = 0$ implies $\mathbf{p}$ is constant). Now, consider an object that, instead of sliding, is *rotating* on a smooth surface but is not moving across the surface (for example, a spinning top). We know that, if friction is negligible, the spinning motion will continue for a very long time. There is a definite parallel here: in the absence of outside influences, a moving object will continue in its state of uniform motion and a spinning object will continue in its state of uniform rotation.

The description of spinning motion can be made quantitative by introducing the concept of *angular momentum*. If an object of mass m moves with a velocity v (and, hence, a linear momentum $p = mv$) in a circular path with a radius r, then the angular momentum is defined to be (see Fig. 5-15).

$$\boxed{\text{angular momentum} = L = mvr} \qquad (5\text{-}5)$$

That is, the *angular* momentum is equal to the product of the *linear* momentum, $p = mv$, and the radius of the path.

How is it possible to change the state of rotation (that is, the angular momentum) of the object in Fig. 5-15? We can only do this by exerting a push or a pull along (or opposite) to the direction of the linear momentum vector $\mathbf{p}$. Such a push or a pull will speed up or slow down the particle and therefore change the quantity mvr. By referring to the way that the forces were applied in the situations illustrated in Fig. 5-8a and Fig. 5-9, we can see that a force that acts in line with $\mathbf{p}$ amounts to the application of a torque around the center of motion O. In fact, the only way to change the angular momentum of an object (or system of objects) is by the application of a torque. We can state this conclusion in terms of a conservation law:

If no torque acts on an object, the angular momentum of the object will remain constant.

This is the law of *conservation of angular momentum*. Notice that the laws concerning the conservation of linear momentum and angular momentum are very similar. If no outside agency exerts a *force* on an object, the *linear* momentum of the object remains constant; if no outside agency exerts a *torque* on an object, the *angular* momentum of the object remains constant.

To see how these two principles operate and to see the distinction between the action of a force and a torque, let us return to the situation shown in Fig. 5-15. Suppose that the particle is moving with constant speed and constant angular momentum and is restrained by a string that passes through a hole in the surface located at O. What will happen if we pull downward on the string so that the radius of the circular motion is decreased from r to r', as shown in Fig. 5-16? The force that is exerted on the particle to draw it closer to the center of rotation is directed along the line that connects the particle and the center of rotation. This force is similar to the force $\mathbf{F}_2$ in Fig. 5-7 in that it does not exert a torque on the particle. The conservation of angular momentum therefore requires that the quantity mvr remain constant. The mass m of the particle does not change, but r decreases. To compensate for the smaller value of r, the velocity v must *increase*. The new velocity is found by equating the original angular momentum mvr to the final angular momentum $mv'r'$. Thus,

$$v' = \frac{rv}{r'}$$

What about linear momentum in this case? Because the particle moves in a circular path, there must be a force acting on the particle. (If there were no such force, $\mathbf{p}$ would remain constant and the particle would move in a straight line.) This force which is exerted on the particle through the string is the *centripetal* force. The centripetal force causes the linear momentum to change, but this change is in *direction*, not in *magnitude*. Even though this force acts continually on the particle, it exerts no torque and so does not change the angular momentum. When the additional centripetal force is applied to reduce the radius of the path, it produces a change in the *magnitude* of $\mathbf{p}$ (which increases from mv to mv').

The spinning Earth obeys the principle of angular momentum conservation and revolves at a (nearly) constant rate. Actually, frictional effects are present, primarily in the form of *tides*. (The Earth's ocean waters tend to be held in place by the gravitational attraction of the Moon while the Earth rotates; thus, there is a frictional drag exerted on the rotating Earth.) Because tidal friction slows down the rate of the Earth's rotation, the

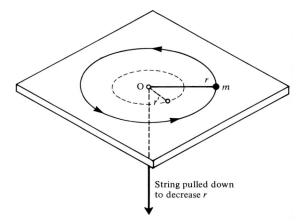

Figure 5-16 When the string is pulled downward to decrease the radius of the circular path, conservation of angular momentum requires that the velocity of the particle increase.

Figure 5-17   By drawing her arms to her sides, a skater increases her rate of rotation because angular momentum is conserved.

length of a day is gradually increasing (at a rate of approximately 10^{-5} s per year).

The conservation of angular momentum is strikingly evident when an ice skater spins with arms outstretched then suddenly brings her arms to her sides. The rotational speed of the skater must increase in order to conserve angular momentum (see Fig. 5-17).

The Angular Momentum Vector

Thus far we have discussed only the *magnitude* of angular momentum. But angular momentum is actually a *vector* quantity and therefore has *direction* as well as magnitude. Angular momentum is a conserved quantity, and if a vector quantity remains constant, this means that both the magnitude and direction are unchanged. We have studied this point in connection with the linear momentum vector. How do we define the direction of angular momentum? In Fig. 5-15 we have an object moving with constant angular momentum. The linear momentum of the object changes direction continually (because there is a centripetal force). Therefore, even though the magnitude of the angular momentum L is proportional to the magnitude of the linear momentum

Angular Momentum and the Formation of Galaxies

Every galaxy of stars in the Universe (including our own Milky Way Galaxy) was formed by the contraction and condensation of a huge cloud of gas and dust. Consider such a mass of material in space. As gravitational attraction acts between every pair of particles, the size of the mass shrinks. In parts of the cloud, local contraction produces embryonic stars while the entire mass continues to shrink. The eventual result is a cluster of stars—a galaxy. In most cases, the original gas-dust cloud will have a certain amount of angular momentum. This angular momentum must remain constant as the cloud contracts, and this can happen only if the rotation speed increases. Because the original cloud is so large (compared to the final size of the galaxy), even a small rotation speed means that the cloud possesses a large amount of angular momentum. As the contraction takes place (see Fig. 5-18), the distances of the star cloudlets from the center decrease and their rotation speeds increase. The stars that lie close to the rotation axis of the galaxy contract readily toward the center, whereas the stars that lie near the plane perpendicular to the rotation axis contract more slowly due to their high rotation speeds. Eventually, the stars are distributed in a disk-shaped cluster that is revolving at a (relatively) high speed. The photographs on page 103 show the basic correctness of this description of the formation of galaxies.

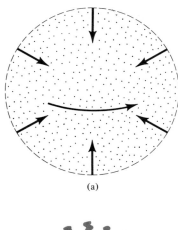

(a)

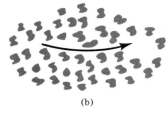

(b)

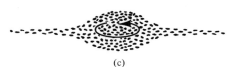

(c)

Figure 5-18 Three stages in the formation of a galaxy. (a) A slowly rotating, but diffuse cloud of gas and dust. (b) The cloud contracts because of gravitational attraction and cloudlets of embryonic stars are formed. The size is smaller, so the rotation speed is greater. (c) The cloudlets have formed stars and further contraction has taken place. Again, the size is smaller and the rotation speed is increased further. The galaxy is now a disk-shaped clustering of stars.

HALE OBSERVATORIES

Disk-shaped galaxies produced by the contraction of a rotating cloud of gas and dust. Top: a spiral galaxy in Ursa Major (identified by the catalog number NGC 3031). Below: a spiral galaxy in Coma Berenices (NGC 4565), seen edge on. This photograph corresponds to the view in the diagram of Fig. 5-18c.

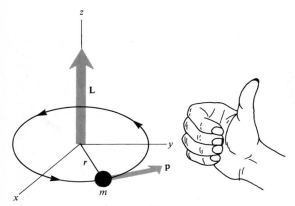

Figure 5-19 Illustrating the right-hand rule for determining the direction of the angular momentum vector **L**. Curl the fingers of the right hand in the direction of motion of the object; then, the thumb points in the direction of **L**.

p, the direction of the vector **p** cannot be the same as the direction of the vector **L**.

We define the direction of the angular momentum vector to be along the *axis* of the rotation. There are two possible directions of this axis; we select one of these for the direction of the angular momentum vector according to the following *right-hand rule* (see Fig. 5-19):

> Curl the fingers of the right hand in the direction of motion of the object. Then, the right thumb points in the direction of **L**.

As the object moves in the circular path with constant speed, both the magnitude mvr and direction determined by the right-hand rule remain constant. Angular momentum is conserved (in a vectorial sense).

The fact that the angular momentum vector maintains a fixed direction in space (in the absence of any applied torque) has many important consequences. The rotating Earth, for example, is essentially free of external torque and its rotation axis always points in the same direction, namely, toward *Polaris,* the *North Star.* In addition to rotating on its own axis, the Earth revolves in an orbit around the Sun. The rotation axis, however, does not point in the same direction as the axis for the orbital motion. As indicated in Fig. 5-20, the rotation axis is inclined at an angle of $23\frac{1}{2}°$ with respect to the perpendicular to the plane of orbital motion. Because the direction of the rotation axis remains fixed in space, the Earth's North Pole tips first toward the Sun and then, six months later, tips away from the Sun. This change causes the occurrence of *seasons.* As shown in Fig. 5-20, the Sun's rays in summer are much more nearly

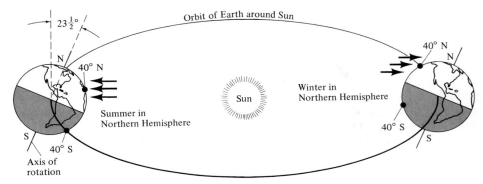

Figure 5-20 The Earth's rotation axis is tilted with respect to the orbital plane. Angular momentum conservation requires the direction of the rotation axis to remain fixed in space and thereby causes the occurrence of seasons.

perpendicular to the Earth's surface in the Northern Hemisphere than they are in winter. When the rays are highly slanted, as they are in winter, the heat delivered to the surface is spread out over a much larger area than in summer and the temperature is correspondingly lower.

SUGGESTED READINGS

Isaac Newton, "The Laws of Motion" (an excerpt from Newton's *Principia*) *Great Experiments in Physics,* M. H. Shamos, ed. (Holt, New York, 1959).

R. E. Peierls, *The Laws of Nature* (Scribner's, New York, 1956).

Scientific American articles:

J. W. Beams, "Ultrahigh-Speed Rotation," April 1961.

G. W. Gray, "The Ultracentrifuge," June 1951.

QUESTIONS AND EXERCISES

1. A boat containing a man is at rest in still water. The mass of the boat is 80 kg and the mass of the man is 100 kg. If the man jumps horizontally from the stern of the boat with a velocity of 2 m/s, how will the boat react?

2. When a bullet is fired from a rifle, the rifle moves in the opposite direction (it *recoils*). How would it be possible to make a *recoilless* rifle?

3. A 10-g bullet is fired from a 3-kg rifle with a muzzle velocity of 900 m/s. What is the recoil velocity of the rifle?

4. A 10 000-kg railway car moves along a horizontal track with a velocity of 6 m/s. The moving car strikes a 5000-kg stationary car and the two cars, coupled together, move along the track. What is the velocity of the pair of cars?

5. A boy on a sled is pulled by a rope to the middle of a frozen lake. The rope breaks and he finds himself stranded. The ice is so smooth that there is no friction between the ice and the sled runners. The boy was on the way home from the grocery store and he has with him a large sack of potatoes. How does the boy move himself and the sled from the middle of the lake? (Assume that he starts from rest; we will not inquire *how* he managed to come to rest!)

6. A ball of clay ($m = 1.5$ kg) is thrown horizontally with a velocity of 30 m/s. The ball strikes and sticks to the back of a cart ($M = 90$ kg) that is originally at rest. Describe the resulting motion.

7. Explain how an astronaut taking a "spacewalk" can propel himself about.

8. Two students are each standing on carts that can roll frictionlessly across a horizontal surface. The students (and their carts) have masses $M_1 = 60$ kg and $M_2 = 80$ kg, and they hold opposite ends of a rope. What will happen as the student with mass M_1 pulls in the rope, hand over hand, at a rate of 1m/s? What will happen as each student pulls in the rope at a rate of 0.5 m/s?

9. Two boys are sliding in a "saucer" with a constant velocity of 4 m/s across smooth ice (that is, frictionlessly). Each boy has a mass of 50 kg and the saucer mass is 5 kg. One

of the boys jumps from the saucer with a horizontal velocity of 2 m/s (with respect to the saucer) at an angle of 135° with respect to the original direction of motion. Describe the subsequent motion of the remaining boy and the saucer.

10. Two automobiles collide at an intersection. One automobile ($m_1 = 2500$ kg) was traveling north with a velocity of 30 km/h and the other ($m_2 = 1500$ kg) was traveling west with a velocity of 40 km/h. The cars lock together upon impact. What is the velocity of the pair after collision?

11. A shell is fired by an artillery piece. As the shell moves along its path, suddenly it explodes into a number of pieces of different sizes that fly off in different directions. Is there any statement that you can make regarding the motion of the shell fragments?

12. Determine the location of the center of mass of the Earth–Moon system. (Use the data in the table inside the front cover.)

13. Four balls are located as follows: $m_1 = 2$ kg at the origin; $m_2 = 4$ kg at $x = 0$, $y = 6$ m; $m_3 = 12$ kg at $x = 4$ m, $y = 6$ m; $m_4 = 6$ kg at $x = 4$ m, $y = 0$. Find the position of the center of mass of the four balls. Proceed by first computing the center of mass of the m_1–m_2 combination and the center of mass of the m_3–m_4 combination. Next, consider the total mass of each pair to be concentrated at its center of mass. Finally, compute the center of mass of the resulting pair, which will be the center of mass of the entire set of balls.

14. In Fig. 5-9, the mass of the ball is 2 kg and $r = 0.4$ m. Make a table of the torque exerted by the gravitational force for angles $\theta = 0°$, 30°, 60°, and 90°. What is the torque for $\theta = -30°$?

15. How can a force of 30 N exert a torque of 30 N-m around one end of a 2-m rod when applied to the opposite end? Make a sketch of the situation.

16. A special type of wrench (called a *torque wrench*) has a built-in scale for reading the amount of torque applied to a bolt head. How do you suppose such a device works? What is the use of a torque wrench?

17. A plank rests on a support and has its left end beneath a fixed protrusion of a wall. The right end of the plank is loaded with two masses, as shown in the diagram. Calculate the force exerted on the triangular support.

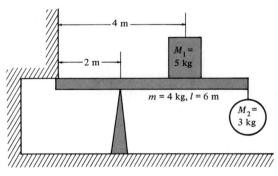

18. A 12-kg block is suspended from the end of a 2-m boom, as shown in the diagram. The mass of the boom is 4 kg. Make a sketch of all the forces acting on the boom and find the magnitude of the tension **T** in the support cable.

19. A pendulum consists of a ball attached to the end of a string. As the pendulum oscillates back and forth, angular momentum is not conserved. Why?

20. A horizontal platform can rotate freely around a vertical axis. A student stands on

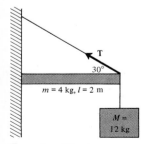

Exercise 18

and even the large-scale motions of stars and galaxies in space. The electrical force, although it is effective in some large-scale phenomena such as lightning strokes and solar flares, is in evidence mainly in effects that take place in the atomic domain. All elastic forces, for example, are the result of attractive electrical forces that exist between atoms and molecules, and the magnetic forces that drive electrical motors and guide electrons in television tubes are due to the motion of electrons within atoms or through wires. Furthermore, radio waves, light, and X rays (in fact, all forms of electromagnetic radiation) are the result of electrical interactions between atoms and electrons.

Gravitational and electrical forces can account for all of the events that we directly observe taking place in Nature. But there are many important phenomena that occur in the subatomic domain which we *cannot* see or experience directly. In this century, as instruments have been developed for examining the processes that take place in atomic nuclei, we have become aware that gravitational and electrical forces are inadequate to account for nuclear phenomena. The study of processes involving nuclear and elementary particle interactions has shown that there are two additional basic forces; these are called the *nuclear* or *strong* force (which is responsible for the binding together of protons and neutrons to form nuclei) and the *weak* force (which is responsible for β decay and similar processes—see Section 20-1). Thus, as far as we know today, there are four and only four fundamental forces that govern all processes in Nature:

(1) gravitational force
(2) electrical force
(3) strong (nuclear) force
(4) weak (β decay) force

The first of the four basic forces to be studied in detail was the gravitational force. We see the gravitational force in operation on and near the surface of the Earth, but the first clues that gravity is more than an Earthbound phenomenon came from observations of the movements of the Moon and the planets. After discussing the gravitational force we will consider the electrical force. We will find that, in some respects, these two forces are very similar in character. Concluding this chapter, we will discuss briefly the nuclear and weak forces.

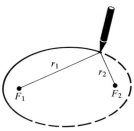

Figure 6-1   A simple way to construct an ellipse. Attach a string to two points, F_1 and F_2 (the *foci*). With a pencil point, keep the string taut while drawing the curve that everywhere has $r_1 + r_2 =$ constant. The Earth's elliptical orbit is nearly circular – the difference between the maximum and minimum distances from the Earth to the Sun is only about 3 percent.

6-2 PLANETARY MOTION

Kepler's Laws

The key ideas that were necessary to advance the heliocentric theory of Nicolaus Copernicus and to lay the groundwork on which Newton was able to construct the theory of gravitation were supplied by Johannes Kepler (1571–1630). Analyzing the extensive records of the Danish astronomer Tycho Brahe (1546–1601), with whom he worked from 1597 until Brahe's death, Kepler tried first one planetary model and then another in an effort to interpret Brahe's data on the motion of the planets, particularly Mars. All of these attempts failed until Kepler at last decided to give up the idea of circular orbits. (Kepler's mystical nature and his fascination with regular geometrical forms had led him to base his astronomy, as had the ancient astronomers, on the "perfect" circle.) Trying other geometrical forms, Kepler decided that the motions of the planets were best described in terms of *ellipses,* and accordingly, in 1609, he published his first law of planetary motion:

> 1. The orbit of a planet around the Sun is an ellipse, with the Sun at one focus. (See Fig. 6-1.)

Furthermore, in his second law, also announced in 1609, Kepler abolished the long-held idea that the planets move with constant speed in their orbits:

> 2. The line joining a planet with the Sun sweeps out equal areas in equal times.

According to this law, the speed of a planet changes during the course of executing its orbit, being fastest when nearest the Sun and slowest when farthest away (see Fig. 6-2).

Finally, in 1619, the third of Kepler's laws appeared:

> 3. The square of the period of a planet's orbit around the Sun is proportional to the cube of the mean distance of the planet from the Sun.

This law, in equation form, becomes

$$\frac{\tau^2}{R^3} = \text{constant} \qquad (6\text{-}1)$$

where τ is the orbital period of the planet and R is the

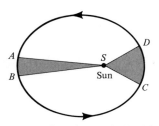

Figure 6-2   According to Kepler's third law, if a planet moves from A to B and from C to D in equal times, then the areas ABS and CDS are equal. The speed is greater for the part of the orbit CD than for the part AB. For the Earth, the orbital speed nearest the Sun is about 3 percent greater than when farthest from the Sun.

TABLE 6-1 PLANETARY DATA

PLANET	AVERAGE DISTANCE FROM SUN (10^6 km)	DIAMETER (km)	MASS[a] (Earth masses)	PERIOD (years)
Mercury	58	4 880	0.054	0.241
Venus	108	12 110	0.814	0.615
Earth	150	12 740	1.000	1.000
Mars	228	6 800	0.107	1.881
Jupiter	778	143 000	317.5	11.86
Saturn	1426	121 000	95.0	29.46
Uranus	2868	47 000	14.5	84.01
Neptune	4494	45 000	17.6	164.8
Pluto	5896	6 000?	0.1?	249.9

[a] Mass of Earth $= 5.98 \times 10^{24}$ kg

average planet–Sun distance. We use this equation in the following way. Because the "constant" in the equation is the same for all planets, we can equate τ^2/R^3 evaluated for one planet to τ^2/R^3 evaluated for any other planet. For example,

$$\frac{\tau^2(\text{Earth})}{R^3(\text{Earth})} = \frac{\tau^2(\text{Mars})}{R^3(\text{Mars})}$$

By substituting into this expression the data from Table 6-1, we can verify the equality for this case (or for any other pair of planets). Notice, however, that it is simpler to make the comparison if we rearrange the equation to read

$$\left(\frac{\tau(\text{Mars})}{\tau(\text{Earth})}\right)^2 = \left(\frac{R(\text{Mars})}{R(\text{Earth})}\right)^3 \qquad (6\text{-}2)$$

Substituting values,

$$\left(\frac{1.881 \text{ y}}{1 \text{ y}}\right)^2 = \left(\frac{228 \times 10^6 \text{ km}}{150 \times 10^6 \text{ km}}\right)^3$$

Calculating these numbers, we find approximately 3.5 in each case.

After many centuries of fruitless searching, the mystery of planetary motion was finally solved. Kepler's three laws, although based solely on observational data and not supported by any fundamental theory, provide a correct description of the motions that take place in the solar system. The underlying theory, the product of Isaac Newton, did not appear until near the end of the 17th century.

Nicolaus Copernicus

The idea that the Sun, and not the Earth, is the center of the solar system dates from the ancient Greek era. But it was not an idea that was widely accepted. It remained for the Polish astronomer, Nicolaus Copernicus (1473–1543), to bring the Sun-centered scheme boldly to public view. Copernicus recognized that the positions of the planets in the sky could be more easily predicted if the calculations were carried out by assuming that the central body of the system was the Sun instead of the Earth. (Although Copernicus transferred the key position in the solar system from the Earth to the Sun, he believed that the Sun was the center of the Universe.) The Copernican model of the solar system explained in a straightforward way two puzzling features of planetary motion that for many centuries had plagued all those who attempted to account for the motions of planets. Mercury and Venus never appear in the sky very far from the position of the Sun. According to the Copernican scheme this is a natural consequence of the fact that these planets move around the Sun in orbits that lie closer to the Sun than the path followed by the Earth. Therefore, as viewed from the Earth, the angular separation of the inner planets from the Sun is always small. Moreover, the orbits of the planets Mars, Jupiter, and Saturn lie outside the Earth's orbit. The moving Earth periodically overtakes these outer planets, causing them to appear, on occasion, to move backward in the sky. Thus, the Copernican model accounts also for the so-called retrograde motion of the outer planets. Copernicus' book explaining his theory of planetary motion was published only shortly before his death, in 1543. However, general acceptance of the idea of a Sun-centered solar system did not take place for more than another century, and it was not until 1835 that Copernicus' book was removed from the list of books banned by the Catholic Church.

6-3 UNIVERSAL GRAVITATION

The Falling Moon

Kepler had given a successful *description* of planetary motion in terms of his three laws, but, by 1666 when Newton turned his attention to the problem, no one had yet provided an *explanation* for the dynamical behavior of the planets. What was the *cause* of the regular and precise movement of the Moon around the Earth and of the planets around the Sun? According to the legend, as Newton pondered this question sitting in the orchard at his Woolsthorpe farm in 1666, he heard the gentle thud of an apple falling to the ground beside him. The apple was a free and unattached object, and it fell toward the Earth. The Moon was also a free and unattached object—should it not also fall toward the Earth? This simple incident and the question it provoked provided the impetus that led Newton eventually to the descrip-

Johannes Kepler (1571–1630). Kepler was born in Württemberg, Germany, and was educated at Tübingen. Even in his early years, Kepler was a staunch supporter of the Copernican system. He became professor of mathematics at Graz, in Austria, but because of religious persecution of Protestants at that time, he moved to Prague in 1597. There, he worked with the astronomer Tycho Brahe attempting to fit Brahe's extensive set of planetary data into a geocentric theory devised by Brahe. Failing this, and inheriting Brahe's voluminous collection of data upon his death, Kepler tried a number of different hypotheses in an effort to account for the planetary motions. Finally, in his stumbling way, Kepler hit upon the correct answer: the planets move in elliptical orbits around the Sun. Although Kepler had a troubled personal life (his first wife went mad and he was not much luckier with his second) and held many mystical beliefs (at one point, financial troubles forced him to astrological fortune telling), he developed a thorough understanding of planetary motions and he appreciated that planetary dynamics were dependent upon gravity. In his *New Astronomy*, Kepler wrote that gravity is a "mutual corporeal tendency of kindred bodies to unite or join together." He argued that if the Moon and the Earth were not restrained "each in its own orbit, the Earth would move up toward the Moon . . . and the Moon would come down toward the Earth . . . and they would join together."

tion of the motion of the Moon and the planets in terms of a theory of universal gravitation.

An object that is projected horizontally near the surface of the Earth will, because of the Earth's gravitational attraction, move in a curved path toward the Earth (Fig. 6-3a). The Moon also departs from straight-line motion and moves in a curved path, always toward the Earth (Fig. 6-3b). The object clearly falls toward the Earth; does not the Moon also "fall" toward the Earth

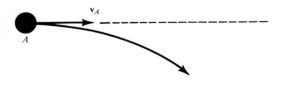

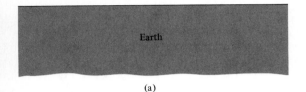

Earth

(a)

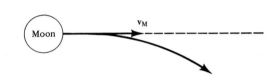

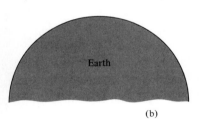

Earth

(b)

Figure 6-3 (a) An object projected with a horizontal velocity v_A falls toward the Earth because of gravity. (b) The Moon also deviates from straight-line motion and "falls" toward the Earth because of gravity.

(actually, *around* the Earth)? If gravity is responsible for the motion of the falling object, is it not also responsible for the motion of the "falling" Moon? So Newton argued.

Newton knew the approximate distance to the Moon ($r_M = 3.84 \times 10^8$ m) and he knew the period of the Moon's rotation around the Earth ($\tau_M = 27.3$ days). In order to maintain this motion, the Moon must experience a centripetal acceleration, $a_c = v^2/r_M$ (see Eq. 3-19). The velocity of the Moon is

$$v = \frac{\text{circumference of orbit}}{\text{period of rotation}}$$

$$= \frac{2\pi r_M}{\tau_M} = \frac{2\pi \times (3.84 \times 10^8 \text{ m})}{(27.3 \text{ days}) \times (86\ 400 \text{ s/day})}$$

$$= 1.02 \times 10^3 \text{ m/s}$$

Therefore, the centripetal acceleration of the Moon is

$$a_c = \frac{v^2}{r_M} = \frac{(1.02 \times 10^3 \text{ m/s})^2}{3.84 \times 10^8 \text{ m}}$$

$$= 2.72 \times 10^{-3} \text{ m/s}^2$$

This acceleration that the falling Moon experiences is far less than the acceleration of an object falling near the surface of the Earth ($g = 9.8$ m/s²). If the same agency—gravity—is responsible for both falling motions, how can the accelerations be so vastly different? Newton drew upon his knowledge of optics to answer this question.

Imagine that two transparent spheres surround a light source. The smaller sphere has a radius of 1 m and the larger sphere has a radius of 2 m; the center of each sphere is located at the position of the source. All of the light that is radiated outward by the source must pass through both spheres. The surface area of a sphere with radius r is $4\pi r^2$. Therefore, the area of the larger sphere ($r = 2$ m) is 4 times the area of the smaller sphere ($r = 1$ m). Consequently, the light that falls on a small section of the larger sphere has an intensity (amount of light per unit area) that is one-quarter as great as the intensity of the light falling on a section of the smaller sphere having the same area. Figure 6-4 shows the situation in detail. The amount of light that falls on a small square of the surface of the 1-meter sphere is distributed over *four* similar squares on the surface of the 2-meter sphere. Moreover, the light will spread out to cover *nine* similar squares at a distance of 3 meters. Compared to the intensity of light falling on the small square at a distance of 1 meter, the intensity on a square of the same size at 2 meters is $\frac{1}{4}$ and the intensity at a distance of 3 meters is $\frac{1}{9}$. We therefore conclude that *the intensity of light varies inversely as the square of the distance from the source;* that is,

$$\text{intensity} \propto \frac{1}{r^2}$$

Newton argued that *gravity* must somehow radiate uniformly outward from a mass (such as the Earth) in the same way that light radiates uniformly outward from a source. Therefore, gravitational force must depend on distance in the same way that light intensity does. That is,

$$\text{gravitational force} \propto \frac{1}{r^2}$$

What distance should be used for r? For an object falling near the surface of the Earth, r is not the distance of the object above the Earth's surface; we know that at a height of 1 m the gravitational force and, hence, the acceleration due to gravity is the same as at a height of 10 m. Newton appreciated the fact that it is the *entire* Earth that attracts a falling object, and he succeeded in proving that the distance r must therefore be measured from the *center* of the Earth (Fig. 6-5). Consequently, if the Moon were located at the Earth's surface, 6400 km from the center of the Earth, the acceleration would be 9.8 m/s². Or, if the Moon were located at a height of 6400 km above the surface (or 12 800 km from the center), the distance r would be *twice* that in the first case and, hence, the acceleration would be *one-fourth* as great, namely, 2.45 m/s². But the center of the Moon is actually located approximately 384 000 km from the center of the Earth—that is, a distance 60 times as far away from the center as an object located on the surface. Therefore, the Moon's acceleration should be

$$a_c = \frac{1}{(60)^2} \times g = \frac{9.8 \text{ m/s}^2}{3600}$$
$$= 2.72 \times 10^{-3} \text{ m/s}^2$$

The value of the Moon's acceleration determined in this way is the same as the value calculated from a knowledge of the Moon's orbit. Newton had succeeded in accounting for the falling motions of the Moon and Earth-bound objects with a single force, the gravitational force of the Earth.

Newton went on to complete his analysis of the

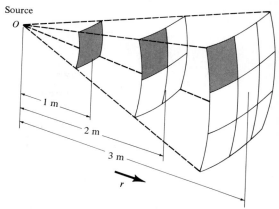

Source
O
1 m
2 m
3 m
r

Figure 6-4 The intensity of light decreases with distance from the source as $1/r^2$.

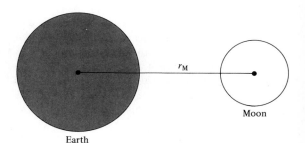

r_M

Moon

Earth

Figure 6-5 In the calculation of gravitational forces between spherical bodies, such as the Earth and the Moon, the distance r is the distance between the *centers* of the bodies. This is equivalent to imagining, for purposes of gravitational calculations, that the entire mass of each body is concentrated at its center.

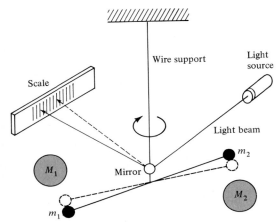

Figure 6-6 Schematic diagram of the apparatus used by Henry Cavendish to measure the value of G. A rod carrying two small masses, m_1 and m_2, is attached to the lower end of a thin wire which is suspended from a fixed support. The apparatus is allowed to come to rest with the large balls, M_1 and M_2, placed some distance away. The equilibrium position is then measured by shining a light beam onto a mirror that is attached to the wire; the point at which the reflected beam strikes the scale is recorded. Next, the large balls are moved close to the small balls, as indicated in the diagram. The force of attraction between M_1 and m_1 and between M_2 and m_2 tends to rotate the rod into the new position shown by the dashed line. In this position, the gravitational force between the balls is exactly balanced by the torsional force in the twisted wire. The angular movement of the rod is determined by noting the change in position of the reflected light beam on the scale. In a separate experiment, the amount of twist in the wire due to known forces is measured. In this way Cavendish measured the gravitational force between known masses separated by a known distance. It was then possible to compute G using Eq. 6-3. The value of G obtained from Cavendish's data is only 1 percent different from the value now accepted (Eq. 6-4).

problem by noting that the *weight* of an object (that is, the gravitational force acting on the object) is proportional to the *mass* of the object, $w = mg$. Thus, the gravitational force exerted on the Moon by the Earth is proportional to the mass of the Moon. But every force is accompanied by an equal and oppositely directed reaction force (Newton's third law). Therefore, the Moon must exert a gravitational force on the Earth and this force is proportional to the mass of the Earth. The conclusion, drawn by Newton, is that the gravitational force that exists between the Earth and the Moon (or, in fact, between *any* two objects) is proportional to the *product* of the two masses. When this result is combined with the $1/r^2$ dependence, the expression for the gravitational force can be written as

$$F_G = G \frac{m_1 m_2}{r^2} \qquad (6\text{-}3)$$

where the constant of proportionality G must be determined by other experiments in which the gravitational force between two objects is directly measured. Because this force is so small for laboratory objects, precision measurements with delicate instruments must be made. The first successful attempt to determine G in a laboratory experiment was made in 1797 by the English physicist and chemist Henry Cavendish (1731–1810); see Fig. 6-6. Modern experiments, using refinements of Cavendish's technique as well as other methods, have determined G to be

$$G = 6.673 \times 10^{-11} \text{ N-m}^2/\text{kg}^2 \qquad (6\text{-}4)$$

Notice that the units of G, N-m^2/kg^2, are necessary in order to give the force in newtons—that is, (meters)^2 cancels with the units of r^2 in the denominator and (kilograms)^2 cancels with the units of $m_1 m_2$ in the numerator, leaving newtons for the dimensions of F_G.

The Mass of the Earth

Newton's law of gravitation (Eq. 6-3) can be used to determine the mass of the Earth. To do this, we write two different expressions for the gravitational force on an object of mass M near the surface of the Earth. The first expression is Eq. 6-3 with $m_1 = M$ and with $m_2 = M_E$, the mass of the Earth:

$$F_G = G \frac{M M_E}{R_E^2}$$

where R_E is the radius of the Earth. The force on the object is also given by the mass M of the object multiplied by the acceleration due to gravity g:

$$F_G = Mg$$

Equating these two expressions for F_G, we have

$$G\,\frac{MM_E}{R_E{}^2} = Mg$$

The mass M occurs on both sides of the equation and therefore cancels. Solving for M_E, we find

$$M_E = \frac{gR_E{}^2}{G}$$

Using the known values of g, R_E, and G, we obtain

$$M_E = \frac{(9.8\ \text{m/s}^2) \times (6.38 \times 10^6\ \text{m})^2}{6.67 \times 10^{-11}\ \text{m}^3/\text{kg-s}^2}$$
$$= 5.98 \times 10^{24}\ \text{kg}$$

Universal Gravitation

The great importance of Newton's gravitational force equation, Eq. 6-3, is not that it successfully accounts for the motion of the Moon around the Earth, but that it applies to *any* pair of objects. Newton developed a *universal* theory of gravitation, and he made applications to the motions of planets and comets with equal success. He showed mathematically from his force equation that every planet (and comet) must exhibit an elliptical orbit, thereby deriving Kepler's first law. And he was also able to prove that Kepler's other two laws follow directly from gravitation theory.

One of the great triumphs of Newton's theory of gravitation was the discovery of the planet Neptune. After Uranus was detected in 1781, subsequent study of its motion revealed that there was no orbit that could satisfactorily describe its motion, even after account was taken of the perturbing influences of Jupiter and Saturn. Concern was expressed that Newton's theory of gravitation, which had proved supremely successful in describing the motion of the other planets, might contain a defect. Two young astronomers, John Couch Adams (1819–1892) of England and Urbain Leverrier (1811–1877) of France, independently hit upon the idea that Uranus was not following its prescribed orbit because it was being disturbed by another planet, as yet undiscovered. Both men proceeded to calculate where this new planet must be in order to account for the observed motion of Uranus. When the calculation was finally put into the hands of Johann Galle at the Berlin Observatory in 1846, it required less than an hour to locate the planet we now call Neptune. This was a monumental tribute to the power of calculational techniques applied to Newton's theory!

More recently, the theory has been used to analyze the motions of artificial satellites as well as the motions of objects outside the solar system, particularly pairs of stars that revolve around one another (*binary* stars). As far as we know today, *every* object in the Universe is subject to the gravitational force law set down by Newton.

The Gravitational Field

How does the Earth exert a force on the Moon through the vacuum of space? How does the Sun maintain the planets in their orbits even though there is no material connecting link between the Sun and the rest of the solar system? The answer, of course; is *gravity*. But what does *that* mean? How does gravity really *work*? We must draw a careful distinction here. It is one thing to give a description — even an absolutely correct description — of *how* Nature behaves, and it is quite a different thing to ask *why* Nature behaves in that particular way.

Newton worked out the mathematical description of the way in which objects gravitationally attract one another. Although relativistic corrections are

Photograph of a cluster of stars in the constellation Hercules. Hundreds of thousands of stars are held together in this cluster by gravitation. Richard Feynman, a physicist at the California Institute of Technology and a winner of a share of the Nobel Prize in physics for 1965, has said that "if one cannot see gravitation acting here, he has no soul."

required in certain circumstances, Newton's gravitation theory is basically a correct description of *how* Nature behaves.

Two-hundred years after Newton's formulation of gravitation theory, a different way to describe gravity was invented. This new method involves the concept of the *gravitational field*. We say that any object (for example, the Earth) sets up a condition in space to which another object (for example, the Moon) responds by experiencing an attractive force. This "condition in space" is the gravitational field. No material medium is necessary for a gravitational field to exist—the field extends even through vacuum. The strength of the gravitational field due to an object is proportional to the mass of the object and the strength decreases with distance away from the object as $1/r^2$. Near a large mass, such as Earth or the Sun, the gravitational field is strong, and far out in space the field is weak.

What have we accomplished by introducing the idea of the gravitational field? Have we not simply exchanged a mysterious *force* for a mysterious *field*? In a sense, *yes!* By using the field concept we have gained nothing in understanding *why* gravity works. But we have gained a mathematical advantage in describing gravity. In complicated situations it is easier to formulate a problem and to carry out calculations using the idea of the field than it is using the gravitational force directly. In this nonmathematical survey of physics, we will not have an opportunity to see the full power of gravitational field theory at work. We will, however, return to the field concept in the discussion of electric and magnetic phenomena (Chapter 12), where we will take advantage of a pictorial method for illustrating electric and magnetic fields.

6-4 SPACE TRAVEL

Achieving Earth Orbit

The launching of artificial satellites into orbit around the Earth and rocket expeditions to Mars and Venus, even manned landings on the Moon, have become almost commonplace in recent years. How do we place these satellites in orbit and launch vehicles on interplanetary tours? Although the navigational planning and the in-flight maneuvering require precise calculations and delicate timing, the laws that govern space travel are exactly those that determine planetary motions and the dynamics of objects in the laboratory. That is, Newton's laws of motion and the universal theory of gravitation are all that we require to understand satellite and rocket problems.

A single-stage rocket (that is, a rocket capable of firing only once and incapable of performing any subsequent maneuvers) cannot be placed into orbit around the Earth. Consider the possibilities. If the rocket is

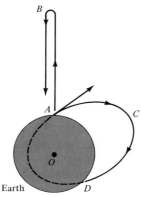

Figure 6-7 A single-stage rocket, whether fired straight up or at an angle to the vertical, will always fall back to the Earth (or will escape from the Earth's gravity if the initial velocity is sufficiently great). A single-stage rocket cannot achieve an Earth orbit.

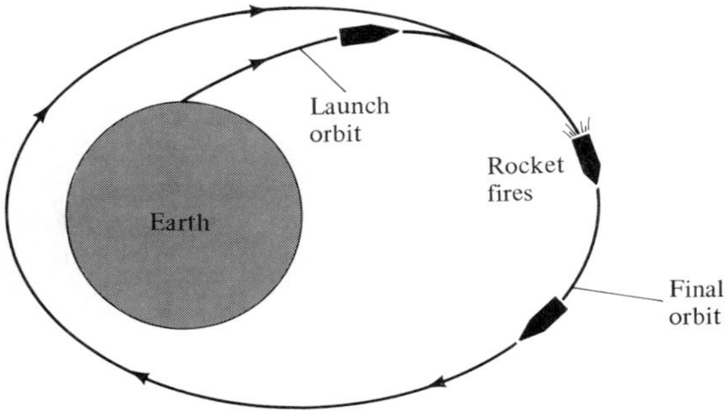

fired straight up, it will simply rise to a certain height (point *B* in Fig. 6-7) and fall back to its launch point (*A*). If the initial velocity of the rocket is sufficiently large (greater than 11.3 km/s near the Earth's surface), it will be able to escape from the gravitational attraction of the Earth and proceed into deep space, never returning to the Earth. In either case, an Earth orbit is not achieved. If the rocket is fired at an angle to the vertical, its path will be elliptical, the same as for any object orbiting another object under the influence of gravity. Because the launch point is on the surface of the Earth, the elliptical path, as shown in Fig. 6-7, always intersects the Earth. Thus, the rocket will rise along an elliptical path to a maximum height at *C* and will then fall back toward the Earth, impacting at *D*.

In order to project a rocket into an Earth orbit, there must be some method of altering the otherwise elliptical trajectory that intersects the Earth. The simplest way of accomplishing this is shown in Fig. 6-8. Here, the launch phase is the same as that sketched in Fig. 6-7, but as the rocket reaches the highest point in its launch orbit (which, if completed, would intersect the Earth), the rocket engine is fired again. The added velocity alters the trajectory and starts the rocket on a new elliptical path that is a true orbit. The exact dimensions of the elliptical orbit are determined by the position and the velocity of the rocket at firing and by the duration of firing. In practice, if the orbit is not precisely the one desired, subsequent brief firings of the rocket engine can supply the necessary corrections to adjust the orbit.

Staging

Instead of carrying the entire rocket into orbit, the usual procedure is to *stage* the launch. The amount of fuel required to boost a payload into orbit is many times more massive than the payload itself. Therefore, it is very inefficient to carry the fuel containers after the fuel has been exhausted. Many rocket systems are of three-stage design, as shown in Fig. 6-9. After launch, the first-stage fuel container is dropped and, later, the second-stage container is also released. The entire third stage may be placed into orbit or the fuel section of this stage may also be jettisoned, leaving only the bare payload in orbit. The empty fuel stages continue on ballistic trajectories and either burn up upon reentering the atmosphere or impact the Earth.

Figure 6-8 A rocket can be placed into Earth orbit by refiring the rocket engine as the vehicle reaches the highest point of the launch orbit. The new orbit is also elliptical, but it does not intersect the Earth and is a true orbit.

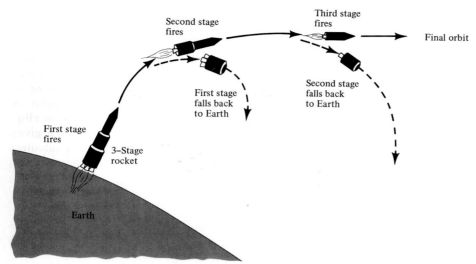

Figure 6-9 Instead of carrying the empty (and massive) fuel containers into orbit, the rocket stages that supplied fuel for launch are released and fall back to Earth while the payload is placed into orbit by the final rocket stage.

Orbit Shapes

Suppose that a rocket is placed into a launch orbit and reaches a certain height above the Earth, indicated by point A in Fig. 6-10. When the rocket engine is fired at this position and in a direction perpendicular to the line connecting the point with the Earth, what kind of orbit will result? If only a small velocity increment is added to the rocket velocity at point A, the launch orbit (which, remember, will intersect the Earth if unchanged) will be altered so that it just misses the Earth. The resulting orbit (labeled 1 in Fig. 6-10) will be an elongated ellipse with its *apogee* at A and its *perigee* close to the Earth (see Fig. 6-11). A larger additional velocity increment will produce a *circular* orbit (orbit 2). A still larger velocity increment will result in another elliptical orbit (orbit 3), but this orbit will have its perigee (instead of its apogee) at A.

Circular and elliptical orbits are the only orbits in which a satellite is *bound* to the Earth (or to any other parent object). If the rocket is given a very large velocity increment at point A, the vehicle can proceed into space in a *parabolic* or *hyperbolic* path, never to return to the Earth. (Such rockets, having escaped from the Earth, usually go into orbit around the Sun.)

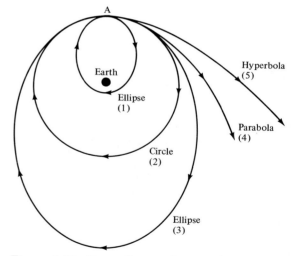

Figure 6-10 Depending on the magnitude of the velocity given a rocket at point A, the resulting orbit can have a variety of shapes. The elliptical (1 and 3) and circular (2) orbits are true Earth orbits, but the parabolic (4) and hyperbolic (5) paths (which result from the rocket being given large velocity increments) continue into space and never return to the Earth.

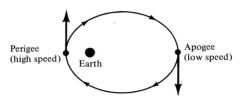

Figure 6-11 The highest point in an elliptical Earth orbit is called the *apogee* and the lowest point is called the *perigee*. Because the satellite has a constant angular momentum, the perigee is the point of highest speed and the apogee is the point of lowest speed (see Fig. 6-2).

The Size of Circular Orbits

What are the requirements on placing a satellite in a circular Earth orbit? As we have seen, the additional velocity increment given a rocket at the apogee of its launch orbit determines the shape of the final orbit. A low or a high velocity increment will result in an elliptical orbit; only a single velocity increment in a given situation will produce a circular orbit. We can calculate the velocity necessary for a circular orbit at a specified height by using Newton's expression for the gravitational force and the formula for the centripetal acceleration (Eq. 3-19).

First, we note that there is only *one* force acting on the rocket after its engine has been shut down; this is the gravitational force,

$$F_G = G\,\frac{M_E m}{r^2} \tag{6-5}$$

where M_E is the mass of the Earth and m is the mass of the rocket. The rocket moves according to Newton's equation, $F = ma$, where F is the gravitational force and a is the only acceleration that the rocket is experiencing, namely, the centripetal acceleration, $a_c = v^2/r$. Therefore, we can write

gravitational force = (mass) × (centripetal acceleration)

or,

$$G\,\frac{M_E m}{r^2} = m \times \frac{v^2}{r} \tag{6-6}$$

The rocket mass m cancels in this equation and is therefore irrelevant in determining the size of the orbit. Solving Eq. 6-6, we find

$$v^2 = \frac{GM_E}{r}$$

from which

$$v = \sqrt{\frac{GM_E}{r}} \tag{6-7}$$

Suppose that the rocket is at a height of 150 km above the Earth (see Fig. 6-12). That is, the distance r in Eq. 6-7, which is measured from the center of the Earth, is equal to the Earth's radius plus 150 km:

$$r = 6380 \text{ km} + 150 \text{ km} = 6530 \text{ km}$$

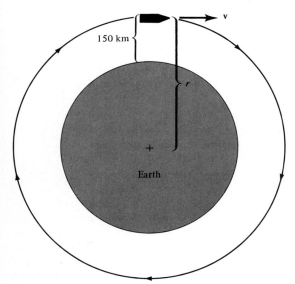

Figure 6-12 The velocity v required for a circular orbit at a distance r from the center of the Earth is specified by the theory of gravitation.

TABLE 6-2 IMPORTANT EVENTS IN THE HISTORY OF SPACE TRAVEL

Robert H. Goddard

1903	First paper on rocket travel published by the Russian school teacher, K. E. Tsiolkovskii. (In the same year, the Wright brothers made their first flight at Kitty Hawk.)
1926	First liquid-fueled rocket launched by Robert H. Goddard from Auburn, Massachusetts.
1944	German V-2 rockets bombard London; the 14-ton missiles reached heights of about 50 miles.
1949	First *staged* rocket fired from White Sands in New Mexico; the first stage was a modified V-2 and the second stage was a U.S. WAC-Corporal rocket. This vehicle reached an altitude of 244 miles.
1957	Launching of the Russian satellite, Sputnik I (October 4). Sputnik II, a half-ton vehicle, carried the first living creature, the dog Laika, into orbit (November 3).
1958	First U.S. satellite, Explorer I, launched from Cape Canaveral (January 31); the Earth's radiation belts were first detected on this flight.
1959	First rocket (the Russian Lunik I) to reach the Moon.
1961	First manned space flight by the Russian Cosmonaut, Yuri Gargarin, in the vehicle Vostok 1 (April 12); Gargarin flew for 1 hour 48 minutes, completing one orbit, and reached an altitude of 203 miles.
1961	First American in space; Alan Shepard made a suborbital flight in Freedom 7 (May 5).
1962	First space vehicle to reach Venus – Mariner 2.
1962	First communications satellite, Telstar I, placed into orbit.
1965	First docking operation in space performed by Astronauts Stafford and Schirra in Gemini 6.
1966	First spacecraft to land on another planet – Venera 3 on Venus.
1968	Astronauts Borman, Lovell, and Anders orbit the Moon in Apollo 8.
1969	First landing of Man on the Moon – Astronauts Neil Armstrong and Edwin Aldrin land in Eagle, the lunar module of Apollo 11.
1971	First spacecraft to go into orbit around another planet – Mariner 9 around Mars.
1972	Last manned Moon mission (Apollo 17).
1973	First space vehicle to reach Jupiter – Pioneer 10.
1973–1974	Skylab missions; numerous scientific experiments and observations carried out in Earth-orbit.
1974	First probe of Mercury – Mariner 10.
1975	American–Soviet joint mission; Apollo–Soyuz linkup.
1976	Viking mission to Mars; search for extraterrestrial life.

Yuri Gargarin

NASA

Eagle (Apollo 11)

The velocity required for a circular orbit at this height is

$$v = \sqrt{\frac{GM_E}{r}}$$

$$= \sqrt{\frac{(6.67 \times 10^{-11} \text{ N-m}^2/\text{kg}^2) \times (5.98 \times 10^{24} \text{ kg})}{6530 \times 10^3 \text{ m}}}$$

$$= 7.8 \text{ km/s}$$

or about 17 500 mi/h. This type of orbit is close to that achieved in the first manned orbital flights.

If the distance r is increased to 42 200 km (26 400 mi), the velocity becomes

$$v = \sqrt{\frac{(6.67 \times 10^{-11} \text{ N-m}^2/\text{kg}^2) \times (5.98 \times 10^{24} \text{ kg})}{4.22 \times 10^7 \text{ m}}}$$

$$= 3.07 \text{ km/s}$$

At this velocity, the period of the motion is

$$\text{period} = \frac{\text{circumference of orbit}}{\text{velocity}}$$

Thus,

$$\tau = \frac{2\pi r}{v} = \frac{2\pi \times (42\ 200 \text{ km})}{3.07 \text{ km/s}}$$

$$= 86\ 400 \text{ s}$$

The period of rotation is therefore just equal to *one day*. That is, the satellite rotates at the same rate as the

Landsat — How It Works for Us

The most widely publicized and best known aspect of the United States' space program has been the Apollo missions to the Moon. Experimental apparatus placed on the Moon by the Apollo astronauts and the rock samples they brought back to Earth have provided us with a truly enormous amount of scientific information concerning the geologic structure and history of the Moon. As important as the manned missions to the Moon and the unmanned flights to Mars and Venus have been in learning about the solar system, NASA has not overlooked the fact that orbiting satellites are ideal instrument platforms for viewing the *Earth*. In July 1972 the space agency placed into orbit the first Landsat satellite at a cost of $200 million. (The Landsat satellites were originally called ERTS — Earth Resources Technology Satellites.) Never before has there been such a rich and immediate reward from a scientific venture.

Landsat I and II are equipped with several high-resolution cameras that photograph the Earth using light from different parts of the spectrum. By comparing the different photographs of the same region, scientists can learn the details of the surface features — for example, the type and quality of vegeta-

Landsat I

tion, geologic features that are clues to mineral deposits, and the sources of air and water pollutants.

During the first year and a half after launch, Landsat I made detailed photographic surveys of about three-quarters of the Earth's land mass and almost the entire United States. The discoveries made during this brief interval have been as varied as they have been valuable.

Item. The color and contour of certain areas of western Canada strongly indicate the presence of undiscovered deposits of nickel-bearing minerals.

Item. All of the strip mines in several states have been pinpointed.

Item. The burned-out regions of California's forest land have been mapped.

Item. Land-use maps have been made for several of our larger cities and for several entire states.

Item. Many geographic errors have been revealed. For example, Landsat photographs showed lakes in Brazil that were as far as 20 miles from the positions on previous maps.

Item. Snow levels in the western mountains were determined, thus allowing accurate predictions of the magnitude of the spring run-offs.

Item. Photographs of Alaska's North Slope show a peculiar alignment of the lakes which, according to geologists, indicate a direct link to the huge petroleum deposits that underlie the entire region.

Item. Two large areas of copper deposits have been discovered in remote parts of Pakistan.

Item. Inventories have been made of Nevada's wheat grass, Arizona's eroded soil, South Dakota's productive soils, and Indiana's diseased trees.

The economic benefits of the photographic maps and surveys made by Landsat I and II defy calculation. Money saved in the mapping of highway and pipeline routes alone will probably equal the cost of launching the Landsat instruments. Soil and water conservation projects will be made considerably easier. Cities and states will be able to plan land use more efficiently. Control of crop diseases will be made more effective. Areas that are ripe for forest fires can be identified. And the list goes on. The Landsat program has already demonstrated an enormous capability to work for humankind and the benefits should continue into the future.

Earth, and to an observer on the Earth the satellite will appear to remain stationary. Such satellites are called *synchronous* because they rotate in time with the Earth. Synchronous satellites are used extensively in the worldwide communications network (see Fig. 6-13).

Artificial Gravity

An astronaut in a satellite orbiting the Earth will "fall" around the Earth at exactly the same rate as will his satellite. Relative to his space capsule, he will be "weightless," and there will be no force pressing him to any side

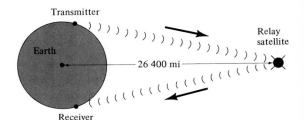

Figure 6-13 Satellites placed in synchronous orbits are used to relay radio communications between points on the Earth that would not otherwise be able to communicate by radio because of the straight-line propagation of high-frequency radio waves.

In 1962 the National Aeronautics and Space Administration launched the *Syncom* satellite, shown here in an artist's conception. This was the first synchronous satellite to be used in the now extensive international radio communication system.

of the capsule. If the satellite is a space laboratory, designed to stay in orbit for an extended period of time, this "weightlessness" may be a severe handicap to the astronaut in carrying out his laboratory duties, making observations, and performing experiments. In order to avoid this problem, an "artificial gravity" can be created in the space station which will permit the inhabitants to function in a near-normal manner. All that is required is to set the space station into rotation around its own axis. Consider a donut-shaped space station, as shown in Fig. 6-14. The outer radius of the station is R and it rotates around a central axis (perpendicular to the plane of the drawing) with a rim speed v. Any point on the outer rim experiences a centripetal acceleration, $a_c = v^2/R$ (see Eq. 3-19). An astronaut therefore exerts a centripetal reaction $ma_c = mv^2/R$ on the outer rim. That is, he is pressed toward the outer rim and exerts a centrifugal force on the "floor" of the space station. (Compare the discussion of centripetal reaction in Section 4-3.) The floor pushes on the astronaut and the astronaut pushes on the floor just as if gravity were acting.

With what speed must a space station rotate in order to produce at its outer rim an artificial gravity equal to the true gravity at the surface of the Earth? Suppose that the station has a radius of 200 ft. Then, solving $a_c = v^2/R$ for v, we have

$$v = \sqrt{a_c R} = \sqrt{(32 \text{ ft/s}^2) \times 200 \text{ ft}}$$
$$= \sqrt{6400} \text{ ft/s} = 80 \text{ ft/s}$$

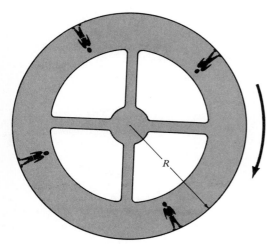

Figure 6-14 The rotation of a space station around its own axis creates an "artificial gravity" for the occupants.

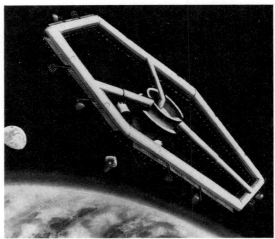

An artist's conception of a space station which can rotate to produce an "artificial gravity."

The period of this rotation is

$$\tau = 2\pi R/v = (2\pi \times 200 \text{ ft})/(80 \text{ ft/s}) = 15.7 \text{ s}.$$

Any large space laboratories that are constructed in the future will probably have some degree of rotation in order to produce an artificial gravity, but, in order to keep the problems associated with high rotation rates to a minimum, the magnitude of the acceleration produced will probably be only a fraction of g.

The Space Program — What Is Its Future?

Between July 1969 and December 1972, the United States launched six space missions that landed men on the Moon. This series of lunar landings stands as the greatest scientific-technological achievement ever made. The Apollo missions have been appropriately called "Man's greatest adventure." Probably not within this century will astronauts again venture outside the immediate vicinity of the Earth. What, then, lies ahead for the space program?

During the next 20 or 30 years the primary emphasis of space missions will be the exploration of the planets with unmanned probes, while at home the astronauts will fly a series of Earth-orbit missions to examine in still greater detail the face of the Earth and its changing conditions. The Skylab missions (flown in 1973 and 1974) were enormously successful, especially in the wealth of data acquired regarding the Earth and the Sun. In addition, nu-

NASA

Skylab photographed from command module.

NASA

Photograph of the city of Chicago taken from Skylab.

NASA

The Space Shuttle vehicle, which is launched piggyback on a fuel container and then maneuvers into orbit. At the end of the mission, it leaves orbit and flies to a landing on Earth.

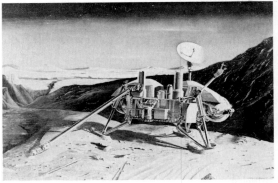

NASA

Viking lander. Instruments in this vehicle will scoop up samples of the Martian surface to test for signs of life.

merous physical and biological experiments were performed to determine the effects of "weightlessness" in various processes (for example, the growing of crystals in a zero-g environment). Thousands upon thousands of photographs of the Earth, the Sun, and the stars were returned from the Skylab missions; the analysis of this gigantic collection of information is just beginning. Many of the Earth photographs will supplement and improve upon the Landsat data (see the previous essay). In the future, similar missions (some involving the recoverable Space Shuttle vehicle) will be flown to add to our growing knowledge of the Earth and the Sun.

The remainder of this century will see a series of spectacular space missions designed to study the Sun, planets, moons, and other objects in the solar system. By 1976, the Sun will have been closely approached by two spacecraft (Helios 1 and 2). Both of these vehicles are instrumented to study various Sun phenomena such as the solar wind, cosmic rays, and the residue of past comets. The Mariner 10 spacecraft passed close to the planet Mercury in 1974, and two later missions (to be launched in 1987) will place probes into orbit around the planet nearest the Sun. Up to 12 probes are planned for the planet Venus, beginning with the launch of a pair in 1978. Soft landings on the surface will be made, and two metal balloons will be released in the Venusian atmosphere to chart wind turbulence and circulation patterns. The Viking missions to Mars, planned for 1976 and 1979, will land on the planet and conduct a search for the first signs of extraterrestrial life.

The outer planets will not be neglected in NASA's plans for the future. Jupiter has already been visited (by Pioneer 10 and 11), and additional missions will orbit the planet for closer observations. Saturn's rings will be studied; planned for 1985 are two spacecraft that would orbit the planet in paths that would carry them through the ring structure. Plans are being formulated for missions to Uranus and Neptune in the late 1980's, but Pluto is too far away for consideration, at least for the next 30 years or so.

Smaller members of the solar system are also scheduled for visits—the moons of Mars, Jupiter, and Saturn, and at least one of the larger asteroids. Comets, too, are tentative targets. A vehicle may be launched to rendezvous with Comet Encke in 1981, and an even more spectacular mission would be a meeting with Halley's comet when it approaches next in 1986.

Because of fluctuations in the NASA budget, not all of these planned missions may survive. But if only a fraction actually take place, we can look forward with anticipation to a era during which our knowledge of the solar system will be enormously advanced.

6-5 THE ELECTRICAL FORCE

Positive and Negative Electricity

In ancient times, electrical phenomena were much less studied and understood than were gravitational phenomena. In the 6th century B.C., Thales of Miletus (640–546 B.C.), a Greek merchant turned astronomer and mathematician, discovered that amber, when rubbed with silk, would attract light objects such as straws and dried leaves. We are all familiar with this electrostatic effect: when a comb is drawn through dry hair, it will attract small bits of paper. Or, if we shuffle across an Acrilan carpet, we become "electrified" and will experience an unpleasant shock upon touching a door knob.

We now know that there are two types of electricity. The basic carriers of negative electricity are *electrons,* the particles that are found in the outer layers of all atoms. The atomic cores, the nuclei, are the seats of positive electricity. When a piece of bulk matter is electrically charged, this is almost always accomplished by adding electrons to or removing electrons from the object. The more massive, positively charged atomic nuclei remain essentially stationary in almost all electrical processes. Thus, an object is given a negative electrical charge by the addition of extra electrons or the object is given a positive electrical charge by the removal of electrons.

In its normal condition, all matter is electrically *neutral.* That is, the negative charge of the atomic electrons is just balanced by the positive charge of the nuclei. The fundamental unit of positive charge is carried by the *proton,* and the magnitudes of the electron and proton charges appear to be exactly equal. (Experiments have shown that if any difference exists, it must be less than 1 part in 10^{19}). Therefore, in any electrically neutral atom,

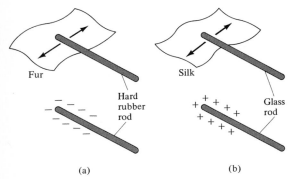

Figure 6-15   (a) A hard rubber rod that is rubbed with a piece of fur becomes *negatively* charged. (b) A glass rod that is rubbed with a piece of silk becomes *positively* charged.

molecule, or piece of bulk matter, the number of electrons and protons is exactly the same. *Changing* the number of electrons results in an electrical charge on the object.

Attraction and Repulsion

A few simple experiments are all that is required to reveal some of the basic properties of electrical charge. If a hard rubber rod is stroked vigorously with a piece of fur (Fig. 6-15a), some of the electrons in the fur are removed by friction and they are transferred to the rubber rod. The rod, therefore, becomes *negatively* charged. Similarly, if a glass rod is rubbed with a piece of silk (Fig. 6-15b), electrons are transferred from the rod to the silk and the glass rod acquires a *positive* charge.

The charge that resides on the rubber rod or the glass rod can be further transferred (at least, partially) to other objects by simply touching a rod to an object. For example, suppose that we suspend two light-weight balls (such as pith balls) from a fixed support by means of threads. If we now touch each of these balls with a charged glass rod, both balls will become positively charged. We now observe that the two balls stand farther apart than they did in the uncharged condition (Fig. 6-16a). That is, a repulsive electrical force exists between the two similarly charged objects — *like charges repel.*

If we touch one of the balls with a charged rubber rod and the other with a charged glass rod, we then find that the balls stand closer together than when uncharged (Fig. 6-16b). That is, an attractive electrical force exists between the two oppositely charged objects — *opposite charges attract.*

Although we do not ordinarily think in electrical terms, many of the forces that we experience every day are due to the attraction or repulsion between electrical charges. A force is required to stretch a rubber band, and when the ends are released it snaps back to its original length. What is the nature of the force that accounts for the *elasticity* of a rubber band? When we stretch a rubber band we are pulling on the long rubber molecules in the band. This pull moves the atoms and electrons from their normal positions. A force is required to move the electrons farther from the positively charged nuclei to which they are attracted. When the band is released, electrical attraction returns the atoms and electrons to

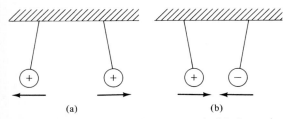

Figure 6-16   (a) Like charges repel. (b) Opposite charges attract.

the positions they previously occupied. Of course, if we pull too hard, the electrical forces can no longer maintain the integrity of the band and it breaks.

Conductors and Insulators

A free atom or molecule of any substance will exist naturally in an uncharged condition with its normal complement of electrons attracted to and bound to the atomic nucleus or nuclei. Some materials—primarily, *metals*—have the interesting property that in the bulk state, some of the atomic electrons are not bound to any particular atom but are free to move around within the material. Such materials are called *conductors*. If, for example, a number of electrons are placed on the surface of a copper sphere, these electrons, because of their mutual repulsion, will almost instantaneously distribute themselves uniformly over the surface of the sphere (Fig. 6-17a). Similarly, if the sphere is given a positive charge by removing some of the electrons, the mutual repulsion of the remaining electrons will cause the positions of the missing electrons to be distributed uniformly over the surface. In the first case, there is a uniform distribution of negative charge over the surface of the sphere. In the second case, there is a uniform distribution of positive charge over the surface of the sphere.

Many materials—such as glass, wood, paper and plastics—do not possess any substantial number of free electrons. These materials are therefore *non*conductors or *insulators*. Because electrons do not move readily in such materials, an electrical charge placed on a sphere of glass, for example, will remain localized on the surface for a considerable period of time (Fig. 6-17b). No material is a perfect insulator, however, and eventually the charge on a piece of glass will "leak off" (that is, it will be conducted to the surroundings) and the glass will again become electrically neutral.

Gases are generally good insulators. But if two large charges (of opposite sign) are brought close together in air, a few electrons may be ripped off the air molecules and a path between the two charges will become temporarily conductive. Electrons will then flow from the negatively charged object to the positively charged object and a *spark* will result. On a large scale in the atmosphere, the same process leads to a lightning stroke (see Section 11-6).

Pure water is a good insulator. The water from most sources, however, contains impurities which raise the

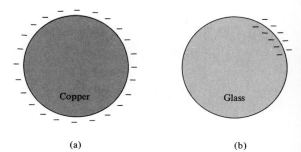

(a)   (b)

Figure 6-17   (a) Because some of the electrons are free to move within a conductor, such as copper, any charge placed on a sphere of a conducting material will rapidly distribute itself uniformly over the surface. (b) Charge placed on the surface of a nonconductor, or insulator, will remain localized.

very small conductivity of pure water to the point that it readily conducts electricity. (It is therefore a poor policy to be in water during a thunderstorm.) On humid days, a thin film of water will collect on most surfaces and will destroy the insulating properties of most materials. Electrostatic experiments can therefore be performed much more easily on dry days than on humid days.

The Earth is a reasonably good conductor and it acts as a giant reservoir for supplying or receiving the electrons that are required to charge or discharge an object. Thus, when we wish to discharge (and render neutral) any charged object, we simply connect a wire between the object and the Earth. (If the amount of charge involved is not too great, the human body—which is a conductor, albeit a rather poor one—can be used as the connecting link.) Electrons flow through the wire from or to the Earth and the charge is drained from the object. In order to prevent the possibility of electrical shocks, the metal shielding that surrounds most electrical devices should be connected by a wire to the Earth (that is, the shield should be *grounded*).

Charging by Induction

One way to give an electrical charge to a neutral object is to bring it into contact with a charged object, thereby causing some of the charge to be transferred. This is how we imagined the balls in Fig. 6-16 were charged by the rods in Fig. 6-15. It is also possible to *induce* a charge on an object without bringing it into contact with a charged object. In Fig. 6-18a, a positively charged ball is brought near the end of a conducting rod. The free electrons in the rod are attracted toward the ball and they concentrate in the end nearer the ball. The opposite end acquires a positive charge in the process. (The rod is said to be *polarized,* with a positive pole at one end and a negative pole at the other end. But the rod as a whole carries *no net charge.*)

Next, we imagine that the rod is cut into two pieces, nearly in contact, as in Fig. 6-18b. The distribution of charge remains the same as in Fig. 6-18a. Finally, we remove the charged ball and further separate the two halves of the rod. The charge now distributes itself over the two parts: one half carries a net positive charge and the other carries a net negative charge. The parts of the rod have been charged by *induction*.

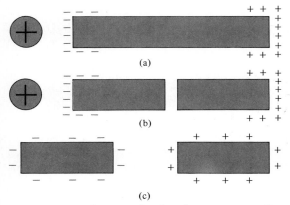

Figure 6-18 The steps in charging two parts of a conducting rod by induction.

The Electron Charge

In order to make measurements of electrical charge, we must have a standard unit of charge, just as we have standard units of length, time, and mass. What do we use for a *standard charge?* An obvious choice is the charge of an electron, the basic unit of negative charge. The units that are used to measure electric charge are *coulombs* (C), and in these units, the magnitude of the electron charge, denoted by the symbol e, is

$$e = 1.602 \times 10^{-19} \text{ C} \qquad (6\text{-}8)$$

The electron charge and the proton charge have the same magnitude but opposite signs:

$$\text{electron charge} = -e$$
$$\text{proton charge} = +e$$

No electrical charge smaller than e has ever been detected and, as far as we know today, no such charge exists in Nature. Every charge that occurs, whether in atomic systems or in bulk matter, is equal to some integer number times e: $2e$, $10e$, or $10^{19} e$. Because the value of e is so small, the discrete nature of electrical charge is not apparent in everyday phenomena. In fact, sensitive laboratory instruments used for measuring electrical charge are generally unable to detect a charge smaller than about 10^{-12} C. Such a charge corresponds to about 6×10^6 electrons and one electron more or less makes no difference whatsoever.

Conservation of Charge

When an object is given an electrical charge, the charge that is *gained* by the object is *lost* by some other object. Electrons are neither created nor destroyed in such a process. Thus, when a hard rubber rod is given a negative charge by rubbing with fur (Fig. 6-15a), the electrons that are gained by the rod are lost by the fur and the fur becomes positively charged. If the rod and the fur are considered to constitute a *system,* then there is no net change in the charge of that system. The conservation of charge is one of the fundamental conservation laws of Nature, on a par with the conservation of linear momentum and of angular momentum:

> The total amount of electrical charge in any isolated system remains constant.

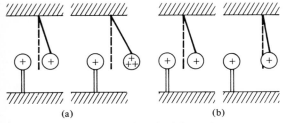

Figure 6-19 (a) The electrical force between two charged objects is increased by increasing the charge on either object. (b) The electrical force between two charged objects decreases if their separation is increased.

Within a system, charge can be transferred from one object to another, but the total charge of that system cannot be altered in the process. No experiment has ever been performed that revealed a violation of the law of charge conservation.

Coulomb's Law

Some simple experiments, such as those sketched in Fig. 6-19, show the following properties of the electrical force that exists between charged objects:

(a) If the amount of charge on either of the two charged objects is increased, the electrical force increases (Fig. 6-19a).

(b) If the distance separating two charged objects is increased, the electrical force decreases (Fig. 6-19b).

In 1785, the French physicist Charles Augustin de Coulomb (1736–1806) studied the characteristics of the electrical force by using a sensitive balance to measure the force between two objects as the charges and distances were varied. He was able to conclude that the electrical force is directly proportional to the product of the two charges, which we label q_1 and q_2; that is,

$$F_E \propto q_1 q_2$$

This behavior of the electrical force is exactly analogous to the way that the gravitational force depends on mass: $F_G \propto m_1 m_2$. Coulomb further showed that the force decreases with the square of the distance between the objects; that is,

$$F_E \propto \frac{1}{r^2}$$

This result, too, is exactly the same as that found for the gravitational force. Consequently, the expression for the electrical force between two charged objects has the same form as Newton's equation for the gravitational force (Eq. 6-3):

$$\boxed{F_E = K \frac{q_1 q_2}{r^2}} \tag{6-9}$$

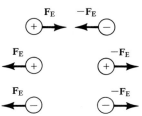

Figure 6-20 The electrical force between two objects is *attractive* if the signs of the charges are *opposite* and the force is *repulsive* if the signs are the *same*. Notice that when a force $\mathbf{F}_E$ acts on one of the charges, a force $-\mathbf{F}_E$ acts on the other charge (Newton's third law).

This result is called *Coulomb's law*. In using this equation, we must remember that the force is *attractive* if q_1 and q_2 have *opposite* signs and is *repulsive* if q_1 and q_2 have the *same* signs (Fig. 6-20).

Because the units of force (newtons), distance (meters), and charge (coulombs) are all specified, the constant K must be determined to give the correct force for a given pair of charges separated by the distance r. The value of K is

$$K = 9 \times 10^9 \text{ N-m}^2/\text{C}^2 \qquad (6\text{-}10)$$

The numerical value of K actually depends on the square of the speed of light. Therefore, the leading factor is not exactly equal to 9. However, for our purposes we will use the approximate value given in Eq. 6-10.

If two charges of $+1$ C are separated by a distance of 1 m, the repulsive force between them would be 9×10^9 N. This is truly an enormous force. (It is approximately equal to the gravitational force on a mass of 10^9 kg, a million metric tons!) Usually, we deal with charges that are much smaller than 1 C. In a typical laboratory experiment, the charge on an object might be 10^{-7} C (0.1 μC). Two such charges, separated by a distance of 10 cm would experience a force

$$F_E = (9 \times 10^9 \text{ N-m}^2/\text{C}^2) \times \frac{(10^{-7} \text{ C}) \times (10^{-7} \text{ C})}{(0.1 \text{ m})^2}$$

$$= 9 \times 10^{-3} \text{ N}$$

which is approximately equal to the gravitational force on a mass of 1 g.

The electrical force is immensely stronger than the gravitational force. Consider a proton ($m_p = 1.67 \times 10^{-27}$ kg) and an electron ($m_e = 9.1 \times 10^{-31}$ kg); in a hydrogen atom these particles are separated by a distance of approximately 5.3×10^{-11} m. Both gravitational and electrical forces exist between the proton and the electron (and both forces are attractive). The magnitude of the gravitational force is

$$F_G = G \frac{m_1 m_2}{r^2}$$

$$= (6.67 \times 10^{-11} \text{ N-m}^2/\text{kg}^2)$$

$$\times \frac{(1.67 \times 10^{-27} \text{ kg}) \times (9.1 \times 10^{-31} \text{ kg})}{(5.3 \times 10^{-11} \text{ m})^2}$$

$$= 3.6 \times 10^{-47} \text{ N}$$

whereas the magnitude of the electrical force is

$$F_E = K \frac{q_1 q_2}{r^2}$$

$$= (9 \times 10^9 \text{ N-m}^2/\text{C}^2)$$
$$\times \frac{(1.60 \times 10^{-19} \text{ C}) \times (1.60 \times 10^{-19} \text{ C})}{(5.3 \times 10^{-11} \text{ m})^2}$$
$$= 8.2 \times 10^{-8} \text{ N}$$

The ratio of these two factors is

$$\frac{F_E}{F_G} = \frac{8.2 \times 10^{-8} \text{ N}}{3.6 \times 10^{-47} \text{ N}} = 2.3 \times 10^{39}$$

Thus, the electrical force between a proton and an electron is stronger than the gravitational force by a factor exceeding 10^{39}! Within atoms, then, gravitational effects are completely overwhelmed by the electrical forces between the constituents of the atoms. It is only in bulk matter, in which electrical effects are neutralized by the presence of large and equal numbers of positive and negative charges, that the gravitational force is apparent and important.

6-6 THE NUCLEAR AND WEAK FORCES

The Attractive Nuclear Force

The two forces that we have discussed so far—the gravitational force and the electrical force—are the only forces needed to account for all of the large-scale phenomena that take place in the Universe as well as all of the microscopic phenomena that take place in the domain of atoms and molecules. But these forces are not adequate to describe *nuclear* effects such as radioactive decay and the occurrence of nuclear reactions and fission. Even to account for the fact that protons and neutrons are bound together in nuclei requires the introduction of a new fundamental force.

We know that all nuclei are extremely small—a typical nuclear radius is only a few times 10^{-15} m. Within the tiny nuclear volume are crowded together a number of protons and neutrons. Each proton exerts a repulsive force on every other proton in the nucleus because they all carry the same electrical charge. And this repulsive force is huge. Consider two nuclear protons that are separated by a distance of 2×10^{-15} m, a typical separation for protons in nuclei. The repulsive electrical force between these protons is

$$F_E = K\,\frac{q_1 q_2}{r^2}$$

$$= (9 \times 10^9 \text{ N-m}^2/\text{C}^2) \times \frac{(1.6 \times 10^{-19} \text{ C})^2}{(2 \times 10^{-15} \text{ m})^2}$$

$$= 58 \text{ N}$$

This force between two *nuclear* protons is approximately equal to the gravitational force on a 6-kg mass at the surface of the Earth! What can hold a nucleus together against this enormous repulsive force? Clearly, a new force of extraordinary strength must be operating within nuclei.

The force that acts at the extremely small distances within nuclei and holds the nuclei together in spite of their tendency to fly apart because of repulsive electrical forces is called the *strong nuclear force,* or sometimes simply the *strong force* or the *nuclear force.* This strong force acts not only between protons and protons (*p-p*) but also between protons and neutrons (*p-n*) and between neutrons and neutrons (*n-n*).

The character of the strong force is decidedly different from the gravitational and electrical forces. Not only is the strength of the strong force vastly greater than the strength of these other forces, but the way in which the strong force depends on distance bears no resemblance to the familiar $1/r^2$ pattern of the gravitational and electrical forces. The strong force is effective only over very small distances. If a proton and a neutron are separated by more than about 10^{-15} m, they no longer exert a strong force on one another. That is, the *range* of the strong force is only about 10^{-15} m (see Fig. 6-21). As the separation of the two particles is decreased below 10^{-15} m, the strong attractive character of the force becomes evident. However, for separations less than 0.1–0.2×10^{-15} m, the strong force becomes *repulsive* (Fig. 6-21). Thus, nuclei are prevented from collapsing.

One of the central problems in our efforts to understand the behavior of nuclei and elementary particles is to learn the details of the strong force. Although much has been discovered, there are still aspects of the strong force that require more study.

The Weak Force

Unstable nuclei exhibit two different types of radioactive decay processes: α decay, in which a helium nu-

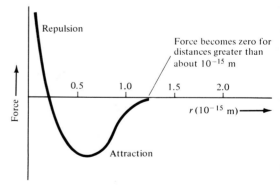

Figure 6-21   The strong nuclear force is attractive for separation distances between about 0.2×10^{-15} m and 10^{-15} m, and it becomes repulsive for distances smaller than about 0.2×10^{-15} m. The dependence of the strong force on distance shown here is only schematic; the strong force actually has several complicating features which cannot be shown in a diagram such as this.

TABLE 6-3 RELATIVE STRENGTHS OF THE FUNDAMENTAL FORCES BETWEEN VARIOUS PAIRS OF PARTICLES

FORCE	APPROXIMATE STRENGTHS AT DISTANCES NEAR 10^{-15} m			
	$e-\nu^a$	$e-p$	$p-p$	$p-n$ $n-n$
Strong nuclear	0	0	1	1
Electrical	0	10^{-2}	10^{-2}	0
Weak	10^{-13}	10^{-13}	10^{-13}	10^{-13}
Gravitational	0	10^{-41}	10^{-38}	10^{-38}

^{a} The symbol ν stands for "neutrino."

cleus is ejected, and β decay, which involves the emission of an electron and a neutrino (see Section 20-1). Because only protons and neutrons are emitted during α decay, this process is governed by the strong force (and the electrical force). The particles emitted during β decay—electrons and neutrinos—do not interact via the strong force and so this force is not adequate to describe the β decay process. A fourth fundamental force—the *weak force*—is required to account for the way in which electrons and neutrinos (as well as muons) interact with one another. The weak force has an extremely short range, certainly less than 10^{-15} m, but as far as we now know, the range may even be considerably smaller.

Table 6-3 shows a comparison of the strengths of the four basic forces for various pairs of particles. The strength of the strong force has been arbitrarily chosen to be unity. Notice that the gravitational force is by far the weakest force, even compared with the weak force.

SUGGESTED READINGS

G. Gamow, *Gravity* (Doubleday, Garden City, New York, 1962).

A. D. Moore, *Electrostatics* (Doubleday, Garden City, New York, 1968).

Scientific American articles:

J. Christianson, "The Celestial Palace of Tycho Brahe," February 1961.

C. Wilson, "How Did Kepler Discover His First Two Laws?" March 1972.

QUESTIONS AND EXERCISES

1. The orbital speed of the Earth around the Sun is approximately 30 km/s. What are the orbital speeds of Venus and Mars? (Use the data in Table 6-1). Do you expect the orbital speed of Jupiter to be greater or smaller than any of these three values?

2. Calculate the gravitational force that the Earth exerts on the Moon by multiplying the mass of the Moon by its acceleration. Calculate the same quantity using Newton's gravitational equation (Eq. 6-3) and compare the results.

3. Two identical spherical objects are separated by a distance of 1 m. If the gravitational force between them is 6.7×10^{-9} N, what is the mass of each object?

4. In the experiment performed by Henry Cavendish in 1798, he measured the force between two lead balls with masses of 49.5 kg and 0.775 kg. When the centers of these balls were separated by a distance of 0.2 m, what was the gravitational force?

5. In Jules Verne's *A Trip Around the Moon,* published just after the Civil War, the author describes a huge cannon that fires a 10-ton projectile, containing three men and several animals, to the Moon. On the flight, as the vehicle coasts unpowered (a ballistic trajectory), the passengers walk around normally within the vehicle on the side nearest

the Earth. As they approach the Moon, they walk on the side nearest the Moon. Is this really what happens on a ballistic Earth – Moon trip?

6. The weight of an object is $w = mg$. But the *weight* is just the gravitational force, $F_G = GMm/r^2$, where M is the mass of the Earth. Equate these two forces and solve for g at the surface of the Earth. Use the known values of G, M, and r, and compute the acceleration g. Compare the result with the measured value, $g = 9.8$ m/s².

7. Use the expression for g obtained in the preceding exercise and substitute values of M and r appropriate for the Moon, thereby obtaining the value of the acceleration due to gravity at the surface of the Moon.

8. If the Earth had only one-half of its actual radius and only one-eighth of its actual mass, what would be the value of g at the surface? (Refer to Exercise 6 above.)

9. An astronomer claims to have discovered a new planet midway between Mars and Jupiter with a period of 2.50 years. What is your opinion of his claim and why?

10. Use the data in the Table 6-1 and verify Kepler's third law for Venus, Jupiter, and Saturn.

11. Use Eq. 6-7, properly identifying the symbols, to calculate the velocity of the Earth in its orbit around the Sun.

12. What is the velocity of a satellite in a circular orbit 5 km above the surface of the Moon? Why is such a low-altitude orbit practical around the Moon but not around the Earth? (Near the surface of the Moon, $g = 1.62$ m/s², about one-sixth of the Earth value.)

13. What is the rotation period of an Earth satellite that is in a circular orbit 150 km above the Earth's surface? Compare this period with that for the first manned orbital flight of Vostok 1 (see Table 6-2).

14. In Fig. 6-18a, suppose that the positively charged end of the rod is momentarily connected to the ground by a wire. After the wire is disconnected, the charged ball is removed. Describe what happens. What is the final condition of the rod? (This is another way to charge an object by induction.)

15. Someone proposes that there are *three* different kinds of electrical charge instead of two, and he provides you with three objects which, he claims, each carry a different kind of charge. What experiments would you perform to test his assertion?

16. Will the electrical force between a proton and an electron always be greater than the gravitational force by a factor of 2.3×10^{39} regardless of the separation of the two particles? Explain.

17. An experimenter attempts to measure the electrical force between two objects each of which carry a charge of 0.2 C by bringing the objects together to a separation of 0.5 m. What value for the force does he obtain? Is this a practical experiment? Explain.

18. A gram of copper contains approximately 10^{22} atoms. Suppose that two 1-g spheres of copper are separated by a distance of 3 m and that one electron has been removed from every copper atom in each sphere. What is the electrical force between the two spheres?

19. Two identical copper spheres carry charges of $+4 \times 10^{-5}$ C and -12×10^{-5} C, respectively. If the spheres are separated by a distance of 1 m, what is the electrical force between them? If the spheres are now brought together so that they touch and then are returned to their original positions, what is the electrical force? (Indicate whether the force is attractive or repulsive in each case.)

20. Two electrical charges, $q_1 = +3 \times 10^{-4}$ C and $q_2 = +8 \times 10^{-4}$ C are separated by a distance of 1.6 m. A charge $q = +4 \times 10^{-4}$ C is located midway between q_1 and q_2. Calculate the force (magnitude and direction) on each of the three charges.

21. Three electrical charges are located as follows: $q_1 = +2 \times 10^{-4}$ C at $x = 0$, $y = 0.8$ m; $q_2 = -3 \times 10^{-4}$ C at $x = 0$, $y = 0$; $q_3 = +6 \times 10^{-4}$ C at $x = 0.6$ m, $y = 0$. What is the force on q_2?

22. What forces act between the following pairs of particles: ν–p, ν–n, e–n, ν–ν?

23. A proton moves directly toward another proton, starting from a distance of 10^{-10} m. Describe the forces that the protons experience as the separation distance decreases to 10^{-16} m.

24. Use the information in Table 6-3 and in the example in Section 6-6 to estimate the magnitude of the weak force that exists between two protons separated by a distance of 10^{-15} m.

7
ENERGY

In Chapter 4 we found that several of our intuitive ideas regarding *force* correspond closely with the precise physical definition of this quantity. The same is true of some of our notions concerning the concepts of *work* and *energy*. We say that "a person who has a lot of energy can do a large amount of work." The statement made by the physicist is: "Energy is the capacity to do work." And we say that "a person who eats a good meal will have a lot of energy." The corresponding scientific statement is: "The stored chemical energy in foodstuffs can be utilized in biological systems to do work." In this chapter we will develop these and other ideas concerning work and energy.

The importance of the energy concept was not fully understood until the middle of the 19th century. By this time it was realized that energy takes many forms: motional energy, chemical energy, heat energy, electromagnetic energy, biological energy, and so forth. The great unifying principle that places all of the various forms of energy on an equal basis and makes the energy concept of truly universal significance was first stated in 1847 by the German physicist Hermann von Helmholtz (1821–1894). Helmholtz's important contribution was to realize that energy can be converted from one form to another and transferred from one object to another *without loss*. That is, the total energy of a system remains the same even though energy is being changed from one form to another within the system or is being exchanged between objects that are part of that system. Thus, Helmholtz discovered a principle—the principle of energy conservation—that, along with the principles of linear momentum conservation, angular momentum conservation, and charge conservation, represents one of the foundation stones of modern science.

The importance of the energy conservation principle can hardly be over-emphasized. *Every* process in Nature, physical as well as biological, takes place in accordance with this principle. The discovery of the principle of energy conservation has been one of the giant steps forward in our efforts to understand the way in which Nature behaves.

It must be emphasized that energy occurs in many forms. We see changes between various forms of energy taking place in all kinds of everyday situations. The chemical energy in gasoline is changed into motional energy of an automobile. Electrical energy is changed into light and heat by a light bulb. And the radiant energy in light is converted by plants into the chemical energy in foods. Only when we realize the many ways in which energy can manifest itself, can we understand fully the fundamental importance of the energy conservation principle.

In this chapter we will be concerned primarily with mechanical forms of energy. In later chapters we will extend the discussion of thermal energy, electrical energy, elastic energy, and mass-energy.

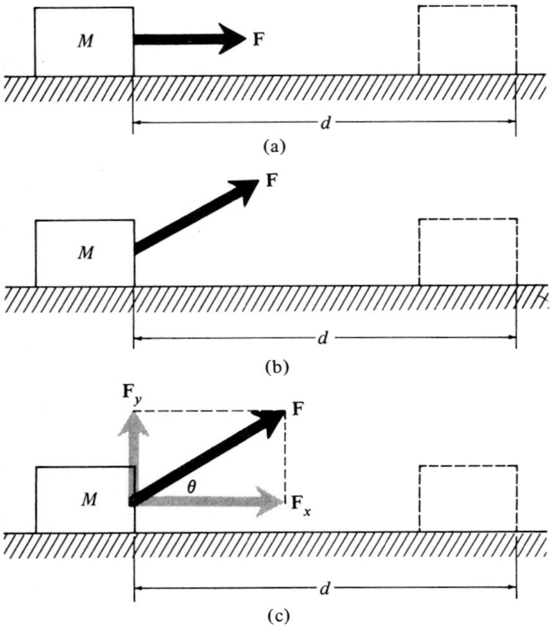

7-1 WORK

(Force) × (Distance)

When we push an object across a rough surface, continually exerting a force in order to overcome the effect of friction, we are conscious of the fact that we have exerted our muscles and we say that we have done *work*. The amount of work that is done depends on how much force was exerted and on how far we moved the object. Increasing either the applied force or the distance through which the object is moved increases the amount of work done. That is, the work done is proportional both to the applied force and to the distance through which the force acts (Fig. 7-1a). The equation which expresses this statement is

$$W = Fd \tag{7-1}$$

Figure 7-1 (a) The work done by the force **F** is $W = Fd$. In (b) and (c) the work done by the force **F** is $W = Fd \cos \theta$. In each case the force is assumed to be constant during the motion.

where

> F = force (in newtons)
> d = distance (in meters)
> W = work done (in newton-meters)

We give to the unit of work the special name *joule:*

$$1 \text{ joule (J)} = 1 \text{ newton-meter (N-m)} \qquad (7\text{-}2)$$

The unit of work is named in honor of the English physicist James Prescott Joule (1818–1889), whose experiments—particularly those concerning heat energy—greatly clarified the concepts of work and energy. It was largely because of Joule's careful experiments that Helmholtz was able to formulate the energy conservation principle.

It is important to realize that Eq. 7-1 for the work done is valid only for the case in which the direction of movement is the same as the direction of the applied force. (Equation 7-1 is also valid if the force is applied directly *opposite* to the direction of motion; this would be the case if a force were exerted to slow down a moving object.)

In Fig. 7-1b we see a case in which the direction of $\mathbf{F}$ is not the same as the direction of movement. We assume that $\mathbf{F}$ remains constant throughout the movement. We analyze this case by separating $\mathbf{F}$ into horizontal and vertical components, $\mathbf{F}_x$ and $\mathbf{F}_y$; this is done in Fig. 7-1c. Now, $\mathbf{F}_x$ *is* in the direction of motion and this component of $\mathbf{F}$ does an amount of work equal to $F_x d = F d \cdot \cos \theta$. The component $\mathbf{F}_y$ is perpendicular to the direction of motion; the block undergoes no displacement in the y-direction. That is, $\mathbf{F}_y$ does no work. Consequently, the total amount of work done by the force $\mathbf{F}$ is

$$\boxed{W = F_x d = F d \cos \theta} \qquad (7\text{-}3)$$

Notice that this equation is valid also for the case shown in Fig. 7-1a. If $\mathbf{F}$ is in the direction of movement, then $\theta = 0°$ and $\cos \theta = 1$; then, the general result given in Eq. 7-3 becomes the same as Eq. 7-1.

If all of the work done in moving the block is contributed by $\mathbf{F}_x$, what is the function of $\mathbf{F}_y$ in the process? Although $\mathbf{F}_y$ does no work, it does have an effect on the amount of work done by $\mathbf{F}_x$ in the event that friction exists between the block and the surface across which it moves. According to Eq. 4-8, the magnitude of the frictional force is proportional to the normal force on the block. If $\mathbf{F}_y$ were zero (this is the case of Fig. 7-1a), the

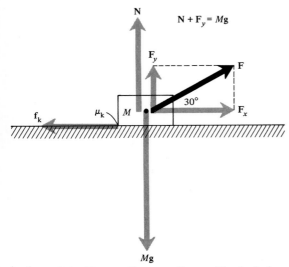

Figure 7-2 Force diagram for a block being pulled by a force with a vertical component.

normal force would be equal in magnitude to Mg, the weight of the block. But when $\mathbf{F}_y$ is present, the weight of the block is balanced by the sum, $\mathbf{N} + \mathbf{F}_y$. Then, the magnitude of the normal force is $Mg - F_y$. Because the normal force is smaller in this case, the frictional force is also smaller. That is, when the applied force has a vertical component, the friction is reduced and the block can be moved at a constant velocity with a smaller horizontal force.

In Fig. 7-1c, suppose that $\mu_k = 0.5$, $M = 5$ kg, $\theta = 30°$, and $d = 20$ m. What is the magnitude of $\mathbf{F}$ and how much work is done in moving the block at constant velocity? The force diagram is shown in Fig. 7-2. First, we calculate the frictional force,

$$f_k = \mu_k N = \mu_k (Mg - F_y)$$

where we have made use of the fact that $N = Mg - F_y$. If the block is moved with constant velocity, this frictional force is exactly balanced by $\mathbf{F}_x$, so we can write

$$f_k = F_x = \mu_k (Mg - F_y)$$

We also know that

$$F_x = F \cos 30°; \qquad F_y = F \sin 30°$$

Substituting these expressions into the equation for F_x, we have

$$F \cos 30° = \mu_k (Mg - F \sin 30°)$$

Solving for F,

$$
\begin{aligned}
F &= \frac{\mu_k Mg}{\cos 30° + \mu_k \sin 30°} \\
&= \frac{(0.5) \times (5 \text{ kg}) \times (9.8 \text{ m/s}^2)}{0.866 + (0.5) \times (0.500)} \\
&= 22.0 \text{ N}
\end{aligned}
$$

And finally for the work done, we find

$$
\begin{aligned}
W = F_x d &= (F \cos 30°) \times d \\
&= (22.0 \text{ N}) \times (0.866) \times (20 \text{ m}) \\
&= 381 \text{ J}
\end{aligned}
$$

Notice that if a horizontal force just sufficient to balance the frictional force (not reduced by any vertical pull) were applied, the work done would be

$$
\begin{aligned}
W = Fd &= \mu_k Mgd \\
&= (0.5) \times (5 \text{ kg}) \times (9.8 \text{ m/s}^2) \times (20 \text{ m}) \\
&= 490 \text{ J}
\end{aligned}
$$

and we see that the work done is much larger in this case. In both situations, the work is done against the frictional force.

In the preceding example we have seen how the application of a vertical force decreases the frictional force and thereby decreases the amount of work that must be done against sliding friction. We use this idea in many ways. If we wish to slide a heavy box across a rough floor, we have learned from experience that it is usually easier to exert some vertical force (perhaps sufficient to raise one end) instead of pushing or pulling the box in a strictly horizontal direction. Various "air cushion" devices have been designed to take advantage of this effect by reducing sliding or rolling friction. Figure 7-3 shows, in a schematic way, an air cushion system. The surface over which the block is to move is perforated with a number of small holes. An air compressor forces air upward through these holes. The air currents, pushing vertically on the block, provide a cushion of air on which the block moves. In this type of air cushion system, the block does not make contact with the surface. Therefore, the ordinary sliding friction between the block and the surface is completely eliminated. There remains, of course, the friction due to the air.

7-2 WORK DONE AGAINST VARIOUS FORCES

Work Done in Lifting

It is important to remember that work is always done when an object is moved against some resisting force. In the previous examples the resisting force was the force of friction. But now suppose that we raise a block into the air. In this case we encounter negligible friction due to the motion of the block through the air, but work is still being done. The resisting force against which work is done is the gravitational force.

Suppose that a block with a mass $M = 100$ kg is raised through a distance $h = 30$ m, as shown in Fig. 7-4. How much work is done in this case (neglecting friction in the pulley)? In order to raise the block, the downward force of gravity must be overcome; this force is just the *weight* of the block, $w = Mg$, and the distance moved is the height h. Therefore, the work done is

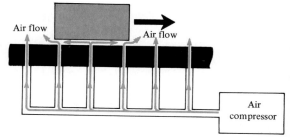

Figure 7-3 The upward flow of air through the perforated plate provides a cushion of air on which the block moves. Because the block is not in contact with the surface, sliding friction is eliminated; only air friction remains.

TABLE 7-1 RANGE OF ENERGIES IN PHYSICAL PROCESSES

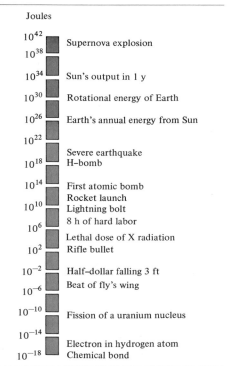

Joules	
10^{42}	Supernova explosion
10^{38}	
10^{34}	Sun's output in 1 y
10^{30}	Rotational energy of Earth
10^{26}	Earth's annual energy from Sun
10^{22}	
10^{18}	Severe earthquake / H–bomb
10^{14}	First atomic bomb
10^{10}	Rocket launch / Lightning bolt
10^{6}	8 h of hard labor
10^{2}	Lethal dose of X radiation / Rifle bullet
10^{-2}	Half–dollar falling 3 ft
10^{-6}	Beat of fly's wing
10^{-10}	Fission of a uranium nucleus
10^{-14}	
10^{-18}	Electron in hydrogen atom / Chemical bond

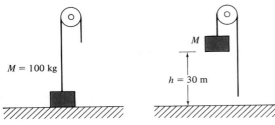

Figure 7-4 The work done in raising the block against the downward force of gravity is $W = Mgh$.

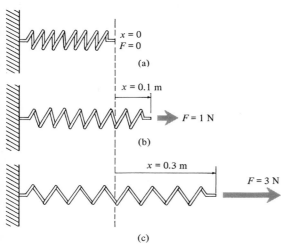

Figure 7-5 The force required to extend a spring by a certain distance is proportional to that distance: $F = kx$.

$$W = Fd = wh = Mgh$$
$$= (100 \text{ kg}) \times (9.8 \text{ m/s}^2) \times (30 \text{ m})$$
$$= 29\,400 \text{ J} = 2.94 \times 10^4 \text{ J}$$

Of course, in any real situation friction is always present. Therefore, to lift the 100-kg block to a height of 30 m would actually require an amount of work greater than 29 400 J. (The extra work would appear in the form of heat in the rope and pulley.)

There is a subtle point here about which a remark should be made. We have stated that the block is raised by the application of an upward force which has a magnitude Mg. But the block is initially at rest. If we apply an upward force exactly equal to the weight of the block, the block will be suspended, but it will not move because the net force acting is zero. Therefore, to raise the block we must apply a force in excess of Mg. If the applied force is constant, the block will have a constant upward acceleration, and when the height h is reached, the block will be in motion and the amount of work done will have exceeded Mgh. Actually, we wish to consider the case in which the block is again at rest at the height h. We can accomplish this by removing the applied force at a height below h such that the block will "coast" upward, coming to rest at h. Because the applied force does not act through the entire height h, the work done does not exceed Mgh. In fact, if we calculate the work done in this case (try it), we find that it does not depend on the extra force applied and is exactly equal to Mgh. Therefore, regardless of the way the upward force is applied, the amount of work required to lift a mass M, initially at rest, to a height h, where it is again at rest, is always exactly Mgh.

Work Done against a Spring

How much work is required to stretch a coiled spring? Most springs, if they are not stretched too far, have the property that the force required to extend a spring by a certain distance is proportional to that distance. This property is illustrated in Fig. 7-5. In Fig. 7-5a we have a spring in its normal condition; the applied force is zero and the extension is zero. In order to extend this spring by 0.1 m, a force of 1 N is required (Fig. 7-5b). For an extension of 0.3 m, a force of 3 N is required (Fig. 7-5c), and so forth. That is, the relationship between the applied force F and the extension x from the normal condition is

$$F = kx \qquad (7\text{-}4)$$

Springs, or any elastic materials, that follow this rule are said to obey *Hooke's law* (named for Robert Hooke, 1635–1703, a contemporary of Newton).

In Eq. 7-4, the quantity k is the *force constant* of the spring and has units of *newtons per meter* (N/m). Each spring will have its own particular value of k. For the spring in Fig. 7-5, the value of k is (1 N)/(0.1 m) or 10 N/m.

Figure 7-6 shows a force–distance graph for a spring that obeys Hooke's law. The force increases linearly with the distance of extension. We can use this graph to determine the amount of work required to extend the spring by an amount x. We follow the same procedure that we used in Section 3-4 to determine the distance traveled by an object, namely, by calculating the area beneath the velocity–time graph (Fig. 3-8). We know that the work done is given by the product of the force applied and the distance moved. For the spring, the force is not constant, but the work done can be found by calculating the area beneath the force–distance graph (Fig. 7-6). We see that the area is $\frac{1}{2}Fx$, and we know that $F = kx$. Therefore, the work done in stretching the spring to an extension x is

$$W = \tfrac{1}{2}kx^2 \qquad (7\text{-}5)$$

Suppose that a force of 0.8 N is required to extend a spring from $x = 0$ to $x = 4$ cm. How much work is done in extending this spring from $x = 0$ to $x = 12$ cm? First, we need to know the force constant:

$$k = \frac{F}{x} = \frac{0.8 \text{ N}}{0.04 \text{ m}} = 20 \text{ N/m}$$

Then, the work done is

$$W = \tfrac{1}{2}kx^2 = \tfrac{1}{2}(20 \text{ N/m}) \times (0.12 \text{ m})^2 = 0.144 \text{ J}$$

Muscular Work

If a man pushes a box across a floor (Fig. 7-7a), he does work—he has exerted a force on the box and the box has moved. However, if the box is against a wall (Fig. 7-7b), no motion can occur and we conclude that the man does no work. But if the man continues to push, exerting a muscular force, he will become tired from the effort—he *feels* as though he has been doing work. Actually, work *is* being done in this case. But the work does not appear as external motion; the work is done

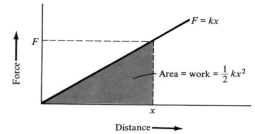

Figure 7-6   The work required to extend a spring by an amount x is equal to the area beneath the force–distance graph: $W = \frac{1}{2}kx^2$.

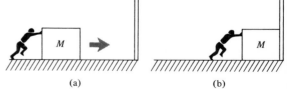

Figure 7-7   (a) The man does work in pushing the box across the floor. (b) The man's push does not result in any motion of the box, but work is still being done within the man's muscles.

internally, in the man's muscles. A force can be exerted when the muscle fibers extend or contract. This muscular action comes about when an electrical nerve signal triggers the flow of ions and electrons through the walls of the muscle cells and causes the fibers to extend or contract. This ionic motion, caused by electrical forces, constitutes *work*. As a result of this work, heat is produced, the temperature of the body rises, and the man eventually begins to perspire. One popular physical fitness procedure involves *isometric* exercises, in which various body muscles exert forces against one another or against fixed objects. No external motion results but the muscles do internal work and are thereby "kept in shape." We will discuss further the topics of heat and body energy in later sections.

7-3 POWER

The Rate of Doing Work

Two men do equal amounts of work by lifting identical boxes from floor level and placing them on a shelf. One of the men works rapidly and the other works slowly. Although the total amount of work performed by each man is the same, the two men have quite different bodily sensations when their tasks are completed. The reason is that the two men have been working at different *power* levels; the faster-working man was converting body chemical energy into work at a more rapid rate than was the slower-working man. It is difficult for the body to maintain a high rate of energy conversion and so the faster working man feels a greater "drain" on his internal energy supply.

Power is the *rate* at which work is done. That is,

$$\text{power} = \frac{\text{work done}}{\text{time}} \tag{7-6}$$

or,

$$P = \frac{W}{t} \tag{7-7}$$

Work is measured in *joules* and time is measured in *seconds,* and so the unit of power is the *joule/second* (J/s). To this unit we give the special name, *watt* (W):

$$1 \text{ J/s} = 1 \text{ W} \tag{7-8}$$

Also, 10^3 W = 1 kilowatt (kW) and 10^6 W = 1 megawatt (MW).

The unit of power is named in honor of the Scottish engineer, James Watt (1736–1819). Although Watt did not invent the steam engine, his improvements of existing designs resulted in the first commercially practical model.

Another widely used unit of power is the *horsepower* (h.p.) which is now defined in terms of the watt:

$$1 \text{ h.p.} = 746 \text{ W}$$
$$\cong \tfrac{3}{4} \text{ kW} \qquad (7\text{-}9)$$

A man lifts 45 boxes (each with $M = 25$ kg) from floor level to a height of 2 m in 5 min. At what average power was the man working? The work required to lift n boxes (each with mass M) through a height h is $W = nMgh$. If this work is done in a time t, the average power is

$$P = \frac{nMgh}{t} = \frac{45 \times (25 \text{ kg}) \times (9.8 \text{ m/s}^2) \times (2 \text{ m})}{(5 \text{ min}) \times (60 \text{ s/min})}$$

$$= 73.5 \text{ W} \cong \tfrac{1}{10} \text{ h.p.}$$

The Kilowatt-Hour

One of the terms we often hear about (or see on electric bills) is the *kilowatt-hour*. Let us see what this term actually means. Because 1 W is equal to 1 J/s, if a 1-W device operates for 1 s, the amount of work performed is 1 J:

$$1 \text{ J} = 1 \text{ W-s}$$

Similarly, if a 1000-W (1-kW) device operates for 1 h, the amount of work performed is

$$W = Pt = (1000 \text{ W}) \times (3600 \text{ s})$$
$$= 3.6 \times 10^6 \text{ J}$$

This amount of work is called 1 *kilowatt-hour* (kWh):

$$1 \text{ kWh} = 3.6 \times 10^6 \text{ J} \qquad (7\text{-}10)$$

A certain ventilating fan is operated by a $\tfrac{1}{2}$-h.p. electric motor. How much does it cost to run the fan continuously for a month if the cost of electricity is 2.5 cents per kWh?

$$\text{cost} = (\tfrac{1}{2} \text{ h.p.}) \times \left(\tfrac{3}{4} \frac{\text{kW}}{\text{h.p.}} \right)$$

$$\times \left(24 \frac{\text{h}}{\text{day}} \right) \times \left(30 \frac{\text{days}}{\text{mo.}} \right)$$

$$\times (\$0.025/\text{kWh})$$

$$= \$6.75 \text{ per month}$$

It is important to realize that the *kilowatt-hour* is a unit of *work* or *energy;* the *kilowatt* is a unit of *power*. Be certain that you understand the distinction between *power* and *energy*. Power is the *rate* at which work is done or energy is used.

7-4 KINETIC AND POTENTIAL ENERGY

Motional Energy

Suppose that we have a block with mass M which rests on a frictionless surface. If a constant force **F** is applied to the block, it will accelerate. Let the force be applied at time zero when the block is at rest. At the later time t, the block has moved a distance d and has a velocity v (Fig. 7-8). From our previous discussions, we know the following:

distance traveled:
$$d = \tfrac{1}{2}at^2 \quad \text{(Eq. 3-12)}$$

final velocity:
$$v = at \quad \text{(Eq. 3-9)}$$

force applied:
$$F = Ma \quad \text{(Eq. 4-2)}$$

work done:
$$W = Fd \quad \text{(Eq. 7-1)}$$

We can express the work done in terms of the final velocity by substituting for F and d:

$$W = Fd = (Ma) \times (\tfrac{1}{2}at^2)$$
$$= \tfrac{1}{2}M(at)^2$$

and since $v = at$, we have, finally,

$$W = \tfrac{1}{2}Mv^2 \qquad (7\text{-}11)$$

That is, an amount of work W has been done on the block and we say that the block has thereby acquired an amount of *energy* equal to $\tfrac{1}{2}Mv^2$. This amount of energy which the block possesses by *virtue of its motion,* is called the *kinetic energy:*

$$\boxed{\text{kinetic energy} = \text{K.E.} = \tfrac{1}{2}Mv^2} \qquad (7\text{-}12)$$

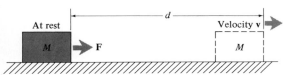

Figure 7-8 A force **F** uniformly accelerates a block from rest to a velocity v in a distance d. The energy imparted to the block is $\tfrac{1}{2}Mv^2$.

Work and energy have the *same* units: *newton-meters* or *joules*.

Notice that the kinetic energy is proportional to the mass and to the *square* of the velocity. Thus, a 3600-lb sedan moving at 40 mi/h has twice the kinetic energy of an 1800-lb sports car moving at the same speed. But if we double the speed of the sports car to 80 mi/h, it now has *four* times its previous kinetic energy and *twice* the kinetic energy of the 40-mi/h sedan.

Potential Energy

In Fig. 7-4 we considered lifting a mass M to a height h. We found that the work done in such a case is $W = Mgh$. The object was originally at rest, and in its final position the velocity is again zero. Thus, no kinetic energy was imparted to the object. But the object has a capability to do work that it did not have in its original position. For example, if we drop the object and allow it to fall through the height h work can be done in driving a stake into the ground (Fig. 7-9). That is, the raised block has the *potential* to do work and we call this capability the *potential energy* of the object:

$$\boxed{\text{potential energy} = \text{P.E.} = Mgh} \qquad (7\text{-}13)$$

An object does not do work as the *direct* result of its potential energy. After all, work involves *movement*. To make use of the potential energy possessed by an object, we must first convert the potential energy into motional energy. If we drop an object from a certain height, we know that it will accelerate due to gravity and will strike the floor with a certain velocity. The greater the height from which the object is dropped, the greater will be its impact velocity and the greater will be the kinetic energy and the amount of work that can be done.

We see in this example an illustration of the definition of the energy concept:

$$\boxed{\text{Energy is the capacity to do work.}}$$

The potential energy of the raised block, Mgh, can be converted into an equal amount of work. The potential energy is first converted into kinetic energy, $\frac{1}{2}Mv^2 = Mgh$, and the kinetic energy is then converted into work by driving the stake into the ground. The net result of the operation is to use the original amount of work to drive the stake.

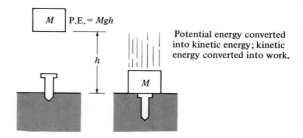

Figure 7-9   The potential energy of a raised block can be converted into kinetic energy which, in turn, is converted into work. The work done is equal to the force that the ground exerts on the stake multiplied by the distance that the stake moves, $Fd = Mgh$.

Conservation of Energy—
Forms of Energy

In the example above, no energy is lost in any phase of the process. Work is done on the block and the block possesses this amount of energy—in the form of potential and/or kinetic energy—until the energy is given up when the block does an equivalent amount of work on the stake. The energy delivered to the stake is finally converted into heat and sound energy. Although energy is converted from one form to another, no energy is lost—*energy is conserved.*

Every process known in Nature takes place in accordance with the principle of energy conservation. But the true importance of this principle cannot be fully understood or appreciated unless it is realized that energy appears in many forms. If we add up all of the energy in its various forms that an isolated system possesses before an event or process takes place and do the same afterward, we always find an exact balance. We can make this calculation only if we know all of the ways in which energy can appear. If we did not realize the existence of potential energy, for example, we would discover many situations in which there is an apparent increase or decrease in energy.

We have discussed only two forms of energy thus far: the energy due to motion—*kinetic energy*—and the energy due to the gravitational attraction of the Earth for an object—*gravitational potential energy.* Both kinetic and potential energy can manifest themselves in other ways. The molecules in every piece of matter—solid, liquid, or gas—are in a continual state of motion. This random, agitated motion constitutes an *internal* kinetic energy or *thermal energy* that an object possesses even though the object as a whole may not be in motion. A change in the internal energy of an object can be brought about by supplying *heat* to the object or by doing *work* on the object. If we do work on an object (for example, by repeatedly hitting a block of metal with a hammer), the molecules are caused to move more rapidly; the internal energy is thereby raised and there is an accompanying increase in *temperature.* Heat considerations are particularly important in processes that involve friction because the energy that is expended in working against frictional forces always appears in the form of heat. Thus, in any real physical process some energy will be "lost" in the heating of the objects involved and their surroundings. We will discuss internal energy and heat more thoroughly in Chapter 9.

The transmission of *sound* from one point to another takes place when the sound source (for example, a vibrating speaker diaphragm) sets into motion the air molecules in its immediate vicinity. These molecules collide with other nearby molecules and further molecular collisions cause the propagation of the sound to other points. Thus, sound is due to molecular motions and constitutes another form of kinetic energy.

If we raise an object near the surface of the Earth to a higher position, the work done in accomplishing this relocation appears as the *gravitational potential energy* of the object. In this case work is done against the attractive gravitational force. Work can also be done against electrical forces and the potential energy that results is called the *electrical potential energy*.

Elastic energy is another form of electrical energy. When a spring is compressed or extended, work is done against the intermolecular electrical forces and electrical potential energy is stored (see Fig. 7-10). All forms of *elastic energy* are basically *electrical* in character.

When gasoline burns or when dynamite explodes, the potential energy stored in the substance is converted into heat or motional energy. When the fuel *methane*, CH_4 (a gas), burns to completion, the chemical reaction is

$$CH_4 + 2\ O_2 \longrightarrow CO_2 + 2\ H_2O \qquad (7\text{-}14)$$

The burning of 1 gram of methane releases approximately 55 000 joules of energy (which can be used to heat some other material). Where does this energy come from? We can represent the oxidation reaction in the following schematic way:

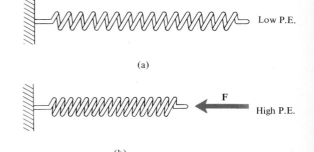

(a)

(b)

Figure 7-10 Work against the intermolecular electrical forces is required to compress (or extend) a spring. A compressed spring therefore possesses more potential energy than a relaxed spring.

$$CH_4 \quad + \quad 2\ O_2 \longrightarrow CO_2 + 2\ H_2O$$

where each short line connecting two element symbols represents a pair of electrons that bind the two atoms together. In order for the reaction to proceed, several atomic bonds must be broken and new ones formed:

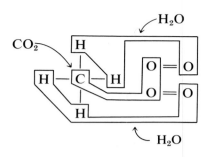

There is a certain amount of electrical potential energy in every molecule; this is due to the arrangement of electrons around the positively charged nuclei. Some arrangements of electrons and nuclei have more potential energy than others. Any system will always tend to move or react in such a way that the potential energy is *decreased*. There is *more* electrical potential energy in the combination $CH_4 + 2\ O_2$ than when the same atoms are in the arrangement $CO_2 + 2\ H_2O$. Thus, when methane is burned to produce carbon dioxide and water, energy is released. All forms of *chemical energy* are basically *electrical* in character.

Calculations Using Energy Conservation

By using the idea of energy conservation, many types of problems can be easily solved. As a simple example, suppose that we wish to find the velocity of a ball as it strikes the ground, the ball having been dropped from a height h. Energy conservation tells us that the kinetic energy at the moment of impact is equal to the potential energy at the instant of release:

$$\tfrac{1}{2}Mv^2 = Mgh$$

The mass M of the ball occurs on both sides of the equation and therefore cancels. We then have

$$v^2 = 2gh$$

and taking the square root,

$$v = \sqrt{2gh} \tag{7-15}$$

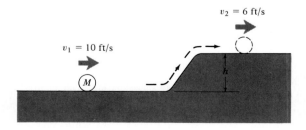

Figure 7-11   Calculate h by using energy conservation.

This result shows that the impact velocity does not depend on the mass of the object; this is the same conclusion that we reached in Section 3-5.

If the ball is dropped from a height of 256 ft, the final velocity will be

$$v = \sqrt{2 \times (32 \text{ ft/s}^2) \times (256 \text{ ft})}$$
$$= \sqrt{16\ 384 \text{ ft}^2/\text{s}^2}$$
$$= 128 \text{ ft/s}$$

which is the same result we obtained in Section 3-5 by directly applying the kinematic formulas.

Next, consider a slightly more complicated situation. A ball rolls (without friction) across a horizontal surface with a velocity of $v_1 = 10$ ft/s (Fig. 7-11). The ball encounters an incline and after rolling uphill to a new horizontal surface at a height h above the original surface, the velocity is $v_2 = 6$ ft/s. What is the height h? Again we use energy conservation and equate the total energy (kinetic plus potential) on the lower surface to the total energy on the upper surface. But the potential energy on the lower surface is zero; therefore,

$$\tfrac{1}{2}Mv_1{}^2 = \tfrac{1}{2}Mv_2{}^2 + Mgh$$

The mass of the ball cancels, and solving for h, we find

$$h = \frac{\tfrac{1}{2}v_1{}^2 - \tfrac{1}{2}v_2{}^2}{g} = \frac{1}{2g}\,(v_1{}^2 - v_2{}^2)$$

$$= \frac{1}{2 \times (32 \text{ ft/s}^2)} \times [(10 \text{ ft/s})^2 - (6 \text{ ft/s})^2]$$

$$= \frac{100 - 36}{64} \text{ ft} = 1 \text{ ft}$$

Energy Differences

How much potential energy can be converted into kinetic energy (or into work) if a 1-kg object falls from a height of 6 m to a height of 5 m? The original potential energy is

$$(\text{P.E.})_1 = Mgh_1$$
$$= (1 \text{ kg}) \times (9.8 \text{ m/s}^2) \times (6 \text{ m})$$
$$= 58.8 \text{ J}$$

And the final potential energy is

$$(\text{P.E.})_2 = Mgh_2$$
$$= (1 \text{ kg}) \times (9.8 \text{ m/s}^2) \times (5 \text{ m})$$
$$= 49.0 \text{ J}$$

Therefore, the potential energy that can be converted into work is

$$(\text{P.E.})_1 - (\text{P.E.})_2 = 58.8 \text{ J} - 49.0 \text{ J}$$
$$= 9.8 \text{ J}$$

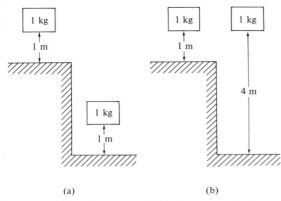

Figure 7-12 (a) Each of the blocks will release the same amount of gravitational potential energy by falling even though they start from different heights. (b) Each of the blocks has a different amount of gravitational potential energy that can be released by falling even though they are at the same height.

We can obtain the same result by making the following observation. The original height is 1 m above the final height. Therefore, if we measure height above the final height instead of above the ground, we can state that the original potential energy is

$$P.E. = Mgh$$
$$= (1 \text{ kg}) \times (9.8 \text{ m/s}^2) \times (1 \text{ m})$$
$$= 9.8 \text{ J}$$

and *all* of this potential energy can be converted into work because the potential energy at the final height is zero ($h = 0$).

Thus, the gravitational potential energy that can be converted into work depends only on the *change* in height that the object undergoes. A 1-kg object that falls through a distance of 1 m can convert into work an amount of potential energy equal to $Mgh = 9.8$ J. It does not matter whether this object starts falling from a height of 1 m, 100 m, or 10 000 m — the only important fact is that the total distance of fall is 1 m. Thus, each of the blocks in Fig. 7-12a will release the *same* amount of potential energy in falling 1 m, even though each block starts from a different height. In the same way, each of the blocks in Fig. 7-12b possesses a *different* amount of potential energy that can be released as work, even though both of the blocks are at the same height.

It does not matter what level we elect to use as the zero level (or *reference* level) for the purpose of measuring heights and computing potential energies. Only the *difference* in potential energy between the initial and final positions is important because only this *difference* can be converted into other forms of energy.

Efficiency in Energy Conversions

A *machine* is any device that can extract energy from some source and convert this energy into useful work. The energy source might be the potential energy in water stored behind a dam, the chemical (electrical) energy in gasoline or coal, or the radiant energy in sunlight. Various machines have been constructed to utilize the energy from these and other sources. No machine, however, can completely convert available energy into useful work. By one means or another, energy always manages to escape to the surroundings in any conversion process. Friction exists in every moving system and the effect of friction is to convert energy from the source into thermal energy, thereby raising the temperature of the surroundings. An operating automobile engine

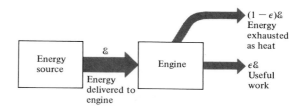

Figure 7-13 The flow of energy through an engine whose operating efficiency is ϵ.

becomes hot and the thermal energy in the engine block cannot be recovered and used to assist in propelling the vehicle. In other situations, such as the explosion of a stick of dynamite, some of the energy is released in the form of light and sound.

Every machine can be characterized by an *efficiency*, which is the ratio of the useful work performed to the amount of energy used in the process:

$$\text{efficiency} = \epsilon = \frac{\text{work done}}{\text{energy used}} \qquad (7\text{-}16)$$

Schematically, the situation is that pictured in Fig. 7-13. An energy source (for example, a tank of gasoline) delivers an amount of energy $\mathscr{E}$ to an engine. A fraction $\epsilon\mathscr{E}$ appears as useful work (for example, in the motional energy of the automobile), and the remainder, $(1 - \epsilon)\mathscr{E}$, appears in the form of heat.

Almost all machines that are used on a wide and regular basis have efficiencies less than 0.5 (50 percent). An automobile engine, typically, has an efficiency of about 25 percent (the figure varies depending on how well the engine is tuned and on the operating speed). The best steam engine has an efficiency of only about 10 percent. In a coal-fired electricity-generating plant, about 40 percent of the chemical energy in the fuel can be converted into electrical energy. Therefore, a power plant that produces 1000 MW (10^9 W) of electrical power, consumes energy at a rate of 2500 MW and exhausts energy in the form of heat to the air or to a river at a rate of 1500 MW. No wonder that thermal pollution is an increasing problem! (See also Section 8-6.)

Under optimal conditions, the human body is actually a rather efficient machine. The maximum efficiency of a muscle for converting chemical energy into mechanical work is about 40 percent. Thus, 60 percent (or more) of the available body chemical energy is expended as heat. This heat which is produced by a working muscle is not wasted, but is utilized in maintaining the body temperature. There are no special body units that exclusively produce heat, but when the body loses more heat than its muscles are producing, the muscles are stimulated into a special action, namely shivering, in order to generate additional heat. Muscle-produced heat plays a dominant role in maintaining proper body temperature.

It is important to realize that energy is not "lost" in any of these processes, no matter how inefficient. The energy lost by one part of a system always appears in some other part of the system or in some other form. Energy is always conserved.

7-5 MASS-ENERGY

Energy in Relativity Theory

In this chapter we have emphasized the importance of the principle of *energy* conservation. Even before the energy principle was appreciated, there had been established, about 200 years ago, another conservation law of equal significance — the conservation of *mass*. This law depends ultimately on the fact that atoms can neither be created nor destroyed in any kind of ordinary chemical or physical process (with the exception of nuclear processes which were unknown until the 20th century). In all everyday processes, conservation of energy and conservation of mass are separately valid. However, the theory of relativity (Chapter 16) shows that the concepts of mass and energy are intimately connected. Interchanges between mass and energy become important in physical processes that take place in nuclei or at extremely high speeds. Einstein demonstrated that the mass m of a system and its energy content $\mathscr{E}$ are related in a simple way; this is the famous Einstein mass-energy equation,

$$\boxed{\mathscr{E} = mc^2} \qquad (7\text{-}17)$$

where c is the velocity of light, 3×10^8 m/s.

Because mass and energy are related and one can be transformed into the other, the proper statement of the conservation principle is that the *mass-energy* of a system remains constant.

The value of c is extremely large and, therefore, the energy content of even a small amount of matter is enormous. If 1000 kg of matter were converted completely into energy, the amount would be

$$\mathscr{E} = mc^2 = (10^3 \text{ kg}) \times (3 \times 10^8 \text{ m/s})^2 = 9 \times 10^{19} \text{ J}$$

or about 2.5×10^{13} kWh, which is approximately the total amount of energy used in the United States in a year!

If matter could be completely converted into energy, we would have no need to be concerned about sources of energy for the world's use — only a few tons of matter per year would be required to meet the worldwide need for energy. Unfortunately, the basic constituents of matter — protons, neutrons, and electrons — cannot be converted completely into energy. Nature allows us only to *rearrange* these particles into forms that have different mass and to extract energy equal to the mass difference. For example, when a nucleus of ^{235}U (92 protons and 143 neutrons) undergoes fission, two less massive nuclei are formed, some neutrons are emitted, and energy is released. A typical fission reaction is

$$^{235}_{92}U \xrightarrow{\text{fission}} {}^{139}_{56}Ba + {}^{94}_{36}Kr + 2n + \mathscr{E}$$

The total number of protons and neutrons is the same after fission as before (check this), but the mass of ^{235}U is slightly greater than the sum of the masses of ^{139}Ba, ^{94}Kr, and two neutrons. This mass difference, when multiplied by c^2, is equivalent to an energy release of approximately 3×10^{-11} J per fission event. The *total* mass–energy of a nucleus of ^{235}U is about 3.5×10^{-8} J. Therefore, only about 10^{-3} or 0.1 percent of the total mass–energy is actually released in the fission process. On the other hand, when a chemical fuel is burned, only about 10^{-10} of the total mass–energy is released. Consequently, the fission process is about 10^7 times more efficient in energy generation than is the burning of chemical fuels.

SUGGESTED READINGS

P. Brandwein, *Energy: Its Forms and Changes* (Harcourt, New York, 1968).

M. Wilson, *Energy* (Time, Inc., New York, 1966).

Scientific American articles:

S. W. Angrist, "Perpetual Motion Machines," January 1968.

S. H. Schurr, "Energy," September 1963.

QUESTIONS AND EXERCISES

1. A 100-kg man climbs a ladder to a height of 5 m. How much work has he done?

2. A pile driver drives a stake into the ground. The mass of the pile driver is 2500 kg and it is dropped through a height of 10 m on each stroke. The resisting force of the ground for this stake is 4×10^6 N. How far is the stake driven on each stroke?

3. A force of 60 N is required to push a 4-kg object up an inclined plane to a height of 3 m. The total distance moved along the plane is 10 m. How much work was done by the force?

4. In the previous exercise, how much work was done against gravity? How much work was done against friction?

5. The coefficient of kinetic friction between a certain block and a surface is 0.6. The mass of the block is 6 kg. What constant force must be applied at an angle of 40° above the horizontal in order to move the block with an acceleration of 2 m/s²?

6. In the preceding exercise, how much work has been done against friction after the block has been moved a distance of 12 m? What is the kinetic energy of the block? What is the total amount of work done?

7. Refer to Fig. 7-1c. Suppose that **F** is directed at an angle of 30° but *downward* from the horizontal. Choose the same values for the various quantities as in the example in the text ($\mu_k = 0.5$, $M = 5$ kg, and $d = 20$ m), and calculate the magnitude of **F** and the amount of work done. Why is the work done greater in this case?

8. A 40-kg boy sits on a 10-kg sled and is pulled across an icy surface ($\mu_k = 0.25$) at constant velocity. The rope used for pulling makes an angle of 35° with the horizontal. What is the magnitude of the pulling force and how much work is done if the sled is pulled for 200 m?

9. A 3-kg block rests on a plank. The coefficients of friction for the block-plank combination are $\mu_s = 0.6$ and $\mu_k = 0.4$. One end of the plank is raised slowly, as in Fig. 4-20. At a certain angle, the block slides down the plane. After the block has moved 2.5 m, how much work has been done against friction?

10. An amount of work equal to 2 J is required to compress the spring in a spring-gun. To what height can the spring-gun fire a 10-g projectile? What is the velocity of the projectile as it leaves the spring-gun?

11. The pumping action of the heart gives to the blood some kinetic energy. Where does this energy originate and what happens to the blood's kinetic energy?

12. What is the kinetic energy of a 1500-kg automobile moving with a speed of 40 km/h?

13. A constant force of 20 N is applied to a 5-kg block and the block is moved a distance of 30 m. The frictional force between the block and the surface is 6 N. What is the final velocity of the block?

14. Classify the energy in the following systems according to *basic* energy forms: (a) water in a storage tower, (b) sonic boom, (c) food, (d) boiling water, and (e) moving automobile.

15. An amount of work equal to 0.8 J is required to stretch a certain spring from $x = 0$ to $x = 0.2$ m. How much additional work is necessary to extend the spring to $x = 0.5$ m?

16. A workman at a construction site wishes to raise a 100-kg load to the top of a 10-story building ($h = 33$ m) within 1 min. He has at his disposal several motors, the output ratings of which are $\frac{1}{2}$ h.p., $\frac{3}{4}$ h.p., 1 h.p., 1.5 h.p., and 2 h.p. Which is the smallest motor that can do the job? Assume that half of the power is expended in overcoming friction.

17. An automobile ($m = 1000$ kg) is moving with a speed of 10 m/s. The driver "steps on the gas" and accelerates to a speed of 30 m/s in 8 s. What power (in h.p.) is required of the engine? What factors have been neglected in this calculation? Do you expect that a real automobile would require a more powerful engine to accelerate at this rate?

18. What is the power of an engine that lifts a 1000-kg mass to a height of 100 m in 3 min?

19. If a force F moves an object at constant velocity v, show that the power expended is $P = Fv$.

20. A man moves a loaded sled ($M = 40$ kg) at a constant velocity of 2 m/s by pulling horizontally on a rope attached to the sled. The coefficient of kinetic friction is $\mu_k = 0.4$. At what power does the man work?

8

ENERGY IN TODAY'S WORLD

One of the vital problems that this country and the world as a whole faces today is concerned with the production of adequate amounts of energy to meet the increasing needs of a technological society. Not only must we seek efficient methods to provide energy to the world population, but we must ensure that these developments do not have undesirable or even disastrous effects on our environment.

In this chapter we will examine our energy requirements and the methods we are now using to meet these needs. We will look at all of the major sources of energy and at the undesirable side effects peculiar to each type of source. No energy source is perfect — the exploitation of each is accompanied by unpleasant environmental consequences. Although we will experience some near-term shortages, the long-term prospect for energy production appears satisfactory. Our primary problem is to find ways of using the available energy sources that are economically acceptable and at the same time do not despoil the Earth on which we live and the atmosphere which we breathe.

8-1 HOW MUCH ENERGY DO WE USE?

An Accelerating Rate of Consumption

Present-day society consumes energy at a fantastic pace. Almost every aspect of modern civilization is

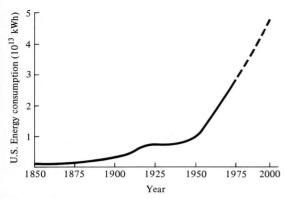

Figure 8-1 Consumption of energy in all forms in the United States since 1850.

geared to the use of energy, and the rate of energy consumption is continually increasing. The magnitude of the industrialized world's appetite for energy can be appreciated by noting that *half* of the energy that has ever been used by Man has been used during the last 100 years. Consequently, the business of providing energy has become one of our primary occupations. In the last 100 years the rate of consumption of energy in the United States has increased by a factor of about 20. (During this same period the population has increased by a factor of 5.) That is, we now use 4 times as much energy per person as we did in the 1870's. Because the rate of energy usage is accelerating, we can expect another doubling in energy consumption before the year 2000 (see Fig. 8-1).

A projection for the future, such as that shown in Fig. 8-1, is based primarily on the pattern established in the recent past and on the expectation that the current trend will extend into future years. We are now beginning to make a serious effort to alter the pattern of expansive and often wasteful usage of energy by instituting conservation programs, by voluntary action and by higher taxation of energy consumption. These measures cannot halt completely the need for expanded energy production facilities in the future, but they may serve to decrease the rate of growth below the steeply rising curve shown in Fig. 8-1. By the year 1990, the energy consumption in the United States may be 15 percent below the projected value in Fig. 8-1.

Reliable figures for energy usage are difficult to obtain, but it has been estimated that the present worldwide consumption of energy amounts to about 6×10^{13} kilowatt-hours (kWh) annually. Of this total, the United States uses approximately one-third, or 2×10^{13} kWh per year. Because the numbers involved in energy consumption are so enormous, let us express all energy quantities in terms of the current U.S. energy consumption per year. We will call this amount of energy 1 *energy unit* (E.U.):

$$\text{Current U.S. energy consumption}$$
$$= 1 \text{ E.U. per year}$$
$$= 2 \times 10^{13} \text{ kWh/y} \qquad (8\text{-}1)$$

$$\text{Current worldwide energy consumption}$$
$$= 3 \text{ E.U. per year}$$
$$= 6 \times 10^{13} \text{ kWh/y}$$

For comparison, 1 E.U. is approximately equal to the

TABLE 8-1   ENERGY CONSUMPTION IN THE UNITED STATES FOR 1974

SOURCE	AMOUNT	10^{12} kWh	PERCENTAGE
Coal	540×10^6 metric tons	4.6	21.1
Oil	6.0×10^9 barrels[a]	10.2	46.8
Natural gas	22×10^{12} cu ft[a]	6.4	29.3
Hydropower		0.3	1.4
Nuclear		0.3	1.4
		21.8	100

[a] These are not metric units, but they are the units in which these fuels are usually measured.

amount of energy obtained from burning 3.5 billion (3.5×10^9) tons of coal.

About one-tenth of the energy now used in the United States is *electrical* energy. That is, the annual consumption of electrical energy is approximately 0.1 E.U., or 2×10^{12} kWh.

Table 8-1 shows the amounts of energy from various sources used in the U.S. in 1974.

We do not use energy very efficiently. The consumption of 2×10^{13} kWh of energy each year in the United States means that the rate at which every person uses energy is approximately 10 kW, on the average. (By way of contrast, the rate at which a person uses food energy amounts to approximately 0.15 kW, averaged over a day.) But only about half of the 10 kW of power is used in performing useful work (see Fig. 8-2). The other half is consumed in converting the energy in the original fuel supply into a useful form of energy and in transporting the energy to the user. We produce about as much waste heat as useful energy in this process.

The Changing Sources of Energy

The ultimate source of almost all the energy that we use today is in the radiant energy that comes from the Sun. All of our primary fuels—wood, coal, oil, and natural gas—are derived from plant life and animal life that grew because of the action of sunlight. The water that drives hydroelectric generating plants is lifted to high land through evaporation and precipitation processes that result from solar heating. The most important source of energy that is not derived directly from the Sun is stored in nuclei and can be released through fission and fusion. (There are other sources of nonsolar energy but these are of little importance at present. The source of *geothermal* energy is the heat produced in the

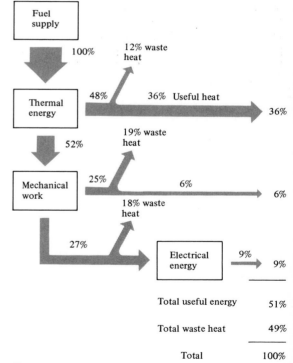

Total useful energy	51%
Total waste heat	49%
Total	100%

Figure 8-2   Flow of energy from chemical fuels to user. The diagram represents an average for all chemical fuels. We lose approximately half of the original energy as waste heat in the process of delivering energy to the user.

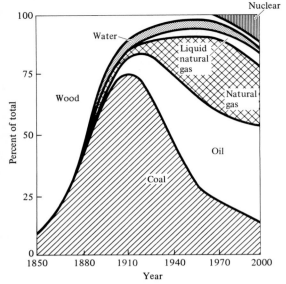

Figure 8-3 Sources of energy in the United States during the period 1850–2000. Notice that the fraction of the total worldwide production of energy by burning coal is now about the same as it was 100 years ago. Even by the year 2000, the impact of the generation of energy by nuclear reactors will be only beginning. (Adapted from Hans H. Landsberg.)

Earth's interior by radioactive decay, and *tidal* energy is due to the rotation of the Earth relative to the Moon.)

Until about 150 years ago, the primary sources of energy were wood, water, and wind (see Fig. 8-3) plus, of course, the heating effect of the Sun's direct rays. We still make use of these sources, but only water power in the form of electricity generated by huge hydroelectric plants is a significant factor in the world energy supply. Most of the energy used today is the result of the burning of the various chemical fuels. In the future, an increasing fraction of the world's energy will be obtained from nuclear fission reactors and, it is hoped, in the not too distant future, fusion reactors will be available to take over the major burden of energy production. Although the use of solar energy (in the form of solar heating and in the generation of electricity from solar cells) may never be our most important source of energy, improved techniques may result in some practical systems for the generation of energy for specialized purposes directly from the Sun's rays.

Figure 8-4 shows a breakdown of the energy sources currently used and proposed. We will discuss many of these sources in turn.

8-2 WATER POWER

A Limited Supply

Historically, the utilization of the energy in the flowing water of rivers and streams by means of water wheels provided the first plentiful and continuous source of energy. Today, we no longer use water power directly but, instead, use the potential energy in water stored behind dams to generate electricity. In a modern hydroelectric plant, water is allowed to pass through conduits and drive huge turbines whose rotating shafts are connected to electrical generators that produce power for the users.

The largest dams produce electrical power in the range of thousands of megawatts. Grand Coulee dam on the Columbia River in the state of Washington produces about 2000 MW of electricity for use in the Northwest. Most hydroelectric plants have considerably smaller outputs. The largest plant in the Tennessee Valley Authority (TVA) system, Wilson dam in Alabama, produces only about one-fourth as much electrical power as Grand Coulee dam. And the largest dam in the Missouri

River system produces only about 130 MW. Altogether, hydroelectric plants account for about 16 percent of the electrical energy used in the United States.

At present the total installed hydroelectric generating capacity in the U.S. amounts to about 60 000 MW. The maximum possible hydropower capacity has been estimated to be 300 000 MW. But it is unrealistic to suppose that this figure will ever be reached. There are too many objections to the huge number of dams that would be necessary to approach the ultimate power figure. More reasonably, we might look forward to a doubling of the present capacity. This situation might be achieved around the year 2000. Other countries, particularly Canada and the U.S.S.R., as well as countries in South America and Africa, will probably continue to develop water power sources well past the time that the United States has found it impractical to do so.

8-3 FOSSIL FUELS

A Transient Energy Supply

Since the beginning of the 20th century, most of the world's energy has been derived from the burning of fossil fuels. At the present time less than 10 percent of the energy used in the United States is obtained from nonfossil sources (see Fig. 8-3). Even though nuclear reactors will supply an increasing fraction of our energy in the future, fossil fuels will continue to be our main source of energy well into the 21st century. (The development of a practical fusion reactor or solar power plant might alter this outlook.)

Approximately 80 percent of our fossil fuels are used directly, in space heating, in transportation, and in industry; only about 20 percent are used in the generation of electricity. By the year 2000 we will be converting a substantially larger fraction of our fossil fuels (primarily coal) into electrical energy as we shift toward a more electrically-oriented economy.

Fossil fuels are produced over long periods of time; but we are using these fuels at a rapid rate. How long can we continue to do this? We have already used approximately 16 percent of the estimated total supplies of oil and natural gas. Fortunately, our supplies of coal are much more extensive; there probably remains 50 times as much coal as has already been mined. Even so, the supply is limited, and at our present rate of con-

TABLE 8-2 KEY EPISODES IN THE DEVELOPMENT OF ENERGY SOURCES

c 40,000 B.C.	Fire used by Paleolithic Man
c 3000 B.C.	Use of draft animals
1st Century B.C.	Water wheel
12th Century	Vertical windmill
16th Century	Large-scale mining, metallurgical techniques developed
18th Century	Steam engines of Savery (1698), Newcomen (1712), Watt (1765)
18th–19th Centuries	Understanding of the energy concept
19th Century	Formulation of the laws of thermodynamics and electromagnetism
1859	First producing oil well, Titusville, Pennsylvania (Drake)
1876	Internal combustion engine (Otto, Langen)
1882	First steam-generated electric plant, New York City (Edison)
1884	Steam turbine (Parsons)
1892	Diesel engine (Diesel)
1896	First alternating-current hydroelectric plant, Niagara Falls, New York (Westinghouse)
1905	Discovery of relationship between mass and energy (Einstein)
1933	Tennessee Valley Authority (TVA) Act
1942	First self-sustaining nuclear fission chain reaction
1945	First nuclear weapons used, Hiroshima, Nagasaki
1946	Atomic Energy Commission established
1952	First nuclear fusion device (H-bomb), Eniwetok Atoll
1957	First U.S. nuclear power plant devoted exclusively to generating electricity, Shippingport, Pennsylvania
1975	Energy Research and Development Administration (ERDA) established
?	First nuclear fusion reactor

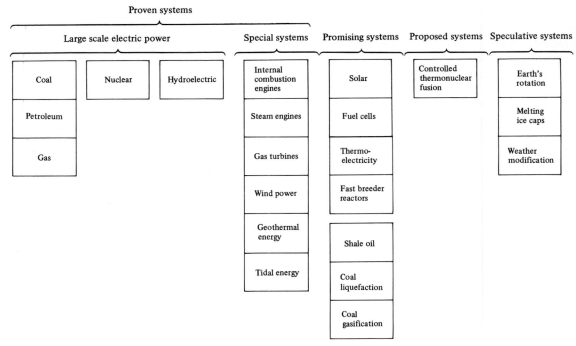

Figure 8-4   Sources of energy.

sumption we will exhaust the world's fossil fuel supply within a few hundred years.

The high rate of utilization of fossil fuels during the modern era is strikingly illustrated in Fig. 8-5 which shows the rate of energy production from fossil fuels on a time scale that extends from 5000 years in the past to 5000 years in the future. In this diagram we can see that fossil fuels play an important role only during a brief interval of the world's history. Within 200 years or so we shall have to rely almost exclusively on other sources of energy.

Estimates of this type are necessarily based on the projected status of future technology. If we are successful in devising methods for utilizing low-grade coal and for extracting oil from shale deposits, we may be able to extend the reserves of fossil fuels. But it is clear that it is imperative to develop other nonfossil energy sources. Nuclear reactors, utilizing the fission and fusion processes, together with solar power plants must eventually assume the primary burden of supplying the world with energy.

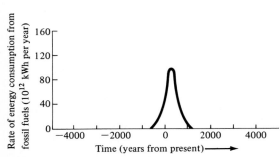

Figure 8-5   Exploitation of fossil fuels during the epoch from 5000 years in the past to 5000 years in the future. (Adapted from M. King Hubbert.)

Supplies of Coal and Oil

Coal occurs widely throughout the world. The largest deposits are in the U.S.S.R. where about 60 percent of the world's coal reserves are located. Coal has been mined in many parts of the United States. There are relatively few deposits of anthracite or *hard* coal; these occur primarily in Pennsylvania, Virginia, Arkansas, and Colorado. Softer coals, bituminous and lignite, are found throughout the eastern and central parts of the country and in the Northern Plains. The estimated coal reserves in the United States amount to about 500 E.U.

Much of the U.S. coal (particularly that in the eastern part of the country) contains a substantial amount of sulfur (3 to 6 percent). When such coal is burned, sulfur oxides are produced and exhausted into the air (unless elaborate and expensive trapping equipment is used); see Section 8-7. Because of the injurious effect that sulfur oxides have on plant and animal life, restrictions have been placed on the large-scale burning of high-sulfur coal, as in coal-fired electrical generating plants. In the northeastern United States, where there is no nearby source of low-sulfur coal, many of the generating plants converted their facilities to burn oil. But now oil is in short supply, and reconversion to coal is necessary. In order to meet the regulations on sulfur oxide emissions, these plants generally mix low-sulfur coal (which, in the Northeast, may have to be transported over large distances) with the more readily available high-sulfur coal. This is one of the factors that has contributed to the increasing cost of electrical energy.

Because oil is formed from marine life, oil deposits are widely distributed, especially in coastal areas and beneath the continental shelves. Oil is also found in inland regions that were once submerged, such as the southwestern United States, parts of the Sahara Desert, and in the regions bordering the Persian Gulf. Large deposits of oil, probably more than occur beneath land areas, are found beneath the continental shelves. The most significant deposits occur off the eastern, southern, and southwestern coasts of the United States, in the North Sea, in the Persian Gulf, and in the waters of southeast Asia and Indonesia. Because of the greater expense of drilling for oil in offshore locations, these deposits have not yet been fully exploited.

In a recent report concerning the U.S. oil and gas reserves, it is estimated that 37×10^9 barrels of oil

PHILLIPS PETROLEUM COMPANY

The petroleum resources that are located beneath the continental shelf are exploited with off-shore drilling equipment such as shown here.

remain in fields now producing and that 113×10^9 barrels are contained in yet untapped fields. The total oil reserves, 150×10^9 barrels, represent a 25-year supply at the present rate of consumption (see Table 8-1). The total reserves of natural gas, 600×10^{12} cubic feet, also represent about a 25-year supply. Other estimates place the reserves 2 or 3 times higher (the primary difference is in the judgment of offshore reserves). Even if the higher figures are correct, it is clear that our supplies are quite limited and that alternate sources of energy must be actively sought and developed.

Petroleum compounds occur not only in liquid form as crude oil but also in solid form in *shale* deposits. (In the United States there are extensive oil shale deposits in Colorado and Utah.) It has been estimated that there is about 1000 times as much hydrocarbon material in oil shales as in crude oil throughout the world. Extracting useful fuel from oil shales poses a variety of special problems which have not yet been solved. At the present time only about 0.01 percent of the known oil shale deposits are classified as "recoverable." If methods can be devised to extract fuels from these shales in an efficient manner, the world's useful reserves of fossil fuels will increase enormously.

The most serious problem involved in the development of shale oil deposits is that all methods now proposed require the use of large amounts of water. It is not clear whether the areas in which these deposits occur are capable of supplying the necessary amounts of water without serious disruptive effects.

8-4 NUCLEAR ENERGY

The New Fuel

The most concentrated form of energy that is available to Man is stored in nuclei. This energy can be released in the processes of *fission* (the splitting apart of heavy nuclei) and *fusion* (the fusing together of light nuclei). Fission reactors have been producing electricity in commercial quantities for only about 20 years. But as our reserves of fossil fuels are depleted and it becomes more and more expensive to extract these fuels from low-grade deposits, nuclear power plants will supply a larger and larger fraction of the energy we use. In the United States in 1968, for example, the usage of nuclear generated electricity amounted to about 900 kWh per

The Point Beach nuclear plant near Two Creeks, Wisconsin consists of two units, each with a capacity of 497 MW, in operation since 1970 and 1972.

person. By the year 2000, it is estimated that this figure will increase to 35 000 kWh per person (and during the same period the population will increase by 50 percent from 200 million to 300 million). Although we may continue to use fossil fuels for certain purposes (for example, natural gas for space heating and petroleum fuels for transportation), it is most likely that during the early part of the 21st century we will be generating electricity almost exclusively from nuclear power plants. These nuclear plants will use uranium and thorium in fission reactions, and when a feasible fusion reactor has been developed, heavy hydrogen (*deuterium*) will probably become the principal fuel.

By 1985, about 30 percent of our electrical energy will be derived from nuclear plants. By the year 2000, the figure will have risen to about 50 percent.

Energy from Uranium

The energy available in a given mass of nuclear fuel is several *million* times greater than in the same mass of a fossil fuel. For example, the fission energy contained in 1 kg of uranium is the same as that contained in 3.4×10^6 kg of coal. A total of about 10 billion tons of coal would be required to produce 3 E.U. of energy (the worldwide usage per year), whereas only 3000 tons of uranium could produce the same amount. The situation at present, however, is not nearly this attractive. There are three major factors that increase the amount of uranium necessary to produce a given number of kilowatt hours of electrical energy:

(1) Present-day reactors use uranium-235 (^{235}U) which has an abundance of only 0.7 percent in naturally occurring uranium.

(2) Only about 2 percent of the theoretical maximum available fission energy is actually extracted from the uranium fuel rods used in today's reactors.

(3) The efficiency of converting fission energy into electrical energy in present-day reactors is about 32 percent.

When all of these factors are taken into account, it is seen that about 70 million tons of natural uranium metal (instead of 3000 tons) is required to produce 3 E.U. of energy with the reactors of today. Fortunately, all three of these factors can be significantly improved:

(1) New types of reactors will be able to utilize the

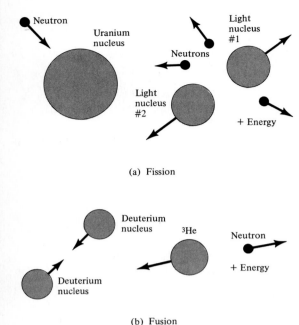

(a) Fission

(b) Fusion

Figure 8-6 (a) When a uranium nucleus is struck by a neutron, a fission reaction is triggered which releases energy and produces several neutrons which can induce other fission reactions. (b) When two deuterium nuclei collide at high speed, a reaction can take place which produces a ^{3}He nucleus and a neutron and releases energy.

isotope ^{238}U (which makes up 99.3 percent of natural uranium).

(2) By recycling the uranium in used fuel rods, the efficiency of extracting fission energy can be increased to about 3 percent. Even more important is the fact that new *breeder* reactors (those that produce new fuel as well as energy) will increase the yield to 80 percent or perhaps even more.

(3) New designs for the cooling systems in reactors will increase the electrical efficiency from 32 to 40 percent (which is the efficiency of coal-powered generating plants).

If all of these improvements can be successfully and simultaneously incorporated into fission reactors, the net result will be that about 9000 tons of uranium metal will produce 3 E.U. of energy.

Because breeder reactors will be able to use thorium as fuel (in addition to natural uranium), the possible sources of energy for the future are considerably enlarged. The most profound effect on the world's energy supply, however, will be the development of a practical power plant that operates on the fusion principle. Energy is released not only when a heavy nucleus (such as uranium) undergoes fission into two lighter nuclei but also when two light nuclei (such as nuclei of deuterium) fuse together and produce a heavier nucleus (Fig. 8-6). There are difficult technical problems involved in constructing a fusion reactor that will operate continually, but progress so far indicates that a successful device could be operating before the end of the century. If so, we might have a network of fusion power stations by the year 2050.

The Nuclear Fuel Supply

As we approach an era in which there will be increasing reliance on nuclear fuels as the principal source of our energy, it is indeed fortunate that the supplies of uranium and thorium seem adequate for several hundred years and that the deuterium supply is truly enormous. The U.S. reserves of high-grade uranium ore amount to only about 2 E.U., but the undeveloped and lower-grade ores raise the total to about 15 E.U. The fuel supply for breeder reactors, however, amounts to about 100 E.U. and 700 E.U., respectively. Taking into account the fact that the rate of energy usage will increase, we will still have sufficient energy from this source alone for several centuries.

The fuel supply picture for fusion power is extremely attractive. The world's oceans constitute a huge source of deuterium (or *heavy hydrogen*) in the form of water, and relatively little energy is required to extract deuterium from water. There are approximately 2×10^{13} tons of deuterium in the ocean waters, equivalent to about 10^{11} E.U. of energy. If we can succeed in the development of a deuterium-fueled fusion power plant, then the source of supply of fuel is ensured for millennia!

In Chapter 20 we will discuss more fully the subject of nuclear power from fission and from fusion.

8-5 SECONDARY ENERGY SOURCES

Geothermal Energy

The interior of the Earth is extremely hot—so hot, in fact, that a part of the core consists of molten iron. From a temperature of about 20 °C at the surface of the Earth, there is an increase to about 1000 °C at a depth of only 40 km. Even for the relatively shallow penetrations of mine shafts, the temperature increases are substantial and limit the depths at which miners can work. Molten rock, liquefied at the extreme temperatures beneath the Earth's crust, is forced to the surface through cracks and fissures and is ejected in the form of lava from volcanoes. Hot water and steam are similarly released at the Earth's surface from hot springs and geysers. It has been estimated that there are 700 000 000 cubic kilometers of superheated water (at temperatures of about 200 °C) beneath the Earth's surface. All of this heated material—rocks, steam, and water—represents an enormous reservoir of energy.

As long ago as 1904, engineers in Larderello, Italy tapped the supply of geothermal energy by drilling special wells into the underground steam supply. This natural steam drives electricity-producing turbines and now the Larderello plant generates 390 MW of electrical power. Other geothermal systems are in operation in New Zealand, the Soviet Union, Japan, Iceland, Mexico, and Kenya. At the Geysers, 90 miles north of

PACIFIC GAS AND ELECTRIC COMPANY
The Geysers steam field near San Francisco.

San Francisco, steam wells drive generators that supply 300 MW of electrical power.

Although at first glance they may seem to represent an ideal form of natural power, geothermal sources are far from being trouble-free and without pollution. Even the purest underground steam contains enough hydrogen sulfide (with its characteristic odor of rotten eggs) to be extremely unpleasant and enough minerals to poison fish and other forms of marine life in streams and rivers into which the condensed steam is discharged. Furthermore, the removal of underground steam and water causes the surface to subside. At the Wairakei plant in New Zealand, for example, the subsidence amounts to about 16 inches per year. This plant discharges 6.5 times as much heat and 5.5 times as much water vapor into the environment as would a modern coal-fired plant with the same electrical output. Some of the difficulties attending the utilization of geothermal power could be overcome if the condensed steam were pumped back underground, but such measures are not yet in general use.

Of even greater potential than underground steam and heated water is the heat energy stored in subsurface rocks. Some of this energy could be recovered and used by pumping water into the region by means of deep wells. Upon being pumped back to the surface, the heated water could be utilized to drive electrical generators in the same way that natural underground hot water is used. Although there is probably 10 times as much energy that could be recovered from heated rocks than is available from natural steam and hot water (a potential of about 600 000 MW), no plants have yet been constructed to tap this energy source. Scientists are now studying a 2- by 5-mile region near Marysville, Montana, where rock at 500 °C lies only a mile below the surface. This relatively small source has the potential of supplying 10 percent of the U.S. electrical needs for 30 years.

There are a sufficient number of potential geothermal sites in the world that, with vigorous development, these could represent a significant energy resource. Geothermal energy will not, at least in the near future, replace the major energy sources now being used. But it has been estimated that by the end of this century, the U.S. could be producing 10^5 MW of geothermal electrical power. This figure represents about 10 percent of the projected electrical power requirements of the U.S. in the year 2000.

Tidal Power

It is possible to extract energy from water in ways other than the damming of rivers. For example, in certain parts of the world tides rise to prodigious heights. On the coasts of Nova Scotia and Brittany (in northern France), and in the Gulfs of Alaska and Siam the tidal variations amount to 40 feet or more. This twice-daily surging of water back and forth in narrow channels represents a potential source of power. Although not of major significance on a worldwide scale, tidal power should be useful in particular areas. The first tidal-powered electric generating plant is on the Rance River in France and is harnessing the power of the English Channel tides which rise to as much as 44 feet at this location. By opening gates as the tide rises and then closing them at high tide, a 9-square-mile pool is formed behind the Rance River Dam. As the tide lowers, the trapped water is allowed to flow out, driving 24 electricity-generating turbines of 13 MW capacity each. Other potential tidal power sites are Cook Inlet in Alaska, San José Gulf of Argentina, and a location on the White Sea near Murmansk in the U.S.S.R. A proposal to develop the tidal power at Passamaquoddy Bay between Maine and Canada has been abandoned as uneconomical.

Solar Energy

The source of energy most readily available to us is sunlight. Mirrors and lenses have been used to concentrate the energy in the Sun's rays into small spots; even an ordinary magnifying glass can be used to start a fire. Every year the Earth absorbs about 22 000 E.U. of solar energy. All of this energy, plus a small additional amount that is conducted to the Earth's surface from the interior, is eventually reradiated into space. (If this energy were not reradiated, the Earth's surface would soon become as hot as the interior.) If we could utilize only 0.1 percent of the incident solar energy, there would be more than enough to satisfy the entire world's energy requirements.

Panels containing hundreds of semiconductor solar cells are used on satellites and space vehicles to produce the electrical energy needed to operate the various pieces of equipment aboard. But the diffuseness of the solar energy supply is a great handicap in terms of Earth-bound uses of this energy. Enormous areas of ab-

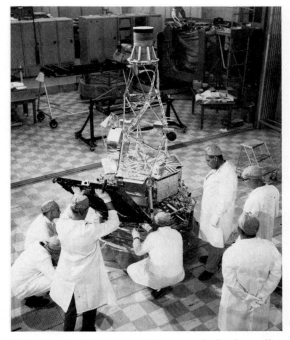

NASA technicians assemble a panel of solar cells.

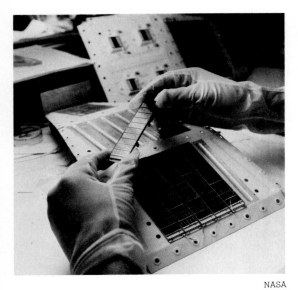

A strip of semiconductor solar cells.

sorbers are necessary to produce any substantial amount of useful energy. In the relatively cloudless desert regions of the southwestern United States, for example, the rate at which solar energy reaches the Earth's surface during the 6 to 8 hours around midday is about 0.8 kW per square meter. The energy incident per square meter per year amounts to about 2000 kWh. The most practical way to utilize this energy is to allow it to vaporize water into steam which can then be used to drive conventional electricity-producing turbines. There are formidable problems in this type of development, however, such as finding the proper surface for absorbing (and not reradiating) the sunlight and devising an efficient method for converting the energy in the absorbing material into steam energy.

Recent advances suggest that these difficulties can be overcome and that a 1000-MW "solar farm" could be constructed which would consist of about 50 square kilometers (12 500 acres) of solar converters. However, such a solar farm would probably cost several times more than a nuclear plant with the same power output and the cost of electricity delivered to the user would be 3 to 4 times today's cost. Thus, the utilization of solar energy to produce electrical power on a large scale is not now economically feasible. But as the costs of conventional fuels rise, as construction becomes more expensive for conventional facilities, and as more environmental controls are required, solar energy may appear as a viable alternative to conventional sources of energy.

Solar Heating

Although it appears that the large-scale generation of electricity from solar radiation is at best many years in the future, the heating (and cooling) of homes and businesses with the Sun's rays may become widespread within a much shorter time. By 1975, only a few dozen homes in the United States had been constructed with solar heating systems. But as fuel costs rise and increasing emphasis is placed on "clean" energy, it seems probable that more and more new construction will incorporate some sort of arrangement for the utilization of solar radiation. It has been estimated that in the next 5 to 10 years, perhaps 10 percent of the new homes will be at least partially heated by solar radiation.

One type of home solar heating system is shown schematically in Fig. 8-7. Solar radiation is incident on a

Figure 8-7 Schematic diagram of a simple home solar heating system.

collector which is placed on the south-facing slope of the roof. Water in the transfer loop is heated and is pumped to the heat reservoir (also water) which receives a portion of the heat. When heat is required in the house (as sensed by a thermostat), warm water from the heat reservoir is pumped through a coil in the heating duct. A fan forces warm air throughout the house. If the water temperature in the reservoir is not sufficiently high to provide adequate heating, an auxiliary supply adds heat to the reservoir water.

During extremely cold weather, especially on sunless days, a large amount of auxiliary heating will be required. Therefore, in a typical installation, an average of only about one-third to one-half of the necessary heating could be supplied by solar radiation. Solar heating therefore offers the possibility of substantial savings in fuel costs. But it must be remembered that these savings come only at the price of increased construction costs. According to some estimates, the special equipment required for an effective solar heating system may add up to 10 percent to the construction cost for a new house. In order to offer real economic advantages to a homeowner, it will be necessary to install a more complex system, one that provides *cooling* as well as *heating*. It is this type of solar heating-cooling system that the experts envisage coming into the new construction market in substantial quantities within a few years. Perhaps 10 percent of the new homes will be so equipped. The resulting impact on total fuel consumption will not be large, but it will provide a useful saving. The most significant effect of home solar systems will be to relieve the peak-load situations that occur in hot weather when everyone turns on air conditioners and power cutbacks or "brownouts" are sometimes experienced.

8-6 THE STORAGE OF ENERGY

Pumped Storage

One of the problems associated with the generation of electricity is that the demand for power fluctuates. During the day the power requirements, especially for commercial purposes, are much greater than during the nighttime hours. If there were some way to *store* electrical energy, the generating plants could be operated at capacity during the night, storing up energy to be re-

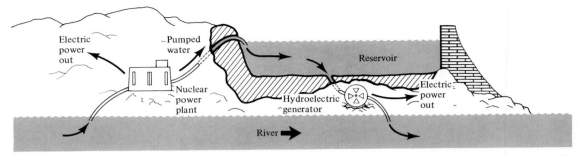

Figure 8-8 Schematic of a pumped storage system. Water is pumped into the reservoir during off-peak hours by electrical pumps powered by the output of the generator station. The water is later allowed to flow through a hydroelectric generator, producing electrical power at times of high demand.

leased when the demand increases. But how can electrical energy be stored? For small-scale uses, we have *batteries;* however, for the large energy requirements of homes and industry, batteries are completely impractical. One solution seems to be the pumped storage of water. At night, when the demand for power is low, instead of decreasing the operating level of the plant, some of the output power is used to pump water from a river or lake into a storage reservoir at a high elevation (Fig. 8-8). This water represents stored energy which can be recovered by allowing the water to return to the original height, turning a turbine generator on the way. By pumping during the night and adding the hydroelectric power to the plant's output during the day, the generated power can be more closely matched to the demand.

Although this scheme appears quite feasible and sensible, the plan requires the construction of an enormous water reservoir on a hill or plateau that is located near a river or lake. One of the proposals to construct a pumped storage system (on the Hudson River) has been challenged because of the environmental damage that would result from the construction of the reservoir. The largest pumped storage facility now in operation is located at Luddington, Michigan, and is operated jointly by the Consumers Power Company and the Detroit Edison Company. This plant uses Lake Michigan as the lower reservoir; the upper reservoir is a manmade lake 1.5 km by 3 km. The maximum power output is approximately 1900 MW and the energy storage capacity is 15 000 000 kWh.

Hydrogen As a Fuel

One of the more interesting recent ideas regarding energy storage on a longer time scale is the proposal to make widespread use of hydrogen gas as a fuel. Ever since the disastrous accident in 1937 when the hydrogen-filled airship Hindenburg was consumed by flames, hydrogen has been considered too dangerous for public use. During the intervening years, however, we have developed the techniques for handling hydrogen with safety. In the space program, for example, liquid hydrogen and liquid oxygen have long been used as the propellants in our most powerful rockets. The most serious problem associated with the introduction of hydrogen as a major fuel is probably one of public acceptance.

Hydrogen offers an attractive possibility for the storage of electrical energy. Most hydrogen in use today is produced by *electrolysis*—an electrical current is passed through water, and it dissociates into hydrogen and oxygen. The component elements can be recombined into water with the release of energy directly in the electrical form in devices called *fuel cells,* or the hydrogen gas can be burned and the heat used in the same way that the heat from the burning of natural gas is used.

By operating electrical generating plants at full capacity (which is the condition of maximum efficiency), electrical power could be supplied to meet the immediate requirements and any excess power could be used to electrolyze water into hydrogen and oxygen. Hydrogen, instead of electricity, could be delivered to homes and factories where it could be burned or where fuel cells could produce electricity on the spot as needed.

Hydrogen possesses a number of advantages as a fuel for space heating. Natural gas is the cleanest of the fossil fuels. When natural gas is burned, only carbon dioxide and water (and sometimes carbon monoxide) are produced. On the other hand, when hydrogen burns, *only* water is formed:

$$2 \, H_2 + O_2 \longrightarrow 2 \, H_2O$$

Hydrogen is therefore the cleanest possible combustible fuel. (An added benefit is that *pure* water is produced and could assist in meeting the local demand for water.)

The energy content of hydrogen gas is only about one-third of that of natural gas—about 95 kWh per

1000 ft^3 for hydrogen, compared to about 300 kWh per 1000 ft^3 for natural gas. But hydrogen burns with a hotter flame, and because no noxious fumes are produced, it can be burned in an unvented space. (A home furnace using hydrogen gas could perhaps be operated without a flue or chimney. There might be some problems arising from the production of nitrogen oxides in the heated air.) Hydrogen could be stored in central depots and routed to homes and factories through the underground pipeline system now used to deliver natural gas.

The Storage of Solar Energy

Is it possible to store solar energy? This is done, of course, in the formation of coal and oil, but these are long-term processes which cannot be speeded up. One of the ideas for the storing of solar energy is to allow solar radiation to promote the biochemical processes that convert our organic waste materials into useful chemical fuels (such as methane and hydrogen gas). This attractive idea is still in the development stage, but several pilot projects are under way. It appears unlikely that the fuels derived from wastes will ever amount to more than 1 or 2 percent of our national requirements, but at least there is the possibility that our wastes can be put to some good use.

Another way in which solar energy is stored is in the growth of plants such as food substances (which we use as fuel in our bodies) and trees (which we use hardly at all as fuel). In 1972 the domestic harvesting of wood amounted to 130×10^6 tons. Most of this wood is used as lumber or converted into paper products. If it were all burned, the thermal energy output would be about 0.4×10^{12} kWh. At a conversion efficiency of 40 percent, the annual wood crop could yield 0.16×10^{12} kWh of electrical energy, or about 10 percent of the amount used in the United States each year. Because of the large effect on our forest lands, it does not appear likely that we will attempt to double our wood harvest in order to supplement our electrical output by only 10 percent. (It should be noted, however, that wood contains very little sulfur so that one of the major problems associated with the burning of coal is not a factor in the use of wood as a fuel.) On the other hand, some electrical generating plants are being modified to use waste materials (primarily wood and paper trash) as fuels. In St. Louis a project is under way to utilize essentially all of the area's solid wastes in the generation of electricity. Scheduled for operation by mid-1977, the plant will be able to handle 2.5 to 3 million tons of solid wastes annually and to produce an average power of about 300 MW. In addition to the production of useful amounts of power, the trash-burning system will help alleviate the trash disposal and land-fill problems.

A much more practical and efficient way to utilize forest products as fuel is to convert them to *methanol* (methyl or wood alcohol, CH_3OH). Methanol is a clean-burning, inexpensive, easily transported fuel. It can be used to supplement (and perhaps eventually even substitute for) our gasoline and other liquid fuel supplies. In fact, tests have shown that methanol mixed with gasoline in proportions up to 15 or 20 percent and used in standard automobile engines without modification will increase the performance (reduce acceleration time and improve mileage per gallon) and decrease the pollutants in the exhaust gases. Moreover, methanol can be produced from wood or coal or from almost any chemical fuel, and sold at prices below the current inflated prices for gasoline. It has been estimated that with proper management of our commercial forests, sufficient methanol could be produced to generate all of the electrical power that we now use. Because of the many attractive features of methanol, it would seem prudent and economical to begin shifting toward methanol as a primary liquid fuel.

8-7 ENERGY AND THE ENVIRONMENT

What Is the Cost?

Every month we pay our fuel bills (or *energy* bills). We receive accountings for our use of electricity, oil, and natural gas in our homes and for the gasoline used in our automobiles. And there are indirect charges that we pay for the energy used in manufacturing processes and for the transportation of goods. At the present time, the average per person consumption of energy in the United States amounts to about 100 000 kWh annually. (Only half of this amount represents useful energy because of the wastage—see Fig. 8-2.) Because our rate of using energy is increasing (doubling in about 20 years), it is becoming more expensive to generate sufficient energy to meet the demands. Consequently, the cost to the

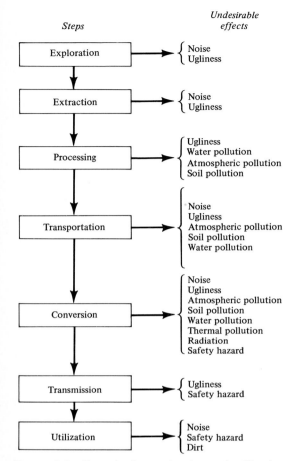

Steps	Undesirable effects

Figure 8-9 Steps in the production and utilization of energy together with the associated effects on the world. Each step also involves thermal pollution. (After Ali Bulent Cambel.)

energy-user per kWh consumed is increasing at a rate of about 10 percent per year. Thus, as we manufacture and sell more goods, add more devices for our comfort and entertainment, and make more use of transportation facilities, we can look forward to larger and larger energy bills as well as to shortages of some of our fuels.

We pay for the energy we use not only in terms of the direct and indirect charges for electricity and fuel consumption but also in terms of the effects that energy production has on our world (Fig. 8-9). It is not possible to place a dollar value on many of the side effects associated with energy production. What is the value of the health impairment caused by automobile exhaust fumes? What value do we place on the destruction of farm land caused by the strip mining for coal? What value is associated with the loss of seaside beaches because of oil spills washing ashore?

Modern society cannot exist without the production and utilization of energy. And as long as we continue to use fuels, there will necessarily be undesirable side effects. We must pay a price for energy. How much are we willing to pay?

The Effects of Water Power

We often think of electrical energy generated by hydro-electric power plants as the least offensive of the various energy-producing systems in use today. But there are serious problems associated with the construction of giant dams on natural waterways. A case in point is the Aswan High Dam on the Nile River in southern Egypt. For thousands of years, the annual floods of the Nile have been carrying silt from the African highlands to revitalize the soil along the banks of the Nile, virtually the only cultivatable land in Egypt. The flood waters flushed away the soil salts that had been accumulated during the previous year and annually dumped 130 million tons of rich sediment into the Mediterranean Sea, adding to the food chain of marine life and helping to maintain the proper salinity in the entire eastern end of the Mediterranean.

The presence of the Aswan High Dam has changed all this. Without the Nile sediment flowing into the Mediterranean, the plankton and organic carbons, vital to the marine life, have been reduced by one-third. The number of fish in the area has been drastically diminished, with some species having been forced into other waters to feed. The annual catch of sardines has been reduced by 20 percent. No one yet knows what will be the amount of ecological damage to the eastern Mediterranean.

The silt-free waters of the Nile now flow much more rapidly than the sediment-laden waters of the old river. As a result, parts of the riverbed are being carried away, undermining hundreds of bridges across the river.

In addition, the Aswan High Dam has triggered a variety of health problems. The still waters of Lake Nasser behind the dam are becoming breeding grounds for disease-carrying mosquitoes. The population of *bilharzia,* a parasite carried by water snails, had formerly been limited by the periods of dryness between the annual floods. But there are no longer alternate periods of flood and dryness, and the water snails, which flourish in the constantly placid irrigation canals, are on the increase. As a result, the incidence of *bilharziasis,* a debilitating intestinal disease, has risen to the point that more than half the population is infected.

Although substantial health and economic problems have been generated by the Aswan High Dam, there have been undisputed benefits. The 10 000-MW capacity of the dam's 12 hydroelectric generators (only a few of which have been placed into service) will provide the electrical power that the Egyptian economy desperately needs. New agricultural lands will be opened by irrigation from the backed-up waters in Lake Nasser, and a new fishing industry will also develop in the lake. But the lake is not filling as rapidly as anticipated, apparently because of unexpected underground losses and a higher evaporation rate than was calculated (due to the neglect of the high wind conditions).

The losses incurred by the construction of the Aswan High Dam will surely be permanent. Will the gains also be permanent or will they turn out to be only temporary? Has too high a price been paid for this new source of energy?

Although the problems associated with the Aswan High Dam are exceptionally severe, they are by no means unique. The construction of any dam alters the downstream ecology as well as that in the lake area behind the dam. What will be the effect of the loss of silt in the downstream portion of the river? How will the silt that builds up behind the dam be removed? What is the value of the land that is submerged by the dam's lake? These and other similar questions must always be

The old and the new at the Aswan High Dam in Egypt.

answered before intelligent decisions can be reached whether to construct a dam and where to locate a dam so that the damage is minimized.

Fossil Fuels — Multiple Threats

Most of the energy that is generated throughout the world at the present time is derived from the burning of fossil fuels: coal, natural gas, and petroleum products. The fact that *combustion* is necessary in the utilization of these fuels presents a number of problems that are different from those encountered with hydroelectric or nuclear reactor power sources in which combustion does not occur.

Because we are using combustible fuels at an ever-increasing rate, and because combustion involves the absorption of oxygen and the production of carbon dioxide, are we not in danger of depleting the world's supply of oxygen and upsetting the oxygen–carbon-dioxide balance that is necessary for plant and animal life? There are a multitude of problems associated with the burning of fossil fuels, but, fortunately, this is not one. All of the fossil fuels that have ever been burned have used only 7 out of every 10 000 oxygen molecules available to us. If the burning of these fuels continues at the present accelerating rate (increasing by 5 percent per year), then by the year 2000 we shall have consumed only about 0.2 percent of the available oxygen supply. Even the combustion of all of the world's known reserves of fossil fuels would use less than 3 percent of the available oxygen. Thus, the use of fossil fuels does not present us with the spectre of exhausting our oxygen supply.

Although we need not be concerned about depleting the atmospheric oxygen, there are numerous environmental problems associated with the utilization of fossil fuels. These problems can be separated into several categories:

(1) *Extraction of the fuel from the Earth.* The most plentiful fuel source in the world is *coal*. The highest quality coal (*anthracite*) generally occurs sufficiently far underground to require deep-mining techniques. The costs associated with this type of mining have risen to the point that many mines are being closed because they are uneconomical to operate. Consequently, in recent years there has been increased interest in the mining of lower quality coal that lies close to the surface. *Strip-*

Figure 8-10   Strip mining—the peeling back of the landscape to reveal the buried seams of coal.

mining techniques have been developed which allow the recovery of coal that was once considered to be of little value (Fig. 8-10). Huge machines have been constructed, such as the Consolidation Coal's "Gem of Egypt," that can take 200-ton bites out of the Earth to uncover the seams of coal. Strip mining now accounts for almost half of the coal production in the United States.

Although strip mining is providing the coal necessary to run our electricity-generating plants, huge amounts of landscape are suffering in the process. More than 3000 square miles of land (about $2\frac{1}{2}$ times the area of the state of Rhode Island) have been stripped in the United States. It is estimated there are 71 000 square miles of land in the U.S. that can be profitably strip mined! This area is nearly equal to the size of the states of Maryland, New Jersey, New York, and Connecticut combined.

Strip mining affects more than just the land that is mined. Unless careful measures are taken, adjoining property will suffer from landslides, erosion and sedimentation, and deterioration of water quality due to chemical effects. It has been estimated that strip-mining operations have affected from 3 to 5 times the area that has actually been mined.

The most significant aspect of the land problem is the fact that on only about one-third of this area has there been any attempt at reclamation. Some mining companies are making substantial efforts to reclaim the stripped land (and new laws are forcing others to do so), but the effects of strip mining will be unpleasantly visible for many years.

CONSOLIDATION COAL COMPANY

The "Gem of Egypt," the major tool of the Consolidation Coal Company's operation in Egypt Valley near Belmont, Ohio.

The strip mining for coal causes serious and continuing environmental problems. But strip-mined coal appears to be our best hope (perhaps our only hope) to meet the short-term fuel shortages.

The extraction of oil from the ground does not tend to desecrate the land the way that strip mining does. (But the drilling rigs and pumping stations that dot the countryside in many locations do not contribute a great deal to the scenery.) The most serious environmental problem associated with oil-well drilling occurs at offshore sites. Much of the world's oil reserves are located under the continental shelves—off the coasts of North America and Saudi Arabia, in the North Sea, and near the Indonesian islands. Because of the many technical difficulties inherent in offshore drilling, if a rupture occurs or if the drilling opens a crack in the rock that contains the oil deposit, a major leakage of oil into the water can occur before the damage is repaired or the crack is sealed. Leaks of these types have occurred in the Gulf of Mexico and off the coast of southern California. The release of substantial amounts of oil into the water can be injurious to the marine life and can foul the beaches when the oil washes ashore.

(2) *Transportation of the fuel.* Much of the world's oil is transported to refineries via sea. The tremendous size of the modern ocean-going tankers (some are capable of carrying more than 300 000 tons of oil) has rendered them extremely slow to answer controls. (Consider the momentum of a 300 000-ton tanker traveling at 15 knots!) The possibility of collision with another ship or with reefs and rocks in narrow waters presents a substantial hazard, as in the case of the Torrey Canyon accident in 1967. When such an accident does occur, rupturing the oil tanks, an enormous oil spill can result, endangering marine life and polluting beaches and harbors.

Leakage from offshore drilling operations and spills from damaged tankers do not represent the major sources of oil pollution in the world's ocean waters. More than two-thirds of the oil dumped into the seas by Man is from the crankcases of automobile and other engines. It has been estimated that as much as 350 million gallons of used crankcase oil is dumped into sewers and eventually runs into the seas each year. Most of this oil could be re-refined and used again, but the practice is not usually followed. Legal restrictions and substantially

UPI

Torrey Canyon accident in 1967 in which 100 000 tons of oil were spilled. When this oil washed ashore it despoiled many miles of English beaches. It was this accident that first focused attention on the problem of oil spills in the ocean waters.

increased oil prices are beginning to force the recycling of used oil.

Oil is also transported overland via pipelines. In some cases there are serious problems associated with this mode of delivery. The most economical method of transporting oil from the huge fields at Prudhoe Bay on the arctic coast of Alaska is by pipeline to the southern coast of Alaska where the oil is loaded onto tankers for further shipment. Because the proposed pipeline passes through large areas of untouched land, conservationists fear that the presence of the pipeline will upset the ecological balance of the region. Migratory animals may be forced to use new routes because of blockage by the pipeline. If the line were to break, sizable areas could be soaked by oil before the appropriate valves could be closed. Because the oil must be pumped through the pipeline at high speeds, friction will heat the oil to about 170 °F. The hot pipes could conceivably melt their way through the permafrost layer, then sag and rupture. Even if rupture did not occur, the melting of the permafrost might cause irreversible changes in the local ecology. Designers of the system maintain that safety features will eliminate the possibility of catastrophic accidents. The Alaskan pipeline will be in full operation by 1980 and we will then know the real price of the opening of the Prudhoe Bay oil fields.

Natural gas is also transported via pipeline. In the United States there are almost a quarter of a million miles of underground pipes that deliver natural gas from the sources (primarily in Texas, Louisiana, and Oklahoma) to the industrial and urban users. Although this system is relatively trouble-free, serious leaks and explosions have occurred on occasion, and several deaths each year result from these incidents.

(3) *Combustion of the fuel.* When the fuel has been delivered to the user—coal to steam-generator electrical plants, natural gas to homes, and gasoline to automobiles—the energy content can be utilized only through the combustion process. The burning of fossil fuels releases a variety of noxious gases and particulate matter into the atmosphere (see Table 8-3). The major contributors to this atmospheric pollution are coal and oil products; natural gas is by far the least offensive of the fossil fuels. One of the major problems is the presence of sulfur in coal and oil. Depending on the source, the sulfur content can be several percent and, upon

TABLE 8-3 TYPES OF AIR POLLUTANTS RELEASED BY THE
COMBUSTION OF FOSSIL FUELS

TYPE	AMOUNT RELEASED ANNUALLY IN THE U.S. (millions of tons)	MAJOR SOURCE
Carbon monoxide	151	Automobiles
Sulfur oxides	33	Power plants
Hydrocarbons	37	Automobiles
Nitrogen oxides	24	Automobiles; power plants
Particulate matter	35	Industrial plants

combustion, various oxides of sulfur (particularly SO_2) are produced. As shown in Table 8-3 these sulfur oxides are a major source of air pollution. During the great London smog catastrophe in December 1952 (which resulted in 3900 deaths), the SO_2 concentration reached 1.5 parts per million. (The danger level is considered to be 0.1 ppm).

When SO_2 is released into the atmosphere, it combines with water vapor and forms sulfuric acid. It is this sulfuric acid that is injurious to plant and animal life. Excessive amounts of SO_2 in the atmosphere have been directly linked to the high incidence of several types of respiratory ailments. Recently it has been found that atmospheric sulfuric acid is eating away the limestone facings of many monuments and public buildings in urban areas. The Acropolis in Greece will have to be moved in order to survive the present-day pollution. The Lincoln Memorial in Washington, D.C., is also being attacked and a major project will be required to prevent its surface from decomposing. Restorations expert Kenneth Eisenberg has said of the Lincoln Memorial, "It's like a giant Alka-Seltzer tablet. You can almost hear it fizz when it rains."

The sulfur can be removed from coal and oil, but in many cases a major effort is needed to reduce the content to a level that is consistent with the new Federal regulations. Some of the coal mines in the Eastern U.S. have been closed because the sulfur content is too high and because to remove the sulfur is not now economically feasible. The loss of this coal has placed an even greater burden on our oil supply. The Environmental Protection Agency (EPA) has estimated that the annual cost of air pollution damage to health, vegetation, and property values to be more than $16 billion. How much does air pollution cost *you* each year?

The combustion of natural gas produces far less in the way of pollutants than does either coal or oil. One of the ways of solving the coal and oil problem is to convert these fuels into cleaner-burning gases and liquid hydrocarbons. With the sulfur removed and particulate matter prevented from entering the atmosphere, a primary source of air pollution would be largely eliminated. A major effort is being mounted to perfect methods for coal gasification and liquefaction. Perhaps within a decade or so we will no longer burn coal but will instead use coal-gas or liquid fuel obtained from coal.

The burning of gasoline in internal combustion engines is the major source of carbon monoxide, nitrogen oxides, and hydrocarbons in the atmosphere (see Table 8-3). In addition, about 200 000 tons of lead per year are released into the atmosphere from automobile gasolines. It is alarming to note that about 15 000 tons of these various pollutants are introduced into the air *daily* over Los Angeles County. These compounds and the products of the photochemical reactions in which they engage produce the noxious mixture known as *smog*. There seems to be no escape from the health hazards of smog until some effective way is found to remove the pollutants from automobile exhaust gases or until some practical substitute for the internal combustion engine is developed.

Environmental Effects of Carbon Dioxide and Carbon Monoxide

The production of carbon dioxide is a necessary consequence of every combustion process. Therefore, even if we were to eliminate all of the sulfur from our hydrocarbon fossil fuels and if the combustion of these fuels could be made perfect, we would still release huge quantities of carbon dioxide into the atmosphere. It has been estimated that the carbon dioxide content of the atmosphere has increased by 10 percent in the last 50 years and that by the year 2025 the content will be almost double the value that prevailed in the early 19th century before the large-scale use of fossil fuels began.

Carbon dioxide molecules strongly absorb radiant energy of the type emitted from the surface of the Earth. By reradiating this energy at the lower temperature of the upper atmosphere, carbon dioxide reduces the heat energy lost by the Earth to space. (The absorption and reradiation of energy by atmospheric carbon dioxide is called the *greenhouse effect*.) It has been argued that the continued burning of fossil fuels will result in a steady increase in the Earth's surface temperature. Indeed,

—— *piceaque gravatum*
Fædat nube diem [1] ;

It is this horrid Smoake which obfcures our Church and makes our Palaces look old, which fouls our Cloth and corrupts the Waters, fo as the very Rain, and refre ing Dews which fall in the feveral Seafons, precipitate t impure vapour, which, with its black and tenacious qu lity, fpots and contaminates whatever is expofed to it.

—— *Calidoque involvitur undique fumo* [k] ;

It is this which fcatters and ftrews about thofe black a fmutty *Atomes* upon all things where it comes, infinuati itfelf into our very fecret *Cabinets*, and moft precio *Repofitories*: Finally, it is this which diffufes and fpread: Yellowneffe upon our choyceft Pictures and Hanging which does this mifchief at home, is [l] *Avernus to Fou* and kills our *Bees* and *Flowers* abroad, fuffering nothing our Gardens to bud, difplay themfelves, or ripen; fo

[j] Claud. de rap. Prof. l. i.   [k] Ovid.
[l] A lake in Italy, which formerly emitted fuch noxious fumes, that birds, wh attempted to fly over it, fell in and were fuffocated; but it has loft this bad quality many ages, and is at prefent well ftocked with fifh and fowl.

Excerpt from John Evelyn's book, *Fumifugium: or The Inconvenience of the Aer and Smoake of London Dissipated,* first published in 1661 and reprinted in 1772. Air pollution has been with us for a long time!

there was a general increase in temperature between 1860 and 1940, but since 1940 there has been a slight lowering of temperature for the world as a whole. The problem of atmospheric carbon dioxide is extremely complex, and arguments regarding the inevitability of temperature increases based only on the absorption characteristics of the carbon dioxide molecule are too simplistic. An increase in the temperature of the Earth's surface and lower atmosphere has the compensating effect of increasing evaporation and cloudiness. Because clouds reflect some of the incident sunlight, increases in cloudiness tend to decrease the surface temperature. Furthermore, the release of particulate matter into the atmosphere from fuel burning increases the number of condensation sites around which water droplets can form. The result is an increase in the amount of rain, hail, and thunderstorms which lead to a lowering of the temperature.

The amount of atmospheric carbon dioxide is regulated by the presence of the ocean waters which contain 60 times as much carbon dioxide as the atmosphere and which absorb a large fraction of the carbon dioxide released by the burning of fuels. Also, the increased level of carbon dioxide in the atmosphere actually stimulates the more rapid growth of plants. This increased utilization of carbon dioxide further reduces the atmospheric excess. The carbon dioxide stored in plants will eventually be returned to the atmosphere when the plants decompose. But forests account for about one-half of the plant growth in the world and the long lifetime of trees will hold this extra carbon dioxide and distribute its return over a long period of time.

By examining some of the various aspects of the carbon dioxide problem, we see that the world's climate and the world's ecology are influenced to an important extent by changes in the amount of atmospheric carbon dioxide. Apparently, Nature has been kind enough to provide compensatory effects so that our use of fossil fuels will not precipitously alter the climatic features of our world. However, we do not yet completely understand the role that carbon dioxide plays in our environment and we must continue to examine the possible consequences of increased consumption of fossil fuels.

What about carbon monoxide? Are Man's activities, particularly the burning of fossil fuels, provoking a serious imbalance of carbon monoxide in the atmosphere? Apparently not. A recent study shows that about 3.5 billion tons of CO from natural sources enter the atmosphere each year, mostly the result of decaying plant matter. On the other hand, only about 0.27 billion tons of CO are produced by Man. The injection of this amount of excess carbon monoxide into the atmosphere does not yet constitute a serious disturbance of the *average* value. However, in local situations, such as city streets that carry heavy automobile and truck traffic, the carbon monoxide concentration can reach health-affecting levels.

Very little is known about the natural way by which carbon monoxide is destroyed. Therefore, the process of CO removal from the atmosphere cannot now be identified with any degree of certainty.

Thermal Pollution

All electric generating plants (except for hydroelectric plants) produce electricity by driving huge turbine generators with steam. The steam is condensed in a cooling system and is cycled back to the heating unit for reuse. The "cooling system" can be water that is pumped from some nearby reservoir (a river, lake, or bay) or it can be a cooling tower in which the heat is dissipated into the atmosphere. Each kilowatt-hour of electric energy generated by a modern fossil fuel plant requires the equivalent of about 1.5 kWh of heat to be rejected at the condenser. Nuclear power plants, because of their lower efficiencies, present thermal pollution problems that are about 40 percent greater. If the heated water is discharged into a flowing river, the effect will be to increase the water temperature by a few degrees in the vicinity of the plant. If the water is discharged into a static reservoir, such as a lake, the effect can be even more severe. In either case, the change in the water temperature will affect the oxygen content of the water and will influence the growth rate of aquatic plants and animals. The ecological balance in the water system will therefore be disturbed.

In order to reduce as far as possible the undesirable effects of heat rejection by power plants, both nuclear and conventional, it will probably become necessary to equip these plants with cooling towers. (Several of the newer plants are so equipped.) By dissipating most of the excess heat into the atmosphere instead of the water system, the damage to the aquatic life will be considerably lessened. But the use of cooling towers will mean a more expensive operation and it will also mean a change in the local atmospheric conditions (for example, an increase in fog formation). Although either system tends to alter the natural conditions, on balance it often seems preferable to reject as much of the heat as possible into the atmosphere instead of into rivers, lakes, and bays.

Thermal pollution is generated by the energy *user* as well as by the energy *producer*. Almost all of the energy we use is eventually converted into heat, by friction, by electrical resistance heating (Section 11-3), and in combustion processes. Most of this waste heat is dissipated into the air where it contributes to the general atmospheric heating. In large cities, where energy consumption is concentrated, the air temperature is usually several degrees higher than in the surrounding rural areas. This increased temperature is an important factor in the production of urban smog.

In order to gauge the magnitude of the urban waste heat problem, consider the situation in Los Angeles County. The population of the county is approximately 7 million persons. Assuming an average rate of energy use of 10 kW by each person (which is the national average), the total for the county is 7×10^{10} W. The area of Los Angeles County is 4069 mi^2 or approximately 10^{10} m^2. Therefore, the average rate of energy usage (and, hence, heat production) is about 7 W/m^2. Solar radiation reaching the surface of the Earth, averaged over a day, is about 200 W/m^2. Thus, the artificially produced heat in this urban area is about 3 percent of that received from the Sun, and this figure will increase substantially as the rate of energy usage continues to climb. It has been estimated that by the year 2000, the rate of release of thermal energy by the 56 million people who will then live in the Boston–Washington corridor will be about 32 W/m^2, a significant fraction of the solar energy input.

Even if we discover ways to eliminate the other problems associated with energy production and usage, thermal pollution will still be with us. And we do not know what the long-term consequences of this subtle form of pollution will be.

Nuclear Power and Radioactivity

An increasingly important fraction of the energy generated in this country is being produced by nuclear power plants. Just as for all other types of energy sources, there are significant hazards and pollution problems associated with nuclear energy. These are: (1) thermal pollution caused by the discharge of heated water into waterways, (2) radioactive emissions during operation, (3) the necessity to dispose of the radioactive wastes produced by nuclear fission in the fuel rods, and (4) the possibility of the release of substantial amounts of radioactivity into the atmosphere due to an accident. We will return to a discussion of these problems in Chapter 20 after we have had an opportunity to develop the necessary background material.

The Limitations of Energy Consumption

Society today requires huge amounts of energy in order to function. Every source of energy entails certain hazards to the environment. Hydroelectric plants disturb the balance of river ecology; fossil fuels give rise to atmospheric pollutants; nuclear power generators produce radioactivity problems; and all energy production and usage contributes to thermal pollution. Because we require energy in increasing amounts in order to satisfy worldwide needs, we must learn to cope with these mounting problems. Although we will probably experience some severe short-term difficulties, in the long term, we are not really limited in our energy consumption by the supply of fuels; instead, the limitation is really the degree to which we can safely alter our environment. The main long-range problem that we face is how to increase effectively the production of energy and at the same time maintain the deleterious side effects at a livable level.

SUGGESTED READINGS

R. E. Lapp, *The Logarithmic Century* (Prentice-Hall, Englewood Cliffs, New Jersey, 1973).

J. M. Fowler, *Energy and the Environment* (McGraw-Hill, New York, 1975).

Scientific American articles:

J. Barnes, "Geothermal Power," January 1972.

A. M. Squires, "Clean Power from Dirty Fuels," October 1972.

QUESTIONS AND EXERCISES

1. Examine the ways in which energy is used in the various processes that lead from the discovery of a deposit of iron ore to the use of a nail in the building of a house.

2. An average person requires approximately 3000 Calories of food energy per day. (1 Cal = 1.16×10^{-3} kWh.) Examine the lighting in a room of your home and estimate the

amount of electrical energy used per day to operate the lights. Compare this electrical energy with your food energy requirements.

3. In the United States we use about 0.1 E.U. of electrical energy each year. If the average cost of electricity is $0.03 per kWh, what is the annual electrical power bill of the U.S.? What is the average monthly cost of electricity for a family of 4? (If this seems higher than your monthly electric bill, remember that you pay for much electricity indirectly in the form of manufactured goods and services.)

4. The *plant efficiency factor* of an electrical generating facility is defined to be the ratio of the actual amount of electrical energy delivered to the amount that could have been delivered if the plant had operated full time at maximum capacity. Usually, the factor is computed on the basis of a year's operation. (Shut-downs and operations at below peak capacity during low-load hours make the plant efficiency factor always less than 100%.) Compute the plant efficiency factor for all U.S. facilities taken together by considering that in 1974, the installed capacity was about 470 000 MW and the delivered electrical energy amounted to 1.9×10^{12} kWh. Do you think your result is reasonable? Explain.

5. If energy truly is conserved, why do we have any concern about an "energy crisis"? Why not simply convert energy from one form to another depending upon the needs of the moment? (To answer these questions, trace the history of energy starting with the potential energy stored in a water reservoir behind a dam. What is the final step and why can the energy in this form *not* be used further?)

6. The construction of the Glen Canyon Dam on the Colorado River in a remote section of northern Arizona has flooded 200 000 acres of canyonlands. Conservationists strongly objected to this destruction of natural canyons. But the dam has formed 200-mile-long Lake Powell and now visitors may tour the partially submerged canyons by boat. Whereas previously only very few persons ever saw the original canyons, thousands now see the lake region every year. Comment on whether the environmental price paid for the Glen Canyon Dam was too high.

7. There are approximately 300 000 miles of overhead high-voltage electrical transmission lines in service in the United States. The rights-of-way on which the familiar steel towers are placed average 110 feet in width. How much land is used for these transmission lines? Compare this area with that of the state of Connecticut.

8. It will be possible to pump approximately 2×10^6 barrels of oil per day through the Trans-Alaska Pipeline from Prudhoe Bay to southern Alaska. (a) The energy equivalent of one barrel of oil is 1700 kWh. What fraction of the U.S. energy requirements will pass through this pipeline? (b) A barrel of oil has a mass of approximately 310 pounds. How many 100 000-ton tanker loads would be required annually to transport the amount of oil carried by the pipeline? From the standpoint of oil logistics, comment on the advantages of the pipeline transport system compared to the tanker transport system.

9. It has been estimated that 71 000 square miles of the United States could be profitably strip mined. Suppose that, instead of strip mining, this area were covered with some kind of solar energy system that would absorb 20 percent of the energy in sunlight and transform it into electricity. Compute the amount of energy that such a system would produce annually and compare the figure with the present worldwide rate of energy consumption. (Would 71 000 square-miles of strip-mined land have any less visual appeal than 71 000 square miles of solar cells?)

10. Parallel with the increase in energy consumption during the last 100 years or so has been a dramatic increase in world population. Does this mean that we are "locked in" to a system that requires the generation of huge amounts of energy or could we return to a situation in which the expenditure of energy resources is far less than it is today?

9

HEAT

In this chapter we continue the discussions begun in Chapter 7 and concentrate on *heat* as a form of energy. Because energy occurs in a variety of different forms in different situations, we have developed a number of ways of describing the various forms of energy. Thus, the terms and the equations that we use to describe heat phenomena appear rather different from those that we use for describing motional energy or mass-energy. Do not be confused or misled by these differences—the thermal energy in a hot bar of iron is just as real and important a form of energy as the motional energy in a falling hammer. Indeed, if we pound the iron bar with the hammer, we can convert some of the motional energy into thermal energy.

As we proceed with the discussion of heat, we will keep before us the idea that heat is a manifestation of the microscopic action of atoms and molecules.

9-1 THERMAL ENERGY

The Microscopic View

It is easy to visualize the ideas of kinetic and potential energy. The kinetic energy of an object depends on its motion and the potential energy depends on its position. But how can we visualize thermal energy or heat? We appreciate the fact that there is a connection between *heat* and *temperature*. We know that we must supply heat to increase the temperature of a room on a chilly day. And we know that when work is done against friction, the temperature of the objects involved will rise. If you rub your hands together vigorously for a few seconds, you will readily be able to sense the increase in temperature.

What actually happens when you rub your hands together or when you slide a block back and forth over a rough surface? You are exerting a force and something is moving; therefore, work is being done. The energy that you expend is transferred to the hands or to the block and the material over which it slides. The greater the amount of work that is done, the greater will be the temperature increase. Temperature is therefore an indicator of the amount of energy that is transferred. The *thermal energy* that an object possesses does not depend on the motion of the object or on its position. (The object remains "hot" even after its motion ceases.) Is thermal energy therefore some new kind of energy, completely different from the familiar kinetic and potential energies? Not at all. The reason is easy to see when we recall that all matter is composed of molecules. When a block slides over a rough surface, the irregularities in the surface of each material tend to "snag" on the irregularities in the surface of the other material. The sliding motion therefore displaces the molecules of both materials and causes them to move about in an agitated fashion. Work is done in changing the state of motion of the molecules, and the energy transferred to the materials is in the form of motional energy of the molecules.

Thermal energy is therefore the *internal* energy of an object. If the molecules that make up the object are moving slowly, we say that the thermal energy is low; if the molecules are moving rapidly, we say that the thermal energy is high. That is, thermal energy can be thought of as a form of kinetic energy that a body possesses by virtue of molecular motions. Thus, thermal energy is not really a new form of energy—it is simply kinetic energy at the microscopic level.

The First Law of Thermodynamics

In Chapter 7 we described the behavior of objects by considering only the kinetic energy and the potential energy. We applied the energy conservation principle without including the *thermal* energy. Were we in error in proceeding in this way? Not really. If an object undergoes a certain change in state of motion or position in such a way that its internal energy *does not change*, then the constant value of the thermal energy can be ignored in computing the total energy of the object. Remember, only energy *changes* are physically meaningful, and if the thermal energy remains constant during a process, it need not be included in the calculation.

It is often the case, however, that the thermal energy of an object *does* change during some physical process. Such a change is reflected in the increase or decrease of the temperature of the object. When thermal energy changes do take place, it is necessary to include these changes when applying the energy conservation principle. If energy conservation is expressed in a way that includes thermal energy changes, it is usually called the *first law of thermodynamics*. There is no new physical content in this law—it is only an extension of the established law of energy conservation.

In our introduction to the subject of thermal energy, heat, and temperature, we have introduced no new physical ideas. Thermal energy is nothing more than microscopic kinetic energy, and the first law of thermodynamics is only an extension of the familiar energy conservation principle. As we proceed with the developments in this chapter, we will use some new terms and some new units, such as *absolute temperature, Calories,* and *specific heat*. The reason is mainly historical tradition—these terms and units have always been used in discussion of thermodynamics and it is convenient to continue to do so. But the change in the style of approach to *thermal* problems compared to *mechanical* problems should not overshadow the fact that we are dealing here with nothing more than energy in a slightly different form.

We will begin by discussing the concept of temperature and its measurement. Next, we will establish the connection between heat and temperature. Finally, in Chapter 10, we will return to the subject of thermodynamics and discuss the relationship between the microscopic concept of thermal energy and the properties of bulk matter.

Temperature

Temperature is a familiar concept, indicating the degree of "hotness" or "coldness" of an object. If we hold an object near a flame (and add heat to it), we know that the object becomes "hotter" and the temperature rises. We also know that if we place a warm object on a block of ice, it will become "colder" (because heat is extracted

Figure 9-1   The level of mercury in a thermometer increases with temperature.

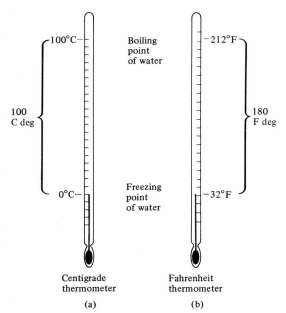

Figure 9-2   Two types of thermometer scales in common use. (a) Centigrade scale. (b) Fahrenheit scale.

from the object), and the temperature decreases. Although these qualitative ideas concerning temperature are correct, we need a method for precisely defining the temperature of an object.

One way to measure temperature is to make use of the fact that the volume of a definite mass of a liquid such as mercury (or alcohol) depends upon the temperature. As the temperature is increased, mercury expands and the volume increases. Therefore, if a small amount of mercury is sealed in a narrow glass tube with a reservoir at one end, the expansion of the mercury when the temperature is raised will cause the level of the liquid in the tube to rise. Similarly, if the temperature is lowered, the contraction will cause the level to fall (Fig. 9-1). This is the operating principle of the mercury *thermometer*. Somewhat less expensive and more common are thermometers that use colored alcohol as the liquid.

For a thermometer to be useful, we must establish some kind of temperature *scale*. We fix two points on the scale by choosing the temperature of boiling water and the temperature of freezing water. In the temperature system commonly used in the United States (but nowhere else in the world), we call the freezing point 32 degrees and we call the boiling point 212 degrees. This is known as the *Fahrenheit* temperature scale, and the fixed points are abbreviated 32 °F and 212 °F (Fig. 9-2). Between these two points the scale is divided into 180 equal parts. This temperature scale was devised by the German physicist Gabriel Daniel Fahrenheit (1686–1736) who constructed the first practical mercury thermometer in 1720.

Throughout most of the world, and in all scientific matters, the centigrade or *Celsius* temperature scale is used. In this system the temperature range between the freezing and boiling points of water is divided into 100 equal parts. Each such part is called *1 centigrade* (or *Celsius*) *degree*. The freezing point is designated 0 °C and the boiling point is 100 °C (Fig. 9-2). This temperature scale was first used by the Swedish astronomer Anders Celsius (1701–1744) in 1742.

We can deduce the relationship between the Fahrenheit and centigrade temperature scales in the following way. As shown in Fig. 9-2, a change of 100 centigrade degrees (C deg) corresponds to a change of 180 Fahrenheit deg. Therefore, the Fahrenheit degree is $100/80 = \frac{5}{9}$ of the centigrade degree. A change in temperature of 20 deg on the centigrade scale corresponds to a change of $\frac{9}{5} \times 20 = 36$ deg on the Fahrenheit scale. Because the freezing point corresponds to 32 °F and to zero on the centigrade scale, the conversion between the two scales is accomplished by using the relations,

$$T_F = \tfrac{9}{5}T_C + 32° \qquad (9\text{-}1a)$$

$$T_C = \tfrac{5}{9}(T_F - 32°) \qquad (9\text{-}1b)$$

where T_F stands for the Fahrenheit temperature and T_C stands for the centigrade temperature.

[Notice the distinction between the terms *degrees centigrade* (°C) and *centigrade degrees* (C deg). When we write 20 °C, this means a particular temperature. But 20 C deg means a temperature *interval* of 20 degrees on the centigrade scale; for example, the temperature range from 50 °C to 70 °C covers 20 C deg.]

The *absolute* (or *Kelvin*) temperature scale was devised by William Thomson, Lord Kelvin (1824–1907), the great Scottish mathematical physicist of the Victorian era. The size of the Kelvin degree (K deg) is the same as that of the centigrade degree, but the zero of the Kelvin scale is placed at *absolute zero*, the lowest temperature that any physical system could ever attain (but, which in practice, can never actually be attained). Absolute zero occurs at −273 °C, so that the centigrade and absolute temperature scales are related according to

$$T_K = T_C + 273° \qquad (9\text{-}2)$$

In Section 10-2 we will see why the absolute temperature scale is useful and important, and we will discuss the significance of the concept of *absolute zero*.

A comparison of temperatures on the three scales is shown in Fig. 9-3. Some of the temperatures (in °K) found in the Universe are shown in Table 9-1.

Heat

In this section so far we have introduced three new terms. To avoid any confusion, let us review what we mean by each of these terms:

Thermal energy: The internal energy of an object associated with the agitated motion of the constituent molecules.

Temperature: The degree of "hotness" or "coldness" of an object. The temperature of an object is an indication of how rapidly the molecules are moving. We have established a method for measuring temperature and assigning temperature values. But we have not yet indicated precisely how the temperature of an object is related to the state of motion of the molecules that make up the object; we will do this in Chapter 10. Notice, however, that the temperature of an object does

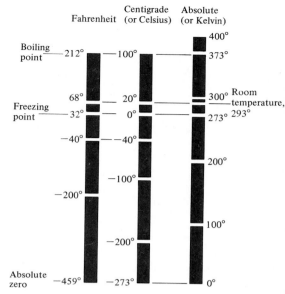

Figure 9-3   Comparison of the three temperature scales.

TABLE 9-1   SOME TEMPERATURES FOUND IN THE UNIVERSE

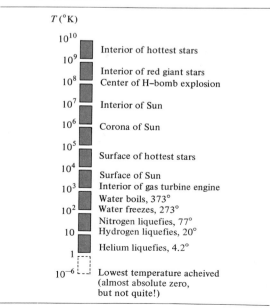

not signify the total internal energy content of the object: a large piece of iron clearly has a greater total internal energy than a small piece of iron at the same temperature.

Heat: Thermal energy in transit. We supply heat to an object in order to increase its thermal energy and to raise its temperature; we remove heat from an object in order to decrease its thermal energy and to lower its temperature. We have already seen that only *changes* in the potential energy of an object are physically meaningful. In the same way, only *changes* in the internal energy of an object are important. We keep account of these changes in terms of the *heat* supplied to or removed from an object.

We must now establish a method for measuring heat just as we have done for temperature. Water is used as the basic substance to define the *temperature* unit, and it is also used to define the unit of *heat*. The amount of heat required to raise the temperature of 1 kg of water by 1 C deg is called *1 Calorie* (Cal). For a temperature rise greater than 1 C deg or for a mass of water greater than 1 kg, the amount of heat required is correspondingly greater. For example, it requires 15 Cal of heat to raise the temperature of 1 kg of water from 30 °C to 45 °C. And to raise the temperature of 5 kg of water by 3 deg from 50 °C to 53 °C requires $5 \times 3 = 15$ Cal.

In addition to the Calorie (Cal), we sometimes see the *calorie* (cal) used with a lowercase *c*. One calorie is defined to be the amount of heat required to raise the temperature of 1 gram of water by 1 C deg. Thus,

$$1 \text{ Cal} = 10^3 \text{ cal} = 1 \text{ kcal} \qquad (9\text{-}3)$$

The Calorie is used to specify the energy content of foods. Some typical values are shown in Table 9-2. Most Americans consume about 3000 Calories per day, 40–45 percent of which is in the form of fats.

TABLE 9-2   ENERGY CONTENT IN CALORIES OF SOME FOODS

FOOD	CALORIES	FOOD	CALORIES
Apple, small	65	Doughnut	240
Bacon, 1 slice	35	Egg	75
Banana, medium	85	Fish, 4 oz.	140
Beef, lean, 4 oz.	190	Ham, 4 oz.	250
Bread, 1 slice	70	Jello, $\frac{3}{4}$ cup	110
Cake, chocolate, 1 slice	200	Milk, 1 glass	165
Carrots, $\frac{1}{2}$ cup	30	Potato, medium	90
Cheese, 1 slice	135	Veal, 4 oz.	200

The Second Law of Thermodynamics

If you place an ice cube in a glass of water, heat is transferred from the water to the ice cube, thereby cooling the water and melting the ice cube. Why is this so? Why did heat not flow from the ice cube to the water, thereby warming the water and making the ice cube colder? There would be no change in the total energy of the system if the energy gained by the water equals the energy lost by the ice cube. Therefore, the process could take place without violating the law of energy conservation. But the process does *not* take place. Think of some other situations involving hot and cold objects in contact. The result is always the same: heat always flows from the hotter object to the colder object. Unless work is done on the system by an outside agency, heat never flows from a colder to a hotter object. This is a new physical idea, quite different from the first law of thermodynamics. The principle that governs the direction of heat flow is the substance of the *second law of thermodynamics*.

A *refrigerator* is a device which extracts thermal energy from an object and lowers its temperature. Water can be made to freeze in a refrigerator even though the room in which the refrigerator exists is at a higher temperature. Thus, heat flows from the water at a lower temperature into the room at a higher temperature. But work is being done by the refrigerator during the process. Electrical energy is used to drive the motor that operates the refrigerator, and there is no violation of the second law of thermodynamics. The operation of a refrigerator in a room actually causes the temperature of the room air to increase slightly.

9-2 THERMAL EFFECTS ON SIZE

Linear Expansion

We have already mentioned that a mercury thermometer indicates temperature changes by virtue of the fact that mercury expands and its volume increases as the temperature increases. In fact, almost all substances—solids, liquids, and gases—have the property that they expand when heated. We can describe thermal expansion effects in terms of changes in *length* or changes in *volume*. For liquids and gases, only volume expansion is meaningful.

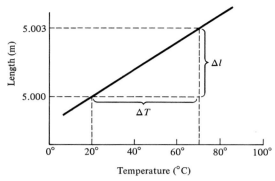

Figure 9-4 Variation of length of an iron bar with temperature. At $T = 20\ °C$ the length is 5.000 m. Raising the temperature to 70 °C produces an expansion of 0.003 m or 3 mm. The coefficient of linear thermal expansion is proportional to $\Delta l / \Delta T$.

Let us consider first the effect of temperature on the length of an iron bar. Suppose that we measure the length of the bar to be 5 m at a temperature of 20 °C (room temperature). If we heat the bar to 70 °C and again measure the length, we find a value of 5.003 m. That is, the length has increased by 3 mm. If we repeat this experiment with iron bars of different lengths and for different temperature changes, we find that the change in length Δl is proportional to the original length l and to the temperature change ΔT. Writing the proportionality constant as α, we have

$$\Delta l = \alpha l\ \Delta T \qquad (9\text{-}4)$$

The length l' at any temperature is then given by

$$l' = l + \Delta l \qquad (9\text{-}5)$$

The quantity Δl can be either positive or negative. If the temperature decreases, ΔT is negative and Δl is negative. Then, l' is less than l; this means that the bar is shorter at the lower temperature. Equations 9-4 and 9-5 taken together mean that the graph of l' versus T is a straight line, as shown for an iron bar in Fig. 9-4.

The quantity α is called the *coefficient of linear thermal expansion*. From the measurements of the length of the iron bar, we can obtain the value of α for iron. Solving Eq. 9-4 for α, we find

$$\alpha = \frac{\Delta l}{l\ \Delta T} = \frac{0.003\ \text{m}}{(5\ \text{m}) \times (50\ \text{C deg})}$$
$$= 12 \times 10^{-6}\ (\text{C deg})^{-1}$$

The unit for α is *per degree centigrade* or $(\text{C deg})^{-1}$. That is, the fractional change in length of an iron bar is $12 \times 10^{-6}\ (\text{C deg})^{-1}$ or 12 parts per million per degree centigrade. For a temperature increase of 1 C deg, a 1-m bar will expand by 12×10^{-6} m; a 2-m bar will expand by $2 \times 12 \times 10^{-6}$ m; a 3-m bar will expand by $3 \times 12 \times 10^{-6}$ m; and so forth. Note carefully that α gives the *fractional* change in length for each degree centigrade of temperature change.

There is a wide range in values of α for different materials. Values for some common substances are given in Table 9-3. Note the relatively large values for aluminum and lead, and the very low values for diamond, fused quartz, and Invar. The composition of the alloy Invar (64% iron, 36% nickel) has been specially adjusted to give a value for α near zero. Invar is used in applications that require the absolute minimum of length change with temperature.

TABLE 9-3 COEFFICIENTS OF LINEAR THERMAL EXPANSION FOR SOME SOLID MATERIALS

MATERIAL	α $(10^{-6}$ C deg$^{-1})$ FOR $T = 20\ °C$
Aluminum	23
Brass	18
Carbon (diamond)	1
Concrete	12
Glass (ordinary)	approx. 8
Gold	14
Invar	approx. 0
Iron	12
Lead	29
Quartz (fused)	0.4

The coefficient of linear thermal expansion for any particular material is actually not constant but has some variation with temperature. Iron, for example, has $\alpha = 11.8 \times 10^{-6}$ (C deg)$^{-1}$ at $T = 20\ °C$, whereas the value is 13.2×10^{-6} (C deg)$^{-1}$ at $T = 125\ °C$. For temperature changes that are not too great, α can be treated as constant for most purposes. A few materials actually have thermal expansion coefficients that are *negative*. At very low temperatures, the nonmetallic element silicon and fused quartz both *contract* when the temperature is increased.

Thermal expansion effects are important in various kinds of construction situations. The steel beams in a building must be cut to length and placed in position in such a way that expansion and contraction do not affect the structural integrity of the building. Highway materials are sometimes subject to large temperature variations. In the Midwest, for example, the temperature extremes for a concrete highway can range from a winter low of $-40\ °F$ ($-40\ °C$) to a summer high of $140\ °F$ ($60\ °C$). (Owing to the absorption of the direct rays from the Sun, the highway surface can be considerably hotter than the air.) The coefficient of linear thermal expansion for concrete is about 12×10^{-6} (C deg)$^{-1}$ (see Table 9-3). If a section of concrete highway is 15 m in length, the change in length between the seasonal low and high is

$$\Delta l = \alpha l\, \Delta T = (12 \times 10^{-6}\ C\ deg^{-1})$$
$$\times (15\ m) \times (100\ C\ deg)$$
$$= 0.018\ m = 1.8\ cm$$

or almost $\frac{3}{4}$ of an inch. This is why gaps filled with tar are left between concrete sections in a highway.

Volume Expansion

When a liquid or a gas is heated, the thermal expansion is most conveniently described in terms of a change in *volume*. In the same way that we expressed linear expansion, we can write for volume expansion,

$$\Delta V = \beta V\, \Delta T \qquad (9\text{-}6)$$

where β is the *coefficient of volume thermal expansion*. Also, in analogy with Eq. 9-5, the volume V' at any temperature is given by

$$V' = V + \Delta V \qquad (9\text{-}7)$$

Values of β for some liquids are given in Table 9-4.

This concrete highway buckled because of thermal expansion.

TABLE 9-4 COEFFICIENTS OF VOLUME THERMAL EXPANSION FOR SOME LIQUIDS

LIQUID	β (10^{-3} C deg^{-1}) FOR $T = 20\ °C$
Alcohol, ethyl	1.12
Alcohol, methyl	1.20
Carbon tetrachloride	1.24
Gasoline	0.95
Glycerine	0.50
Mercury	0.18
Olive oil	0.72
Turpentine	0.97
Water	0.21

Bimetallic Thermometers and Thermostats—How They Work

The fact that a strip of metal will expand and contract as the temperature changes can be used to construct a metallic thermometer. Because a small piece of metal will undergo only a small expansion for a modest increase in temperature, metallic thermometers are not precision instruments. Indeed, the effect is so small that it is not practical to construct a thermometer from a single piece of metal. Instead, two strips of different metals with different expansion coefficients (for example, aluminum and iron—see Table 9-3) are attached lengthwise by welding or riveting. One end is clamped into a stationary support and a pointer is attached to the other end (Fig. 9-5). When such a bimetallic strip is heated, the side which has the greater expansion coefficient (aluminum) will expand more than will the other side (iron). Consequently, the pointer end of the strip will bend toward the side with the smaller expansion coefficient, as shown in Fig. 9-5.

A simple bimetallic strip no larger than a small thermometer will exhibit only a slight movement of the free end. In order to increase the sensitivity, the length of the strip must be increased. This can be accomplished in a small space by coiling the strip, as shown in Fig. 9-6.

A thermostat is a device that controls electrical equipment by turning it on or off at a preset temperature. One type of thermostat is shown in Fig. 9-7. As the temperature increases, the bimetallic strip bends toward the right until, at a certain temperature, contact with the stationary element is lost and the relay circuit is opened. A thermostat connected as in the figure could be used to prevent the operation of electrical equipment at elevated temperatures that might damage the equipment. With the contact point on the other side, the thermostat could be adjusted to turn on some equipment (for example, an air conditioner) when the temperature rises to a certain value.

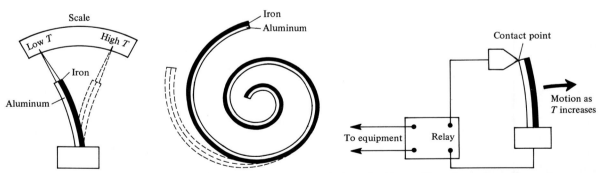

Figure 9-5   (Left) A thermometer consisting of a bimetallic strip and a calibrated temperature scale. The degree of bending shown here has been exaggerated.

Figure 9-6   (Center) Most bimetallic thermometers are in the form of coiled strips in order to increase sensitivity.

Figure 9-7   (Right) A bimetallic thermostat arranged to open a relay circuit when the temperature rises above a preset value.

Notice that the value for mercury is rather low in comparison with the other liquids. Between 0 °C and 100 °C, the fractional volume change, $\Delta V/V$, for mercury is

$$\frac{\Delta V}{V} = (0.18 \times 10^{-3} \text{ C deg}^{-1}) \times (100 \text{ C deg}) = 0.018$$

That is, the volume of mercury in a thermometer expands by only 1.8 percent as the temperature rises from the freezing point of water to the boiling point. This small volume change is made to correspond to a substantial change in the level of mercury in the thermometer tube by making the volume of the mercury reservoir at the bottom of the thermometer much larger than the volume of the tube (as shown in Fig. 9-2).

All gases expand in nearly the same way when heated. That is, the coefficient of volume thermal expansion is essentially the same for all gases. For 0 °C (and for normal pressure),

$$\beta(\text{gases}) = 3.66 \times 10^{-3} \text{ (C deg)}^{-1} \qquad (9\text{-}8)$$

We will return to the subject of the thermal behavior of gases in Chapter 10.

The density of a substance is the *mass per unit volume:* $\rho = M/V$. Because the mass remains constant, when a substance expands due to heating, the density decreases, and vice versa. The fractional change in density, $\Delta \rho / \rho$, is the same as the fractional change in volume, $\Delta V/V$, but the sense is opposite. That is,

$$\frac{\Delta \rho}{\rho} = -\frac{\Delta V}{V} = -\beta \, \Delta T \qquad (9\text{-}9)$$

Thus, a 1 percent *increase* in volume corresponds to a 1 percent *decrease* in density.

If a liquid or a gas is heated at one spot, the density of the heated portion becomes less than that of the surrounding liquid or gas. The heated portion is therefore buoyed up and rises through the more dense medium. (This is the reason for the familiar expression, "Hot air rises.") As the heated liquid or gas rises, it loses heat to its cooler surroundings. This transfer of heat through the buoyant motion of a liquid or gas is called *convection.* We will discuss this and other heat transfer processes in Section 9-4.

Most liquids (and gases) exhibit a uniform increase in volume and decrease in density with increasing temperature. Water, however, shows a variation of density with temperature that is most unusual. If we begin with a

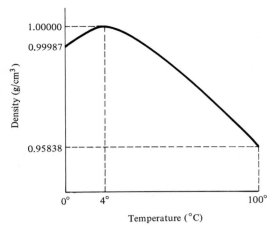

Figure 9-8 The variation of the density of water with temperature. Notice that the density is maximum at 4 °C. Both scales have been distorted in order to show the effect clearly.

certain mass of water at 0 °C, we find that as heat is supplied, the density *increases* until a temperature of 4 °C is reached. That is, in the temperature interval from 0 °C to 4 °C, the density of water changes *opposite* to that expected for a normal liquid. For temperatures above 4 °C, the density decreases in the normal way. Figure 9-8 shows schematically the density of water as a function of temperature.

Although the density of water at 4 °C is greater by only a tiny amount compared to the density at 0 °C (see Fig. 9-8), this difference has important consequences. Suppose that we consider an expanse of water such as a pond or lake. When the air temperature falls, as during a winter night, the surface layer of water becomes cooled and sinks. The water temperature decreases from the top of the pond to the bottom. This general variation of water temperature with depth remains until the surface layer is cooled below 4 °C. Now, this layer has a density *less* than the layer immediately below; therefore, the coldest part of the water remains on *top*. Here it is further cooled until it freezes. Thus, the unusual variation of water density with temperature ensures that ponds and lakes freeze at the top and not at the bottom. (If the pond consisted of any other liquid, it would freeze from the bottom up.) The surface layer of ice tends to insulate the water beneath from the cold air and so a large fraction of the pond remains as liquid water. If ice first formed at the bottom, there would be no insulating effect and the pond would freeze completely during severely cold weather. If ponds and lakes became solid ice, imagine the consequences for marine life!

The process of freezing decreases the density of water still further. The density of ice is 0.917 g/cm³. That is, a certain mass of water will increase in volume by about 8 percent upon freezing. Consequently, ice always floats on liquid water.

9-3 HEAT CALCULATIONS

Specific Heat

If we supply 1 Cal of heat to a 1-kg sample of water, the temperature of the water will increase by 1 C deg. But if we supply the same amount of heat to a 1-kg sample of iron, we find that the temperature of the iron increases by 8.4 C deg. How can we account for such a discrepancy in the temperature increase in two samples of matter with the same mass? First recall that the mass of a water molecule is 18 AMU, whereas the mass of an iron atom is 56 AMU. That is, the mass of the fundamental unit of iron is about 3 times the mass of the fundamental unit of water. Therefore, in a 1-kg sample of water there are about 3 times as many molecules as there are iron atoms in a 1-kg sample. We must remember that temperature is a measure of molecular motion. If we supply heat to equal-mass samples of water and iron, there are more molecules of water to set into motion and, consequently, more heat is required to raise the temperature a given amount.

By this reasoning we can account for a factor of 3 difference between the temperature increase of iron compared to water. But the actual difference is a factor of 8.4. The remainder is due to the fact that the water unit is a *molecule,* whereas the iron unit is an *atom.* In a piece of iron, the atoms are arranged in a regular crystal lattice. When heat is supplied to an iron crystal, the only effect is that the atoms vibrate more rapidly around their normal positions in the lattice. The fundamental unit of water, on the other hand, is a more complicated molecular structure. When heat is supplied to a sample of water, not only do the molecules move more rapidly, but some of the heat is used to make the atoms vibrate *within* each molecule. Therefore, an additional amount of heat is required to raise the temperature of water by 1 C deg compared to the amount required for the same temperature increase in an equal mass of iron. Molecules of different types have different ways in which they can vibrate internally. Therefore, a sample of NH_3, for example, will require a different amount of heat than a sample of water for the same temperature rise.

The amount of heat in Calories required to raise the temperature of 1 kg of a substance by 1 C deg is called the *specific heat* of that substance. Thus, the specific heat of water is 1 Cal/kg-C deg. By taking into account the number of molecules per unit mass and the effects of any internal molecular vibrations, we can understand the range of specific heat values that we find for other materials. Some typical values are given in Table 9-5.

How do we use the specific heat value for a material to predict the temperature rise when we supply a certain amount of heat? The appropriate formula is one that we can write down with only a little thought. The amount of heat Q that must be supplied to a sample to raise its temperature must be proportional to the temperature change ΔT. Furthermore, if we increase the mass M of

TABLE 9-5   SPECIFIC HEATS OF SOME MATERIALS NEAR ROOM TEMPERATURE

SUBSTANCE	SPECIFIC HEAT (Cal/kg-C deg)
Air	0.17
Aluminum	0.219
Copper	0.0932
Ethyl alcohol	0.535
Glass (typical)	0.20
Gold	0.0316
Iron	0.119
Lead	0.0310
Mercury	0.0333

the sample, then we must also increase the heat supplied; that is, Q is proportional to M. Finally, a material with a large specific heat c requires more heat than a material with a low specific heat for the same temperature change; that is, Q is proportional to c. Putting together these three statements, we can write

$$\boxed{Q = cM \, \Delta T} \qquad (9\text{-}10)$$

where

$$Q = \text{heat transferred (in Cal)}$$
$$c = \text{specific heat (in Cal/kg-C deg)}$$
$$M = \text{mass (in kg)}$$
$$\Delta T = \text{temperature change (in C deg)}$$

How much heat is required to raise the temperature of a 2-kg piece of iron from 20 °C to 35 °C? From Table 9-5, we find $c = 0.119$-Cal/kg-C deg for iron. Therefore,

$$Q = (0.119 \text{ Cal/kg-C deg}) \times (2 \text{ kg})$$
$$\times (35 \text{ °C} - 20 \text{ °C})$$
$$= 3.57 \text{ Cal}$$

If we supplied this same amount of heat to a 2-kg sample of water, the temperature rise would be

$$\Delta T = \frac{Q}{cM} = \frac{3.57 \text{ Cal}}{(1 \text{ Cal/kg-C deg}) \times (2 \text{ kg})}$$
$$= 1.8 \text{ C deg}$$

The Mechanical Equivalent of Heat

Heat is simply another form of energy, and so there is a relationship connecting the unit of heat (the Calorie) and the unit of energy (the joule). One way to obtain this relationship is to do a measurable amount of mechanical work on a certain mass of water and determine the increase in temperature of the water. The mechanical work done can be directly measured in joules and the temperature change of the water can be used to calculate the number of Calories of heat supplied to the water. James Prescott Joule performed such experiments in the 1840's and was able to determine the *mechanical equivalent of heat*.

One of Joule's experiments is shown schematically in Fig. 9-9. A beaker contains a certain mass of water and work can be done on this water by the paddlewheel which is turned as the block M falls through the height h. The work done (assuming no frictional losses) is

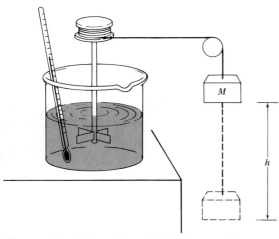

Figure 9-9 Schematic of Joule's experiment to determine the mechanical equivalent of heat. The falling mass M turns the paddle wheel which heats the water. The temperature rise is measured with a thermometer.

Mgh, and by measuring the mass and the temperature rise of the water, the heat equivalent of this work can be determined. Joule's results were actually quite close to the value accepted today:

$$1 \text{ Calorie} = 4186 \text{ joules} \qquad (9\text{-}11)$$

Suppose that a 2-kg block of copper is dropped from a height of 30 m onto some surface. Assume that half of the heat generated in the impact is retained by the block (and half by the material of the surface). What is the temperature rise of the block?

The amount of heat produced in the copper block is one-half of the initial potential energy of the block:

$$\begin{aligned} Q &= \tfrac{1}{2} Mgh \\ &= \tfrac{1}{2} \times (2 \text{ kg}) \times (9.8 \text{ m/s}^2) \times (30 \text{ m}) \\ &= 294 \text{ J} \\ &= (294 \text{ J}) \times \left(\frac{1 \text{ Cal}}{4186 \text{ J}}\right) \\ &= 0.070 \text{ Cal} \end{aligned}$$

Then, using Eq. 9-10,

$$\begin{aligned} \Delta T &= \frac{Q}{cM} \\ &= \frac{0.070 \text{ Cal}}{(0.0932 \text{ Cal/kg-C deg}) \times (2 \text{ kg})} \\ &= 0.38 \text{ C deg} \end{aligned}$$

9-4 HEAT TRANSFER

Conduction

There are three different processes by which heat can be transferred from one location or object to another. The most familiar of these is *conduction,* whereby heat is transferred from a hotter object to a cooler object with which it is in contact. Heat from an electric stove element is conducted through the bottom of a cooking pan to the food inside. And if a soft-drink can is placed on a piece of ice, heat will flow from the can to the ice, thereby cooling the can and its contents.

Not all materials conduct heat equally well. Special insulating materials (poor heat conductors) are placed in the walls and roofs of buildings to decrease the outward flow of heat in winter and to decrease the inward flow of heat in summer. Generally, metals are good conductors of heat, whereas such materials as glass, wood, as-

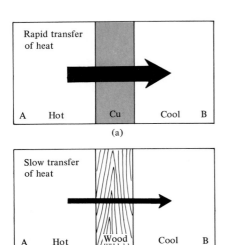

Figure 9-10 Heat is conducted much more rapidly through copper (a good heat conductor) than through wood (a poor heat conductor or good insulator).

bestos, and brick are poor conductors. If a piece of copper is placed between two blocks of material at different temperatures (Fig. 9-10a), heat will be rapidly conducted from the hotter material (A) to the cooler material (B). On the other hand, if a piece of wood is substituted for the copper (Fig. 9-10b), the transfer of heat from A to B will be quite slow. For equal thicknesses of copper and wood, and for the same temperature difference between A and B, heat will flow about 3000 times faster through the piece of copper than through the piece of wood!

Have you ever noticed that a piece of metal at room temperature feels cool to the touch, whereas a piece of wood at the same temperature feels much warmer? The reason is that the metal conducts heat much more rapidly from your hand (which is at body temperature, 37 °C) than does the wood. The rapid outflow of heat from your hand results in a sensation of coolness. Similarly, on a very cold day you can handle a piece of wood with bare hands but to touch a piece of metal is uncomfortable. If the temperature is very low, heat could be conducted from your hand to the metal sufficiently rapidly that the skin will actually freeze to the metal. Injuries of this type are frequently suffered by unwary persons.

Convection

As we have already mentioned, heating a certain mass of a liquid or a gas will cause the volume to increase and the density to decrease. If this mass is free to move, it will rise through the more dense surrounding liquid or gas. Figure 9-11 shows the way in which water will flow in a tank that is heated at one end. As the hot water flows upward and around the tank, it encounters cooler water which is warmed by the flow of heat from the hot water. As a result, the tank of water as a whole is raised in temperature, not simply the region near the flame. The process by which heat is distributed throughout a medium by the flow of a liquid or a gas is called *convection* or *convective heating*.

Convection is a common method of household heating. If hot water is piped to a radiator, the air adjacent to the radiator coils is heated and convective flow distributes the heat to the room. More common than ordinary convective heating in modern construction is *forced convection* in which a fan or blower is used to distribute the air more quickly and more uniformly.

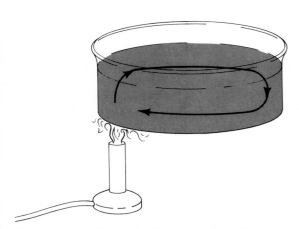

Figure 9-11 If a tank of water is heated at one end, the warm water will rise and will be pushed to the opposite end of the tank where it cools and flows downward and back toward the heated end. This process is called *convection*.

Radiation

If you hold your hand near (but not touching) an electric light bulb, you will feel a sensation of heat when the switch is turned on. This heating effect is not due to conduction or convection. Instead, it is due to *radiation,* the third process by which heat can be transferred.

When an object is heated, the constituent atoms become more and more agitated. The higher the temperature, the faster the atoms move and the more violent are the collisions between atoms. In many of these collisions, some of the kinetic energy of the atoms is transformed into electrical potential energy of the atomic electrons. Immediately following each such a collision, the electrons spontaneously return to their previous energy condition and the excess energy is emitted in the form of *electromagnetic radiation.* (We will discuss electromagnetic radiation more thoroughly in Chapters 14, 17, and 18.) At temperatures below about 1000 °C, the energy radiated by a heated object consists primarily of long wavelength radiation (radio waves and infrared radiation); such an object emits little or no visible light. At 2000 °C, a sufficient amount of the radiated energy consists of visible light so that the object is seen to glow with a dull red color. As the temperature is increased further, the color changes to cherry red, then to yellow-white, and finally blue-white. For a temperature of 6000 °C (the temperature of the Sun's surface), more yellow light is emitted than any other color. At 20 000 °C (a temperature that cannot be achieved on Earth but which is found for some particularly hot stars), the color is bluish-white. Thus, the color of a heated object is indicative of its temperature. Indeed, we make use of this fact in some instruments (called *pyrometers*) for determining high temperatures.

Experiments have shown that the amount of radiant energy emitted by an object depends on the *fourth power* of the absolute temperature, $T_K{}^4$. A surface of area A (in m²) at a temperature T_K emits energy at a rate P (in J/s or watts) given by

$$P = (5.67 \times 10^{-8})\, A e T_K{}^4 \text{ watts} \qquad (9\text{-}12)$$

The quantity e is called the *emissivity* of the particular material. The emissivity, which is a number between 0 and 1, indicates how effective a radiator the material is. A perfect radiator would have $e = 1$. Black substances usually have emissivities above 0.9, whereas white or shiny materials have very low values ($e = 0.2$ or less).

For example, candle soot has $e = 0.95$, and polished aluminum has $e = 0.05$. Any material which is a good *emitter* of radiation is also a good *absorber*. Thus, a surface covered with candle soot will absorb most of the radiation incident upon it; the absorbed radiation will increase the temperature to a value at which the emitted power is equal to the absorbed power. A polished aluminum surface, on the other hand, will reflect most of the incident radiation and will absorb (and then emit) relatively little.

An electric light consists of a fine tungsten wire which is heated by passing through it an electrical current. A typical filament has a surface area of about 1 cm² (or 10^{-4} m²) and is heated to a temperature of 3000 °K. The emissivity of tungsten at this temperature is 0.34. What is the radiated power of such a bulb? Substituting into Eq. 9-12,

$$P = (5.67 \times 10^{-8}) \times (10^{-4}) \times (0.34) \times (3000)^4$$
$$= 156 \text{ watts}$$

which is a value close to the power rating of many bulbs. It should be noted that this power is not all in the form of visible light. In fact, only about 2 percent of the energy emitted by a tungsten-filament light bulb is in the visible region; most of the energy is in the form of infrared radiation. (This is why an ordinary light bulb makes such a good warming element.)

It is important to understand that of the three ways by which heat can be transferred, only radiation does not require any material medium to be effective. Heat radiation (indeed, *all* forms of electromagnetic radiation) can propagate through empty space. This is exactly the way that we receive energy from the Sun. In Chapter 14 we will discuss additional details concerning electromagnetic radiation.

SUGGESTED READINGS

S. C. Brown, *Count Rumford—Physicist Extraordinary* (Doubleday, Garden City, New York, 1952).

J. Tyndall, *Heat: A Mode of Motion* (Appleton, New York, 1915).

Scientific American articles:

F. W. Dyson, "What Is Heat?," September 1954.

G. Y. Eastman, "The Heat Pipe," May 1968.

1. Convert the following temperatures to °C: (a) −20 °F, (b) 80 °F, (c) 98.6 °F, (d) 263 °K.

2. The daily food intake of a man is 3000 Cal. Suppose that the man's working efficiency is 10 percent. (That is, 10 percent of the food energy can be converted into useful work.) The man works by lifting 10-kg boxes from a floor onto a shelf that is 2 m high. How many boxes can he lift per day? This calculation should give you some appreciation of the tremendously large amount of energy contained in foodstuffs.

3. If you want to reduce your weight, is it better to decrease your food intake by 10 percent (300 Cal) or to exercise by running up three flights of stairs (total height of 10 m) 50 times a day? (Assume that your mass is 80 kg and that the only work done is in lifting your mass 50 times through 10 m.) Is exercise really a very effective method of weight control?

4. A certain river bridge (steel construction) has a length of 1.5 km. What is the total length of the gaps that must be left in the sections (measured at −20 °C) to allow for expansion between temperatures of −20 °C and 40 °C?

5. In Table 9-3 we see that the coefficient of linear thermal expansion for concrete is the same as that for iron. Why is this important in highway or building construction? (Concrete structures are usually reinforced with iron rods.)

6. When a 100-m length of chromium wire is cooled from 20 °C to 0 °C it is found that the wire is 1 cm shorter. What is the linear expansion coefficient for chromium?

7. When an iron rim is mounted on a wagon wheel, the rim is heated before it is placed around the wheel. Why?

8. A bar of aluminum and a bar of iron each have a length of 2.5 m at a temperature of 20 °C. If both bars are heated to 60 °C, which will be longer and by how much?

9. The hole in an iron plate is not quite large enough to allow an iron rod to be inserted. Should the plate be heated by an amount ΔT or should the rod be cooled by the same ΔT? What happens to the size of the hole as the plate is heated?

10. One liter (10^3 cm³) of methyl alcohol is poured into a flask at 0 °C. What must be the capacity of the flask so that it will not overflow when the temperature is raised to 20 °C?

11. Suppose that you have a cubical block of some material with a linear expansion coefficient α. If the block is heated, each of the three dimensions will expand by an amount $\Delta l = \alpha l \, \Delta T$. What is the new *volume* of the block? Remember that Δl is small compared to l. Show that the volume expansion coefficient is just three times the linear expansion coefficient: $\beta = 3\alpha$.

12. If you filled your 20-gal automobile gasoline tank at 0 °C, how much gasoline would spill out when the temperature increases to 25 °C? (Use the result in Exercise 11 to determine how much the volume of the *tank* expands. Assume the tank is made of iron.)

13. If a thick piece of glass is heated until the temperature is uniform throughout and then is plunged into cold water, the glass will crack or shatter. Explain why.

14. Why is it important to protect water pipes so that they do not freeze?

15. If ice has formed on the top of a deep lake, what is the temperature of the water at the bottom of the lake? Explain.

16. A certain alloy consists of aluminum and copper in equal amounts. If 40 Cal of heat is absorbed by a 2-kg block of this alloy, what will be the amount of temperature increase?

17. How much heat is required to raise the temperature of 2 kg of lead from room temperature to the melting point (327 °C)?

18. How many kilowatt-hours of energy would be required to raise the temperature of a 30 000-gal swimming pool by 5 C deg? (One gallon of water has a mass of approximately 4 kg.)

19. An electric fan is turned on inside a closed and insulated room. Will the air temperature be raised or lowered? Explain.

20. A 1-kg mass of clay ($c = 0.19$ Cal/kg-C deg) is thrown against a wall with a velocity of 40 m/s and sticks to the wall. If no heat is lost to the wall, what is the temperature rise of the clay?

21. Will you grow fat if you drink only *hot* water? How many Calories will be released to your body by the cooling of a quart (approximately 1 kg) of water at 120 °F to body temperature (98.6 °F).

22. A 5-kg mass of water, originally at 20 °C is heated to the boiling point. How much heat is required?

23. In his 1845 paper on the mechanical equivalent of heat, Joule remarked on the expected rise in temperature of the water as it cascades over Niagara Falls. The Falls are 160 feet high; what should be the difference in temperature of the water between the top and the bottom of the Falls? Assume that all of the potential energy possessed by the water at the top of the Falls is converted into heat when it reaches the bottom.

24. Why are the handles of cooking utensils made of wood or plastic instead of the same metal as the utensil?

25. Explain how convection is important in the flow of air in the atmosphere and in the flow of water in the oceans.

26. Explain how convection is important in the cooling of a large body of water such as a lake.

27. Two pieces of aluminum are exposed to direct sunlight. One piece is polished and the other is painted black. Which piece will have the higher temperature? Why?

28. The surface temperature of the Sun is approximately 6000 °K. How much power is radiated from each square meter of the Sun's surface? (Use $e = 1$.) The radius of the Sun is 7×10^8 m. What is the total power radiated by the Sun?

29. The distance from the Earth to the Sun is approximately 1.50×10^{11} m. Use the result in Exercise 28 and calculate how much solar power is intercepted by the disc of the Earth. (The radius of the Earth is 6.38×10^6 m.)

10

LIQUIDS AND GASES

Thus far, we have been concerned primarily with the behavior of solid objects. We now come to the study of *fluids,* substances that do not have rigid structure or form. In this category we identify two different states of matter, namely, *liquids* and *gases*. The basic physical characteristic that distinguishes solids, liquids, and gases is the way in which the molecules of the materials interact with one another. In solids the intermolecular forces are strong and the molecules are held tightly together so that, on average, there is little or no relative motion of the molecules. The forces are weaker in liquids and the molecules readily slip and slide past each other; that is, liquids *flow*. In gases, on the other hand, the intermolecular forces are quite weak and the molecules have almost no interaction. Consequently, gas molecules are free to move in all directions, each molecule moving independently of the others.

In this chapter we will find that liquids and gases have many similar properties. Indeed, in any physical effect that depends only upon the *mobility* of the molecules in the substance (as, for example, in the transmittal of pressure to all parts of a container), liquids and gases behave in the same way. Differences arise when the *degree* of molecular mobility is important (as, for example, in the process of evaporation). As we discuss these two states of matter, we again must keep in mind the fact that all matter consists basically of atoms and molecules.

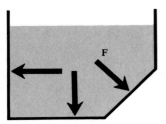

Figure 10-1   The force due to the pressure of a static fluid is always perpendicular to the surface on which it acts.

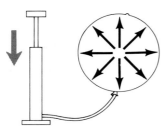

Figure 10-2   The force exerted by a gas on a balloon is everywhere perpendicular to the surface.

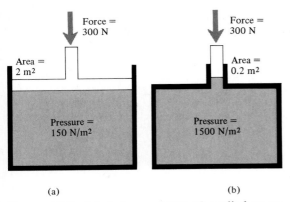

(a)   (b)

Figure 10-3   (a) A force of 300 N applied to an area of 2 m² results in a pressure of 150 N/m². (b) The same force applied to an area of 0.2 m² results in a pressure of 1500 N/m².

10-1 PRESSURE

Force per Unit Area

Fluids can exert forces. The water in a lake exerts a force on the bottom, and the atmosphere exerts a force on the Earth. How can we conveniently describe these forces? Because fluids are mobile substances and deform themselves to conform to the shape of any container, there is no unique point of contact between a fluid and the surface on which it exerts a force. Each small part of the surface receives its share of the total force exerted by the fluid. We can describe this situation in terms of the *force per unit area* exerted by a fluid on a surface. This we call the *pressure:*

$$\text{pressure} = \frac{\text{force}}{\text{area}}$$

$$\boxed{P = \frac{F}{A}} \qquad\qquad (10\text{-}1)$$

The dimensions of pressure are *newtons per square meter* (N/m²).

Any force — in particular, the force exerted by a fluid — is a vector quantity. In what direction does the force exerted by fluid pressure act? If the fluid is not in motion, the force vector is always *perpendicular* to the surface on which it acts, as shown in Fig. 10-1. If the force were not perpendicular, there would be a component of the force parallel to the surface. Some internal rigidity is required for a substance or object to exert a parallel force on another object (Can you see why this is so?), but a fluid has no such rigidity and therefore cannot exert a parallel force. Figure 10-1 shows the case of a liquid in a container, but the situation is the same for a gas in a container such as the balloon shown in Fig. 10-2.

Pascal's Principle

Suppose we have a cylinder that holds a certain quantity of gas. The top surface of the cylinder is a sliding piston with an area of 2 m² (Fig. 10-3a). If a downward force of 300 N is exerted on the piston, the resulting pressure on the piston due to the compressed gas is

$$P = \frac{300 \text{ N}}{2 \text{ m}^2} = 150 \text{ N/m}^2$$

If the area of the piston were smaller, the same force would cause the pressure to be even larger. As shown in Fig. 10-3b, a force of 300 N applied to an area of 0.2 m² produces a pressure of 1500 N/m².

We calculate the pressure on the piston by dividing the force exerted on the piston by the area of the piston. But there is no difference between the piston surface and any other part of the container. If the pressure exerted on the piston by the gas is P, this same pressure is exerted on every other part of the container.

This important idea was first formulated by the French mathematician and scientist, Blaise Pascal (1623–1662). *Pascal's principle* can be stated in the following way:

> A pressure applied to any portion of the surface of a confined fluid is transmitted undiminished to all points within the fluid.

Thus, in Fig. 10-3b, every part of the container surface experiences a pressure of 1500 N/m².

Figure 10-4 illustrates the application of Pascal's principle in a rock crusher. A fluid is confined in a vessel with two arms; each arm is sealed with a tight piston. The left arm has a small area (0.002 m²). A force of 50 N applied to the piston in this arm produces a pressure of 25 000 N/m². This pressure is transmitted to the arm at the right which has an area of 4 m². The force exerted by the piston on the rock is

$$F = P \times A$$
$$= (25\ 000\ \text{N/m}^2) \times (4\ \text{m}^2)$$
$$= 10^5\ \text{N}$$

Thus, by exerting a force of only 50 N on the small piston, a rock-crushing force is produced on the large piston.

Pressure within Static Fluids

What pressure does a fluid at rest exert on the bottom of a container in which the fluid stands at a height h (Fig. 10-5)? First, we know that the *force* exerted on the bottom of the container is simply the gravitational force or *weight* of the fluid, $F = mg$. The volume of the fluid is $V = Ah$, and if the density is ρ, the mass is $m = \rho Ah$. Therefore,

$$F = mg = (\rho Ah)g = \rho gh \times A$$

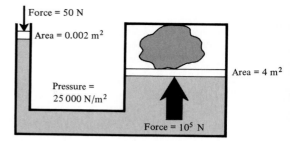

Figure 10-4 The application of Pascal's principle to a rock crusher.

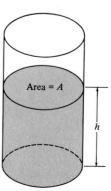

Figure 10-5 The pressure of the fluid on the bottom of the container is ρgh.

and the pressure is

$$P = \frac{F}{A} = \rho g h \qquad (10\text{-}2)$$

What is the pressure at the bottom of a lake that is 200 m deep?

$$P = \rho g h = (10^3 \text{ kg/m}^3) \times (9.8 \text{ m/s}^2) \times (200 \text{ m})$$
$$= 1.96 \times 10^6 \text{ N/m}^2$$

Notice that it was necessary to express the density of water as 10^3 kg/m³ (instead of 1 g/cm³) in order to obtain the pressure in N/m².

Atmospheric Pressure

The Earth's atmosphere exerts a pressure over the entire surface of the Earth. Under normal atmospheric conditions the pressure at sea level amounts to 1.013×10^5 N/m². This pressure is called 1 *atmosphere* (atm):

$$1 \text{ atm} = 1.013 \times 10^5 \text{ N/m}^2 \qquad (10\text{-}3)$$

The value of the atmospheric pressure is equal to the *weight* of the column of air above one square meter of the Earth's surface. Because the acceleration due to gravity g does not vary greatly over the extent of the Earth's atmosphere, we can compute the mass of the air above 1 m² of the Earth's surface by solving the weight equation, $w = mg$, for the mass m:

$$m = \frac{w}{g} = \frac{1.0 \times 10^5 \text{ N}}{9.8 \text{ m/s}^2}$$
$$\cong 10^4 \text{ kg}$$

This is equivalent to approximately 10 tons of air above every square meter of the Earth's surface! The reason we are not crushed by the weight of this huge air mass is that the air within our bodies exerts an outward pressure on the body tissues equal to the inward pressure of the atmosphere.

Measuring Atmospheric Pressure

A barometer (Fig. 10-6) is a device for measuring atmospheric pressure. A long glass tube, sealed at one end, is filled with mercury; the tube is then inverted and the open end placed in a reservoir of mercury. A vacuum space develops at the top (closed) end of the tube. The downward pressure of the atmosphere on the surface of

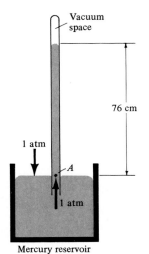

Figure 10-6 A mercury barometer. A pressure of 1 atm will support a column of mercury 76 cm high.

the mercury reservoir is 1 atm, and this pressure is transmitted to the base of the mercury column. There is no downward pressure on the top of the mercury column because there is vacuum over the column. We have an equilibrium situation. At point A, the downward force due to the weight of the mercury column is exactly balanced by the force of atmospheric pressure. (Can you see why?) That is, at point A, the pressure due to the mercury column is exactly 1 atm. If we solve Eq. 10-2 for h, we can compute the height of the mercury column:

$$h = \frac{P}{\rho g} = \frac{1.013 \times 10^5 \text{ N/m}^2}{(1.36 \times 10^4 \text{ kg/m}^3) \times (9.8 \text{ m/s}^2)}$$
$$= 0.760 \text{ m} = 76.0 \text{ cm}$$

That is, a pressure of 1 atm will support a column of mercury 76.0 cm high. Deviations of the height of the column from the normal value of 76.0 cm reflect changes in the atmospheric pressure due to local weather conditions.

Because local conditions can change, the usual practice in reporting scientific data is to correct the results so that they correspond to "normal" or "standard" conditions, namely, a pressure of 1 atm and a temperature of 0 °C. We frequently see the abbreviations NTP (meaning "normal temperature and pressure") or STP (meaning "standard temperature and pressure"); the two designations are equivalent.

10-2 BUOYANCY

Archimedes' Principle

Suppose that you are standing in a swimming pool with the water level at your chest or shoulders. If you now jump up, you will readily notice that as you settle downward you do so very gently. When your feet again touch the bottom, the sensation is far less than if you were to perform the same exercise out of the water. The reason the impact on the bottom of the pool is so small is that the water exerts a *buoyant force* on your body. This force is upward and therefore cancels a portion of the normal downward force due to gravity (your normal *weight*). In fact, if we choose to call your "weight" the net downward force on your body, then this weight is smaller when you are in water than when you are completely out of the water. This is always true. Any object

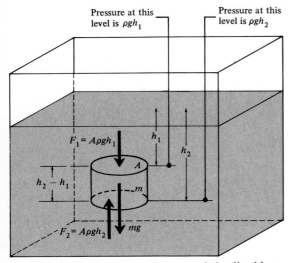

$F_1 = A\rho g h_1$

h_1

h_2

$h_2 - h_1$

A

m

$F_2 = A\rho g h_2$

mg

Figure 10-7 The buoyant force of the liquid on the object is equal to the weight of the liquid displaced by the object.

will have a smaller weight when immersed in water or any other fluid than when in the normal condition in air.

We can calculate the buoyant force on an object in the following way. Consider the cylindrical object, shown in Fig. 10-7, which is immersed in a tank containing a liquid that has a density ρ. (The fact that we consider an object with a regular geometrical shape is of no consequence; the argument can be extended to an object with *any* shape.) What are the forces acting on this object? There is no horizontal motion, so there can be no net horizontal force. (All of the horizontal forces sum to zero.) Therefore, we can confine our attention to the vertical forces. First, there is the downward force due to gravity, mg. Next, there is a downward force due to the liquid pressure on the top surface of the cylinder. This surface is at a depth h_1 in the liquid, and so the pressure is $\rho g h_1$ (Eq. 10-2). The area of the cylinder top is A; therefore the force is

$$F_1 = A\rho g h_1$$

Finally, there is an upward force due to the liquid pressure on the bottom surface of the cylinder. This surface is at a depth h_2 in the liquid, and so the pressure is $\rho g h_2$. The corresponding force is

$$F_2 = A\rho g h_2$$

The net downward force (that is, the *weight*) is

$$\begin{aligned} w &= mg + F_1 - F_2 \\ &= mg - (F_2 - F_1) \\ &= mg - A\rho g(h_2 - h_1) \end{aligned}$$

Notice that $A(h_2 - h_1)$ is the volume V of the cylinder and that ρV is the mass M of an equal volume of the liquid. The product $A\rho g(h_2 - h_1) = Mg$ is the weight w_l of the liquid that has been displaced by the cylinder. The expression for the weight of the cylinder in the liquid now becomes

$$w = mg - w_l \qquad (10\text{-}5)$$

The quantity w_l is the amount by which the normal weight mg of the cylinder has been decreased because of the buoyant effect of the liquid. Thus,

$$\begin{aligned} \text{buoyant force} = w_l &= V\rho g \\ &= \text{weight of liquid displaced} \\ &\qquad \text{by the object} \qquad (10\text{-}6) \end{aligned}$$

The idea that an object immersed in a fluid is buoyed up by a force equal to the weight of the displaced fluid is called *Archimedes' principle*.

Look again at Eq. 10-5. What will happen if the buoyant force w_l is larger than mg? This will be the case whenever the mass m of the object is smaller than the mass M of an equal volume of the fluid. This means that the density of the object is less than the density of the fluid. In such a case the net downward force on the object (that is, w) will be negative; in other words, the net force will be *upward*. Consequently, the object will rise in the fluid. When the object begins to emerge from the surface, the volume of fluid displaced by the object decreases and the buoyant force also decreases. The object will rise to a level such that the buoyant force exactly equals the downward force due to gravity. Then, in this condition of equilibrium, the object *floats*, with part of its volume submerged and part above the surface. The smaller the density of the object, the larger will be the fraction of its volume that is exposed. If the density of the object is only slightly less than the density of the fluid, only a small part of the volume will appear above the surface. This is the case for icebergs whose density is approximately 0.9 times that of sea water. Therefore, an iceberg floats with only 10 percent of its volume exposed. (It is for this reason we say that we see only the "tip of the iceberg.")

The Measurement of Density

We know that the density of an object is equal to its mass divided by its volume. If the object has an irregular shape, this will not affect the measurement of its *mass* but it will make extremely difficult a determination of its *volume*. We can use the idea of buoyant forces to assist us in measuring the density of such an object. We need to make two measurements: we weigh the object when it is immersed in a fluid of known density (we will use water, with $\rho = 1$ g/cm³), and we weigh the object in air (Fig. 10-8). Our measurements give us values for w_1 (the weight immersed) and w_2 (the weight in air):

$$w_1 = mg - w_l, \qquad w_2 = mg$$

Combining these expressions, we have

$$w_l = w_2 - w_1$$

The buoyant force w_l is equal to $V\rho_w g$, as we found in Eq. 10-6. (We will use ρ_w for the density of water and ρ_o for the density of the object.) Thus,

$$V\rho_w g = w_2 - w_1$$

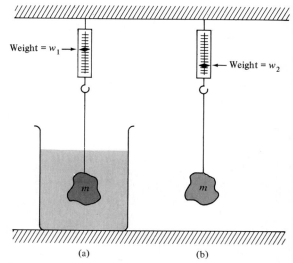

Figure 10-8 The density of the object can be determined by measuring its weight when immersed (a) in water and (b) in air.

from which

$$Vg = \frac{w_2 - w_1}{\rho_w}$$

Now the density of the object is

$$\rho_o = \frac{m}{V} = \frac{mg}{Vg} = \frac{w_2}{(w_2 - w_1)/\rho_w}$$

so that we have, finally,

$$\rho_o = \frac{w_2}{w_2 - w_1} \rho_w \qquad (10\text{-}7)$$

Suppose that we weigh (in air) an irregular piece of some metal and find $w_2 = 20.35$ N. Next, we suspend the metal in water and weight it again, as in Fig. 10-8a, finding $w_1 = 12.90$ N. What is the density of the metal? Using Eq. 10-7, we immediately write

$$\rho_o = \frac{20.35 \text{ N}}{20.35 \text{ N} - 12.90 \text{ N}} \times (1 \text{ g/cm}^3)$$
$$= 2.73 \text{ g/cm}^3$$

and we identify the metal as aluminum (see Table 1-9). Notice that we obtained the density without determining explicitly either the mass or the volume. Notice also that we did not bother to express the density of water in kg/m³ because we were satisfied to have the result for ρ_o expressed in g/cm³.

It is important to understand that the effects discussed in this section and in the preceding section pertain equally well to gases and liquids. (For example, the buoyant force on a balloon is equal to the weight of the displaced air.) In the following sections we will concentrate on some of the important properties of *gases*.

10-3 THE GAS LAWS

Boyle's Law

In 1662, the Irish chemist Robert Boyle (1627–1691) studied the relationship between the pressure and the volume of a confined gas. Boyle placed a certain quantity of gas in a cylinder and closed the cylinder with a tight-fitting piston. He could measure the force exerted on the piston and, hence, he could determine the pressure of the gas in the cylinder; and he could measure the volume of the gas at various pressures (see Fig. 10-9). Boyle discovered that the pressure and the volume are related in a particularly simple way. With the tempera-

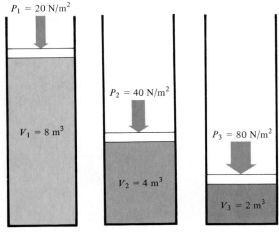

Figure 10-9 Robert Boyle discovered the relationship that connects the volume and the pressure of a confined gas at constant temperature: $PV =$ constant.

ture T held constant, Boyle found that the product of pressure and volume always remains constant. That is,

$$PV = \text{constant} \qquad \text{(when } T \text{ is constant)} \qquad (10\text{-}8)$$

This relationship is called *Boyle's law.*

For a series of three different conditions such as those shown in Fig. 10-9, we have

$$P_1V_1 = P_2V_2 = P_3V_3 \qquad \text{(when } T \text{ is constant)} \qquad (10\text{-}8a)$$

In order to *halve* the volume of a gas, the pressure must be *doubled,* and so forth.

For a particular sample of gas, the "constant" in $PV =$ constant is the same for all pressure–volume combinations. But for a different quantity of gas, the "constant" will have a different value.

The Law of Charles and Gay-Lussac

An extension of Boyle's law was made in 1802 by the French physicists Jacques Charles (1746–1823) and Joseph Louis Gay-Lussac (1778–1850) who, independently of one another, discovered the way in which the volume of a gas varies with temperature for constant pressure.

In order to see the way in which the volume and temperature of a gas are related, let us examine a specific case. Suppose that we confine a certain quantity of gas in a cylinder. We adjust the conditions until the temperature is 0 °C and the volume of the gas is 273 cm³ (see Fig. 10-10). For all of the remaining operations, we maintain the *same pressure.* If we lower the temperature

---

Robert Boyle (1627–1691). Boyle, the fourteenth child of the first Earl of Cork, was a child prodigy. At the age of eight he entered Eton and at eleven he toured Europe. Because he believed in the power of experiment as opposed to unsupported theory, Boyle is often regarded as the father of modern chemistry. He was the first to give a clear definition of a chemical element, but he believed that all matter consisted of atoms (not of molecular units). Boyle was the first to demonstrate that air is necessary for the transmission of sound. He lived simply throughout his life and was one of the original fellows of the Royal Society.

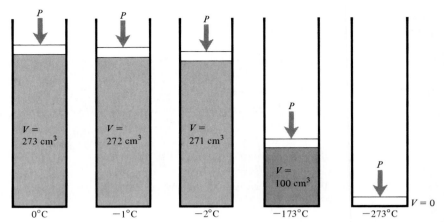

Figure 10-10   For constant pressure, the volume of a gas decreases at the same rate that the temperature is lowered. The temperature $T = -273\,°C$ represents "absolute zero."

to $-1\,°C$, we find that the volume of the gas decreases to 272 cm³. At a temperature of $-2\,°C$, the volume is 271 cm³. That is, for each centigrade degree that the temperature is lowered, the volume decreases by 1 cm³ or $\frac{1}{273}$ of the original volume. When $T = 173\,°C$ is reached, we find $V = 100$ cm³. If the gas continues to behave in this way, we would expect that a temperature of $-273\,°C$ would result in *zero* volume. We cannot actually reach this extreme condition because (a) any real gas would liquefy at a temperature above $-273\,°C$ and (b) the temperature $-273\,°C$ can never be attained in any real system. In spite of these practical difficulties, the temperature $T = -273\,°C$ has an important significance because it represents the lowest temperature that is *conceivable* (even if it cannot actually be *attained*).

The above results show that when the temperature is lowered from $0\,°C$ to $-1\,°C$ (that is, by $\frac{1}{273}$ of the range from $0\,°C$ to $-273\,°C$), the volume changes by $\frac{1}{273}$. In other words, at constant pressure, *the fractional change in temperature is the same as the fractional change in volume*. This means that a graph of volume versus temperature is a *straight line* (Fig. 10-11). Furthermore, the straight line, when extended to $T = -273\,°C$ corresponds to $V = 0$. Because $T = -273\,°C$ represents the "absolute zero" of temperature, we can use a temperature scale which sets $T = 0°$ at this point. This is just the *absolute* or *Kelvin* temperature scale discussed in Section 9-1. The size of the degree on this scale (deg K) is the same as the centigrade degree (deg C). Therefore, the point $0\,°C$ corresponds to $273\,°K$, and, in general, the absolute temperature (T_K) is

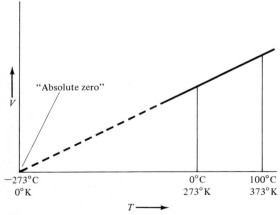

Figure 10-11   The variation of volume with temperature at constant pressure is a straight line, indicating that $V/T_K = $ constant. The point $T = -273\,°C$ is the "absolute zero" of temperature.

given by

$$T_K = T_C + 273° \qquad (10\text{-}9)$$

where T_C is the centigrade temperature.

Using the absolute temperature, we can now write a simple relation that represents both the measurements given above and the graph of Fig. 10-11. We have

$$V \propto T_K$$

or

$$V = (\text{constant}) \times T_K$$

which can be written as

$$\boxed{\frac{V}{T_K} = \text{constant} \qquad (\text{when } P \text{ is constant})} \qquad (10\text{-}10)$$

This is the law discovered by Charles and by Gay-Lussac. Notice the following important point regarding this law. We want an expression that states: "When the temperature is doubled (at constant pressure), the volume also doubles." If we give the temperature in degrees centigrade (or in degrees Fahrenheit), the statement is not true—the volume does not double if we increase the temperature from 10 °C to 20 °C. Equation 10-10 is valid only if we express the temperature in *degrees absolute*.

The Ideal Gas Law

The laws of Boyle and of Charles and Gay-Lussac can be combined into a single equation,

$$\boxed{\frac{PV}{T_K} = \text{constant}} \qquad (10\text{-}11)$$

When the temperature is held constant, Eq. 10-11 reduces to $PV = \text{constant}$, which is Boyle's law. And when the pressure is held constant, we have $V/T_K = \text{constant}$, which is the Charles–Gay-Lussac law.

The Boyle and the Charles–Gay-Lussac gas laws, and therefore also Eq. 10-11, are only approximately correct. Nevertheless, they are reasonably accurate descriptions of the way in which any real gas behaves at ordinary pressures and at temperatures above the liquefaction temperature (see Table 10-1). Equation 10-11 is therefore strictly valid only for an imaginary ideal gas; accordingly, this expression is called the *ideal gas law*.

TABLE 10-1 LIQUEFACTION TEMPERATURES FOR SOME GASES AT STANDARD PRESSURE

GAS	T_K (°K)
Chlorine (Cl_2)	238.6
Xenon (Xe)	166.1
Krypton (Kr)	120.8
Oxygen (O_2)	90.2
Argon (Ar)	87.5
Nitrogen (N_2)	77.4
Hydrogen (H_2)	20.7
Helium (He)	4.2

For a series of different conditions, the ideal gas law can be written in the form

$$\frac{P_1 V_1}{T_{K,1}} = \frac{P_2 V_2}{T_{K,2}} = \frac{P_3 V_3}{T_{K,3}} \qquad (10\text{-}11a)$$

Suppose, for example, that we have 3 m^3 of a certain gas at a pressure of 2 atm and at a temperature of 300 °K. If we compress the gas to one-third of its original volume ($V_2 = 1$ m^3) and at the same time increase the pressure to $P_2 = 9$ atm, what will be the final temperature? Solving Eq. 10-11a for $T_{K,2}$, we have

$$\begin{aligned} T_{K,2} &= \frac{P_2 V_2}{P_1 V_1} \times T_{K,1} \\ &= \frac{(9 \text{ atm}) \times (1 \text{ m}^3)}{(2 \text{ atm}) \times (3 \text{ m}^3)} \times (300 \text{ °K}) \\ &= 450 \text{ °K} \end{aligned}$$

(Notice that because we are taking the *ratio* of two pressures, we are permitted to express the pressures in atm instead of N/m^2. Would it also be permissible to express the volumes in liters instead of m^3?)

10-4 KINETIC THEORY

Boyle's Law

The idea that the properties of gases can be explained in terms of the microscopic actions of component molecules developed slowly through the 18th and 19th centuries. As early as 1738, the Swiss mathematician Daniel Bernoulli (1700–1782) pointed out that an explanation of Boyle's law, $PV = $ constant, can be given by appealing to the motion of gas particles (which we now know to be molecules). Bernoulli argued that the pressure exerted by a gas on the walls of a container is due to the repeated impacts of gas molecules on the walls. If the volume of a confined gas is decreased, the reduction in the space available to the molecules means that the molecules will collide with the walls more frequently. If the volume is *halved,* the number of times that each molecule will strike the walls per second will be *doubled.* Thus, when the volume is *decreased* by a certain factor, the pressure will *increase* by exactly the same factor. The result is that the product of pressure and volume, PV, remains constant. Bernoulli's simple argument therefore leads directly to the statement of Boyle's law.

A Microscopic Description of Gases

Bernoulli's explanation of Boyle's law in terms of molecular impacts represented the first crude attempt to deal with the properties of gases using a theory based on the microscopic constituents of gases. In the 19th century this idea was expanded upon and developed into a complete theory of the behavior of gases. In this theory—which is called *kinetic theory*—we use Newton's laws of dynamics to describe the way in which gas molecules interact with the walls of the container that encloses a gas.

In order to sketch the theory, we must first inquire into the connection between temperature and the motion of the molecules in a gas. What do we expect for the velocity of a gas molecule in a typical situation? We know that when *sound* is propagated through air, sound energy is transmitted by the successive collisions of air molecules. The speed with which sound travels should therefore be related to the speed of the air molecules. We cannot expect that the speeds are exactly the same, but the speed of sound should be indicative of the magnitudes of molecular velocities with which we must deal. Under normal atmospheric conditions the speed of sound in air is easily measured to be 330 m/s. We therefore expect to find molecular velocities in gases at room temperature that are similar to this value. (Actually, the speed with which sound is propagated must be *less* than the speed of the gas molecules. Can you see why?)

Can we assume that in a gas sample at a particular temperature *every* molecule will have the same velocity? Suppose this were the case. As the molecules move about, they collide with one another. In many of these encounters, one molecule will be speeded up and the other will be slowed down (but with energy always conserved!). Therefore, a situation will rapidly develop in which the velocities of the gas molecules will be distributed over a range. A few of the molecules will have very small velocities and a few will have very large velocities; most will have velocities with intermediate values.

If we perform an experiment to measure the velocities of the molecules in a sample of a gas at a particular temperature, we find, as expected, a range of velocities for the gas molecules. If we plot the number of molecules with the same velocity versus the velocity, we obtain a *velocity distribution* curve, as shown in Fig. 10-12. Repeating the experiment with different gas temperatures,

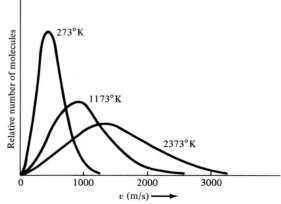

Figure 10-12 Relative numbers of nitrogen molecules with various velocities for three gas temperatures. The average velocity increases with temperature.

TABLE 10-2 AVERAGE VELOCITIES OF NITROGEN MOLECULES AT VARIOUS TEMPERATURES

TEMPERATURE (°K)	AVERAGE VELOCITY (m/s)
273	492
373	575
1173	1020
1773	1255
2373	1450

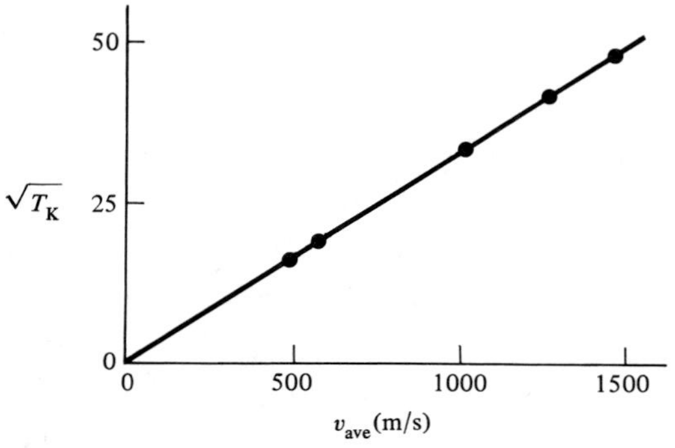

Figure 10-13 The average molecular velocity v_{ave} varies *linearly* with $\sqrt{T_K}$. That is, $v_{ave} \propto \sqrt{T_K}$.

we find that the velocity distribution curve changes shape and that the peak of the curve shifts to higher velocities as the temperature is raised. From each of these curves we can obtain the *average velocity* of the molecules. For the case of nitrogen gas, we find the average velocities given in Table 10-2. We note that the average velocity for standard temperature (0 °C or 273 °K) is 492 m/s, somewhat greater than the speed of sound, as we predicted. A velocity of 492 m/s is almost 1000 mi/h! But a molecule does not travel very far in a gas before its direction of motion is changed by a collision with another molecule. This is why the speed of sound is less than the molecular velocity.

How are we to interpret these results for the average molecular velocities? If we plot the average velocity v_{ave} versus the *square root* of the absolute temperature T_K, we find that all of the points fall on a straight line (Fig. 10-13). This is an important result and tells us that the average velocity is directly proportional to $\sqrt{T_K}$:

$$v_{ave} \propto \sqrt{T_K} \tag{10-12}$$

If we perform the same experiment with other gases, we find that the average velocity depends on the molecular mass M of the gas. For example, at a temperature of 273 °K, the average velocity of a nitrogen molecule ($M = 28$ AMU) is 492 m/s, but the average velocity of a hydrogen molecule ($M = 2$ AMU) is 1840 m/s. By studying several different types of molecules, we are able to conclude that v_{ave} is inversely proportional to $\sqrt{M}$:

$$v_{ave} \propto \frac{1}{\sqrt{M}} \tag{10-13}$$

Combining Eqs. 10-12 and 10-13, we have

$$\boxed{v_{ave} \propto \sqrt{\frac{T_K}{M}}} \tag{10-14}$$

If we square this result, we obtain $Mv_{ave}^2 \propto T_K$. Because this is a proportionality, we can supply a factor of $\frac{1}{2}$ and write

$$\tfrac{1}{2}Mv_{ave}^2 \propto T_K \tag{10-15}$$

Identifying $\frac{1}{2}Mv_{ave}^2$ as the *average kinetic energy* ($\overline{\text{K.E.}}$), we reach the conclusion that the temperature of a gas is directly proportional to the average kinetic energy of the gas molecules:

$$\boxed{\overline{\text{K.E.}} \propto T_K} \tag{10-16}$$

The Significance of Temperature in the Kinetic Theory

From the experiments on molecular velocities, we can draw several important conclusions:

(a) The *average velocity* of gas molecules (as well as the shape of the velocity distribution curve) depends on the *temperature* of the gas sample. Moreover, the average velocity of a particular gas depends *only* on the temperature. Changing the pressure or the volume of the sample does not influence the average molecular velocity as long as the temperature remains constant.

(b) Different types of gas molecules at the same temperature will have *different* average velocities, depending on the molecular masses. The lighter molecules will have the higher velocities and the heavier molecules will have the lower velocities. If the gas sample is a mixture of different molecular types, each individual type will maintain its own particular average velocity. In air, for example, the nitrogen and oxygen molecules have different average velocities. Nitrogen has $M = 28$ AMU and oxygen has $M = 32$ AMU; at 0 °C, v_{ave} (nitrogen) = 492 m/s and v_{ave} (oxygen) = 460 m/s.

(c) The *average kinetic energy* of the molecules in a gas sample is directly proportional to the *absolute temperature*. Although the average velocity of the molecules in a gas at a particular temperature depends on the molecular mass, the average kinetic energy does not. *All* gas molecules, regardless of type, have the same average kinetic energy at the same temperature. In a mixture of hydrogen and nitrogen gases, the average velocity of the hydrogen molecules will be considerably higher than the average velocity of the nitrogen molecules, but the average kinetic energy of the two molecular species will be exactly the same.

Kinetic Theory and the Ideal Gas Law

We can now complete the sketch of the kinetic theory by examining in more detail Bernoulli's idea that the pressure of a gas is due to the impacts of the gas molecules on the walls of the container. The result of this analysis, when combined with the previous result regarding kinetic energy and temperature, yields the relationship connecting pressure, volume, and temperature. We will then have a description of the ideal gas law according to the kinetic theory.

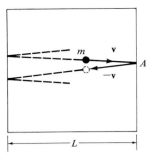

Figure 10-14 When a gas molecule strikes the wall of a container, it exerts a force on the wall.

The essential features of the situation can be seen in the simplified case shown in Fig. 10-14. A gas molecule of mass m is incident on a wall of the container with a velocity $\mathbf{v}$. The impact occurs at point A and the molecule rebounds with a velocity $-\mathbf{v}$. (The molecule loses no kinetic energy in the collision and so the initial speed is equal to the final speed.) The *change* in velocity is, therefore, $v - (-v) = 2v$. The time required for the molecule to travel between the opposite walls of the container is L/v. One collision with a wall occurs for every transit across the container. Consequently, the molecule experiences an average *acceleration* which is equal to the change in velocity per unit time:

$$\text{acceleration} = a = \frac{\Delta v}{\Delta t} = \frac{2v}{L/v} = \frac{2v^2}{L} \quad (10\text{-}17)$$

The average *force* exerted on the wall is equal to the mass of the molecule multiplied by the acceleration:

$$F = ma = \frac{2mv^2}{L} \quad (10\text{-}18)$$

and the *pressure* is equal to the force per unit area. If the container shown in Fig. 10-14 is cubical, the area of each wall is L^2. Therefore,

$$P = \frac{F}{A} = \frac{2mv^2/L}{L^2} = \frac{2mv^2}{L^3} \quad (10\text{-}19)$$

But the *volume* of the container is $V = L^3$, and Eq. 10-19 can be written as

$$P = \frac{2mv^2}{V} = \frac{4 \times (\frac{1}{2}mv^2)}{V} \quad (10\text{-}20)$$

We have already established that the average kinetic energy (which in this simplified situation is $\frac{1}{2}mv^2$) is proportional to the temperature. Therefore, Eq. 10-20 can be expressed as

$$P \propto \frac{\overline{\text{K.E.}}}{V} \propto \frac{T_K}{V} \quad (10\text{-}21)$$

or

$$\frac{PV}{T_K} = \text{constant} \quad (10\text{-}22)$$

which is the ideal gas law equation.

We see, therefore, that the consequences of the idea that a gas consists of rapidly moving molecules are in full accord with the experimental results of Boyle and of

Charles and Gay-Lussac expressed in the ideal gas law. It is rather remarkable that a theory so simply constructed is capable of producing a result that so closely corresponds to the behavior of real gases. The success of the kinetic theory constitutes one of the strongest links in the chain of reasoning that has led to the modern molecular theory of matter.

Improvements in the Kinetic Theory

Although the ideal gas law equation closely describes the behavior of real gases, it is really only an approximate law and deviations from its predictions are observed for all real gases. In order to bring kinetic theory into better agreement with experimental results, several modifications in the theory have been made. One of the crucial assumptions in kinetic theory (which has been implicit in our discussions) is that the molecules do not interact with one another. But all molecules consist of distributions of electric charge and when two such distributions come into close proximity, there will be an interaction through the electric force. Because molecules are complex assemblies of electric charge instead of simple point charges, the electric force that exists between two molecules is likewise complex. The description of these attractive intermolecular electric forces was developed by the Dutch physicist Johannes van der Waals (1837–1923), and they are now known as *van der Waals forces*. (These forces are definitely *not* new fundamental forces; they represent only the particular forms that the electric force takes in the interaction of molecules of various types.)

The incorporation of van der Waals forces and other refinements into the kinetic theory has had the result that this theory is now the most accurate that we have for the description of the properties of bulk matter.

10-5 CHANGES OF STATE

Evaporation

All matter consists of molecules, and we have seen that the molecules of a gas are in continual, rapid motion. What about liquids — are the molecules in liquids also in motion? Indeed they are. But because the inter-

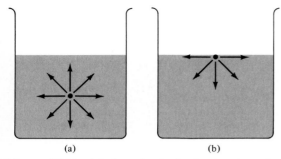

Figure 10-15 (a) A molecule in the interior of a liquid is acted upon by attractive van der Waals forces from all sides. (b) A molecule at the surface is prevented from evaporating by the predominantly downward forces unless it has an exceptionally high velocity.

molecular forces in liquids are greater than those in gases, the effects of molecular motions in liquids are influenced to a much greater degree by the intermolecular van der Waals forces.

Consider a liquid. If there were no van der Waals forces in operation, the molecules at the surface would be constrained in no way and because of their motion they would fly off into the surrounding air. The liquid would quickly dissipate and would, in fact, become a gas. But the existence of the van der Waals forces prevents this rapid dissipation and maintains a high density for the substance—the substance remains a *liquid*.

A molecule in the interior of a real liquid is acted upon by the attractive van der Waals forces due to the surrounding molecules (Fig. 10-15a). Although the molecule is in motion, the forces acting on all sides prevent the molecule from moving very rapidly away from its original position. But movement does occur, and the molecule is as likely to wander in one direction as in any other. A molecule at the surface (Fig. 10-15b), is also acted upon by van der Waals forces but none of these forces act *upward* because there are no liquid molecules above the surface layer. Consequently, a surface molecule is prevented from moving upward, separating from the liquid, and dissipating into the air *unless the molecule has a velocity significantly greater than the average velocity*. Thus, only those surface molecules that happen to be moving upward with exceptionally high velocities can overcome the downward attractive forces and escape from the liquid. This process by which a liquid is converted into a gaseous vapor is called *evaporation* or *vaporization*. Because relatively few of the molecules have sufficient velocities to escape, the evaporation of a liquid is generally a slow process.

Only the faster moving molecules leave a liquid during the process of evaporation, and so the residual liquid contains molecules with a lower average velocity. The temperature of an evaporating liquid is therefore lowered; that is, evaporation *cools* a liquid. This is a familiar phenomenon: on a hot day we perspire, and the evaporation of the perspiration cools our bodies.

Evaporation changes the state of a substance—from liquid to gas. Molecules can also escape from the surfaces of solids, but this process is exceedingly slow and, except for a few materials, the vaporization of solids at room temperature is immeasurably slow. One of the exceptional materials is solid carbon dioxide (*dry ice*) which passes directly from the solid to the gaseous state. This process is called *sublimation*. At temperatures below the normal freezing point (0 °C), ice will also sublime: water vapor will be formed but liquid water is never evident.

Vapor Pressure

In different liquids at the same temperature the molecules have different velocity distributions and the strengths of the van der Waals forces are different. Consequently, the rate of evaporation of different liquids, even under the same conditions, is generally different. A beaker of alcohol exposed to air, for example, will evaporate much more quickly than will a beaker of water (Fig. 10-16). Alcohol is more *volatile* than water.

What happens if we confine the vapors by placing covers over the beakers of alcohol and water? Evaporation will continue to take place, but now the molecules, instead of dissipating into the air, strike the cover and rebound toward the liquid (Fig. 10-17). If the enclosed volume above the liquid is not too great, a situation will rapidly develop in which the number of molecules leaving the surface per unit time is just equal to the number returning to the liquid. That is, an *equilibrium* condition is reached in which the *evaporation* rate is equal to the *condensation* rate.

In the equilibrium condition at any particular temperature, there will be a definite amount of vapor in every unit volume of the confined space above the liquid. In order to measure the amount of vapor, we first evacuate the confined space with a vacuum pump. When the pump is switched off, the evaporation process will reestablish the equilibrium. There are now no air molecules in the space, but because there is essentially no interaction among molecules in a gas, the number of vapor molecules per unit volume is the same as in the previous condition with the air present. We next determine the pressure in the enclosed volume by using some sort of pressure gauge (such as a mercury barometer). This pressure is called the *vapor pressure* of the liquid at the particular temperature. We can state the vapor pressure in newtons per square meter (N/m^2) or in terms of the number of centimeters of mercury that this pressure will support (cm Hg). At 20 °C, for example, the vapor pressure of water is 1.75 cm Hg whereas the vapor pressure of methyl alcohol (CH_3OH) is 9.4 cm Hg.

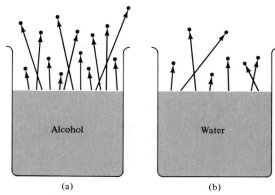

Figure 10-16 Because more alcohol molecules than water molecules escape from the surface per unit time, the evaporation rate of alcohol is greater than that of water.

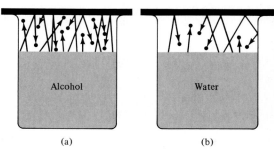

Figure 10-17 In an enclosed space, an equilibrium condition is reached in which the number of molecules escaping from the surface per unit time is equal to the number returning to the liquid. At the same temperature, the *vapor pressure* of alcohol (a) is greater than that of water (b).

TABLE 10-3 VAPOR PRESSURE OF WATER
AT VARIOUS TEMPERATURES

TEMPERATURE (°C)	VAPOR PRESSURE (cm Hg)
0	0.46
20	1.75
40	5.53
60	14.9
80	35.5
100	76.0

Humidity

On dry summer days it is possible to remain reasonably comfortable (even without air conditioning) in spite of a high temperature. The reason is that our perspiration readily evaporates and our bodies are thereby cooled. In fact, the evaporation may proceed so rapidly that we are not even aware of perspiring. On muggy days, however, a high temperature can be particularly oppressive; the evaporation rate is slow and we feel continually drenched. (It's not the heat, it's the humidity.) The difference between the two situations lies in the amount of water vapor in the air. If the amount of water vapor corresponds to the equilibrium vapor pressure at that temperature (for example, 1.75 cm Hg at 20 °C), the air is completely *saturated* and can absorb no additional water. This condition is called *100 percent humidity*. There is no net evaporation at 100 percent humidity. The situation is the same as that shown in Fig. 10-17b; just as many water molecules evaporate per second as are returned to the source from the saturated air.

At a temperature of 20 °C and at a pressure of 1 atm, each cubic meter of air contains approximately 2.5×10^{25} molecules. If the humidity is 100 percent, the air contains about 5.8×10^{22} water molecules per cubic meter. Fewer water molecules per cubic meter means that the humidity is less than 100 percent and the air can therefore absorb additional water. At a humidity of 50 percent, for example, the number of water molecules per cubic meter is 2.9×10^{22} and the air is capable of doubling its water content.

The vapor pressure of water (and all other liquids as well) increases with temperature (see Table 10-3). Therefore, as the temperature is raised, the capacity of air to hold water vapor increases. A humidity of 40 percent at 25 °C indicates a much higher level of water vapor in the air than the same humidity at 15 °C. Similarly, the humidity of air that is saturated at 20 °C will drop sharply if the temperature increases to 30 °C.

Boiling

If we supply heat to an open beaker of water, the temperature rises and vaporization from the surface takes place at an increasing rate. When a temperature of 100 °C is reached, the vapor pressure has increased to 76.0 cm Hg, which is normal atmospheric pressure (see

Table 10-3). Under these conditions the vaporization ceases to be a strictly surface phenomenon and now takes place throughout the volume of water. Because the vapor pressure of the water is equal to atmospheric pressure (and the water is also at this pressure), bubbles form within the liquid as the vapor pushes aside the water: that is, the water boils. As additional heat is supplied, the temperature does not increase further because this would imply an increase in the vapor pressure and the vapor pressure cannot exceed atmospheric pressure. Therefore, the boiling process continues to completion at *constant temperature*.

At normal atmospheric pressure, water boils at a temperature of 100 °C. If the pressure exceeds 1 atm, the vapor pressure of the water must be increased to this new higher pressure before boiling can occur. Thus, the boiling temperature is *increased*. Conversely, if the pressure is lowered, so is the boiling temperature. On high mountains, where the atmospheric pressure can be considerably lower than at sea level, the boiling temperature of water can be several degrees below 100 °C. Cooking times based on the temperature of boiling water must be lengthened at mountain elevations compared to sea level. At an elevation of 3400 feet, the normal atmospheric pressure is 68 cm Hg and water boils at 97 °C (207 °F).

Heat of Vaporization

When a liquid vaporizes, the escaping molecules have velocities that are higher than the average velocity of the liquid molecules. Therefore, the vaporization process removes energetic molecules and leaves behind the less energetic molecules. That is, vaporization causes the temperature of the liquid to *decrease* (if it is thermally insulated from its surroundings). Owing to the lower temperature, the vapor pressure will decrease and eventually the vaporization process will cease. In order to maintain the liquid at constant temperature during vaporization, heat must be continually supplied from an outside source. For example, the conversion of water at 100 °C to steam at the same temperature requires 540 Calories per kilogram. This quantity of heat is called the *heat of vaporization*.

Because there is no change in temperature in the process of converting water to steam at the boiling point, why is energy required? The answer is to be found by again examining the intermolecular forces in the two states. In the gaseous state the molecules are far apart and the van der Waals forces are weak. In the liquid state, however, the molecules are close together and the van der Waals forces are stronger. Therefore, to increase the average separation of the molecules, *work* must be done on the liquid. Thus, an input of energy is required to change the state of a substance even though there is no change in the temperature of the material.

Melting

The structures of the tightly packed molecules in solid materials take two forms. In *amorphous* materials, such as glass or plastic, the molecules, although close together, have no regular pattern. In a *crystalline* material, on the other hand, the molecules are joined together in a repeating array (see Section 19-4). Most pure substances that have well-defined molecular structures occur as crystals. For example, diamond, quartz, most metals, and many other materials are crystalline. (Glass is a mixture of substances, and the molecules of a plastic material, such as lucite, are not all the same — these substances are therefore amorphous.)

When an amorphous solid is heated, it softens gradually; there is no single temperature that can be said to separate the solid and liquid states. Crystals, however, have a sharp and well-defined *melting point*. For example, the melting point of ice (0 °C) is as well defined as is the boiling point of water (100 °C). In the same way that energy is required to separate the molecules of a liquid to produce a gas, energy must be expended to change the crystal structure of a solid into the mobile molecular system that is characteristic of a liquid. To transform 1 kg of ice at 0 °C into water at 0 °C requires 80 Cal. This amount of heat (80 Cal/kg) is called the *heat of fusion* of water. Conversely, when 1 kg of water freezes, 80 Cal of heat is liberated to and absorbed by the surroundings.

Figure 10-18 illustrates the energy relationships among the three states of matter. Notice that energy is *released* whenever matter in one phase condenses to a phase with a smaller molecular mobility (gas to liquid, liquid to solid, and gas to solid). An input of energy is required to make the phase transition proceed in the opposite direction.

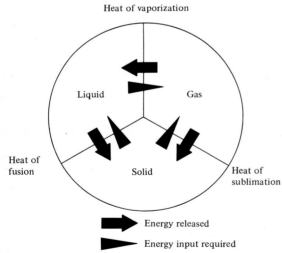

Figure 10-18　Energy flow in the changes of state of matter.

SUGGESTED READINGS

J. Tyndall, *Heat: A Mode of Motion* (Appleton, New York, 1915).

M. Zemansky, *Temperatures, Very Low and Very High* (D. Van Nostrand, Princeton, New Jersey, 1964).

Scientific American articles:

B. J. Alder and T. E. Wainwright, "Molecular Motion," October 1959.

M. B. Hall, "Robert Boyle," August 1967.

QUESTIONS AND EXERCISES

1. In Fig. 10-4, suppose that a force of 275 000 N is required to crush the rock. What force on the small piston is required?

2. The density of sea water is 1.025 g/cm³. What is the pressure at the bottom of an ocean trench that is 8.5 km below the surface? Express the result in N/m² and in atm.

3. The fluid pressure at the bottom of a cylinder filled with glycerin ($\rho = 1.26$ g/cm³) is 3.09×10^4 N/m². What is the height of the column of glycerin?

4. A mercury barometer, under standard atmospheric conditions, stands at a height of 76.0 cm. If the barometer were filled with oil instead of mercury, at what height would the oil stand? (The density of the oil is 0.8 g/cm³.)

5. What is the maximum height to which water could be raised from a well with a vacuum pump at the top of the well? If the well were 100 ft deep, how could water be pumped to the surface?

6. A certain irregular object is weighed first in air with the result, $w_{air} = 0.980$ N. Next, it is weighed while immersed in water: $w_{water} = 0.855$ N. Finally, the object is weighed while immersed in oil: $w_{oil} = 0.880$ N. (a) What is the density of the object? Identify the material. (b) What is the density of the oil?

7. A helium-filled, spherical research balloon has a diameter of 12 m. The total mass of the balloon, helium, gondola, and payload, is 800 kg. Will this balloon "get off the ground"? (The density of air is 1.293 kg/m³.)

8. A cubical block of wood ($\rho = 0.72$ g/cm³) has a mass of 250 kg. If this block is placed in a pool of water, what will be the height of the top surface above the water?

9. Suppose you are attempting to make a precision measurement of the mass of an object by comparing with a standard mass using a beam balance (Fig. 4-8). If the volume of the standard mass is significantly different from that of the object, will it be necessary to make any correction to your result? Calculate the effect in a particular case. Let the standard be 1 kg of iron and let the object be made of aluminum.

10. What is the Fahrenheit temperature of absolute zero?

11. The temperature of a confined gas is held constant. When the volume is changed, what effect does this have on the *density* of the gas?

12. When pumping a tire with a bicycle pump, the cylinder of the pump becomes hot. Why does this happen? (There are *two* reasons.)

13. A 6-liter volume of gas is at a pressure of 15 atm and a temperature of 27 °C. If the temperature is increased by 200 C deg while the volume is increased to 45 liters, what will be the new pressure?

14. Gas is stored in a steel container at a pressure of 300 atm and a temperature of 27 °C. If the bursting pressure of the tank is 450 atm, at what temperature will the tank become unsafe?

15. A quart of water is placed in a 10-gallon can and the can is heated over a flame until most of the water has boiled away. The top of the can is then screwed tightly into place. The can is removed from the flame and allowed to cool. Explain in detail what will happen.

16. Consider the following model of a gas: the gas is composed of particles at rest; the particles exert repulsive forces on one another and these forces are inversely proportional to the distance between particles. Using this model, what features of gases can be explained? (This model is due to Newton but he did not seriously propose it as a model for real gases.)

17. Examine a coffee percolator and sketch its construction. Using the diagram, explain the operation of the percolator.

18. Use the fact that the boiling temperature of water varies with pressure to explain the operation of a *pressure cooker*. Should any precautions be exercised in using such a utensil?

19. A quantity of gas is maintained at constant pressure. If the volume is increased by a factor of 4, how is the average velocity of the molecules affected?

20. The amount of *lift* that an airplane wing produces (at a particular forward speed and angle of attack) depends on the density of the air through which it moves. Do you expect an airplane to operate more efficiently in summer or winter, at the same atmospheric pressure? Explain.

21. Why is it unwise (perhaps even dangerous) to inflate your automobile tires to a high pressure before beginning a long, high-speed trip, especially during the summer?

22. A gas at a certain temperature is a mixture of the following: nitrogen (N_2), water vapor (H_2O), carbon dioxide (CO_2), argon (Ar), and helium (He). Order these gases according to *decreasing* average velocity.

23. At what temperature will the average velocity of a nitrogen molecule be equal to that of a hydrogen molecule at 0 °C?

24. What is the average velocity of helium atoms when the gas is at a temperature of 373 °K? (Use the information in Table 10-2 for nitrogen molecules.)

25. Explain why a simple fan makes one feel cooler on a hot day. Does a fan have an effect if the humidity is 100 percent?

26. Explain in detail why boiling takes place at constant temperature.

27. Make a graph of the data in Table 10-3. Estimate the boiling point of water at the top of Mt. Everest where the normal atmospheric pressure is 24 cm Hg.

28. A 5-kg quantity of steam at 100 °C condenses to water at the same temperature. How much heat is liberated in this process?

29. How much heat is required to convert 20 kg of water at 40 °C into steam at 100 °C?

30. How much energy is released when 2 kg of steam at 100 °C is converted into ice at 0 °C?

31. Suppose that your air conditioner breaks down and you decide to cool your kitchen by leaving open the refrigerator door. What will happen and why?

11

ELECTRICITY

How many things can you think of that require electricity for their operation? The list is almost endless. Truly, electricity drives the modern world. Compare the mode of life today with that of only a hundred years ago. Almost every feature of our life style has been altered through the widespread use of electricity and electrical devices. Transportation, communications, entertainment, manufacturing, even agriculture and medicine, have been profoundly affected by electrically operated machines, instruments, and gadgets. Almost every day sees the introduction of some new electrical device that allows us to work more efficiently or enhances our comfort.

How does electricity *work?* Where does electrical energy come from and where does it go? Is there any danger in using electricity? In this chapter we will discuss these and other questions. We will present the basic ideas behind the electrical concepts of voltage, current, resistance, and power. And we will see how electricity operates in several everyday situations. In the next several chapters we will enlarge the discussions to include electric and magnetic fields, electromagnetic waves, and electromagnetic radiation. With these ideas in mind, we will then be able to understand the way in which electrical phenomena influence the behavior of matter in the microscopic domain—in molecules, atoms, and nuclei.

11-1 ELECTRONS, IONS, AND CURRENT

The Flow of Electric Charges

One of the most interesting and important classes of bulk matter is the group of materials that are conductors of electricity. A *conductor* is any material through which electric charges can be made to move with relative ease. All substances will conduct electricity to some extent, but those through which it is difficult to force the movement of electric charges are classified as *insulators* (see Section 6-5).

The best conductors are metals, particularly silver, copper, gold, and aluminum. In these materials the movement of electric charge is in the form of free electrons which can flow readily through the substance. Liquids can also be conductors, and in these materials the movement of electric charge is in the form of *ions,* not electrons. When salt (sodium chloride, NaCl) is dissolved in water, for example, the compound dissociates into positive and negative ions, Na^+ and Cl^-. Because these ions can move through the water, the solution (an *ionic* solution) is actually a good conductor.

Under certain circumstances, gases can become conductors. (We know, for example, that there is a violent flow of electric charge in a lightning stroke.) Gases usually consist of electrically neutral atoms and molecules. But if an electron is separated from an atom or molecule, by the action of light or in the development of a lightning stroke, both the electron and the ion can move. In fluorescent lights and in neon signs the flow of electricity is in the form of moving electrons and ions.

The free electrons that exist in a metallic conductor constitute a kind of "electron gas" within the material. These electrons move freely through the substance with quite high speeds. Indeed, in a metal at normal room temperature the average electron speed is about 10^6 m/s! Similarly, ions are always in motion in an ionic solution, although the average speeds are much lower. (The movement of an ion is more sluggish than that of an electron because the mass of an ion is much greater than that of an electron.)

Under ordinary conditions the motion of electrons in a metal is completely *random,* just as the motion of atoms in a gas is random. If we consider the electron movement in a metal wire and focus attention on the electrons traversing a particular cross section of the wire (Fig. 11-1), we find that there are just as many electrons moving to the *right* through the area as there are moving to the left. That is, there is *no net flow* of electrons through the area. If we connect the ends of the wire to a battery, forming a closed electrical *circuit,* the situation is altered substantially. Now the electrons are attracted toward the positive terminal of the battery and are repelled from the negative terminal. As a result, there is a net movement of electrons through any cross-sectional area of the conductor (from right to left in Fig. 11-2). That is, a *current* flows in the wire.

Figure 11-1 In an isolated piece of conductor, just as many electrons move to the right through a particular cross-sectional area as move to the left.

The Direction of Current Flow

When a wire is connected to the terminals of a battery, the electrons move away from the negative terminal and toward the positive terminal. Thus, the direction of *electron* flow is from negative to positive. However, by convention we define *current flow* (as distinguished from *electron flow*) as moving from *positive to negative* (see Fig. 11-2). That is, electric current flows in a wire in the same direction that a *positive* charge would move. (Actually, the positive charges in the wire — the atomic nuclei — do not move; only the electrons move. In ionic solutions and in gases, however, there is movement of both positive and negative charges.) This is merely a *convention* that is usually followed; the *physics* does not depend on which direction we elect to say that the current is flowing. The movement of negative charge (electrons) to the *left* in Fig. 11-2 is entirely equivalent to the movement of an equal amount of positive charge to the *right*.

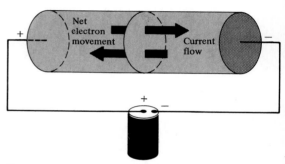

Figure 11-2 When a battery is connected to a conductor, there is a net movement of electrons toward the positive terminal of the battery. By convention, the direction of current flow is opposite to the motion of the electrons.

André-Marie Ampère (1775–1836), in whose honor the unit of electrical current is named. Ampère overcame the shock of seeing his father executed by the Jacobins during the French Revolution and the handicap of severe financial hardships to become professor of mathematics at the Polytechnic School of Paris and then professor of physics at the College de France. Upon hearing in 1820 that Oersted in Denmark had discovered a connection between electricity and magnetism, Ampère quickly carried out a series of experiments and outlined the theory required to explain the phenomenon. By 1825 he had formulated a complete theoretical description of the way in which magnets interact with electrical currents. This work entitles Ampère to be called the father of modern electrodynamics.

Whenever reference is made to "current flow," it will always mean that the direction is from positive to negative. If it is intended to refer to the motion of the electrons, the term "electron flow" will be used.

The Ampere

The measure of electric current is the amount of charge that passes a given point per unit time. If 1 coulomb (C) of charge passes a point in 1 second, the current is defined to be 1 *ampere* (A). Thus,

$$\text{current } (I) = \frac{\text{charge } (q)}{\text{time } (t)}; \qquad I = \frac{q}{t}$$
$$1 \text{ ampere} = \frac{1 \text{ coulomb}}{\text{second}} \qquad (11\text{-}1)$$

An electron carries a charge of 1.6×10^{-19} C; therefore, the number of electrons required to total 1 C is $1/(1.6 \times 10^{-19}) = 6 \times 10^{18}$. This number of electrons must pass a given point each second for the current to be 1 A. But in a typical conductor, such as copper, there are approximately 10^{23} free electrons per cubic centimeter. Consequently, only a very small net speed is required in order that 6×10^{18} electrons pass a given point each second. Ordinary household electrical wire is usually rated for a current of 15 A; at this current the electron drift speed is only about 1 mm/s.

Although a small electron drift speed can give rise to a large current, it must be remembered that this movement is in addition to the random motion that continues with speeds of about 10^6 m/s even when a current is flowing.

Because electrons drift so slowly in a conductor, how is it possible to deliver electricity from *here* to *there* in a reasonable time? When you turn a light switch "on," the light immediately glows—there is no delay as the electrons move slowly along the wires. The reason is that electric *current* is quite unlike *sound*. The propagation of sound from one point to another takes place by molecules banging into each other at successive points along the transmission path. The speed of sound is therefore limited by molecular speeds (see Section 10-4). But when an electrical switch is closed, an *electric field* (see Section 12-1) is propagated through the circuit with the speed of light. This field exerts a force on the electrons and causes them to move practically instantaneously *throughout* the circuit. The electrons do collide with the atoms in the wire and this gives rise to electrical *resis-*

tance (as we will discuss later in this chapter), but this does not alter the fact that all of the electrons begin to move along the wire at the same time. Thus, any device in the circuit that requires current for its operation is activated as soon as the switch is closed.

11-2 VOLTAGE, WORK, AND POWER

Potential Difference and the Volt

Electrons and ions are set into motion and give rise to electric currents because of the action of electric forces. Let us see how we can provide a measure for the tendency of electric charges to move in various situations. Suppose that we have two objects that carry opposite charges, as shown in Fig. 11-3. A positive charge q located at point A will be attracted toward the negatively charged object and repelled by the positively charged object. Therefore, to move the positive charge q from A to B requires that work be done. If $q = 1$ coulomb (C) and if the work required for the movement is $W = 1$ joule (J), we say that the *potential difference* between points A and B is 1 *volt* (V). That is, the potential difference between two points is a measure of the *work per unit charge* required to move a charge from one point to the other. Thus,

$$\text{potential difference} \atop \text{or voltage, } V = \frac{\text{work (W)}}{\text{charge } (q)}$$

$$V = \frac{W}{q} \qquad (11\text{-}2)$$

and the definition of the volt is

$$1 \text{ volt} = \frac{1 \text{ joule}}{\text{coulomb}} \qquad (11\text{-}3)$$

If a potential difference V exists between points A and B in Fig. 11-3, then an amount of work $W = qV$ must be done to move the positive charge q from A to B. Notice also that if the charge moves from B to A, this amount of work can be recovered. For example, if q were released at point B, it would be accelerated toward the negatively charged object and upon reaching point A it would have a kinetic energy equal to qV.

One way to produce a potential difference between two points is through the use of a *battery* (see Section 11-4). We can view the action of a battery in the following way. A battery is a source of *potential energy*. When

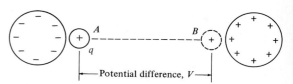

Figure 11-3 If 1 joule of work is required to move 1 coulomb of positive charge from A to B, the *potential difference* between the two points is 1 volt.

Batteries are often thought of as environmentally "clean"—they produce energy with no noxious fumes or dirty smoke. (Of course, energy was used to manufacture the batteries and this energy *did* involve noxious fumes and dirty smoke. And there is the problem of what to do with batteries when they are worn out. Batteries are not "biodegradable"; those that contain toxic materials, such as compounds of mercury, are particularly troublesome.) In spite of at least some attractive features, batteries do not represent a practical source of large amounts of energy—batteries are one of the most expensive forms of energy in common use!

the battery does work in a circuit, this potential energy decreases. The decrease in potential energy when 1 coulomb of charge is transferred through a circuit from the positive terminal to the negative terminal is related to the *potential difference* between the battery's terminals. If the potential energy decrease is 1 joule for 1 coulomb of charge, then, according to Eq. 11-3, the potential difference (or the *voltage*) of the battery is 1 volt.

Chemical reactions within a battery produce the energy and provide the force that causes current to flow. We call this force the *electromotive force* or EMF. In an electric generator, an EMF is produced by fields that change with time (see Section 12-6).

Batteries with various voltages are used in a wide variety of applications. Ordinary flashlight batteries have a potential difference between their terminals of 1.5 V. Automobile batteries are either 6 V or 12 V. The small batteries used mainly for portable radios are 9-V batteries. The voltage rating of a battery is not indicative of the amount of electrical energy stored in the battery: a 6-V automobile battery can deliver a much greater amount of energy than a 9-V radio battery.

Electric Power

Electrical energy is energy in its most useful form. Electrical energy can be conveniently transported over large distances through wires and, at its destination, can readily be converted into mechanical energy by means of motors, into heat with space heaters and ovens, or into light by means of light bulbs. Whenever we operate electrical devices, we are billed by the power company according to how much electrical energy has been used. It is therefore important to inquire how we determine the *rate* at which electrical energy is used by a device.

GENERAL ELECTRIC

A high-voltage test facility at the General Electric Company. The long columns of insulators separate the points of high voltage from one another and from ground.

TABLE 11-1 THE COST OF USING APPLIANCES

APPLIANCE	ESTIMATED AVERAGE ANNUAL kWh USED[a]	ANNUAL COST (based on 3¢ per kWh)
Air-conditioner (window)	1389	$ 41.67
Electric blanket	147	4.41
Carving knife	8	0.24
Clock	17	0.51
Clothes dryer	993	29.79
Clothes washer (automatic)	103	3.09
Dishwasher	363	10.89
Hair dryer	14	0.42
Humidifier	163	4.89
Range	1175	35.25
Refrigerator-freezer (frostless, 14 cu. ft.)	1829	54.87
Shaver	18	0.54
Television (color)	502	15.06
Toaster	39	1.17
Toothbrush	5	0.15
Vacuum cleaner	46	1.38
Water heater (quick recovery)	4811	144.33

[a] From *Changing Times, The Kiplinger Magazine,* November 1972.

According to Eq. 11-3, 1 joule of energy is required to drive 1 coulomb of charge through a potential difference of 1 volt; that is,

$$1 \text{ J} = (1 \text{ V}) \times (1 \text{ C})$$

In order to express the *rate* at which energy is consumed, we divide this equation by the *time:*

$$1 \frac{\text{J}}{\text{s}} = (1 \text{ V}) \times \left(1 \frac{\text{C}}{\text{s}}\right)$$

On the right-hand side, we can identify 1 C/s as 1 A (Eq. 11-1). Furthermore, 1 J/s, being energy per unit time, is *power* and is equal to 1 watt (W), as defined in Eq. 7-8. Therefore, we have

$$1 \text{ W} = (1 \text{ V}) \times (1 \text{ A}) \qquad (11\text{-}4)$$

or,

$$1 \text{ watt} = 1 \text{ volt-ampere}$$

If an electrical device requires a current of I amperes at a potential difference of V volts, the power consumption is P watts:

$$P = VI \qquad (11\text{-}5)$$

For example, a certain household electrical heater operates from a 120-volt line by drawing a current of 8.3

amperes. The power consumption is

$$P = (120 \text{ V}) \times (8.3 \text{ A}) = 1000 \text{ W} = 1 \text{ kW}$$

The total amount of *energy* used is obtained simply by multiplying the rate of energy use (that is, the *power*) by the time of use. (Refer to Section 7-3.) If the 8.3-A device in the previous paragraph were operated for 1 h, the energy used (or, equivalently, the *work* done) would be

$$W = Pt = (1000 \text{ W}) \times (1 \text{ h})$$
$$= \left(1000 \frac{\text{J}}{\text{s}}\right) \times (3600 \text{ s})$$
$$= 3.6 \times 10^6 \text{ J}$$

or,

$$W = (1 \text{ kW}) \times (1 \text{ h}) = 1 \text{ kWh}$$

The kilowatt-hour (kWh) is the usual unit by which electrical energy is sold. The price per kWh varies considerably, depending on the distance of the consumer from the power plant, the cost of fuel used by the plant, and the quantity of energy used. A large user near a hydroelectric plant will generally pay the lowest rate; most household consumers pay about 3 cents per kilowatt-hour.

11-3 ELECTRICAL RESISTANCE

Ohm's Law

How much current will flow through a particular wire if we place a voltage V across the ends? We can see the answer to this question if we take advantage of the analogy between the flow of electrons through a wire and the flow of water through a pipe. Suppose that we have a container with a spout near its bottom. If we place some water in this container and then exert a force F on a piston at the top, water will flow from the spout at a certain rate (Fig. 11-4a). If the diameter of the spout is decreased, the flow will be restricted (Fig. 11-4b). That is, the smaller spout offers a greater *resistance* to the flow of water. Or, if the length of the original spout is increased, the flow rate will also be decreased because friction in the longer pipe makes it more difficult to force the water through the pipe (Fig. 11-4c). The resistance of a pipe to fluid flow increases in direct proportion to its length and in inverse proportion to its cross-sectional area.

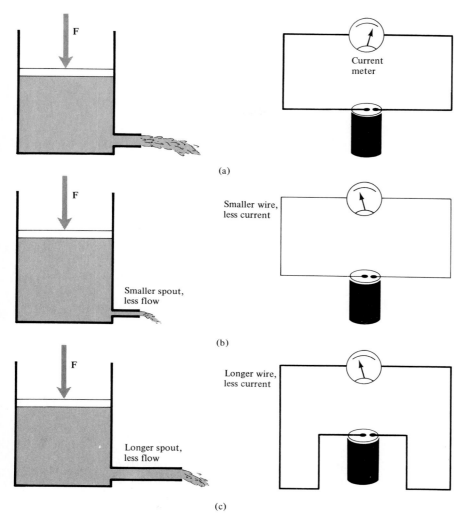

Figure 11-4 Hydraulic and electrical systems behave in similar ways. If the size of the spout (wire) is reduced or if the length is increased, the flow of water (electrical current) is reduced.

The behavior of electric current flow in a wire is exactly the same, with regard to size and length, as the flow of water in a pipe. The *electrical resistance* of a wire increases as the diameter is made smaller or as the length is increased (right-hand portion of Fig. 11-4). If a current of 1 A flows in a wire that is connected to a source of EMF with a potential difference of 1 V, the wire is said to have a resistance R of 1 *ohm*. (The symbol for *ohm* is a capital Greek omega, Ω.)

In 1826, the German physicist George Simon Ohm (1787–1854) discovered that the amount of current flowing in a circuit is directly proportional to the EMF in the circuit and inversely proportional to the total resistance of the circuit; that is,

$$\text{current} = \frac{\text{voltage}}{\text{resistance}}$$

or,

$$\boxed{I = \frac{V}{R}} \qquad (11\text{-}6)$$

This result is known as *Ohm's law*.

What is the resistance of a light bulb which operates at a power level of 150 watts when connected to a 120-volt household line? To determine the current flow, we use Eq. 11-5:

$$I = \frac{P}{V} = \frac{150 \text{ W}}{120 \text{ V}} = 1.25 \text{ A}$$

Then, using Eq. 11-6 to find R,

$$R = \frac{V}{I} = \frac{120 \text{ V}}{1.25 \text{ A}} = 96 \ \Omega$$

Ohm's law is not a law of Nature in the sense that the law of universal gravitation is. Ohm's law is only an approximate description of the way in which many, but not all, materials behave in ordinary electrical circuits. When subjected to very high voltages, all materials will exhibit discrepancies with the simple equation expressing Ohm's law. Nevertheless, the law is sufficiently precise in most circumstances that it is regularly used in the design of all types of electrical circuits. But we must not lose sight of the fact that the predictions of Ohm's law are not valid in *all* situations.

The Resistance of Materials

Electrical *resistance* is an intrinsic property of matter, just as density and specific heat are intrinsic properties. Metals have the lowest resistance of all materials, but even within this group the resistance varies significantly. Generally, a metal wire with a cross-sectional area of 1 mm² and 100 m long will have a resistance of a few ohms (see Table 11-2). Silver is the best conductor of electric current (that is, it has the *lowest* resistance), with copper and gold not far behind. (In Table 11-2, notice the very high resistance of the nonmetal, carbon.) Because it is the least expensive of the good conductors,

TABLE 11-2 ELECTRICAL RESISTANCE OF VARIOUS METALS AT NORMAL ROOM TEMPERATURE

METAL	RESISTANCE (for a 1 mm² wire, 100 m long), Ω
Aluminum	2.8
Copper	1.8
Gold	2.4
Iron	10
Mercury	95.8
Nickel	7.8
Silver	1.6
Tungsten	5.6
For comparison:	
Carbon	3500

copper is almost universally used in the manufacture of wires and conducting cables. Owing to the fact that copper is sometimes in short supply and because the cost is increasing, aluminum is being used as a substitute in certain applications. The melting temperature of aluminum is much lower than that of copper and therefore aluminum can be used for electrical wiring only in low-temperature situations.

The electrical resistance of most materials increases with temperature. The reason is that the agitation of the atoms is more severe at higher temperatures and, therefore, the flowing electrons are more likely to collide with the atoms. These collisions slow the rate of drift of the electrons and decrease the current for a given applied EMF. An exception to this rule is the case of carbon. When this material is heated, the atoms become more agitated, but they also give up more electrons to the pool of free electrons. The net result is to *increase* the current flow as the temperature is raised. Increasing the temperature of a carbon rod from 0 to 2500 °C causes the electrical resistance to drop to one-quarter of its original value. Carbon rods are frequently used as electrodes in arc lamps, such as those in commercial movie projectors and in searchlights.

Because the resistance of a wire of a particular material depends on its size (cross-sectional area) and upon its length, we can use the information in Table 11-2 to find the resistance values for wires of different sizes and lengths. Table 11-2 shows that a copper wire with a cross-sectional area of 1 mm² and a length of 100 m will have a resistance of 1.8 Ω. If the length of the wire is increased to 200 m, the resistance will be *increased* to 3.6 Ω. Similarly, if the cross-sectional area is increased to 2 mm², the resistance of a 100-m length will be *decreased* to 0.9 Ω. We can express these statements as a proportionality:

$$\text{resistance} \propto \frac{\text{length}}{\text{area}} \quad \text{or} \quad \frac{\text{length}}{(\text{diameter})^2}$$

For a copper wire we can write

$$R = (1.8 \ \Omega) \times \frac{(\text{length}/100 \ \text{m})}{(\text{area}/1 \ \text{mm}^2)}$$

Thus, a copper wire 50 m in length and with a cross-sectional area of 0.6 mm² will have a resistance

$$R = (1.8 \ \Omega) \times \frac{(50 \ \text{m}/100 \ \text{m})}{(0.6 \ \text{mm}^2/1 \ \text{mm}^2)}$$

$$= 1.5 \ \Omega$$

A carbon arc. A current of 100 A or more flows through the gap between the carbon rods.

Electrical Heating

When current flows in a circuit, energy is expended in driving the current. Where does this energy go? If the circuit contains a light bulb, some of the energy appears as light. But an electric light bulb is very inefficient in converting electrical energy into light: most of the energy (about 98 percent) appears as *heat*. Similarly, an electric motor converts only a small fraction of the electrical energy into mechanical energy. As everyone knows, glowing light bulbs and running electric motors are *hot*.

When electrons flow through a wire, they collide frequently with the atoms of the wire material. These collisions cause the atoms, which are normally jiggling about, to be further agitated, thereby heating the wire and raising its temperature. If the temperature is increased sufficiently, some of the energy will be radiated away as light. Most electric light bulbs are constructed from thin tungsten wire which has a high resistance (and also remains structurally strong at high temperatures). The wire is quickly heated to incandescence when it is connected to a source of EMF.

The power that is expended in a circuit and which appears mainly as heat is given by Eq. 11-5. We can obtain another form for this expression by using Ohm's law (Eq. 11-6) and substituting IR for V:

$$P = VI = I^2R \tag{11-7}$$

Thus, we see that for a given resistance R, the power expended increases as the *square* of the current. A

TABLE 11-3 SUMMARY OF ELECTRICAL QUANTITIES

QUANTITY	DEFINITION	FORMULA	UNITS
Current (I)	$\dfrac{\text{charge } (q)}{\text{time } (t)}$	$I = \dfrac{q}{t}$	1 A = 1 C/s
Potential difference (V)	$\dfrac{\text{energy or work } (W)}{\text{charge } (q)}$	$V = \dfrac{W}{q}$	1 V = 1 J/C
Power (P)	$\dfrac{\text{energy or work } (W)}{\text{time } (t)}$	$P = \dfrac{W}{t}$ $= VI$ $= I^2R$	1 W = 1 J/s $= 1$ V-A
Resistance (R)	Ohm's law	$R = \dfrac{V}{I}$	R in ohms (Ω)

heating element will deliver 4 times the heat when 10 A flows through it compared to the heat output at 5 A. (Actually, the factor will be slightly greater than 4 because the resistance will increase somewhat with temperature.)

The definitions of electrical quantities and the corresponding formulas are summarized in Table 11-3.

Alternating Current

When a current I flows through a resistance R, such as the light bulb shown in the circuit of Fig. 11-5, the electrical energy expended is I^2R. The amount of energy delivered to the bulb, which appears in the form of light and heat, does not depend on the *direction* of the current flow. If we could arrange a switching system that would quickly reverse the direction of the current at regular intervals, we would see no difference in the light output of the bulb and would feel no difference in its heating effect. Current which reverses its direction in this way is called *alternating current* (AC); current which flows always in the same direction is called *direct current* (DC).

Electric *generators,* which convert mechanical energy into electrical energy, can be constructed to produce either AC or DC. And electric motors, which convert electrical energy into mechanical energy, can also be constructed to operate from AC or from DC lines. Essentially all of the electricity that is now generated for widespread commercial purposes is AC because alternating current can be transported large distances through wires with much less energy loss than can direct current. The reason is based on the properties of a device called a *transformer* (which will be discussed in more detail in Section 12-6). A transformer has the ability to raise or lower the voltage in a circuit while maintaining a constant value for the power (that is, the product VI). A power source that delivers 10 A at 100 V can be changed with a transformer to a 1-A source at 1000 V or to a 1000-A source at 1 V. Because the heating losses in a power line depend on I^2R, it is advantageous to reduce the current I to as small a value as practicable. (Not much can be done about lowering the resistance R of the power lines; however, see the discussion of superconductors in Section 19-5.) By using a transformer, the voltage of the power source can be *stepped up* to a very high value (often as high as

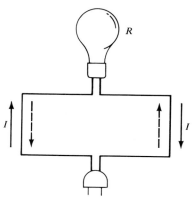

Figure 11-5 The energy (I^2R) delivered to the light bulb does not depend on the direction of current flow. Household current is AC (alternating current) and undergoes 60 complete cycles of reversal each second.

600 000 V) with a consequent lowering of the current necessary to transport energy at a given power level. At the consumer end of the line, the voltage is *stepped down* in a series of transformers to the 220-V or the 110-V level used in industry and households. At the lower voltages, higher currents can be drawn from the line to operate various electrical devices.

In the United States commercial electric current is reversed in direction 60 times each second (and is called *60-cycle AC*). In Europe the practice is to use 50-cycle AC. For specialized applications, other rates of reversal are used; for example, in the aircraft industry, electric generators usually operate at 400 cycles.

The Heating Effect of Alternating Current

Most generators that produce alternating current provide a current that varies *sinusoidally*. That is, a graph of current versus time has the same shape as a graph of the sine function versus angle. Figure 11-6a shows such a current–time graph. Here, *positive* values mean that the current flows in one direction, and *negative* values mean that the current flows in the opposite direction. Now, the heating effect of a current depends upon the *square* of the current, and this is always positive. Figure 11-6b shows a graph of (current)² versus time and corresponds to the current graph in Fig. 11-6a. Because the current varies with time, we must compute the heating effect by taking the average of (current)². As shown in Fig. 11-6b, this average value of (current)² is one-half of the peak value of (current)². What DC current would give the same heating effect as an AC current with a peak value of 1 A? This DC current must have a value when squared equal to $0.5 \times (1 \text{ A})^2$. Thus, the DC or *effective* AC current must be $(1 \text{ A})/\sqrt{2}$ or 0.707 A. In general, for alternating current,

$$I_{\text{eff}} = \frac{1}{\sqrt{2}} I_{\text{peak}} = 0.707 \, I_{\text{peak}} \qquad (11\text{-}8)$$

Suppose that an alternating current with a peak value of 2 A flows through a 6-Ω resistor. What is the power expended in the resistor?

$$P = I_{\text{eff}}^2 \, R = \left(\frac{1}{\sqrt{2}} I_{\text{peak}}\right)^2 R = \frac{1}{2} I_{\text{peak}}^2 \, R$$

$$= \frac{1}{2} \times (2 \text{ A})^2 \times (6 \, \Omega) = 12 \text{ W}$$

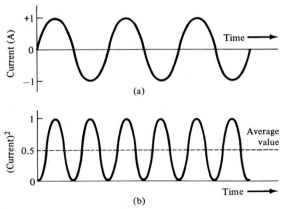

Figure 11-6 (a) Alternating current varies sinusoidally with time. (b) The average value of (current)² is one half of the peak value.

Bodily Resistance and
Electrical Shock

If you simultaneously touch both of the terminals of a small 9-V radio battery, you will feel practically nothing. However, if you touch the terminals with your tongue, you will feel the strong tingling sensation of a low-level electrical shock. Why is the bodily response so different in the two cases? The severity of an electrical shock depends on the amount of current that flows in the body, and this, in turn, depends on the magnitude of the EMF and on the electrical resistance of the body. Depending on the circumstances, the resistance of the body can be as low as a few hundred ohms or as high as a few hundred thousand ohms. If you touch the terminals of a 9-V battery with perfectly dry fingers, the bodily resistance will be about 100 000 Ω, so that a current of only 10^{-4} A will flow (10^{-4} A $= 0.1$ milliampere $= 0.1$ mA). Such a low current will not even be noticed. However, if your fingers are covered with perspiration, the resistance will be low—perhaps 1000 Ω—and the current can be as large as 10 mA. This is not a dangerous current, but you will know that you have touched a battery!

Dry skin is a good insulator and the resistance is sufficiently high to limit the current flow to a low value if the voltage is not too great. But if the skin is moist, the resistance is lowered because of the presence in the moisture of ions that easily conduct electrical current. This is particularly true if the moisture is perspiration because perspiration contains sodium chloride (salt) which forms ions when in solution. Moisture (with its ions) serves to promote the flow of current by providing a highly conductive path that is in intimate contact with both the skin and the electrical terminal.

If you happen to poke a finger into an electrical socket, you will receive a nasty shock. But if you happen to be standing in water or even on a damp floor, the resistance of the conducting path will be so low that the shock could be fatal. Many people are killed each year in just such accidents. It is therefore a good policy *never* to handle anything electrical (even appliances that are claimed to have "good" insulation) when in contact with any wet or moist surface or object.

Electrical current passing through the body attacks the central nervous system (which is the body's electrical network). In particular, an electrical shock will impair the nerve system that controls breathing. A person suffering a severe electrical shock will frequently suffocate. (The best emergency treatment for a shock victim is the application of artificial respiration.) Death will almost always result if a current of 0.1 A (100 mA) passes directly through the heart, and a current half as great will be fatal in some cases.

In order for current to pass through the body, there must be a potential difference between two parts of the body. Touching *one* terminal of a battery will produce no effect whatsoever (if there is no conducting path that connects with the other terminal). Similarly, if you were to jump off the ground and take hold of a sagging high-voltage wire, you would suffer no ill result. However, if you happened simultaneously to touch the ground or another wire at a different potential, the consequences would be most unpleasant (probably a job for your friendly mortician).

11-4   ELECTRIC CIRCUITS

Series Circuits

In order to understand the way in which various electrical systems operate, it is necessary to investigate the current flow in circuits that consist of more than a single element. The simplest type of circuit is the *series circuit,* shown in Fig. 11-7, in which three resistive elements (called *resistors* and indicated by sawtooth lines) are connected together in series. These resistors can represent any type of circuit component that has resistance: light bulbs, heating elements, or pieces of some material (such as carbon) designed to have specific resistance values. We assume that the lines connecting the battery and the resistors have negligible resistance compared to R_1, R_2, and R_3.

The potential difference between the battery terminals is V, and because the connecting wire is considered to have no appreciable resistance, the potential difference between the top of R_1 and the bottom of R_3 is also V. The *same* current I flows through each of the resistors. (Where else would the electrons go?) The magnitude of this current is determined by the voltage V and the *total* resistance R_t of the circuit:

$$I = \frac{V}{R_t} \tag{11-9}$$

What is the value of R_t? As the current flows, the electrons pass through the three resistors in turn. Each

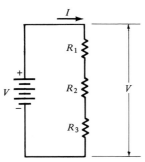

Figure 11-7   A *series* circuit consisting of three resistors connected to a battery. Notice that we have indicated the battery by a special symbol usually employed in circuit diagrams.

resistor contributes its own resistance to the circuit. The net effect is the same as if there were a single large resistor with a resistance equal to the sum, $R_1 + R_2 + R_3$. That is,

$$R_t = R_1 + R_2 + R_3 \quad \text{(series circuit)} \quad (11\text{-}10)$$

Parallel Circuits

Suppose that we connect the resistors of Fig. 11-7 in a *parallel* arrangement, as shown in Fig. 11-8. The basic equation, $I = V/R_t$, is still valid, but now R_t refers to the total resistance of the parallel circuit of resistors. The total current I that flows in the circuit is distributed among the three resistors. Because there are now three different paths for the current flow, the total resistance of the parallel circuit is *less* than the resistance of any one of the individual resistors.

The current flowing through any individual resistor is equal to the voltage across that resistor divided by its resistance value. But notice that the same voltage V is across each resistor. Therefore, we can write

$$I_1 = \frac{V}{R_1}; \qquad I_2 = \frac{V}{R_2}; \qquad I_3 = \frac{V}{R_3}$$

Next, we note that the total current I is equal to the sum of the currents through the individual resistors:

$$I = I_1 + I_2 + I_3$$

Substituting for each current,

$$\frac{V}{R_t} = \frac{V}{R_1} + \frac{V}{R_2} + \frac{V}{R_3}$$

Dividing this equation by V, we have, finally,

$$\frac{1}{R_t} = \frac{1}{R_1} + \frac{1}{R_2} + \frac{1}{R_3} \quad \text{(parallel circuit)} \quad (11\text{-}11)$$

The current flow through a parallel circuit is therefore given by

$$I = \frac{V}{R_t} = V\left(\frac{1}{R_1} + \frac{1}{R_2} + \frac{1}{R_3}\right) \quad (11\text{-}12)$$

To see how these equations for R_t are used, consider three 6-Ω resistors. If the resistors are connected in series, the total resistance will be

$$R_t \text{ (series)} = 6 + 6 + 6 = 18 \ \Omega$$

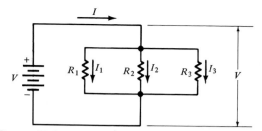

Figure 11-8   A *parallel* circuit consisting of three resistors connected to a battery.

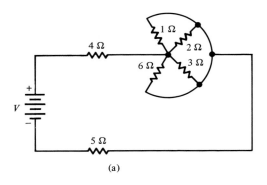

(a)

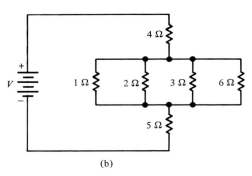

(b)

Figure 11-9 Two equivalent ways to represent the same series–parallel circuit.

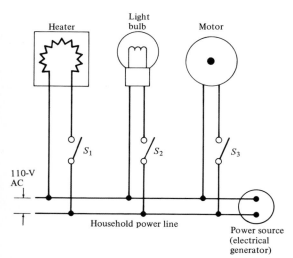

Figure 11-10 Three electrical devices are connected in parallel to a 110-V household power line. Each device is controlled by its own switch and each operates independently of the others.

But if the same resistors are connected in parallel, R_t will have a *smaller* value:

$$\frac{1}{R_t \text{ (parallel)}} = \frac{1}{6} + \frac{1}{6} + \frac{1}{6} = \frac{3}{6} = \frac{1}{2}$$

So that

$$R_t \text{ (parallel)} = 2 \ \Omega$$

Notice that this value of R_t is smaller than that of the individual resistors in the group (6 Ω).

Many circuits that appear to be complicated can actually be analyzed quite easily by breaking them down into combinations of simple series and parallel circuits. For example, consider the system of resistors shown in Fig. 11-9a. By redrawing this circuit in more conventional form, we have the arrangement shown in Fig. 11-9b. The first step in calculating the total resistance is to consider the group of four parallel resistors. The total resistance of this combination is

$$\frac{1}{R} = \frac{1}{1} + \frac{1}{2} + \frac{1}{3} + \frac{1}{6} = \frac{6 + 3 + 2 + 1}{6} = \frac{12}{6} = \frac{2}{1}$$

Therefore, $R = \frac{1}{2} \ \Omega$, and the four resistors can be replaced by an equivalent $\frac{1}{2}$-Ω resistor. We now have a simple series circuit consisting of resistors of 4 Ω, 0.5 Ω, and 5 Ω. Hence, the total resistance in the circuit is

$$R_t = 4 + 0.5 + 5 = 9.5 \ \Omega$$

Most household circuits involve parallel connections. Electrical power, from the power company's electrical generator, enters the home via two wires across which there is an AC voltage of approximately 110 V (peak). Individual household lights and appliances are connected in parallel to the 110-V line, as shown in Fig. 11-10. Here, three circuit elements—a heater, a light bulb, and a motor—are all connected to the same household power line. These devices are all connected in parallel and each device has its own switch. If switch S_1 is closed, the heater will be connected to the power line, but the other two devices will not be operating. The light bulb can be turned on only by closing switch S_2, and the motor can be made to operate only by closing switch S_3. The reason for connecting the devices in parallel is that light bulbs, heaters, motors, and other electrical appliances are designed to operate at a certain *voltage*. If they are connected in parallel, as in Fig. 11-10, the voltage across each device will always be

Three-Way Switches—How They Work

Have you ever wondered how the switches are wired to allow you to turn a stairway light off or on from either the top or bottom of the stairs? Ordinary household switches are used in this "three-way" system, and the diagrams in Fig. 11-11 show how they are connected. There are four possible combinations of switch positions. The first two diagrams show the way that the upstairs switch controls the light when the downstairs switch is in the "UP" position. The last two diagrams show the situation for the downstairs switch in the "DOWN" position. In each "ON" case, trace the flow of current from one side of the power line, through the switches and the light bulb, to the other side of the line. In each "OFF" case, notice that there is no complete path connecting the bulb to the line.

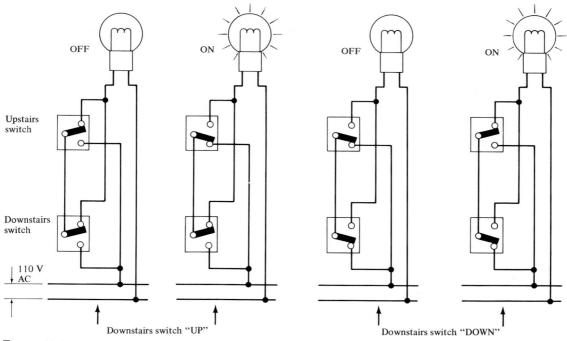

Figure 11-11

equal to the voltage of the power line (which the power company maintains at 110 volts).

Shorts and Overloads

The resistance of a short length of wire is so low that, for most purposes, it can be considered to be zero. If we connect points A and B in the series circuit of Fig.

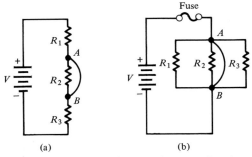

Figure 11-12　Two short circuits. (a) If points A and B in the series circuit are connected by a wire, the resistor R_2 is shorted out of the circuit. (b) If points A and B in the parallel circuit are connected by a wire, the *entire* circuit (R_1, R_2, and R_3) is shorted out. In such a case the current would become very large—until the fuse opens and protects the battery from overloading and burning out.

11-12a with a short wire, the wire will be in parallel with R_2. Because the wire has zero resistance, all of the current in the circuit will flow through the wire and none will flow through R_2. It is as if R_2 is not in the circuit at all—R_2 has been *shorted out*. With the wire in place, *more* current will flow in the circuit because the total resistance has been reduced from $R_1 + R_2 + R_3$ to $R_1 + R_3$.

In an actual series circuit, if one element is shorted out, the increase in current might be sufficient to damage the remaining elements. In a parallel circuit the consequences can be more severe. If the resistor R_2 in Fig. 11-12b is shorted out by connecting points A and B with a short wire, this is equivalent to a direct short across the terminals of the battery. If the circuit actually had *zero* resistance, an unlimited current would flow. But, in reality, there is always in the circuit some small resistance which limits the current to a very large (but not an infinite) value. Even so, the battery will quickly be drained of its capacity to deliver current—it will *burn out*. If a short occurs in a household circuit, a dangerous overload situation can develop. A battery will simply burn out as the result of a short, but the power company can continue to supply current even in the event of a short. It is therefore possible for a large current to flow in a shorted household circuit until the wires become sufficiently hot that they melt and break the circuit. This may not occur, however, before the hot wires have started a fire. For this reason, household circuits (as well as many other kinds of circuits) are protected with *fuses* (see Fig. 11-12b). In the event that a short occurs, the current will increase only to the rated current of the fuse. At this point the fuse (which is made from a material with a low melting point) will melt and open the circuit, thus preventing further current flow. (Most modern electrical systems contain *circuit breakers* instead of fuses. The function is the same, but a circuit breaker can be reset when the fault has been corrected, whereas a burned-out fuse must be replaced.)

Fuses and circuit breakers also protect circuits from *overloads*. Most individual household circuits (every house has several circuits) are designed to carry a maximum of 15 or 20 A. If too many appliances are connected to one circuit, the current flow can exceed the current rating of the circuit. In such a situation the fuse will burn out or the circuit breaker will open, thereby preventing damage.

11-5 ELECTRIC CURRENTS IN SOLUTIONS

Electrolysis

If salt (NaCl) is dissolved in water, we know that the compound will dissociate into sodium and chlorine ions, Na^+ and Cl^-. If we place into this ionic solution a pair of electrodes that are connected to the terminals of a battery, there will be a potential difference across the solution. The ions, which are mobile in the solution, will respond to this potential difference and a current will flow. The Na^+ ions will move toward the negative electrode and the Cl^- ions will move toward the positive electrode, as shown in Fig. 11-13.

What happens when the ions reach the electrodes? When a Na^+ ion reaches the negative electrode, the ion absorbs an electron from the supply of free electrons in the electrode material. The ion then becomes a neutral atom and is deposited on the electrode. The reaction at the negative electrode (called the *cathode*) is

$$Na^+ + e^- \longrightarrow Na \quad \text{(cathode)} \quad (11\text{-}13)$$

At the positive electrode (called the *anode*), a Cl^- ion gives up an electron to the electrode material. Two chlorine atoms combine to produce a molecule of gaseous chlorine which then bubbles away. The anode reaction is

$$2\,Cl^- - 2\,e^- \longrightarrow Cl_2 \quad \text{(anode)} \quad (11\text{-}14)$$

Notice how the current flows in the two parts of the circuit. In the solution, positive and negative ions constitute the current; free electrons do not move through the solution. In the external part of the circuit, electron movement *is* responsible for the current. These electrons are given up to the anode by the Cl^- ions, move around the external circuit, and are then taken up by the Na^+ ions. This process of the dissociation of molecules, the migration of ions, and the deposition or release of material at the electrodes is called *electrolysis*.

Pure water is not a good conductor. But when a small amount of sulfuric acid, H_2SO_4, is added to water, a complex system of conducting ions is formed. Passing a current through a sulfuric acid solution produces a series of ionic reactions which we can schematically represent in the following way. The basic current-carrying ion is H_3O^+ which is formed when a hydrogen ion,

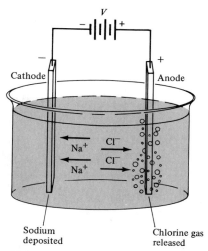

Figure 11-13 The motion of Na^+ and Cl^- ions causes current to flow through the sodium chloride solution. Sodium is deposited on the cathode and chlorine gas is released at the anode.

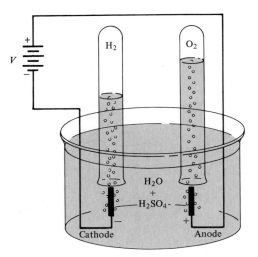

Figure 11-14 The electrolysis of water is accomplished by passing a current through a solution of water and sulfuric acid. Hydrogen gas is released and collected at the cathode; oxygen gas is released and collected at the anode.

H^+, from H_2SO_4 attaches itself to a water molecule:

$$H_2SO_4 + H_2O \longrightarrow H_3O^+ + HSO_4^- \quad (11\text{-}15)$$

The H_3O^+ ions (called *hydronium* ions) move toward the cathode where they pick up electrons to form H_2O and hydrogen:

$$H_3O^+ + e^- \longrightarrow H_2O + H \quad (11\text{-}16)$$

Two hydrogen atoms formed in this way combine to produce a hydrogen molecule, $H + H \rightarrow H_2$. Hydrogen gas is therefore liberated at the cathode (Fig. 11-14).

At the anode, the HSO_4^- ions interact with water molecules and give up electrons to the anode. This produces oxygen atoms, hydronium ions, and re-forms sulfuric acid which is then available to participate again in the reactions:

$$HSO_4^- + 2\ H_2O - 2\ e^- \longrightarrow$$
$$H_2SO_4 + H_3O^+ + O \quad (11\text{-}17)$$

Two oxygen atoms combine to form O_2, and oxygen gas is therefore liberated at the anode (Fig. 11-14).

Notice that the sulfuric acid participates only indirectly in the electrolysis of water by contributing to the formation of hydronium ions. The sulfuric acid itself is not decomposed in the process and by evaporating away the residual water, the original amount of acid could be recovered. Only water molecules are decomposed in this process. Electrolysis is one of the principal methods used for the commercial production of hydrogen. (Oxygen, which can be produced in other ways, is collected as a by-product.)

Electroplating

Many of the metallic materials that we see and use every day do not consist entirely of the metal on the surface. Certain types of metal products are given coatings of a different metal in order to improve their durability or their visual appeal. If iron is left exposed, water vapor in the air will cause corrosion which will weaken the iron and eventually destroy it. The metal chromium, however, resists corrosion. Therefore, iron and steel products (for example, automobile bumpers and trim) are frequently given protective coatings of "chrome." Dinnerware and eating utensils are often coated with silver (these products are called *silverplate*). Coatings of these types are produced in electrolysis reactions and the process is called *electroplating*. For example, suppose that we wish to plate a steel fork with silver. We suspend the fork in a solution of silver nitrate, $AgNO_3$, and connect it to the negative terminal of a battery (Fig. 11-15). Thus, the fork becomes the cathode. The anode is an electrode of pure silver. In solution, the silver nitrate dissociates into Ag^+ and NO_3^- ions. When the electrical circuit is completed, the Ag^+ ions move toward the cathode (the fork) where they pick up electrons, become neutral atoms, and are deposited. At the anode, silver atoms shed one electron each and go into solution as Ag^+ ions. The silver that is plated onto the fork is continually replenished by silver from the positive electrode, and the silver nitrate in solution is not depleted.

Some metals do not adhere well to certain other metals. In such cases it is necessary to provide an intermediate coating of a third metal. For example, if silver is deposited directly onto an iron surface, it will tend to peel off in time. Therefore, in order to silverplate an iron object, it is first necessary to give the object a coating of copper (which adheres well to iron) and then to plate the silver onto the copper.

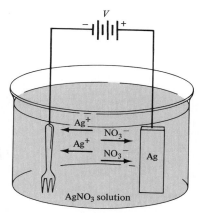

Figure 11-15   A solution of silver nitrate, $AgNO_3$, and a pure silver anode are used for silverplating objects.

Faraday's Law of Electrolysis

When a current is passed through an ionic solution such as that shown in Fig. 11-15, it is easy to see that the number of silver atoms deposited on the cathode depends on several factors. First, more atoms will be deposited per second if the current is high. And more atoms will be deposited if the current flows for a longer time. That is, the total number of atoms deposited depends on the product of *current* (I) and *time* (t). According to Eq. 11-1, the product $I \times t$ is equal to the *charge q*. Thus, the number of silver atoms plated onto the cathode is proportional to the total charge q that is transported through the electrolyte by the silver ions.

Suppose that we have two electrolytic cells, one in which the ions are Ag^+ and one in which the ions are Mg^{++}. If the same current is passed through each cell for the same period of time, how will the number of deposited atoms compare in the two cases? The silver ion, Ag^+, carries a single charge, but the magnesium ion, Mg^{++}, carries a double charge. Therefore, one electron must be given up by the cathode to neutralize and deposit each silver ion, but two electrons are required for each magnesium ion. Therefore, for equal amounts of charge transported through the solution, *twice* as many silver atoms will be deposited compared to magnesium atoms. Thus, the number of atoms deposited at the cathode is *inversely proportional* to the ionic charge of the particular ions. Altogether, we can state

$$\text{number of atoms deposited} \propto \frac{\text{total charge transported}}{\text{ionic charge}}$$

Usually, we are interested, not in the number of atoms, but in the *mass* of material deposited. Therefore, we write

$$\text{mass deposited} \propto \frac{(\text{charge}) \times (\text{atomic mass in AMU})}{\text{ionic charge}}$$

Using symbols and writing the proportionality factor as $1/F$, the equation for the mass is

$$m = \frac{q}{F} \times \frac{\text{atomic mass}}{\text{ionic charge}} \qquad (11\text{-}18)$$

This equation is the expression of *Faraday's law of electrolysis,* discovered in 1834 by Michael Faraday. The constant F in the equation is called the *faraday* in his honor. Although Eq. 11-18 is a straightforward result of the atomic and electric nature of matter, this was not known in Faraday's time. In order to discover that the electrodeposition process follows the law expressed by Eq. 11-18, Faraday made thousands of measurements, using different electrodes, different electrolytes, and different currents. Thus, even before electrons and ions were known, Faraday's exhaustive series of experiments had led to the correct description of electrolysis.

On the basis of his charge and mass measurements, Faraday obtained a direct experimental value for the constant F:

$$F = 96\ 500 \text{ coulombs/mole} \qquad (11\text{-}19)$$
$$= 1 \text{ faraday}$$

We can calculate the value of the faraday by using the values of the electron charge ($e = 1.602 \times 10^{19}$ C) and Avogadro's number ($N_0 = 6.024 \times 10^{23}$ atoms/mole). We ask the question: How much charge is required for the electrolytic deposition of one mole of a substance? If we consider an ion such as Ag^+ which carries a single charge, then one electron charge is required to deposit each atom. One mole of the material contains N_0 atoms. Therefore, the charge that is required to deposit one mole of material is equal to the number of atoms (N_0) multiplied by the charge (e) carried by each ion that becomes a deposited atom. This charge is 1 faraday:

$$1 \text{ faraday} = N_0\, e$$
$$= (6.024 \times 10^{23} \text{ atoms/mole})$$
$$\times\, (1.602 \times 10^{19} \text{ C/atom})$$
$$= 96\ 500 \text{ C/mole} \qquad (11\text{-}20)$$

In order to deposit one mole of a substance whose ionic charge is *two* (such as magnesium, Mg^{++}), a charge of 2 faradays is required, and so forth.

Suppose that we pass a current of 20 amperes through the electrolytic cell shown in Fig. 11-15 for a period of 1 hour. How much silver will be deposited on the cathode? The total amount of charge transported by the ions is

$$q = I \times t$$
$$= (20 \text{ A}) \times (3600 \text{ s})$$
$$= 72\ 000 \text{ C}$$

The atomic mass of a silver atom is 107.9 AMU, which means that silver has a mass of 107.9 grams per mole. Therefore, using Eq. 11-18, the mass of the deposited silver is

$$m = \frac{72\ 000 \text{ C}}{96\ 500 \text{ C/mole}} \times \frac{107.9/\text{mole}}{1}$$
$$= 80.5 \text{ g}$$

The Lead Storage Battery

A common way to store electrical energy is in the form of *batteries*. These devices are useful in a variety of applications requiring portability. They are generally quite reliable and will perform well if the energy requirement of the application is not large. When the energy stored in an ordinary flashlight or transistor radio battery is exhausted, it is not possible to recharge the battery and it must be discarded. Certain types of batteries, however, can be reactivated by passing electrical current through them in the direction opposite to the direction of the current delivered by the battery. The most common battery of this type is the lead storage battery used to start automobile and other internal-combustion engines.

A lead storage battery consists of a pair of lead plates immersed in a sulfuric acid solution (Fig. 11-16). The anode is a plate made from *spongy* lead, that is, lead which is highly porous so that there is a large surface area in contact with the acid solution. The cathode is also a lead plate but this plate contains many holes into which is pressed lead dioxide, PbO_2.

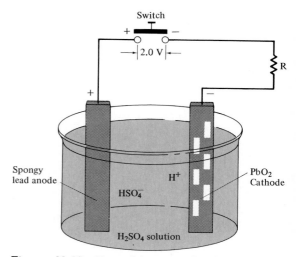

Figure 11-16 Essential parts of a lead storage battery. The voltage between the lead plates is 2.0 volts. Three or six such cells are connected together to make standard 6-V or 12-V batteries.

When the switch in the battery circuit (Fig. 11-16) is closed, reactions take place at the anode and the cathode; these reactions cause a current to flow through battery and through the load resistance R:

anode reaction:

$$Pb + HSO_4^- \longrightarrow PbSO_4 + H^+ + 2e^- \quad (11\text{-}21)$$

cathode reaction:

$$PbO_2 + HSO_4^- + 3H^+ + e^- \longrightarrow PbSO_4 + 2H_2O \quad (11\text{-}22)$$

(Simplified notation is used here. The positive ions are actually H_3O^+ instead of H^+. The net effect is the same.) If we add these two reactions together, we can write a single chemical equation which represents the overall process of extracting energy from the battery:

$$Pb + PbO_2 + 2\ HSO_4^- + 2\ H^+ \longrightarrow 2\ PbSO_4 + 2\ H_2O + energy \quad (11\text{-}23)$$

As the battery is discharged, the sulfuric acid is depleted and lead sulfate accumulates on the electrodes. The amount of sulfuric acid decreases and the amount of water increases until the acid solution is too weak to support additional reactions. The battery is then completely discharged.

The "charge" remaining in a lead storage battery can be determined by measuring the density of the acid solution. Sulfuric acid has a density of 1.8 g/cm³ and the acid solution in a fully charged battery has a density of approximately 1.28 g/cm³. As a battery is discharged, the amount of sulfuric acid in solution decreases and the density is lowered. When the battery is completely discharged, the density of the solution is approximately 1.13 g/cm³. Service station operators use a device (called a *hydrometer*) which sucks up some battery solution into a glass tube where the density can be read from a scale. If the density is below about 1.2 g/cm³, the battery needs recharging.

For liquids, the term *specific gravity* is sometimes used to indicate the density. The specific gravity of a liquid is the ratio of the density of the liquid to that of water. Because the density of water is 1 g/cm³, the specific gravity of a liquid is numerically equal to its density in g/cm³, but without any units attached. The specific gravity of sulfuric acid, for example, is 1.8.

By connecting the terminals of a discharged lead storage battery to a source of direct current which flows in the direction opposite to that of normal battery current flow, the battery can be recharged. The charging reaction is

$$2\ PbSO_4 + 2\ H_2O + energy \longrightarrow$$
$$Pb + PbO_2 + 2\ H_2SO_4 \quad (11\text{-}24)$$

When fully charged, the voltage between the terminals of a lead storage battery is 2.0 volts. Most automobile starters require either 6 V or 12 V to function, so automobile batteries consist of 3 or 6 individual lead cells connected together to provide the proper voltage. With care, a lead storage battery will last many years in an automobile. It will eventually "die" due to the slow mechanical disintegration of the electrodes.

Fuel Cells—How They Work

Almost all of the electrical energy we use is produced in generating plants that employ some kind of heat cycle. In a conventional power plant, fossil fuels are burned to heat water into steam, and high-pressure steam drives turbines that are connected to electrical generators. Any kind of system that converts the chemical energy of fuels into heat, then into mechanical energy, and finally into electrical energy is necessarily quite inefficient because there are so many steps in which heat energy can escape to the surroundings. We have already mentioned that conventional power plants are only about 40 percent efficient in converting the chemical energy of fuels into electrical energy.

We would have a much more efficient system if we could by-pass the intermediate steps and convert chemical energy directly into electrical energy. On a small scale, it is possible to accomplish this direct conversion in devices called *fuel cells*. When water is electrolyzed into hydrogen and oxygen, electrical energy is consumed. But this energy is stored in the released gases as chemical energy. If we could reverse the process, we could recover the stored chemical energy directly as electrical energy, without the necessity of burning the hydrogen in oxygen. One way that this can be done is shown in Fig. 11-17. The process that takes place in this apparatus is essentially a controlled oxidation of hydrogen. The two electrodes are hollow cylinders of porous carbon and are immersed in a potassium hydroxide solution. Hydrogen gas is passed through the cathode and oxygen gas is passed through the anode. The reactions occur in the pores of the electrodes as the gases diffuse into the $K^+ + OH^-$ electrolyte. The anode and cathode reactions are

$$\begin{aligned}
&\text{anode:} \quad 2\ H_2 + 4\ OH^- - 4\ e^- \longrightarrow 4\ H_2O \\
&\text{cathode:} \quad O_2 + 2\ H_2O + 4\ e^- \longrightarrow 4\ OH^- \\
&\rule{7cm}{0.4pt} \quad (11\text{-}25) \\
&\text{net reaction:} \quad 2\ H_2 + O_2 \quad \longrightarrow 2\ H_2O + energy
\end{aligned}$$

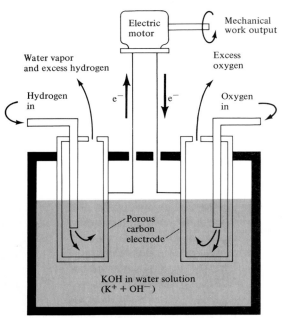

Figure 11-17   A hydrogen–oxygen fuel cell.

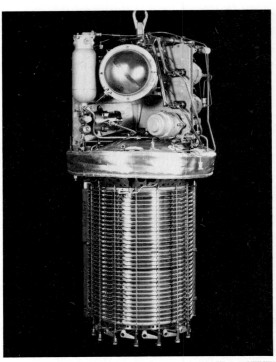

An experimental hydrogen–oxygen fuel cell.

The net reaction is exactly equivalent to the combustion of hydrogen. Electrical energy is delivered to the external circuit by the transfer of electrons from the anode to the cathode within the cell. The amount of energy released in a fuel cell is equal to the heat energy that would be produced in the combustion process.

A fuel cell is by no means a perfect converter of chemical energy. Heat is generated in all fuel cells and the overall efficiency of the best fuel cells is about 45 percent, only slightly higher than that of fossil-fuel power plants. The interest in fuel cells is based on the following considerations.

(1) Fuel cells are compact and portable sources of electrical energy. They were used, for example, to power the on-board equipment during the Apollo missions to the Moon.

(2) Fuel cells can maintain their relatively high efficiency down to low power levels (about 25 kW), whereas conventional steam-turbine plants can approach 40 percent efficiency only at very high power levels (greater than 100 MW). Consequently, fuel cells should be useful as auxiliary sources of electrical energy where high power levels are not required.

(3) When the losses in the transmission of electrical power are included, the overall efficiency of a conventional generating plant drops from about 40 percent to about 34 percent. The use of fuel cells as on-site sources of electrical energy would eliminate these transmission losses. As we pointed out in Sec-

tion 8-6, safe methods for handling and transporting hydrogen are now available. By moving hydrogen instead of electrical energy, a net increase in the efficiency of utilizing energy could be realized. Perhaps we will someday pipe hydrogen into our homes and businesses to generate electrical power as we need it.

Fuel cells will probably never compete with the huge generating plants in the production of electrical energy at high power levels. Even looking 10 years into the future, the maximum power output from a fuel cell will probably not exceed 100 kW. The advantage of fuel cells lies in their ability to produce electrical energy with relatively high efficiency at power levels appropriate for on-site applications.

11-6 ATMOSPHERIC ELECTRICITY

High Voltage in the Atmosphere

Electrical currents, in the form of moving electrons and ions, are continually flowing in the atmosphere. Ultraviolet light from the Sun and cosmic rays from space rip electrons from molecules and atoms in the air, producing free electrons and positively charged ions. But a current will flow only if a potential difference exists between two points connected by a conducting path. The primary source of the potential difference that causes electric current flow in the atmosphere is found in thunderstorms. The violent convective action in a thunderstorm propels positively charged ions high into the atmosphere, even into the ionosphere. Because there are more ions per unit volume at high altitudes than near the surface of the Earth, electrical charge is conducted freely from the tops of clouds into the ionosphere, whereas current flow at low altitudes is inhibited by the lack of sufficient ions and electrons. Observations made with aircraft flying above thunderstorms show that the electrical current flowing upward from the top of a thundercloud averages about 1 ampere. Thus, thunderstorms play an essential role in supplying electrical charge to the upper atmosphere.

There is always thunderstorm activity somewhere on the Earth and so there is a continuous pumping of positive charge into the upper atmosphere. Once at the high altitudes where the supply of electrons and ions is plentiful, the excess charge is quickly conducted horizontally and distributed uniformly in a charged layer surrounding the Earth. The average potential difference

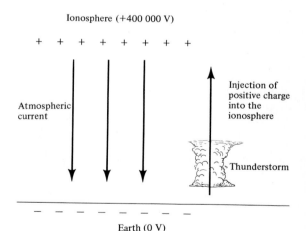

Figure 11-18 Positive charge is injected into the ionosphere by the convective action of thunderstorms. This results in a potential difference of about 400 000 V between the ionosphere and the ground, giving rise to a downward flow of atmospheric electrical current. (The ions and electrons that conduct the current are not indicated.)

between the ionosphere and the Earth's surface is about 400 000 volts!

With the ionosphere at $+400\ 000$ V relative to the ground, and with a conducting path of electrons and ions in between, there is a continual downward flow of current throughout the atmosphere (see Fig. 11-18). In a column with a cross-sectional area of 1 m², the downward atmospheric electrical current amounts to only about 4×10^{-12} A; the total current flowing from the ionosphere to the Earth is approximately 2000 A under normal conditions. The total power (volts $\times$ amperes) represented by the atmospheric current is about 10^9 W $= 1000$ MW, which is close to the power output of a modern commercial generating plant. Therefore, even though there is a natural electrical generator in the Earth's atmosphere, the supply of power is too limited to be of any practical use.

Lightning

The most spectacular evidence of atmospheric electricity is the occurrence of *lightning*. Some lightning strokes take place within clouds (cloud-to-cloud lightning), whereas others strike the Earth (cloud-to-ground lightning). About 100 cloud-to-ground strikes occur each second somewhere on the surface of the Earth. Some of these strikes are relatively mild but 10 000-ft crashers sometimes occur. All of these strikes are potentially dangerous—trees and buildings can be damaged, forest fires can be started, and human lives can be lost.

The details of the way in which lightning strokes occur are complicated and not completely understood, but the general features can be described in the following way. In an active thunderstorm cloud, the updrafts carry water droplets upward where, upon reaching the freezing level, they are converted into ice particles (hail pellets). When these ice particles eventually fall, they collide with the upward-moving water droplets. These collisions produce a separation of positive and negative charge in a way that is not fully understood. The result is that the rising droplets carry positive charges upward while the falling precipitation particles carry negative charges downward. The thundercloud then develops a charge distribution similar to that shown in Fig. 11-19. The highest concentration of negative charge is in the region of strong updrafts and the positive charge is localized in the region above 10 km (about 35 000 ft).

In an ordinary cloud, the moisture is carried only high

Lightning flashes over the George Washington Bridge. Tall structures are frequently hit by lightning strokes (the Empire State Building is hit an average of 23 times a year), but the average home should not be struck more often than about once every 1000 years.

enough that it condenses in the colder air and falls as rain. In a thundercloud, however, the updrafts are so strong that the water droplets are lifted above the freezing level where they solidify, forming hail.

When sufficient charge has been separated in the cloud (generally, about 20 coulombs), the air cannot retain its normal insulating property and an electrical *breakdown* occurs in which a relatively small electron current flows from the base of the cloud toward the ground. A quick succession of surging electrons in these *leader* strokes opens up an ionized path that extends to within 20 or so meters of the Earth. At this point a *streamer* of positive charge advances from the Earth to meet the leader. When this occurs, there exists a conducting path linking the cloud with the ground and along this path a huge current flows (up to 200 000 A) which constitutes the main lightning stroke. A large fraction of the charge in a thundercloud is dissipated in each lightning stroke, but the convection currents are so violent and the charge separation process is so efficient, that only 15–20 seconds are required to recharge the cloud and prepare it for another lightning stroke.

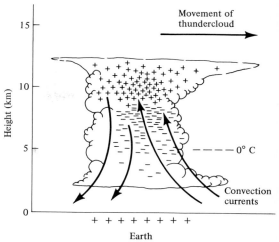

Figure 11-19 Schematic diagram of the charge distribution in an active thundercloud. The convection currents sweep positive charge to the top of the cloud (and even further into the ionosphere). Notice that even though the Earth carries an overall negative charge (see Fig. 11-18), in the vicinity of a thunderstorm, the charge is positive.

SUGGESTED READINGS

E. T. Canby, *A History of Electricity* (Hawthorne, New York, 1963).

I. B. Cohen, Ed., *Benjamin Franklin's Experiments* (Harvard University Press, Cambridge, Massachusetts, 1941).

Scientific American articles:

H. Ehrenreich, "The Electrical Properties of Materials," September 1967.

M. Josephson, "The Invention of the Electric Light," November 1959.

QUESTIONS AND EXERCISES

1. Copper contains 8.2×10^{22} free electrons/cm³. These electrons drift with a net speed of 1.5 mm/s in a copper bar with a cross-sectional area of 1 cm². What current flows in the bar?

2. An electron moving in a wire suffers a collision with an atom of the wire after moving, on the average, a distance L (which is called the *mean free path* between collisions). If the mean free path is *decreased,* will the resistance change? Explain. What could cause the mean free path to decrease?

3. A current of 6 A is drawn from a 120-V line. What power is being developed? How much energy (in J and in kWh) is expended if the current is drawn steadily for a week?

4. A current of 5 A flows through a 3-Ω resistor. What is the potential difference across the resistor? What power is developed in the resistor?

5. A typical color television set requires a current of about 4 A for operation. If you use such a set for an average of 4 h per day, what is the annual cost for electric power (at 3¢/kWh)?

6. If it is necessary for a circuit to carry a large current, why is it better to use thick wire instead of thin wire?

7. Examine the filament of an electric light bulb. Is the wire thick or thin? Why?

8. A 200-m length of copper wire with a diameter of 2 mm has a resistance of 1.1 Ω. What is the resistance of 50 m of copper wire that has a diameter of 1 mm?

9. What is the resistance of a 75-m length of nickel wire that has a cross-sectional area of 1.8 mm²? (Use Table 11-2.)

10. A certain amount of copper is in the form of a cube. The resistance between one face of the cube and the opposite face is R. If this copper cube is now shaped into a long, thin wire, will the resistance between the ends be equal to, larger than, or smaller than R? Explain.

11. Why do birds perched on a high-voltage line suffer no ill effects?

12. The wingspan of birds is a factor in determining the spacing between high-voltage power lines. Why?

13. In some strings of Christmas tree lights, when one bulb burns out, the entire string of bulbs go out. In other strings, only the single bulb goes out. Explain the difference in the two types of light strings.

14. What current flows through the filament of a 60-W light bulb when connected to a 120-V line? What is the resistance of the bulb?

15. What is the total resistance of n identical resistors R connected in series? In parallel?

16. Three identical light bulbs are connected in *parallel* to a power line. If one of the bulbs burns out, what effect will this have on the other two? How will the situation differ if the three bulbs are connected in *series*?

17. Show, using diagrams, five different ways to connect four 1-Ω resistors. Calculate the total resistance of the combination in each case.

18. What is the total resistance between the terminals of the system of resistors shown below? What would be the resistance if the points A and B were connected by a short length of wire? Can you think of an explanation for this result? [*Hint:* Consider the voltage drop along each arm of the circuit when a battery is connected to the terminals.]

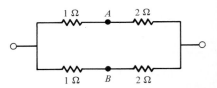

19. What is the total resistance of the resistor system shown below? If the 6-Ω resistor were removed from the set, what would be the resistance?

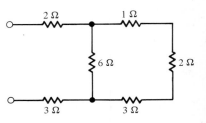

20. What total current flows in the circuit shown below? How much power is dissipated in the 8-Ω resistor?

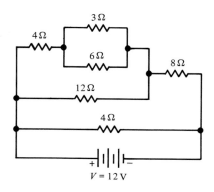

21. Suppose that you need a 2-Ω resistor. If you only have resistors with values of 1, 3, 4, 5, and 6 Ω, how would you make the required resistor?

22. What is the total resistance of the circuit shown below? To what value will the total resistance change if point A is shorted to point B?

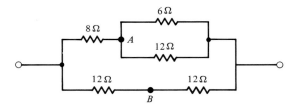

23. Suppose that you have a light bulb connected to a certain voltage source. If you add an identical bulb in series with the first bulb, what will happen to the light output of the first bulb? Why?

24. A fellow student claims that he has devised an electrical circuit which has the property that if any one of the resistors is removed, the total resistance of the circuit *increases*. Do you believe his claim? Why?

25. Four identical resistors R are connected to form a square with one resistor on each side. What is the total resistance between opposite corners of the square?

26. The output voltage of a certain electrical generating station is 100 000 V. Power is delivered to a users' network through a transmission line that has a resistance of 0.2 Ω and carries a current of 1000 A. What is the power output of the generating station? What fraction of the power is lost due to heating effects in the transmission line?

27. Resistors of 2 Ω and 4 Ω are connected in series with a 20-V battery. How much power is dissipated in the resistors? What would be the power dissipation if the resistors were connected in parallel?

28. Express the power dissipated by an AC current in a resistor R in terms of R and the peak voltage V_{peak} across the resistor. If the peak voltage on an AC line is 120 V, what power is expended in a 40-Ω resistor?

29. A current of 15 A is passed through an electrolytic solution containing Mg^{++} ions for a period of 2 h. A total of 13.6 g of magnesium metal is deposited on the cathode. From this information calculate the atomic mass of magnesium.

30. A total charge of 96 500 C is passed through a sample of molten sodium chloride.

How many grams of sodium are deposited at the cathode? How many liters of chlorine gas are liberated at the anode?

31. The useful life of an automobile battery before recharging is necessary is usually given in terms of ampere-hours (A-h). A typical 12-V battery has a rating of 60 A-h. This means that a current of 60 A can be drawn from the battery for 1 h, or 30 A for 2 h, and so on. Suppose that you forget to turn off the headlamps of your automobile. Each lamp consumes 36 watts of power. How long do you have to remember your oversight before your battery is "dead"?

32. If all of the current flowing from the ionosphere to the Earth within a certain column of air could be utilized, what would the cross-sectional area of the column have to be in order to power a 100-W light bulb?

12

ELECTRIC AND MAGNETIC FIELDS

When two material objects interact by gravitation, they exert forces on one another even though they may be separated by complete vacuum. In Section 6-3, we expressed this fact by saying that each mass sets up a condition in space—a *gravitational field*—to which the other mass responds. By introducing the idea of a gravitational field, it becomes easier to visualize the action of one mass on another. The field picture allows us to describe in a convenient manner the way in which gravitation acts.

Electric and magnetic forces are similar to gravitation in that they also require no material medium for their actions to be effective. These forces, too, can be conveniently described in terms of *fields*. In this chapter we will concentrate on the properties of electric and magnetic fields considered individually, and in Chapter 14 we will discuss the combined field—the *electromagnetic field*—and the way in which it carries light, radio waves, and other radiations.

12-1 THE ELECTRIC FIELD

Lines of Force

How can we describe the electric field in the space that surrounds an electric charge? We know that another electric charge placed in the vicinity of the original or *source charge* will experience an electrical force. If we measure this force (magnitude and direction) at a large

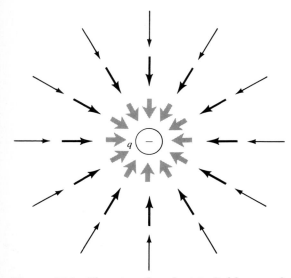

Figure 12-1 Mapping the electric field around a negative source charge q by plotting the vectors that represent the force on a positive charge at various positions around q. (The thinner arrows represent smaller forces.)

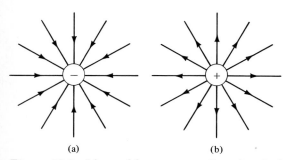

(a)   (b)

Figure 12-2 Lines of force surrounding electrical source charges. The direction of a line of force is always the same as the direction of the electrical force on a *positive* charge. (a) The lines of force surrounding a negative source charge are directed *toward* the charge. (b) The lines of force surrounding a positive source charge are directed *away from* the charge.

number of points around the source charge, we can construct a kind of map of the electrical force field that is due to the source charge. Figure 12-1 shows some of the vectors that represent the force on a *positive* charge in the vicinity of a *negative* source charge q. Notice that all of the vectors are directed *toward* the source charge (because the force between a positive and a negative charge is *attractive*). Notice also that the magnitudes of the force vectors far from the charge q are smaller than those close to q (because the electric force varies as $1/r^2$).

Because we can make such measurements at every conceivable position around the source charge, plotting the results in the same way, we could consider the job finished — the set of force vectors completely describes the field of the source charge. It proves convenient, however, to carry the mapping procedure one step further by connecting together the force vectors to form continuous lines, as shown in Fig. 12-2a.

The lines that are constructed from the force vectors are called *lines of force* or *electrical field lines*. Notice that the map of the lines of force also completely describes the field. First, the direction of the force on the positive charge at any point is the same as the direction of the line of force passing through that point. Second, the force is large where the lines of force bunch together, and the force is small where the lines are sparse. That is, the magnitude of the force is proportional to the *density* of the lines in any small region. The lines of force from a spherical source charge spread out into space in exactly the same way as the light rays from a light source. We have already seen (Fig. 6-4) that the intensity of light falls off with distance as $1/r^2$. The density of the lines of force (Fig. 12-2a) follows this same prescription, showing that the strength of the electric field also decreases as $1/r^2$, a result we have already established.

The direction of the force on a charge placed in the vicinity of a certain source charge depends upon whether that charge carries a positive or a negative electrical charge. To avoid confusion, we will always draw the lines of force that would be mapped by a *positive* charge. Therefore, the lines of force in the vicinity of a negative source charge are directed *toward* the source charge (Fig. 12-2a), and in the vicinity of a positive source charge, they are directed *away from* the source (Fig. 12-2b).

The basic idea that underlies modern field theory was

conceived by the English physicist Michael Faraday (1791–1867), who was probably the most gifted experimental scientist who ever lived. Faraday's view of the world was very mechanistic and he preferred to consider the electrical force between two charged objects as taking place via spidery *lines of force*. To Faraday, the lines of force were *real,* but today we view the lines as only a convenient way to describe field phenomena.

Fields for Different Charge Combinations

Consider a pair of objects that carry electrical charges of opposite sign and which are placed close together (Fig. 12-3). The force on a small positive charge (a *test charge*) located near these charges is the vector sum of the repulsive force due to the positive source charge and the attractive force due to the negative source charge. As the test charge is moved from the vicinity of the positive charge to the vicinity of the negative charge, the force vector (and, hence, the field line) goes smoothly from the positive charge to negative charge (Fig. 12-3). That is, *electric field lines begin on positive charges and end on negative charges.*

Figure 12-4 shows a pair of parallel metal plates that carry equal and opposite electrical charges distributed uniformly over their surfaces. Electric field lines connect the positive and the negative charges. Notice that between the plates the field lines are straight and uniformly spaced. That is, the electric field in this region is *uniform* — the strength and the direction of the field are everywhere the same.

We believe that the Universe is electrically neutral, that it carries no net charge. An object can be given an electrical charge, of course, but only by giving an equal and opposite charge to another object. Therefore, *all* electric field lines begin on some positive charge and terminate on some negative charge; there are no lines "left over" because there are no unpaired charges in the Universe.

Polarization

We can map the electric field from a collection or distribution of charges by using the standard test-charge procedure, but there is an even more revealing method that depends on the phenomenon of *polarization*. Suppose that we have two oppositely charged objects, as

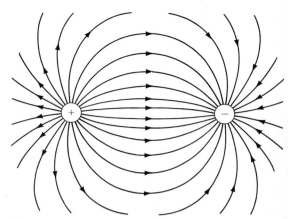

Figure 12-3   Electric field lines begin on positive charges and terminate on negative charges.

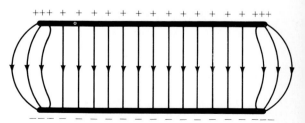

Figure 12-4   The electric field between a pair of charged parallel plates is uniform (except near the edges).

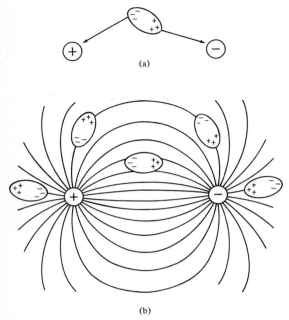

(a)

(b)

Figure 12-5 The action of a polarizable object in an electric field. (a) The source charges produce a separation of charge in the object (*polarization*) and (b) the object then tends to align with the electric field lines.

shown in Fig. 12-5. Somewhere in the vicinity of these source charges we place an object that carries no net electrical charge but in which some of the atomic electrons are free to move about. These electrons are attracted toward the positive charge and are repelled by the negative charge. As a result the object becomes *polarized,* with an excess of electrons at one end and a deficit at the other (Fig. 12-5a). The source charges act further on the polarized object, tending to draw the object into alignment with the field lines in its vicinity. The same effect occurs for other locations of the polarized object (Fig. 12-5b).

If we scatter around a number of polarizable objects in an electric field (a common technique is to use elongated seeds or hair clippings suspended in oil), the objects will orient themselves along the field lines and will provide a visual picture of the entire electric field. The photographs on page 275 show the field configurations of Figs. 12-3 and 12-4 revealed by this method.

Electric Field Strength

If we wish to calculate the electrical force on a charge due to one or two other charges, this is easily done by using Coulomb's law, $F_E = Kq_1q_2/r^2$. But how do we calculate the force on a charge due to a large number of positive and negative charges distributed over a pair of plates? How can we ever sum up the individual force vectors? By using the field concept instead of Coulomb's law, this problem is easily solved.

First, what do we mean by the *strength* of the electric field? If we place a test charge q in a field and the field exerts a large force on the charge, we say that the field strength is high. Similarly, if the force is small, we say that the field strength is low. That is, the strength of an electric field is directly proportional to the force that the field exerts on a test charge. It would make no sense to have a field strength that changes whenever we use a different test charge to determine the strength. We can avoid this difficulty if we define the strength of an electric field as the *force per unit charge* exerted on the test charge. That is, the strength E of the field is

$$E = \frac{F_E}{q} \qquad (12\text{-}1)$$

By writing this equation as $F_E = qE$, we can calculate

the force F_E on a charge q in any field that has a strength E.

An electric field has *direction* as well as *strength* (or magnitude). We know that the electric force on a charge is a vector quantity, so we could have written Eq. 12-1 as a vector equation: $\mathbf{E} = \mathbf{F}_E/q$. That is, the direction of the vector $\mathbf{E}$ that describes the electric field is the same as the direction of the force $\mathbf{F}_E$ on the test charge. The *electric field vector* $\mathbf{E}$ completely describes the electric field at a particular location.

The electric field between the pair of parallel plates in Fig. 12-4 is uniform: $\mathbf{E}$ has the same direction and the same strength everywhere between the plates. But in Fig. 12-3, the field lines curve and bunch together near the source charges. In this case, $\mathbf{E}$ has different directions and different strengths at various positions around the source charges. Remember, at any particular point, $\mathbf{E}$ has the same direction as the line of force passing through that point.

If we connect the terminals of a 6-V battery to a pair of parallel plates, as in Fig. 12-6, we know that the potential difference (or voltage) across the plates is 6 volts. But what is the strength of the field between the plates? If the field idea is to be really useful, there must be a simple way to relate the field strength E to the voltage V across the plates.

First, remember that in Section 11-2, we found that voltage is work per unit charge; that is (see Eq. 11-2),

$$\text{voltage } (V) = \frac{\text{work } (W)}{\text{charge}(q)}$$

How much work must be done to move a positive charge q from the negatively charged plate in Fig. 12-6 to the positively charged plate? Work is equal to (force) $\times$ (distance) and the force is the electric force $F_E = qE$. If the distance between the plates is d, we can write the following steps:

$$V = \frac{W}{q} = \frac{F_E \times d}{q} = \frac{(qE) \times d}{q}$$
$$= E \times d$$

That is, $V = E \times d$, and if we divide by d, we can write the electric field strength as

$$E = \frac{V}{d} \qquad (12\text{-}2)$$

The expression $E = V/d$ for the electric field strength is valid only for the case of a *uniform* field such as that

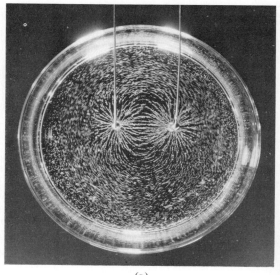

(a)

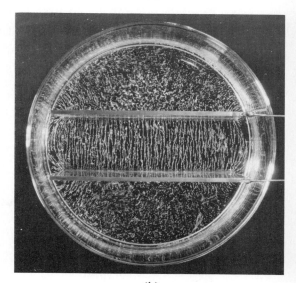

(b)

Electric field patterns as revealed by polarized seeds suspended in oil. (a) Two point charges carrying opposite charges (compare Fig. 12-3). (b) Two parallel plates carrying equal and opposite charges that are distributed uniformly (compare Fig. 12-4).

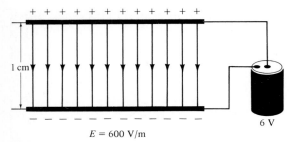

$E = 600$ V/m

Figure 12-6   A voltage of 6 V placed across a pair of parallel plates separated by a distance of 1 cm produces a uniform electric field of 600 V/m.

Robert A. Millikan, winner of the 1923 Nobel Prize in physics for his experiments on the properties of the electron.

between a pair of parallel plates. The equation is not correct for a nonuniform field such as that shown in Fig. 12-3.

In Fig. 12-6 we have 6 volts across a pair of plates separated by 1 cm. Therefore, the strength of the uniform field between the plates is

$$E = \frac{V}{d} = \frac{6 \text{ V}}{1 \text{ cm}} = \frac{6 \text{ V}}{0.01 \text{ m}} = 600 \text{ V/m}$$

Notice that the units of E are *volts per meter*.

It is important to understand the difference between *voltage* and *electric field strength*. A battery, for example, has a characteristic *voltage;* but if the terminals of a particular battery are connected to a pair of parallel plates, the *electric field strength* in the region between the plates depends on the separation of the plates. If the plates are far apart, the field strength will be low because the distance d occurs in the denominator of the expression, $E = V/d$. On the other hand, if the plates are brought close together, the field strength can be made quite high. It is the field strength (and *not* the voltage) that determines, for example, whether *sparking* will occur. In dry air, sparking between a pair of plates or electrodes will occur if the voltage and separation are such that the field strength exceeds about 3 million volts per meter (3 MV/m). Thus, a voltage of 4000 V across a 1-mm gap ($E = 4$ MV/m) will produce a spark, whereas 400 000 V across a 1-m gap ($E = 0.4$ MV/m) will not.

12-2 THE ELECTRON CHARGE AND THE ELECTRON VOLT

The Measurement of the Electron Charge

In 1911 the American physicist Robert A. Millikan (1868–1953) performed a beautifully simple but highly significant experiment in which he established the *discrete* nature of electrical charge and made, for the first time, a precise measurement of the charge on the electron. Millikan set up a pair of parallel plates separated by a distance d (Fig. 12-7). A voltage V was placed across the plates. Into the field between the plates he sprayed tiny droplets of oil. Some of these droplets became negatively charged by friction in the process of spraying. The droplets could be viewed by means of a microscope in the side of the chamber that protected the apparatus from air currents. By adjusting the voltage V, Millikan found that a given droplet could be *suspended* between the plates, with the downward gravitational force mg just balanced by the upward electrical force qE (see Fig. 12-7). Equating the magnitudes of the two forces and using Eq. 12-2 for E, we have

$$\text{force down} = \text{force up}$$

$$mg = qE = \frac{qV}{d}$$

Solving for the charge q,

$$q = \frac{mgV}{d} \qquad (12\text{-}3)$$

Which gives the charge q in terms of measurable quantities. Millikan found that the charges on various droplets, determined in this way, were not of arbitrary sizes. Instead, he found that every charge was an integer number times some basic unit of charge; that is, $q = Ne$, where $N = 1, 2, 3, \ldots$. This basic unit of charge is the *electron charge,*

$$q_e = -e = -1.60 \times 10^{-19} \text{ C} \qquad (12\text{-}4)$$

The Electron Volt

When discussing atomic and nuclear phenomena, it is not particularly convenient to express energies in terms of *joules* because of the large size of this unit compared with ordinary atomic and nuclear energies. Instead, we use a new unit — called the *electron volt* (eV) — which is of tractable size. Suppose that a charged particle, starting from rest, is accelerated by an electrical force through a voltage of 1 volt. If the particle carries a charge of 1 coulomb, it will acquire, by virtue of the increase in velocity, a kinetic energy of 1 joule because 1 volt = 1 joule/coulomb (see Eq. 11-2). That is,

$$\boxed{\text{K.E.} = q \times V} \qquad (12\text{-}5)$$

If the particle is an electron, the energy acquired in accelerating through 1 volt will be considerably less due to the very small size of the electron charge. For an electron, the energy is

$$\text{K.E.} = (1.60 \times 10^{-19} \text{ C}) \times (1 \text{ V})$$
$$= 1.60 \times 10^{-19} \text{ J}$$

This energy we call 1 *electron volt* (1 eV):

$$1 \text{ eV} = 1.60 \times 10^{-19} \text{ J} \qquad (12\text{-}6)$$

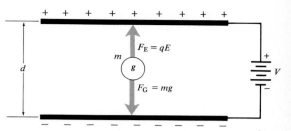

Figure 12-7 Schematic of Millikan's apparatus for determining the electron charge e. The charged droplet can be suspended by adjusting E so that $qE = mg$. Oil droplets were used instead of water droplets because oil does not evaporate as rapidly as water. Actually, Millikan used a dynamic method (instead of the static technique described here), in which he measured the rates of fall of the droplets for different voltages V, but the distinction is not important here.

Electrostatic Precipitation—How It Works

All types of industrial operations, including the burning of fossil fuels for power production, chemical processing, smelting, and so forth, release substantial amounts of particulate matter in a plant's exhaust gases. In the United States, about 40 million tons of exhaust particles are produced each year. (in addition to the even larger amount of gaseous exhausts). Filtering and trapping methods, when utilized, are very efficient in preventing this particulate matter from entering the atmosphere. But at the present time about half of the particulate matter produced in the United States is still exhausted to the atmosphere and contributes to air pollution.

One of the best (and least expensive) methods for removing particles of matter from exhaust gases is that of *electrostatic precipitation*, invented in 1905 by the American chemist Fredrick G. Cottrell (1877–1948).

The Cottrell precipitator makes use of the phenomenon of *corona discharge*. Consider a cylindrical metal tube with a wire along the axis, as shown in Fig. 12-8. A high-voltage power supply is connected between the tube and the central wire so that the tube is positive and the wire is negative. If the power supply voltage is 200 000 V and the radius of the tube is 0.1 m, the *average* electric field strength in the tube will be 2 MV/m, a value below the sparking condition. But notice that in the cylinder the field lines are concentrated near the wire—the density of lines is large near the wire and small near the cylinder wall. This means that the electric field is not uniform; the field strength decreases radially outward. Near the wire the field strength is quite high—higher than the sparking point—and, consequently, a kind of continuous sparking takes place in this region. This phenomenon is called *corona discharge*, in which the gas becomes highly ionized and a greenish glow is emitted by the ionized atoms and molecules. The electrons produced in the corona region are attracted to the positively charged walls and move

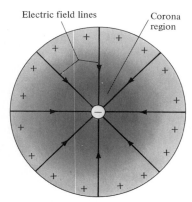

Figure 12-8   If there is a high voltage between the wall of a metal cylinder and a central wire, the electric field strength will be higher near the wire than near the wall. Corona discharge is confined to the high-field region.

GENERAL ELECTRIC

Corona discharge is visible surrounding this high-voltage electrode at a General Electric test facility.

outward. When these electrons strike particulate matter, they often attach themselves to the particles, and the negatively charged particles are forced to the cylinder wall.

Figure 12-9 shows the way in which this effect is employed in the precipitation of particular matter in an industrial exhaust stack. In most cases, the exhaust fumes contain droplets of liquid matter. The electrified droplets collect on the wall and then simply run down into a collecting hopper. If the exhaust gases contain solid particles, the stack wall is vibrated periodically to loosen the particles and they fall into the hopper. A Cottrell precipitation system, when properly designed, is amazingly effective — the efficiency for the removal of particulate matter can be 99 percent. It must be remembered, however, that only particles, not gases, are removed by a precipitator. These devices are not effective in removing undesirable gases such as sulfur dioxide (SO_2). More complicated and more expensive methods are required to remove noxious gases from exhaust stacks.

The basic principle of the Cottrell precipitator has been applied in many other areas. For example, in the dry-copy process (Xerography), electrified carbon particles are attracted to the copy paper which has been selectively charged (by an optical system) in those regions where darkening is required. The paper is then subjected to a rapid heating which fuses the carbon dust to the paper, making a permanent copy.

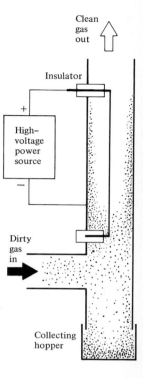

Figure 12-9 Particulate matter in exhaust gases becomes negatively charged in the corona region near the wire and is then attracted to the cylinder wall. The solid or liquid particles are collected in the hopper for disposal. In practice, an industrial exhaust stack contains several stages of precipitation, the net result of which can be a particle removal efficiency of 99 percent.

As we will see, typical energies encountered in an atom are about 10 eV, and in nuclei we find energies of a few million eV (or MeV):

$$1 \text{ MeV} = 10^6 \text{ eV} = 1.6 \times 10^{-13} \text{ J} \qquad (12\text{-}7)$$

The electron volt is a unit of *energy* and can be used to measure the energy of any object whether or not that object carries an electric charge. We often express the kinetic energies of neutrons, for example, in terms of eV or MeV.

12-3 MAGNETISM AND THE EARTH'S MAGNETIC FIELD

Lodestones and the Compass

In the 5th century B.C. the Greeks were aware that certain natural rocks, called *lodestones*, had the curious property that they attracted each other and also at-

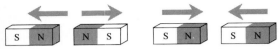

Figure 12-10 (a) Like magnetic poles repel; (b) unlike magnetic poles attract.

Figure 12-11 Cutting a magnet produces two magnets with N and S poles in the same orientation as the original magnet.

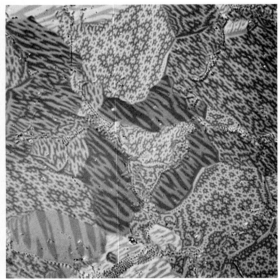

A piece of iron has been polished and then etched with acid to reveal its structure. This microphotograph shows that the iron is composed of microcrystals, each of which is a *magnetic domain*.

tracted bits of iron. These lodestones were found in abundance in Magnesia in Asia Minor and they became known as *magnets*. (Lodestones consist primarily of the mineral *magnetite*.) About 1100 A.D. European navigators discovered that needle-shaped pieces of lodestone, when suspended freely, take up a north–south direction. This observation led to the development of a practical magnetic compass which was soon widely used for navigation purposes, particularly on the open sea.

The end of a compass magnet that is *north-seeking* is called the *N pole* of the magnet and the opposite or south-seeking end is called the *S pole*. Magnets have the familiar property that *like poles repel* and *unlike poles attract*. That is, if two magnets are brought together with their N poles nearest, they will experience a repulsive force (Fig. 12-10a), whereas if one magnet is reversed so that one N pole and one S pole are nearest, the magnets will be attracted toward one another (Fig. 12-10b). Consequently, a freely suspended magnet will always align itself so that its N pole points in the direction of the S pole of another nearby magnet. From this behavior we are led to conclude that the Earth itself acts as a giant magnet, causing the N pole of every compass magnet to point toward the S pole of the Earth. Furthermore, the Earth's S pole must be located near the geographic *north* pole. (Actually, the geomagnetic and geographic poles do not coincide; see Fig. 12-14.)

Permanent Magnets

If it is not disturbed by external influences (for example, by mechanical vibrations or excessive temperatures), the magnetism of a bar magnet will survive indefinitely—it is a *permanent magnet*. When such a magnet is cut in half, as in Fig. 12-11, the result is not a separation of the N pole from the S pole; instead, we find that the two halves are themselves complete magnets with N and S poles in the same orientation as the original magnet. Further division of the magnet produces the same result: the *individual* pieces are all *complete magnets*. If we continue the examination of the magnet down to the microscopic level, we find that the magnet iron consists of an extremely large number of tiny crystalline aggregates of iron atoms, called *magnetic domains,* each of which is individually magnetic. In magnetized iron the domains are aligned with their N poles predominantly in the same direction (Fig. 12-12a), thereby producing the overall magnetism of the bar.

In ordinary, unmagnetized iron, the domains are distributed with their poles pointing in random directions (Fig. 12-12b), so that the individual magnetism of the domains cancel, producing zero net magnetism for the bar as a whole. If an unmagnetized iron bar is stroked with one pole of a magnet, the domains are drawn into alignment and a permanent bar magnet is formed. On the other hand, if a permanent magnet is given a series of sharp blows with a hammer, the domains can be thrown out of alignment and the iron will become demagnetized. Similarly, if the temperature of a magnet is raised, the domains become agitated and are jiggled about, losing their alignment. Above a temperature of about 770 °C, an iron magnet loses its magnetism. (The melting point of iron is 1535 °C.)

The fact that a magnet cannot maintain its magnetism at an elevated temperature allows us to answer one of the questions regarding the Earth's magnetism. There is ample evidence that the interior of the Earth consists of iron. Except for the relatively small inner core, this iron is in the molten state and does not possess any permanent magnetism. Therefore, the Earth's magnetism cannot be due to permanently magnetized iron in the interior. We must seek another and different source. First, we examine some of the features of the Earth's magnetic field, for herein lies the clue to the Earth's magnetism.

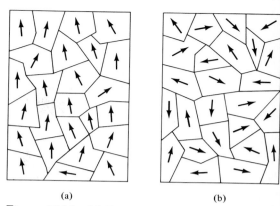

Figure 12-12 (a) In *magnetized* iron, the individual domains have their N poles aligned predominantly in the same direction. (b) If the domains are oriented at random, the iron is *unmagnetized*.

The Earth's Magnetic Field

The compass is a useful device because it reacts to a *magnetic field*. Just as the Earth's *gravitational field* acts on objects and causes them to be attracted toward the Earth, the Earth's magnetic field acts on magnetized objects and causes them to align with the direction of the field. By recording the orientation of a compass at various positions in the vicinity of a bar magnet, we can obtain a picture of the magnetic field in the same way that a test charge is used to probe an electric field. Figure 12-13 shows a series of magnetic field lines obtained in this way. Notice that these lines point in the same direction as the N pole of the compass magnet used to map the field. The magnetic field (at a particular place) is completely described by the field vector **B,** whose direction is the same as the field lines and whose magnitude is defined in terms of the force exerted by the field on a charged particle (as discussed in a later paragraph).

By using compass magnets and other devices, we can

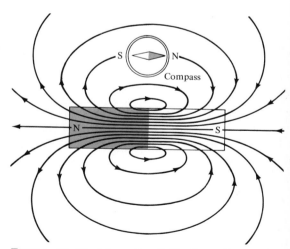

Figure 12-13 Magnetic field lines for a bar magnet mapped by means of a compass. Notice the direction of the field and the direction of the compass magnet.

The Earth's Changing Magnetism

The Earth's magnetic field is not a static affair. As early as 1634 it was noticed that at a fixed place the direction of a compass needle changes slowly with time. At London, for example, a compass needle pointed 11° east of true north in 1580; the direction changed gradually to 24° west of north in 1812; and at the present time a compass points 9° west of north. Thus, the Earth's geomagnetic pole wanders slowly with respect to the geographic pole.

CARNEGIE INSTITUTION

From the 17th century onwards, studies have been made to obtain an accurate description of the Earth's magnetic field. Edmund Halley (1656–1742) was the first to draw a magnetic chart of the Earth based on measurements made during a two-year cruise aboard the *Paramour Pink*. In 1909 the Carnegie Institution of Washington began an extensive mapping of the Earth's magnetic field in the ocean areas using the nonmagnetic brigantine *Carnegie* (see the illustration). In order to eliminate magnetic materials that would perturb the delicate magnetic measuring instruments, this ship was constructed from wooden frames and planking, and was held together with wooden pegs and bolts of copper and bronze (both nonmagnetic metals). The auxiliary engine was built almost entirely of bronze and manganese. Even the crew wore nonmagnetic belt buckles. On her maiden voyage in 1909 the *Carnegie* followed the same route that had been taken by Halley in the *Paramour Pink* 200 years earlier. During the interval between the two voyages, the Earth's magnetic field had shifted sufficiently that if the *Carnegie* had followed the *Paramour Pink's* compass courses, she would have made landfall, not near Falmouth on the southern coast of England as intended, but somewhere along the northwestern coast of Scotland.

An even more spectacular and mysterious effect has been found by analyzing the magnetic properties of rocks that were formed in sedimentary deposits over millions of years of geologic time. As these magnetic rocks hardened, the magnetic axes (that is, the direction from the rock's S pole to its N pole) aligned with the local geomagnetic field direction at the time of formation. By uncovering these rocks and carefully measuring the direction of their magnetic axes, the direction of the Earth's magnetic field can be determined for the geologic period during which the rocks formed. (We must have evidence, of course, that the rocks have not shifted position since the time of formation, but this is frequently the case and so the method is indeed a useful one.) These studies have led to the remarkable conclusion that the direction of the Earth's magnetic field has actually reversed itself many times during the last 80 million years! Apparently, the field undergoes a decrease in strength lasting about 10 000 years, then comes a period of about 1000 years during which the reversal takes place, and finally there is a buildup of the field with the opposite polarity. This same process is repeated with intervals of 50 000 to a few million years. The last reversal occurred approximately 12 400 years ago. Whatever the mechanism within the Earth that is responsible for the generation of the geomagnetic field, it appears to have as a regular feature this continuing field-reversal effect.

also map the Earth's magnetic field. Although we have already concluded that the Earth's field cannot be due to a permanent magnet, we find that the shape of the Earth's magnetic field is quite similar to that of a bar magnet (Fig. 12-14). The *equivalent magnet* that represents the Earth's magnetism does not lie along the Earth's rotation axis but it tilted with respect to the rotation axis by approximately 11°. As a result, the S pole of the Earth is located about 800 miles from the geographic north pole.

With these facts in mind regarding the Earth's magnetic field, let us see what methods, other than permanent magnets, are available for producing similar magnetic fields.

Magnetic Fields Produced by Electric Currents

At the beginning of the 19th century, electric and magnetic effects were believed to be separate and distinct phenomena, without any interconnection. In 1820, however, Hans Christian Oersted (1777–1851), a Danish physicist, discovered that electrical currents can generate magnetism of a temporary nature. Oersted showed that a compass magnet placed in the vicinity of a wire carrying an electric current will be deflected and will take up a direction *perpendicular* to the direction of the wire (Fig. 12-15). Figure 12-16 shows the field lines in a plane perpendicular to a current-carrying wire. This photograph illustrates the *iron-filing technique* for making the field lines "visible." Iron filings are sprinkled on a sheet of paper and the magnetic field induces a magnetism in each of the tiny pieces of iron, causing

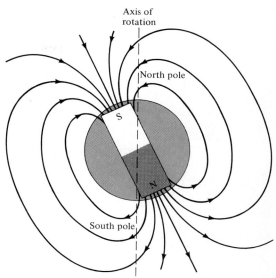

Figure 12-14 The magnetic field of the Earth is similar to that of a giant bar magnet.

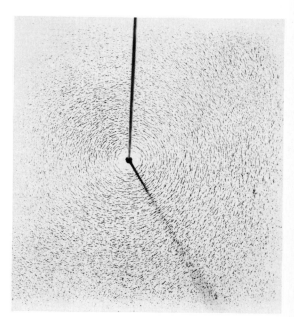

Figure 12-16 The circular field lines surrounding a current-carrying wire shown by the iron-filing technique.

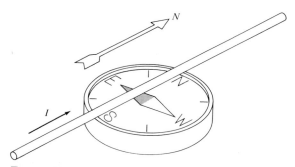

Figure 12-15 Deflection of a compass magnet by a current-carrying wire. A compass placed beneath a wire that carries a current *north* will point to the *west*.

Hans Christian Oersted (1777–1851). Oersted was the son of a poor apothecary in Copenhagen. On a rigid budget, he entered the university in his home city in 1793 to study medicine, physics, and astronomy. He completed his study in pharmacy in 1797 and received the Doctor of Philosophy degree in 1799. For a while he followed in his father's footsteps and operated an apothecary shop, while continuing to study and to lecture, particularly about the electrical sciences. In 1806 he realized his great ambition by being appointed Professor of Physics at Copenhagen University. In the spring of 1820, while lecturing to a small group of students, Oersted noticed that a nearby compass magnet was deviated when a current was caused to flow in an electrical circuit. Oersted's report of his discovery, a four-page privately printed tract, is one of the most important (and rarest) of scientific documents, for it set the stage for development of the entire subject of electromagnetism. The photograph on the left shows the actual compass used by Oersted in his discovery of the magnetic effect of electric current.

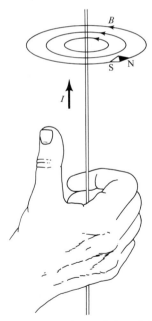

Figure 12-17 Illustration of the right-hand rule for determining the direction of the magnetic field lines surrounding a current-carrying wire. When the thumb of the right hand points in the direction of I, the fingers point in the direction of the field lines.

these miniature magnets to align with the field direction. The field surrounding the wire exists only as long as the current is flowing in the wire.

The direction of the field lines surrounding a current-carrying wire can be determined by using a compass magnet in the same way as for a bar magnet (see Fig. 12-17). The results of such an experiment can be summarized by the following *right-hand rule:*

> Grasp the wire with the right hand with the thumb pointing in the direction of current flow in the wire. The fingers then encircle the wire in the same direction as the magnetic field lines.

If we take a wire that carries a current, form it into a loop, and then map the shape of the resulting magnetic field, we find an interesting and important result: *the magnetic field due to a circular loop of current is essentially the same as that due to a bar magnet* (see Fig. 12-18). Here we have the essential clue to the Earth's magnetism. The molten iron in the Earth's interior must carry electric currents that produce the magnetic field of the Earth. But how are these currents formed and how are they maintained?

The Earth is a rotating mass with a solid crust and a liquid interior consisting of molten iron. Because of the nonuniformity of the interface between the crust and the core, the molten iron does not rotate in exactly the same way as does the crust, and turbulent currents are produced in the liquid interior. The iron is so hot that some of the electrons are separated from every iron atom and the motion of these electrons in the turbulent liquid give rise to electric currents. Although it seems clear that the Earth's magnetism is produced in this general way by electric currents in the core, as yet we have no really satisfactory detailed explanation of the origin of the geomagnetic field.

The Sun's Magnetic Field

The turbulent, rapidly moving ionized gases in the Sun generate a solar magnetic field that is complex and subject to unexpected changes. Various types of measurements have shown that the strength of the Sun's magnetic field varies by a factor of a thousand or so in the vicinity of an active sunspot. Rapid and localized changes in the field accompany the eruptions that we call *solar flares* or *solar prominences*. Charged particles ejected during these violent explosions are guided by the Sun's magnetic field lines and result in the enormous looping arches that rise to heights of thousands of miles and then return to the surface. One spectacular magnetic eruption is shown in the photograph at the bottom of this page.

The Magnetism of the Moon

Before the first lunar landing by the Apollo astronauts, scientists had devised relatively simple descriptions of the Moon's structure. In none of these models was there any suggestion that the Moon possessed (or had ever possessed) a magnetic field. It seemed highly unlikely that the core of the Moon had ever been molten as the Earth's core still is. Therefore, there should never have been any internal currents to produce a Moon magnetism. But the investigations carried out during the Apollo missions revealed a small remnant magnetism in the surface rocks that could have been caused only by exposure to a small magnetic field for a period of at least a billion years. Even though the magnitude of this field is only about $\frac{1}{50}$ of the present Earth field, the way in which the Moon field originated is a complete mystery. None of the several theories that have been developed to account for the Moon's magnetism is entirely satis-

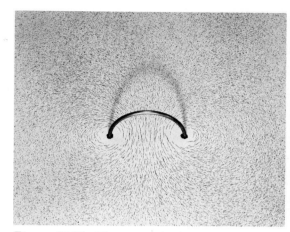

Figure 12-18 The magnetic field lines for a circular loop of electric current. (The other half of the loop extends below the sheet of paper that carries the iron filings.) Notice the similarity to the field shown in Fig. 12-13.

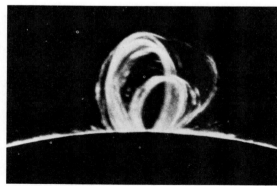

A solar eruption in which the ejected ionized gases are guided by the Sun's magnetic field, forming a gigantic arch. The largest loop extends outward from the surface of the Sun a distance of about 140 000 km.

factory. More information will have to be acquired and further theoretical investigations will have to be carried out before we will have a convincing explanation of the magnetism of the Moon.

Atomic Magnetism

The smallest magnetic systems are found at the atomic level. In a crude way, an atom can be pictured as a number of electrons orbiting around a central nucleus. An electron that revolves around a nucleus is equivalent to a ring of electric current. Consequently, every such electron produces a magnetic field which is similar to that shown in Figure 12-18 for a current loop. In most atoms the individual electron-current magnetic fields cancel with one another so that the atom has no net magnetism. But in certain types of atoms—particularly iron, cobalt, and nickel—the electron fields are not all canceled. These atoms have a net permanent magnetism. When atoms of iron, for example, bind together to form a tiny crystal, the magnetic axes of the atoms are aligned so that the crystal possesses a net magnetism. Such magnetic crystals constitute the *magnetic domains* that are found in all permanent magnets.

TABLE 12-1   RANGE OF MAGNETIC FIELD STRENGTHS IN THE UNIVERSE

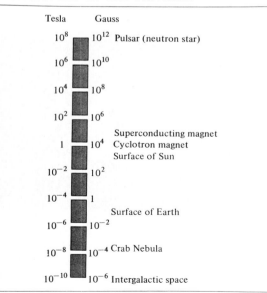

12-4 THE MOTION OF CHARGED PARTICLES IN MAGNETIC FIELDS

Magnetic Field Strength

The strength of an electric field is defined in terms of the force that is exerted on a charged particle in that field. We use the same procedure for defining the strength of a magnetic field. But there is an important difference between the two cases. A stationary charge will experience a force in an electric field (Eq. 12-1), but an electric charge at rest in a magnetic field experiences *no* force. Only when the charge is in *motion* will a magnetic field exert a force on it. This magnetic force increases whenever we increase the magnetic field strength, or the charge on the particle, or the speed with which the particle moves. The equation that describes these features of the magnetic force F_M is

$$F_M = qvB \qquad \text{for} \qquad \mathbf{v} \perp \mathbf{B} \qquad (12\text{-}8)$$

where q and $\mathbf{v}$ are the charge and the velocity, respectively, of the particle, and where $\mathbf{B}$ is the magnetic field vector. The dimensions of B are *tesla* (T); when q is measured in *coulombs*, v in *meters per second*, and B in *tesla*, the force is given in *newtons*. Magnetic field strengths are also frequently given in terms of a unit called the *gauss* (G):

$$1 \text{ tesla} = 10^4 \text{ gauss} \qquad (12\text{-}9)$$

Because charge, velocity, and force have all been defined previously, Eq. 12-8 is therefore the defining expression for the magnetic field strength B.

Experiments show that the force exerted on a moving charged particle is given by Eq. 12-8 only when the velocity vector **v** is *perpendicular* to the field vector **B**. It is only the perpendicular component of **v** that is effective in causing a magnetic force. Therefore, if **v** lies at an angle θ with respect to **B**, then the perpendicular component of **v** is $v_\perp = v \sin \theta$ and the magnetic force is

$$F_M = qvB \sin \theta \qquad (12\text{-}10)$$

The values of F_M for different orientations of **v** and **B** are shown in Fig. 12-19.

The Direction of the Magnetic Force

The *magnitude* of the magnetic force $\mathbf{F}_M$ is given by Eq. 12-10, but in what *direction* does $\mathbf{F}_M$ act? In the electrical case, the force $\mathbf{F}_E$ is always in the same direction as the field vector **E**. In the magnetic case, however, the force $\mathbf{F}_M$ is not in the direction of the field vector **B**, nor is it in the direction of the velocity vector **v**: $\mathbf{F}_M$ *is perpendicular to both* **B** *and* **v**. We can see the relationship among these three vectors in the following simple experiment. Suppose that we place a wire in the field of a permanent bar magnet that has been shaped into a "C," as shown in Fig. 12-20. If the ends of the wire are connected to the terminals of a battery, a current will flow in the wire. As soon as the connection is made, we see the wire pushed to one side by the magnetic force. The flow of a current is the movement of electrical charge, and the direction of current flow in a wire is the same as the direction of motion of *positive* charge (see Section 11-1). Therefore, the effect of the magnetic field on a moving positive charge is the same as on a current-carrying wire.

The relative orientation of the vectors **B**, **v**, and $\mathbf{F}_M$ (for the case of a *positive* charge) is shown in Fig. 12-20 and again in Fig. 12-21. The rule for finding the

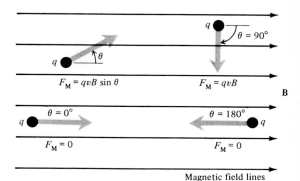

Figure 12-19   Values of the magnetic force F_M on a moving particle that carries a charge $+q$ for different orientations of **v** with respect to **B**. In each case, $F_M = qvB \sin \theta$, but $\sin 90° = 1$, $\sin 0° = 0$, and $\sin 180° = 0$.

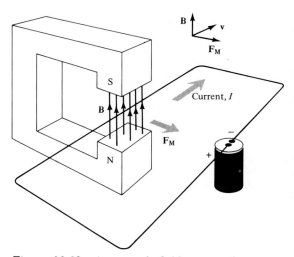

Figure 12-20   A magnetic field exerts a force on a current-carrying wire in the same way that it exerts a force on a moving (positive) charge.

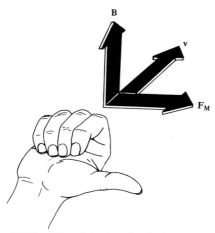

Figure 12-21 The right-hand rule for determining the direction of $\mathbf{F}_M$. If the fingers of the right hand are curled in the direction that carries the vector **v** toward the vector **B**, then the thumb points in the direction of $\mathbf{F}_M$.

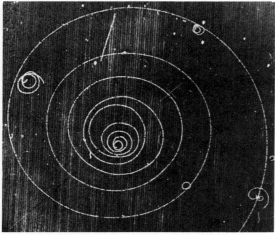

An electron moves in a circular path in a bubble chamber which is located in a magnetic field. The electron loses a small amount of energy in each of the frequent collisions with the atoms of liquid hydrogen in the chamber and so the electron's speed decreases continually. This causes the electron to move in a spiral path with decreasing radius. The tracks of some secondary electrons released in the encounters with hydrogen atoms can be seen near the main track.

direction of the magnetic force on a moving positive charge is illustrated in Fig. 12-21. If the fingers of the right hand are curled in the direction that carries the vector **v** toward the vector **B**, then the thumb points in the direction of $\mathbf{F}_M$.

The direction of the magnetic force on a moving *negatively* charged particle is *opposite* to that on a positively charged particle moving in the same direction relative to the field. Therefore, a proton and an electron will curve in opposite directions when moving in a magnetic field.

The Force on a Current-Carrying Wire

What force does the magnetic field **B** exert on the wire in Fig. 12-20? Within the wire there are many electrons moving with a velocity v. The magnetic force on each ·electron that is within the field region is evB (Eq. 12-9, because $\mathbf{v} \perp \mathbf{B}$). The total force on the wire is equal to the sum of the forces on the individual electrons. Let L be the length of wire that lies within the field region; only the electrons in this section of wire experience a force. Let n be the number of electrons in the wire per unit length. Therefore, the number of electrons acted upon by the magnetic field is nL, and the total force on the wire is

$$F_M = nLevB$$

Now, look at the combination nev. How much charge q will pass a particular point in the wire in 1 second? We can write

$q =$ (number of electrons per unit length, n)
$\times$ (charge of electron, e)
$\times$ (distance each electron moves in 1 s, l)

The distance l is

$$l = (\text{velocity}, v) \times (\text{time}, 1 \text{ s})$$

Therefore,

$$q = nev \times (1 \text{ s})$$

or,

$$\frac{q}{1 \text{ s}} = nev$$

But the charge per unit time is the *current*. Thus,

$$I = nev$$

and the force on the wire becomes

$$F_M = BIL \qquad (\mathbf{v} \perp \mathbf{B}) \qquad (12\text{-}11)$$

If a current of 20 A flows through a wire which has a 10-cm section within a field of 2000 gauss (or 0.2 T), the force on the wire is

$$F_M = (0.2 \text{ T}) \times (20 \text{ A}) \times (0.1 \text{ m}) = 0.4 \text{ N}$$

Equation 12-11 is correct if the direction of the wire is perpendicular to the field direction. In the event that the direction of the wire makes an angle θ with respect to **B**, we must express the force as

$$F_M = BIL \sin \theta \qquad (12\text{-}12)$$

Orbits of Charged Particles in Magnetic Fields

The force exerted on a moving charged particle by a magnetic field is always in a direction that is perpendicular to the direction of motion of the particle. Consequently, the particle experiences an *acceleration* that is always perpendicular to its instantaneous velocity vector. We have already studied just such a case: in Section 3-7 we found that when a particle is accelerated uniformly in a direction perpendicular to **v**, the particle moves in a *circle*. Therefore, a charged particle that moves in a uniform magnetic field will execute a circular orbit. Such a case is shown in the photograph on the opposite page.

We can find the radius of the orbit in the following way. Refer to Fig. 12-22. Here we have a particle with a positive charge q and a mass m that moves with a velocity **v** perpendicular to a magnetic field **B.** (Use the right-hand rule and check that the magnetic force is in the direction indicated.) We know that the magnetic force is

$$F_M = qvB$$

Furthermore, we know that this force is the *centripetal* force which we express as (Eq. 4-7)

$$F_c = \frac{mv^2}{R}$$

Equating these two expressions for the force,

$$qvB = \frac{mv^2}{R}$$

and solving for the orbit radius R, we find

$$R = \frac{mv}{qB} \qquad (12\text{-}13)$$

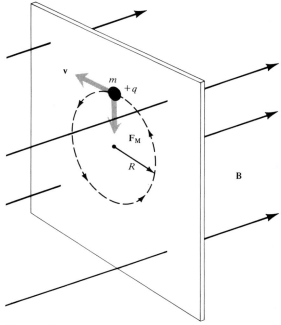

Figure 12-22   A charged particle moving in a uniform magnetic field executes a *circular* orbit.

This result agrees with our expectations. If the mass or the velocity of the particle is increased, it is more difficult for the field to bend the particle and R is increased. If the charge or the field strength is increased, it becomes easier for the field to bend the particle and R is decreased.

Suppose that a proton is accelerated through a potential difference of 10^6 volts and is projected into a magnetic field with a strength of 5000 gauss. What will be the radius of the proton's orbit? First, we need the combination mv (the proton *momentum*) expressed in terms of the kinetic energy. We have

$$\text{K.E.} = \tfrac{1}{2}mv^2 = \frac{1}{2m}\,(mv)^2$$

where we have multiplied and divided by m. Solving for mv,

$$mv = \sqrt{2m\ \text{K.E.}}$$

Therefore, we can express the orbit radius as

$$R = \frac{\sqrt{2m\ \text{K.E.}}}{eB}$$

Next, we note that a proton (charge $= +e$) accelerated through a potential difference of 10^6 volts will have a kinetic energy of 1 MeV or 1.6×10^{-13} J (Eq. 12-7). Then,

$$R = \frac{\sqrt{2 \times (1.67 \times 10^{-27}\ \text{kg}) \times (1.6 \times 10^{-13}\ \text{J})}}{(1.6 \times 10^{-19}\ \text{C}) \times (0.5\ \text{T})}$$
$$= 0.29\ \text{m}$$

Trapping of Charged Particles in a Magnetic Field

A charged particle that moves perpendicular to a magnetic field **B** will execute a circular orbit. But how will the particle move if its velocity vector **v** is not at right angles to **B**? In this case we can imagine **v** to consist of two components, one perpendicular to **B** and one parallel to **B**. The perpendicular component will give rise to circular motion. But there is no magnetic force due to the parallel component (Fig. 12-19) and so this velocity component remains constant. The resulting motion is a combination of the circular motion due to the perpendicular component and a steady motion along the field lines due to the parallel component. The particle therefore spirals along the field line, moving forward at a uniform speed while looping around the line (Fig. 12-23).

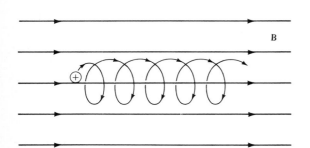

Figure 12-23 If **v** is not perpendicular to **B**, a charged particle will move in a spiral path around a field line.

Many types of magnetic fields are not uniform and the field lines are not straight and regularly spaced. The Earth's field, for example, is not uniform: the field lines bunch together at the poles and spread out in space around the Equator (see Fig. 12-14). When a charged particle moves in a nonuniform magnetic field such as that of the Earth, it performs an interesting motion. Suppose that we follow an electron as it moves in the Earth's magnetic field (Fig. 12-24). Because **v** will not in general be perpendicular to **B**, the electron begins to spiral around a field line. As it moves into the region of increased magnetic field strength near the pole, the path becomes a tighter spiral. The loops become smaller and closer together as the electron moves farther and farther into the high field region. Eventually, a point is reached beyond which the electron cannot penetrate. At this point (called the *mirror point*), the particle is *reflected*—in much the same way that a ball bounces back from a wall—and proceeds back into the low-field region. The electron then moves along the field line to the opposite pole where it is again reflected. Thus, we see that the electron is *trapped* in the magnetic field, bouncing back and forth between the mirror points at the two poles.

The trapping of charged particles (electrons and protons) in the Earth's magnetic field produces the series of gigantic *radiation belts* that encircle the Earth in space. The existence of these belts was discovered by James Van Allen and his co-workers in 1958, during the flight of the artificial satellite, Explorer I. The Earth's radiation belts are sometimes called the *Van Allen belts*.

One of the problems that is associated with the radiation belts is the way in which the gross features of the Earth's magnetic field are influenced by the *solar wind,* charged particles that are ejected from the Sun with extremely high velocities. During the occurrence of solar flares, the intensity of these particles can be quite high. The magnetic field that results from these moving charged particles interacts with and disturbs the Earth's field. On the side nearer the Sun, the Earth's field is compressed, whereas on the "down-wind" side, the field trails off for a considerable distance (see Fig. 12-25). The true magnetic field of the Earth is far from the simple bar-magnet field that we once believed it to be! Because the intensity of the solar wind varies with time, so does the shape of the outer parts of the Earth's magnetic field. These variations also influence the trapping of charged particles in the radiation belts.

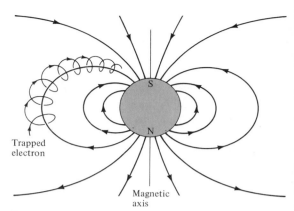

Figure 12-24 An electron trapped in the magnetic field of the Earth.

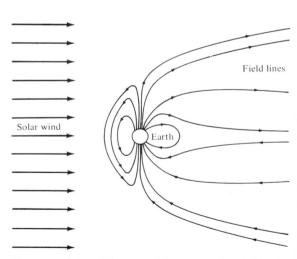

Figure 12-25 The Earth's magnetic field is "blown back" by the rapidly moving charged particles ejected from the Sun. These particles are called the *solar wind.*

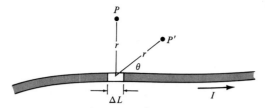

Figure 12-26 The magnetic field strengths at P and P' due to the current flowing in the segment ΔL are given by Eqs. 12-14 and 12-15.

12-5 MAGNETIC FIELDS PRODUCED BY CURRENTS

The Field Due to a Current Element

What is the strength of the magnetic field produced by a current flowing in a wire? Refer to Fig. 12-26. From experiment we can deduce the following. If a current I flows in a wire, then the magnetic field at the point P due to the current I in the small segment of wire with length ΔL is

$$\Delta B = K_{\text{M}} \frac{I\,\Delta L}{r^2} \qquad \text{(at } P) \qquad (12\text{-}14)$$

This expression is correct only if the line that connects ΔL with P is perpendicular to the direction of the segment ΔL (that is, perpendicular to the velocity $\mathbf{v}$ of the charges moving through ΔL). For the point P' in Fig. 12-26, we must include a $\sin\theta$ factor:

$$\Delta B = K_{\text{M}} \frac{I\,\Delta L}{r^2} \sin\theta \qquad \text{(at } P') \qquad (12\text{-}15)$$

The magnetic force constant has the value

$$K_{\text{M}} = 10^{-7}\ \text{N/A}^2 \qquad \text{or} \qquad 10^{-7}\ \text{T-m/A} \quad (12\text{-}16)$$

The value of K_{M} is a consequence of the way in which the ampere is defined (see Exercise 22).

Fields Due to Loops and Straight Wires

The expressions given in Eqs. 12-14 and 12-15 are the contributions to the total field from only a small part of the wire that carries the current. In order to obtain the total field strength in any particular case, we must add the contributions from all of the small segments that make up the actual circuit. A case in which the summing of the individual contributions can easily be made is that of a *current loop*. Figure 12-27 shows a length of wire that has been shaped into a circular loop with a radius r. We wish to determine the strength of the magnetic field at P, the center of the loop. First, notice that the currents entering and leaving the loop flow in two parallel and closely spaced parts of the wire. Because the current flows in opposite directions in these two parts of the wire, the magnetic fields produced are in opposite directions and cancel. The field at the center of the loop is therefore due only to the current flowing in the circular part of the circuit.

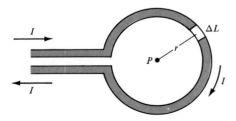

Figure 12-27 The strength of the magnetic field at the center of the loop is $2\pi K_{\text{M}}I/r$.

Every segment of the loop, such as ΔL in Fig. 12-27, is at a distance r from the center and, moreover, the direction from each segment to the center is perpendicular to the direction of the current flow through the segment. Therefore, we can use Eq. 12-14 and simply add all of the ΔB's due to all of the segments ΔL around the loop. The result is the total field B at the center of the loop:

$$B = \text{sum of all } \Delta B$$

We usually denote the phrase "sum of all" by using a Greek capital sigma, Σ. Then, we can write

$$B = \Sigma \; \Delta B = \Sigma K_M \frac{I \; \Delta L}{r^2}$$

Because the quantities K_M, I, and r^2 are all constant, we can write

$$B = K_M \frac{I}{r^2} \; \Sigma \; \Delta L$$

Now, the sum of all ΔL around the loop is the circumference of the loop, $2\pi r$. Therefore,

$$B = K_M \frac{I \times 2\pi r}{r^2}$$

$$= 2\pi K_M \frac{I}{r} \quad \text{(current loop)} \qquad (12\text{-}16)$$

Suppose that a current of 10 A flows in a circular loop with a radius of 10 cm. Then, the field strength at the center is

$$B = \frac{2\pi \times (10^{-7} \text{ T-m/A}) \times (10 \text{ A})}{0.1 \text{ m}}$$

$$= 6.3 \times 10^{-5} \text{ T} \quad \text{or} \quad 0.63 \text{ gauss (G)}$$

If, instead of a loop, we have a long, straight wire carrying a current I, the distances from the individual segments to the point P are no longer the same. In addition, we must use Eq. 12-15 with the $\sin \theta$ factor. Consequently, the calculation of the sum is much more complicated than for the simple loop. We state here only the result:

$$B = 2K_M \frac{I}{r} \qquad \text{(long, straight wire)} \qquad (12\text{-}17)$$

Suppose that a distance of 25 cm from a long, straight wire the field strength is measured to be 5 G. What is the current flowing in the wire? Solving Eq. 12-17 for I,

$$I = \frac{Br}{2K_M}$$

$$= \frac{(5 \times 10^{-4} \text{ T}) \times (0.25 \text{ m})}{2 \times (10^{-7} \text{ T-m/A})}$$

$$= 625 \text{ A}$$

If a long, straight wire that carries a current is placed parallel to another similar current-carrying wire, the magnetic field produced by one wire will act on the moving charges in the other wire. As a result, there will be a force exerted by each wire on the other. This force will be attractive or repulsive, depending on the directions of current flow in the two wires. Measurement of this force can be used to define the ampere (see Exercise 22).

12-6 FIELDS THAT VARY WITH TIME

Electromagnetic Induction

Thus far, we have considered only *static* fields—that is, fields that are constant and do not change with time. There are additional interesting and important effects that take place if we allow the fields to vary with time. Suppose that we remove the battery from the circuit shown in Fig. 12-20. Then, of course, no current will flow through the wire. But now suppose that we pull the wire through the field, as shown in Fig. 12-28. The charges in the wire are now in motion with a velocity **v** which is perpendicular to **B**. (As usual, we consider the motion of positive charges even though only the free electrons actually move through the wire.) The charges therefore experience a magnetic force. This force is *along* the wire (use the right-hand rule) and so a current flows through the wire in the direction shown in the diagram. Thus, a current has been *induced* in the wire because of the motion of the wire in the field. The faster the wire moves, the greater will be the rate at which field lines are cut and the larger will be the induced current. This is the phenomenon of *electromagnetic induction,* discovered by Michael Faraday in 1831.

What will happen in the circuit of Fig. 12-28 if we move the magnet to the left instead of moving the wire to the right? Exactly the same thing! It does not matter whether the field or the wire is "moving" and the other is "stationary." All that is necessary for a current to be induced in the wire is that there be *relative motion* between the wire and the field.

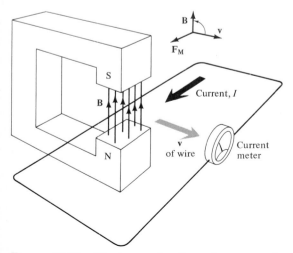

Figure 12-28 Moving a wire through a magnetic field induces a current to flow in the wire.

AC Generators—How They Work

Rotating a coil of wire in a magnetic field is the standard method which is used to generate alternating electrical current (AC). In this way, the mechanical energy of a turbine driven by steam or water power is converted into electrical energy. Basically, the scheme is that shown in Fig. 12-29. Look at the segment of coil labeled OP. This segment is moving to the left in the field $\mathbf{B}$ near the S pole and, according to the right-hand rule, a current is induced to flow in the direction $O \rightarrow P$. Similarly, the bottom segment QR is moving to the right and the induced current flows in the direction $Q \rightarrow R$. Thus, the current induced in each segment flows in the same direction around the coil.

When the segments OP and QR are near the poles of the magnet and are moving horizontally, they cut through the field lines at the maximum rate. At this instant, the current is also maximum. When the coil has rotated to the horizontal position, the segments move instantaneously *along* the field lines. In this condition there is no magnetic force on the free electrons and the induced current is zero. Rotating past the horizontal position, the motion of each segment through the field is in the direction opposite to that in the first half of the cycle. Therefore, the current flows in the opposite direction, reaching a maximum again when the segments are near the poles. The result is a surging of current, first in one direction and then in the other: this is *alternating current* (Fig. 12-30).

In practice, the rotating coil in an AC generator does not consist of a single loop of wire. Instead, a coil of many loops is used so that there are many wire segments cutting field lines, thereby increasing the amount of current generated. Furthermore, permanent magnets are rarely used in generators.

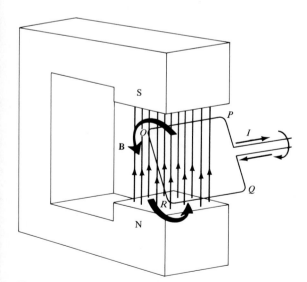

Figure 12-29 An electric generator. As the coil of wire rotates in the field, an alternating electrical current (AC) is induced to flow in the wire.

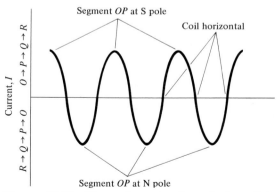

Figure 12-30 Flow of current through the coil in Fig. 12-29.

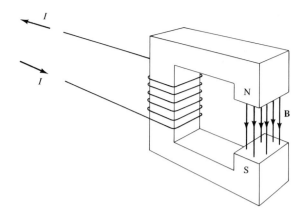

Figure 12-31 An electromagnet. The field lines are carried by the C-shaped piece of iron and produce a strong field in the gap. Use the right-hand rule which gives the direction of the field lines around a current-carrying wire (Fig. 12-17) to verify the direction of the field in this diagram.

Higher field strengths (more field lines that can be cut) are achieved by using *electromagnets*. (Such a magnet is shown in Fig. 12-31). A wire that carries a direct current is wound around a C-shaped piece of iron. The field produced by the current in the wire loops (see Fig. 12-18) is concentrated by the iron, and in the space between the poles there is a strong magnetic field.

Lenz's Law

Refer again to Fig. 12-20. Because of the current in the wire, the wire experiences a magnetic force $\mathbf{F_M}$ and begins to move. As soon as the wire has a velocity relative to the field, there will be a new magnetic force, $\mathbf{F'_M}$, on the charged particles due to their motion in the direction of the original $\mathbf{F_M}$. Application of the right-hand rule shows that $\mathbf{F'_M}$ is in the direction *opposite* to the direction of current flow. That is, there is an induced current that tends to oppose the original current flow. This is a general result—if any electromagnetic change A causes an effect B, then B will always induce a reaction C that tends to oppose A. This principle was discovered by Heinrich Lenz (1804–1865), a German physicist, and is known as *Lenz's law*.

Although Lenz's law is reminiscent of Newton's third law, it is, in fact, a completely distinct statement. Actually, Lenz's law is simply a statement of energy conservation applied to induced currents. (If induced currents *aided* one another, in opposition to Lenz's law, then the currents would grow without the benefit of any energy input to the system. This is a clear violation of energy conservation.)

Transformers

A current will be induced in a wire whenever the magnetic field in its vicinity is changing with time. This will be the case when there is relative motion between the wire and the field. But mechanical motion is not required in order that the wire experience a changing field. Consider the circuits shown in Fig. 12-32. When the switch S is open, no current flows in circuit A and, hence, there is no magnetic field around the wire. What happens when the switch is closed? The field does not instantaneously appear at its full strength. Even though it requires only a small fraction of a second to do so, the field does build up with time from zero field to its final value. During this build-up period, we can imagine the circular field lines growing outward from the wire, as shown in Fig. 12-32. As these field lines "move" outward, they cut across the wire in circuit B. The charges in this wire experience a magnetic force just as they would if the wire were moved through a steady field. Therefore, a current is induced in circuit B and this current will be registered by the current meter. When the field has grown to its final, steady value, there is no

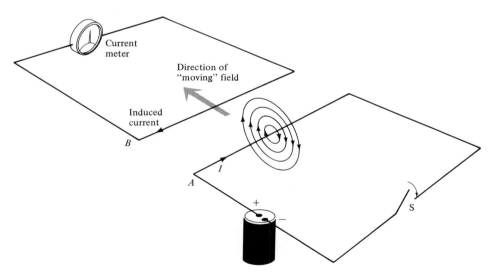

Figure 12-32 When the switch S is closed a current flows through circuit A. As the magnetic field builds up around the wire in circuit A, the field lines "move" outward and cut across the wire in circuit B. This relative "motion" between the wire and the field induces a current in circuit B.

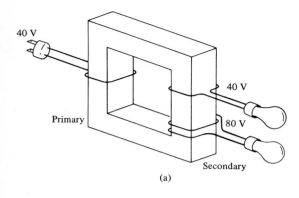

40 V

Primary

40 V

80 V

Secondary

(a)

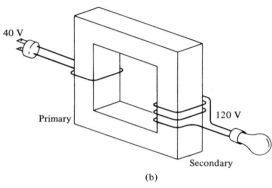

40 V

Primary

120 V

Secondary

(b)

Figure 12-33 Schematic of various coil connections in transformers.

longer any relative motion between the wire and the field lines, and the induced current will decrease to zero.

If the switch is now opened, the field will collapse and a current will again be induced in circuit B as the field lines "move" across the wire in the opposite direction. A current will flow in circuit B (in the opposite direction) until the field has decreased to zero.

If we open and close the switch in circuit A at regular intervals, the induced current in circuit B will flow first in one direction and then in the other. That is, an alternating current will be induced in circuit B. This is basically what happens in the electrical device called a *transformer*.

Figure 12-33 shows the operation of a transformer in a schematic way. The primary coil is connected to a source of AC and is looped around an O-shaped piece of iron which concentrates the field lines and carries them to the secondary coil. As the field changes in the primary, building up first in one direction and then in the other due to the alternating current, a similar alternating current is induced in the secondary. The primary coil is shown as a single loop of wire, but in a practical transformer there are many loops. Figure 12-33a shows two secondary coils, one with a single loop and one with two loops. If the primary is connected to a 40-V source of AC, the induced voltage across the single-loop secondary coil will also be 40 V. The induced voltage across the double-loop secondary, however, will be 80 V because each field line generated by the primary cuts across *two* wires in this secondary. The ratio of the secondary voltage to the primary voltage is the same as the ratio of the number of loops in the secondary coil to the number of loops in the primary coil. Figure 12-33b shows a 3-to-1 ratio of loops and a 3-to-1 ratio of voltages.

In a transformer with more loops in the secondary coil than in the primary, the voltage is increased or *stepped up*. If the primary coil has the greater number of loops, the voltage is *stepped down*. When the voltage is stepped up in a transformer, what happens to the current flow? Only as much power can be extracted from the secondary of a transformer as is delivered to the primary (conservation of energy). In Fig. 12-33b, if the light bulb has a resistance of 60 Ω, the current flowing through the bulb will be (using Ohm's law) $I = V/R = 120 \text{ V}/60 \ \Omega = 2 \text{ A}$. Thus, the power supplied by the secondary to the bulb is $P = VI = (120 \text{ V}) \times (2 \text{ A}) = 240 \text{ W}$. The 40-V source which drives the primary must

Electrical Meters—How They Work

We have seen how the movement of a coil of wire through a magnetic field causes a current to flow—this is the principle of the electric generator (Fig. 12-29). It is also true that when a current flows through a coil of wire located in a magnetic field, the charges are acted upon by a magnetic force which is transmitted to the wire as a whole, thereby causing the wire to move—this is the way in which electrical meters work. Figure 12-34 is a schematic diagram of a meter that is used to measure electrical current and is called a *galvanometer* or *ammeter*. (Only very small currents can be passed through the delicate wires in a galvanometer. Therefore, if large currents are to be measured, most of the current is made to flow through a parallel resistor instead of through the galvanometer meter movement. See Fig. 12-35a.) The basic parts of the meter are a permanent magnet, a wire coil, a spring, a pointer, and a scale. If the meter terminals, A and B, are attached to an external circuit consisting of a battery V and a resistor R, a current I will flow through the wire coil. Note the direction of current flow in the coil relative to the direction of the field lines (not shown in the diagram) which extend from the N pole to the S pole. By applying the right-hand rule, we can see that the vertical sides of the coil will experience forces in the directions of the arrows. These forces produce a torque which causes the coil to rotate in a clockwise sense around the suspension wire. The pointer, which is attached to the coil, moves over the scale. When the torque produced by the magnetic forces is just balanced by that due to the spring S (the lower end of which is attached to the meter case), the pointer comes to rest and the scale indicates the magnitude of the current. The spring S returns the pointer to the zero position when there is no current flow.

If it is desired to know the current flowing in a circuit or in a particular

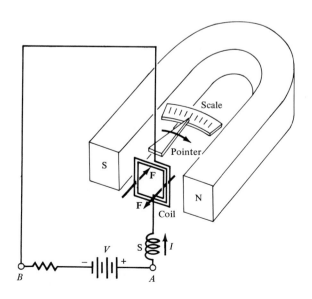

Figure 12-34 Schematic diagram of a galvanometer. The meter connections are A and B; the battery and the resistor represent the external circuit. If large currents are to be measured, the meter is protected by placing it in parallel with a low-value resistor so that most of the current will flow through the resistor, as in Fig. 12-35a.

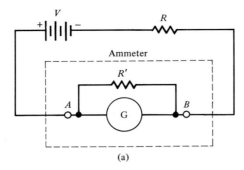

(a)

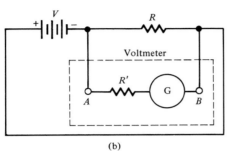

(b)

Figure 12-35 (a) A galvanometer is placed into the circuit consisting of the battery V and the resistor R in order to measure the current. The meter is protected by the parallel resistor R' which has a very low resistance. The meter is calibrated to read directly in amperes for the particular meter resistor R'. The dotted box represents the resulting *ammeter*. (b) The galvanometer is connected to measure the voltage across R. The meter is protected in this case by a series resistor R' which has a very large resistance. The meter is calibrated to read directly in volts and the dotted box represents the resulting *voltmeter*.

part of a complex circuit, the meter is placed directly into the circuit. In order to protect the galvanometer, a resistor with a low resistance value is placed in parallel with the galvanometer, as in Fig. 12-35a. The meter–resistor combination, shown by the dotted outline in the figure, is an *ammeter* and is calibrated to read directly in amperes. The value of R' is made very low so that there will be essentially no change in the current when the meter is inserted into the circuit.

A galvanometer can also be placed in a circuit in such a way that the *voltage* across a particular circuit element can be determined. The method is shown in Fig. 12-35b. Here, the galvanometer is placed in *parallel* with the resistor R. The internal resistance of the meter is low; therefore, if only the meter were placed in parallel with R, most of the current in the circuit would flow through the meter. Thus, the situation would be totally different from that with the meter removed from the circuit. (Moreover, the meter could be damaged.) In order to prevent such a drastic change, a high resistance R' (much greater than R) is placed in series with the galvanometer. Now, there is almost no change in the current through R. The voltage across R is essentially the same as the voltage across R' (because the meter resistance is very low in comparison). This voltage is equal to the value of R' multiplied by the current read on the galvanometer.

Instead of using a galvanometer in this way and multiplying I and R' in order to obtain V, we usually incorporate R' into the meter, thereby converting it into a *voltmeter*, as shown by the dotted line in Fig. 12-35b. The scale is also changed to read directly in volts.

be supplying this amount of power. Therefore, the current in the primary must be $I = P/V = 240$ W/40 V = 6 A. In this transformer, the voltage is stepped *up* by a factor of 3 and the current is stepped *down* by a factor of 3; the product $V \times I$ is the same in the primary as in the secondary (neglecting losses due to heating effects).

By using a transformer, low-voltage AC can be converted into high voltage AC or vice versa. Power from electrical generating plants is stepped up to high voltage (low current) for transmission so that the I^2R heating losses will be minimized (see Section 11-3). At the consumer end of the line, the voltage is stepped down to 220 V or 110 V for household use.

SUGGESTED READINGS

B. Dibner, *Oersted and the Discovery of Electromagnetism* (Blaisdell, Waltham, Massachusetts, 1962).

L. P. Williams, *Michael Faraday* (Basic Books, New York, 1965).

Scientific American articles:

A. Cox, B. Dalrymple, and R. R. Doell, "Reversals of the Earth's Magnetic Field" February 1967.

H. I. Sharlin, "From Faraday to the Dynamo," May 1961.

QUESTIONS AND EXERCISES

1. Can two electric lines of force ever cross? Explain.

2. If a small charge is released from rest at some point in the field shown in Fig. 12-3, will it travel along one of the field lines? Explain. (What do the field lines describe, *force* or *motion?*)

3. What is the strength of an electric field that exerts a force of 3.2×10^{-16} N on an electron?

4. Two parallel plates are separated by a distance of 2 cm. What voltage must be placed across the plates so that the electric field strength between the plates will be 1000 V/m?

5. What is the velocity of a proton that is accelerated, starting from rest, through a potential difference of 10^6 eV? How would you express the kinetic energy of a neutron moving with this same velocity?

6. If a compass magnet (near the surface of the Earth) is suspended on a horizontal axis, it will point down at a certain angle. Explain why.

7. Does it seem that there is any way to separate a magnet into individual N and S poles? Explain.

8. In what direction will a compass magnet point if it is located at the north *geomagnetic* pole? At the north *geographic* pole?

9. Do magnetic field lines have a beginning and an end? (Refer to the various illustrations of field lines. Do they all show the same property?)

10. Make a sketch similar to Fig. 12-18. Choose a direction for the current flowing in the wire loop and use the right-hand rule to label the N and S poles of the field. Use this information and determine the direction that the *electrons* must be flowing in the Earth's core to produce the observed polarity of the Earth's magnetic field.

11. A current-carrying wire lies in the north–south direction. A compass magnet placed immediately *below* the wire points toward the *west*. In what direction are the *electrons* in the wire flowing?

12. An α particle ($^4\mathrm{He}^{2+}$ ion) moves eastward and horizontal to the Earth's surface at the Equator. In what direction does the Earth's field exert a force on the α particle?

13. An electron is projected into a current loop along the axis of the loop. Describe the motion of the electron.

14. An electron moves with a velocity $v = 2.5 \times 10^7$ m/s in a magnetic field of 50 gauss. The velocity vector makes an angle $\theta = 45°$ with respect to **B**. What is the magnetic force on the electron?

15. A current of 5 A flows through a wire which has 0.5 m of its length in and perpendicular to a magnetic field. The wire experiences a force of 0.25 N. What is the strength of the magnetic field?

16. If a 10-MeV proton moves in an orbit with a radius of 2 m in a magnetic field, what is the strength of the field?

17. What is the orbit radius of a 5-MeV α particle ($^4\mathrm{He}^{2+}$ ion) in a magnetic field of 3000 G?

18. A nitrogen ion ($^{14}\mathrm{N}^{3+}$) is accelerated through a potential difference of 1.6 MV and enters a magnetic field of 0.6 T. What is the radius of the orbit?

19. In the example in Section 12-4, we found that a 1-MeV proton will execute an orbit with a radius of 0.29 m in a magnetic field with a strength of 0.5 T. What is the strength of the field in which an electron moving with the same speed will have the same orbit radius?

20. A proton moves with a velocity v in a circular orbit with a radius R in a field $B = 5000$ G. To what value must the field strength be changed if an α particle ($^4\mathrm{He}^{2+}$ ion) moving with a velocity $2v$ is to execute a circular orbit with the same radius?

21. A current of 300 A flows in a long, straight wire. An electron is moving parallel to the wire at a distance of 2 cm with a velocity of 6×10^7 m/s. What is the magnetic force on the electron?

22. Two identical circuits are arranged as shown in the diagram below. Each circuit carries a current $I = 2$ A in the same direction through the 1-m segment that is closest to the other circuit; these segments are spaced 1 cm apart. Consider only the 1-m segments; the magnetic effects due to the other parts of the circuit are small in comparison. Does

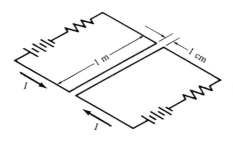

one segment attract or repel the other segment? What is the magnetic force exerted by one segment on the other?

The *ampere* is defined in terms of the force produced in a circuit such as that considered here. If the same current flows in two long, straight, and parallel wires separated by a distance of 1 m, and if the force between the wires is equal to 2×10^{-7} N for each meter of wire length, then the current is defined to be 1 A.

23. A bar magnet is dropped through a horizontal loop of wire with the N pole entering first. Describe the induction in the wire. Use Lenz's law to determine whether the magnet will experience an acceleration greater than or less than g while passing through the loop.

24. A wire is wound around a cardboard tube and the ends are connected to a current meter. What will happen if a bar magnet is thrust into the tube? If the bar magnet is at rest inside the tube and is suddenly pulled out, what will happen? Make a sketch of the first situation, labeling all important quantities.

25. Will a transformer operate on direct current (DC)? Explain.

26. In Fig. 12-33a, suppose that each light bulb has a resistance of 40 Ω. What is the current through each bulb? How much power is being supplied by the secondary? What is the current in the primary?

27. The input to a certain transformer is 110 V at 3.5 A. The output is 15 V at 24 A. What is the *efficiency* of the transformer? (How much of the input power is lost due to heating?)

13

WAVES

Wave motion is one of the most familiar of natural phenomena. Everyone has watched water waves and we have all been fascinated by the way in which they move across the water and crash upon the beach. Our sense of hearing depends upon the fact that our ears are sensitive to the sound waves that travel through air. And, electromagnetic waves, in the form of radio waves, heat radiation, and light, are all around us. We are, in fact, continually immersed in a sea of different wave motions.

In this chapter we will describe how waves are produced and we will discuss some of their important features. In the following chapter we will extend the discussion to electromagnetic radiation and in Chapter 15 we will give more details about the wave character of light.

13-1 WAVE PULSES ON SPRINGS AND STRINGS

Formation of a Wave Pulse

Suppose we have a long loose coil of wire, such as a "slinky" toy, laid out in a straight line. If we grasp one end and quickly move it to one side and then back to its original position, as shown in Fig. 13-1, what will we observe? Here, we deal, not with the motion of a single particle as we have in most of our previous discussions, but with a *collection* of particles that are bound together and which are capable of exerting forces on one another.

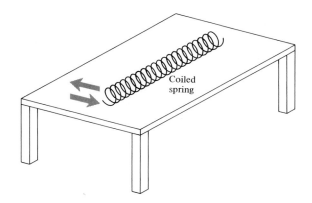

Figure 13-1 The end of a loosely coiled spring which lies flat and straight on a table is quickly displaced to the left and then is returned to its original position. A wave pulse will travel down the spring (Fig. 13-2).

The motion we observe is the *collective motion* of the particles in the spring.

We can imagine that the spring consists of a number of particles linked together in the schematic way shown in Fig. 13-2. As we pull the first particle aside, a force is exerted on the second particle which then begins to move. The movement of this particle away from its original position exerts a force on the third particle, and so on. In Fig. 13-2c, the end particle has been moved to the position of maximum displacement and is stationary. But the second and third particles are in motion and a force is being exerted on the fourth particle. In sequences (d), (e), and (f), the end particle is returned to its original position, where it remains. The motion is transmitted from particle to particle along the line, and a *wave pulse* is propagated down the spring.

Notice that work was done in moving the end particle against the force exerted on it by the second particle. The energy delivered to the end particle remains in the spring as the kinetic energy of the moving particles and as elastic potential energy. (In a real case, friction would eventually damp the motion, and the energy would be transformed into heat.)

The sequence of photographs on page 307 shows the propagation of a wave pulse along a coiled spring. The pulse was formed in exactly the way we have described, namely, by giving a quick displacement to the end. Notice the important point that although the pulse travels along the spring, no part of the spring itself moves very far. The photographs show that a piece of ribbon tied to one loop of the spring moves only a short distance up and down as the wave pulse passes. All types of mechanical wave pulses—whether on springs or strings, on water, or in the air—are characterized by the

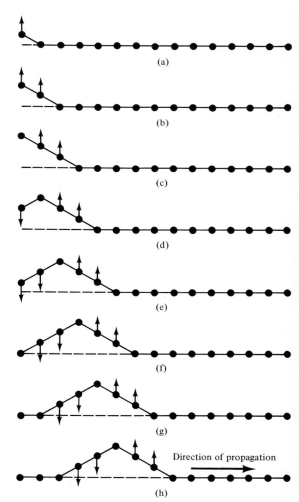

Figure 13-2 Development of a wave pulse by pulling aside the end particle in the line and then returning it to its original position.

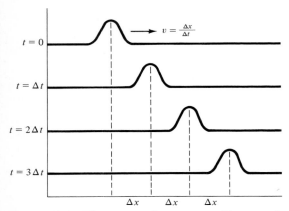

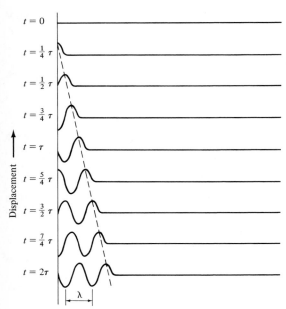

Figure 13-3   The wave pulse moves with a speed $v = \Delta x / \Delta t$.

Figure 13-4   By driving the end of a string up and down in a regular way, a *traveling wave* is formed. The *period* of the wave is τ and the *wavelength* is λ. The dashed line shows the steady movement forward of the first wave maximum.

transfer of motion (and energy) from particle to particle in the medium, but in no case does any part of the medium move any appreciable distance. Waves transport *energy,* not matter.

Wave Speed

The speed of a wave pulse can be determined in the same way that we would determine the speed of a moving object. Figure 13-3 shows a wave pulse at several instants separated by equal intervals of time, Δt. In each such interval, the peak of the wave pulse moves forward a distance Δx. Therefore, according to the usual definition of *speed,* we can write

$$v = \frac{\Delta x}{\Delta t} \qquad (13\text{-}1)$$

The speed of a pulse along a spring or string will be *constant* if the spring or string has uniform properties along its length. If, for example, the size of wire varies along a spring, the speed of a pulse will be affected, slowing down where the wire is thick and speeding up where the wire is thin. (Can you see why the speed changes in this way?)

13-2 TRAVELING WAVES

Period and Wavelength

A wave pulse is formed when we give a single displacement to the end of a spring or string. What will happen if we continue to displace the end in a regular manner, first to one side and then to the other? Figure 13-4 shows that a repeating pattern of wave pulses is formed in this case. Each pulse is exactly the same as any other pulse and each joins smoothly onto the one in front of it and the one behind. The train of pulses moves uniformly along the string and the result is called a *traveling wave.*

The driven end of the string in Fig. 13-4 moves up, and down, and back to the starting position in a time interval indicated by τ. The time τ is the *period* of the driving force and is also the *period* of the traveling wave. That is, at any point along the string, the motion repeats itself in every time interval equal to τ. During each interval τ, the wave form moves forward by a certain distance, indicated by λ in Fig. 13-4. If we examine the shape of the string at any instant, we find that the pattern repeats itself in every distance interval λ. This distance λ is called the *wavelength* of the traveling wave.

Let us look again at Eq. 13-1, keeping in mind the definitions of the period and the wavelength. The wave moves through a distance $\Delta x = \lambda$ in an interval of time $\Delta t = \tau$. Therefore, we can use Eq. 13-1 to express the relationship connecting wave speed, period, and wavelength:

$$\text{wave speed} = v = \frac{\Delta x}{\Delta t} = \frac{\lambda}{\tau}$$

$$= \frac{\text{wavelength}}{\text{period}} \qquad (13\text{-}2)$$

Frequency

The *period* of a wave is the time period between the passage of two successive wave crests past a given point (refer to Fig. 13-4). The *frequency* of a wave is how frequently the wave crests pass a given point. If 100 wave crests pass a point in 1 second, the frequency is 100 per second (100 s^{-1}). In this same case, the time interval between the passage of successive wave crests past the point is $\frac{1}{100}$ of a second; therefore, the period is 0.01 s. That is, the frequency v (Greek *nu*) is the *reciprocal* of the period τ:

$$v = \frac{1}{\tau} \qquad (13\text{-}3)$$

Frequency is usually expressed in *cycles/second,* or simply s^{-1}. This unit of frequency is called the *hertz* (Hz), in honor of the German physicist Heinrich Hertz (1857–1894) who made great contributions to the study of electrical waves: 1 Hz = 1 s^{-1}.

We can now combine Eqs. 13-2 and 13-3 to write the wave speed as

$$v = \frac{\lambda}{\tau} = \lambda v \qquad (13\text{-}4)$$

This is a general expression and is valid for all types of wave motion—waves on springs, sound waves, water waves, and radio waves. For each type of wave there is a particular value of the wave speed v that applies.

Suppose that we measure the wave speed on a particular spring to be 6 m/s (which we can write as 6 m s^{-1}). If we vibrate the end of this spring back-and-forth 3

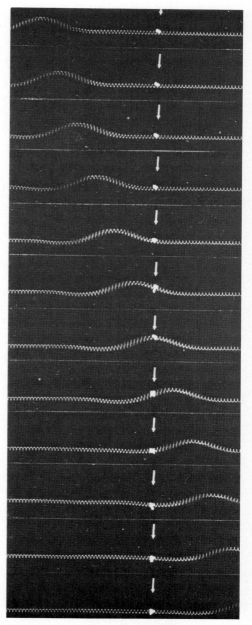

Photographic sequence (taken from frames of a movie) of the propagation of a wave pulse along a coiled spring. The arrow indicates a piece of ribbon tied to one of the coils. Notice that the ribbon moves only a short distance up and down as the pulse passes.

times each second, what will be the wavelength of the wave that travels along the spring? The frequency of the wave is the same as the frequency of the driven end. Therefore, solving Eq. 13-4 for the wavelength, we find

$$\lambda = \frac{v}{\nu} = \frac{6 \text{ m s}^{-1}}{3 \text{ s}^{-1}} = 2 \text{ m}$$

Transverse Waves

Look again at the photographic sequence on page 307 and at Fig. 13-4. In each case notice that the wave propagates to the right but that the particle motions are up and down. Waves that propagate in one direction while the particles of the system or medium are vibrating back and forth in the perpendicular direction are called *transverse waves*. (The direction of particle motion is *transverse* to the direction of wave propagation.)

Water Waves

The surface waves that propagate across deep water are similar to the waves on springs, but water waves are not exactly transverse. Figure 13-5 shows the motions of the water particles that take place when a wave moves on the surface of a body of water. The particles move up and down in small circular orbits as the wave advances. The particles at the top of the wave (the wave *crest*) move in the same direction as the wave, but the particles at the bottom of the wave (the wave *trough*) move in the opposite direction. As in the case of waves on springs, there is no net forward movement of the particles as a water wave propagates.

In an ideal water wave the particles do not undergo any net displacement and the wave cannot transport matter. But real waves are not perfect, and frictional

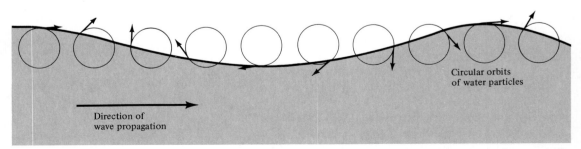

Figure 13-5   The motion of water particles during the propagation of a surface wave in deep water. The particles move in small circular orbits with no net forward motion.

drag effects combine with winds to permit the movement of matter by the waves. Driftwood is continually carried onto beaches by the action of waves, and coastal areas undergo frequent changes as the beach sands are moved by waves.

Ocean Waves

Surface waves on open water build up due to the action of winds. But anyone who has ever seen the sea knows that the waves are not the simple ideal waves we have been discussing. The winds are variable in direction and strength, and ocean waves are therefore complex and irregular, ranging from ripples and chop to giant crashing waves. In describing these waves, one can only refer to *average* properties.

The size of wind-driven waves depends on the speed and the duration of the wind. In a strong, steady wind of 50 km/h, the waves will continue to grow for a day, eventually reaching an average height of about 4 m. For *whole gale* winds (90 km/h), the average height of waves is about 13 meters or 45 feet.

The heights of ocean waves are usually measured from trough to crest. Waves in the North Atlantic with heights greater than 45 feet are fairly common. In 1933, the U.S.S. *Ramapo*, a Navy tanker, was in a Pacific storm which had peak winds of 125 km/h. In one enormous wave, the stern dipped into the trough and the watch officer on the bridge saw the seas astern at a level above the main crow's nest. From the size of the ship and the placement of the bridge and the crow's nest, it was computed that the wave height was at least 112 feet!

Energy in Waves

It is easy to see that a wave with a large height possesses more energy than a wave with a lesser height. How does wave energy depend on the height (or *amplitude*) of the wave? We can deduce the relationship by referring to Fig. 13-6. Here we have two wave patterns (drawn as rectangular pulses for simplicity), the second with twice the amplitude of the first. In order to form the wave pulse, a certain amount of work must be done. If we are considering a water wave, a volume of water must be raised from beneath the normal surface to a height equal to the amplitude of the wave. For the wave with an amplitude of 1 unit, this amounts to 1 unit of work (Fig. 13-6a). If the wave amplitude is increased to 2 units (Fig. 13-6b), an additional equal volume of water must be raised through a height 3 times greater; this

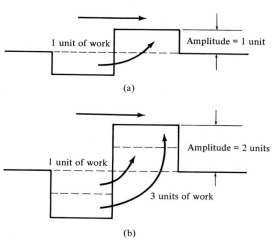

Figure 13-6 The work required to form a wave with an amplitude of 2 units (b) is 4 times that required to form a wave with an amplitude of 1 unit (a). The energy content of a wave is proportional to the *square* of its amplitude.

requires 3 units of work. Therefore, the formation of a wave pulse with an amplitude of 2 units requires 4 times as much work as does a pulse with an amplitude of 1 unit. If we considered a pulse with an amplitude of 3 units, we would find that the work required is 9 times that for the first pulse. That is, the work required to form a wave pulse, and, consequently the energy content of the wave, is proportional to the *square* of the amplitude. This relationship is valid for all types of waves.

13-3 STANDING WAVES

Stationary Wave Patterns

We have been discussing waves that travel on very long springs or strings or across a large expanse of water—ideally, these waves continue to move forward forever. Suppose that we now consider a short string

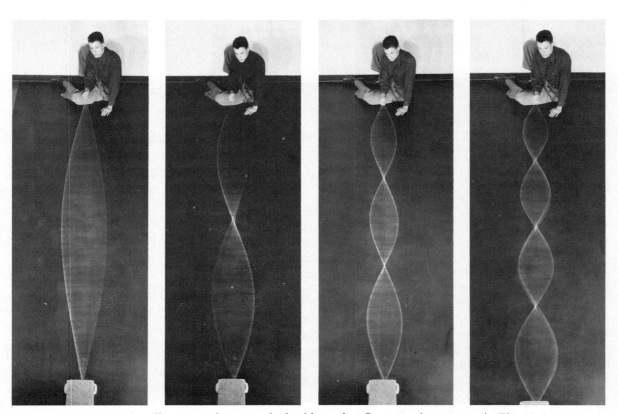

Standing waves in a stretched rubber tube. Compare the patterns in Fig. 13-7.

whose ends are attached to rigid supports. If we grasp the string near one of the supports and move the string back and forth, a traveling wave will be propagated along the string. When the wave reaches the fixed support at the other end, it will be *reflected* and will begin to travel back toward the driven end. We now have *two* waves traveling in opposite directions on the same string. How does this change the vibration pattern of the string?

If the frequency at which the string is vibrated has been poorly chosen, the direct wave and the reflected wave will combine to produce a jumbled wave pattern. By changing the frequency and observing the effect on the string, we can find a number of particular frequencies that produce regular patterns of motion along the string. At these frequencies, certain positions along the string remain stationary (these points are called *nodes*) while the rest of the string vibrates. These regular wave patterns are called *standing waves*. See the photograph on page 310.

The lowest frequency at which a standing wave can be set up produces a wave pattern with nodes only at the fixed ends (Fig. 13-7a). This frequency is called the *fundamental* frequency for the particular string. If the string is vibrated with a frequency that is an integer multiple of the fundamental frequency (and only for such frequencies), standing waves with different patterns will be set up. Figure 13-7 shows the standing wave patterns for the fundamental frequency ν_0 and for the frequencies $2\nu_0$, $3\nu_0$, and $4\nu_0$. The higher frequencies are called *harmonics* or *overtones*. The frequency $2\nu_0$ is the first overtone or second harmonic, $3\nu_0$ is the second overtone or third harmonic, and so forth.

The requirement for a standing wave on a string is that nodes exist at both ends of the string (because the ends are fixed). Therefore, referring to Fig. 13-7, we see that the wavelength of the standing wave bears a definite relationship to the distance L between the fixed supports of the string. In Fig. 13-7d, exactly *two* wavelengths fit between the supports, and in (b) exactly *one* wavelength fits between the supports. Figure 13-7a shows that the distance L is equal to exactly *one-half* wavelength of the fundamental mode of vibration. In general, the wavelength λ of a standing wave is related to the distance L between the supports by

$$n\frac{\lambda}{2} = L, \qquad n = 1, 2, 3, \ldots \qquad (13\text{-}5)$$

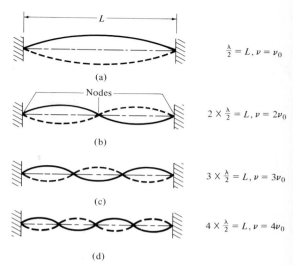

Figure 13-7 Standing waves on a string with fixed ends. The fundamental (a) and the first three overtones are shown. The fundamental frequency is ν_0.

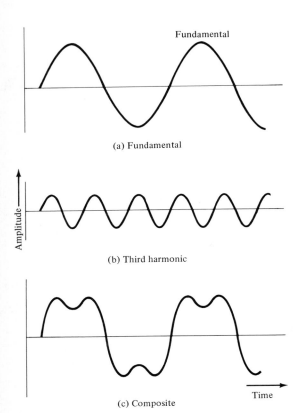

(a) Fundamental

(b) Third harmonic

(c) Composite

Time

Figure 13-8 The composite wave pattern produced by combining the fundamental with the second overtone (or third harmonic). The waves produced by musical instruments usually involve several harmonics and therefore have wave patterns more complex than shown here.

A piano string is a good example of a string that vibrates between fixed supports. When the padded hammer that is connected to a key strikes one of the strings, the string does not vibrate with a pure frequency. Instead, the vibration consists of the fundamental mixed with several of the harmonics.

The Superposition of Wave Patterns

When we discuss complicated waves that consist of vibrations with a number of different frequencies, we usually do so in terms of the variation of the amplitude with *time* instead of the variation with *distance*. A common way of displaying this variation is by connecting the signal (for example, the output of a microphone) to the inputs of a cathode ray oscilloscope. The trace on the oscilloscope face will show directly the time variation of the wave pattern. Figure 13-8 shows a simple case of a composite wave due to the combination (or *superposition*) of vibrations with two different frequencies. Figure 13-8a is the pattern that would be observed for the fundamental alone, and Fig. 13-8b shows the pattern for the third harmonic. Notice that the amplitude of the harmonic is somewhat smaller than that of the fundamental. When these two patterns are combined, the result is that shown in Fig. 13-8c. Notice that the amplitude of the composite wave at any instant is just equal to the *sum* of the amplitudes of the two contributing waves.

If it were not for the production of the harmonic vibrations, the sound produced by a musical instrument would be dull and uninteresting. The quality or *timbre* of the sound is governed by the number and the intensity of the harmonics that are produced. When the same A note is played on a violin and a piano, the sounds are quite different. We have no difficulty in distinguishing between a violin A and a piano A even though the fundamental frequency of both tones is 440 Hz. The reason is that the harmonics generated when a violin string is stroked to produce the note are much more intense than are the harmonics generated by the piano string. Figure 13-9 shows the wave form of the note A produced by a violin and by a piano. The violin note, with its complicated structure, is much richer in harmonics than is the piano wave form.

The sound that we hear and identify as a violin A note can be synthesized by combining the pure notes generated by individual vibrating strings or electrical tone generators. If the mixture of the fundamental and harmonics is the same as that produced by the violin, the resulting tone will sound exactly the same as that from a violin. Indeed, electric organs synthesize tones in just this way. Such organs can generate tones that duplicate a variety of different instruments by combining pure notes in different ways. It is difficult, however, to produce exactly the correct mixture of pure notes, and an artificial violin tone from an electric organ sounds slightly different than a real violin tone. The ear is quite sensitive to the presence of extra harmonics or to those that should be present but are missing.

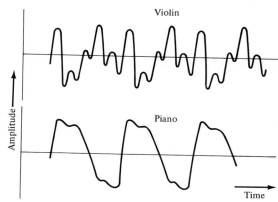

Figure 13-9 Wave forms of the musical note A (440 Hz) when played on a violin and on a piano. The violin sound is much richer in harmonics than is the piano sound. Can you identify the fundamental period in each case?

13-4 SOUND

Longitudinal Waves

The disturbance propagating on the spring pictured on page 307 is a *transverse* wave pulse. This pulse was started by giving a transverse displacement to the end of the spring. What will happen if we give to the spring a quick push and pull *along* the direction of the spring instead of perpendicular to it? In this case, the spring coils near the end are first compressed together and then expanded. The result is a *compressional* pulse that propagates along the spring (Fig. 13-10). Because the motion of the particles in the spring takes place along the direction of the pulse, this kind of disturbance is called a *longitudinal* wave pulse.

If the push–pull driving action on the end of the spring is repeated in a regular fashion, a *longitudinal* or *compressional wave* will propagate along the spring (Fig. 13-11a). The characteristics of this type of wave—wave speed, wavelength, and frequency—are related by Eq. 13-4 just as they are for transverse waves.

A compressional wave in air can be set up by the back-and-forth motion of a piston that is fitted into a tube (Fig. 13-11b). Here, the air molecules in the tube are alternately pressed together and pulled apart by the action of the moving piston. The result is a propagating wave in which the density of the air varies with distance in a regular way—the density pattern is, in fact, exactly the same as the displacement pattern of a transverse wave on a string (Fig. 13-4). Compressional waves in air are what we call *sound waves*. Sound waves are always *longitudinal* waves. (Can you see why transverse sound waves in air are not possible? What could cause air mol-

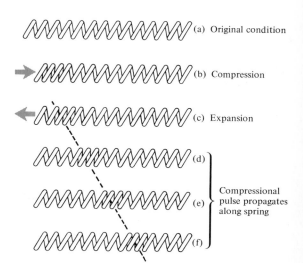

Figure 13-10 The development of a compressional wave pulse in a spring by a quick compression and expansion of the coils near one end.

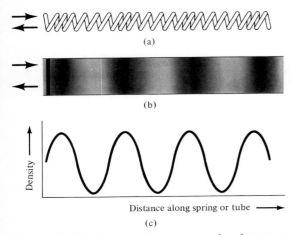

(a)

(b)

(c)

Distance along spring or tube ⟶

Figure 13-11 Propagating compressional waves (a) in a spring and (b) in an air-filled tube. (c) The density of the air in the tube or the degree of compression of the spring coils as a function of distance.

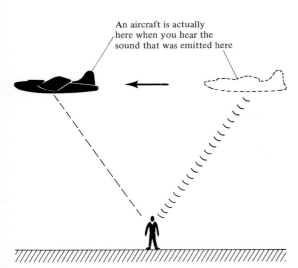

An aircraft is actually here when you hear the sound that was emitted here

Figure 13-12 Because the speed of sound is so much slower than the speed of light, the sound from a rapidly moving aircraft appears to come from a position far behind the actual position.

ecules displaced in the transverse direction to return to their original positions?)

The Speed of Sound

When you see a lightning stroke during a thunderstorm, you do not hear the thunder until several seconds after the flash. When you hear a jet aircraft overhead and attempt to locate it by looking in the direction from which the sound is coming, you find that your line of sight falls a considerable distance behind the aircraft (Fig. 13-12). Both of these effects are due to the fact that sound travels through air with a speed that is extremely slow compared to the speed of light. Thus, the light from a lightning flash reaches your eyes almost instantaneously, but if the stroke is a mile away, there will be a delay of about 5 seconds before the sound of the thunder reaches your ears.

At sea level and at a temperature of 0 °C, the speed of sound in air is approximately 330 m/s or 1100 ft/s. The speed is essentially the same for all frequencies of sound, but the speed does depend on the pressure and the density of the air. At an altitude of 40 000 ft or 12 km (where many jetliners fly), the pressure and density conditions are such that the speed of sound is about 13 percent smaller than at sea level.

Aircraft speeds are frequently stated in units of the speed of sound using *Mach numbers*. A speed of Mach 1 corresponds to the speed of sound (750 mi/h at sea level); Mach 2 corresponds to twice the speed of sound; and so forth. Because of the variation of the speed of sound with altitude, an aircraft flying at Mach 0.9 near sea level actually moves faster than an aircraft flying at Mach 0.9 at a high altitude.

The speed of sound in air and other gases is limited by the fact that the moving molecules must collide with one another in order to propagate the compressional wave (see the discussion in Section 10-4). In liquids and solids, in which the molecules are closer together and interact more strongly with one another, the speed of sound is substantially greater than it is in a gas. Table 13-1 gives the speed of sound in several different materials.

Audible Sound

When a compressional wave in air (that is, a *sound* wave) reaches our ears, it produces vibrations in the membranes of the ears. These vibrations provoke a nervous response and we have the sensation of *hearing* the sound. But not all sound waves are audible. The human ear responds to (*hears*) a sound only if the frequency is in the range from about 16 Hz to about 20 000 Hz. The wavelengths corresponding to these extreme frequencies are

$$\lambda_{\text{long}} = \frac{v}{\nu} = \frac{330 \text{ m/s}}{16 \text{ s}^{-1}} \cong 20 \text{ m}$$

$$\lambda_{\text{short}} = \frac{v}{\nu} = \frac{300 \text{ m/s}}{20\ 000 \text{ s}^{-1}} \cong 0.016 \text{ m} = 1.6 \text{ cm}$$

Sound waves can be produced in air with wavelengths longer than 20 m and shorter than 1.6 cm, but such waves are not audible to humans. For frequencies below 16 Hz, we do not hear a continuous sound; instead, the ear (and, indeed, the body as a whole) detects or *feels* a series of individual pulses. Furthermore, the sensitivity of the human ear to high frequencies tends to decrease with age. A child may be able to hear frequencies above 20 000 Hz, but by middle-age he may be unable to perceive any sound with a frequency above 12 000 or 14 000 Hz. (By the time a person has accumulated a superhigh-fidelity system, he is often unable to appreciate its high-frequency response!)

Sound Intensity

There are two aspects to any sound that we hear: the *pitch* (or frequency) and the *loudness* or (intensity). The *intensity* of a wave is a measure of the amount of energy per second it can deliver to unit area of a surface. (Intensity is measured in watts per square meter.) The frequency of a sound wave depends on the vibration rate, and the intensity depends on the amplitude. These two characteristics of a wave are independent — that is, we can change either the frequency or the intensity of a wave without altering the other (see Fig. 13-13).

When a sound is *heard*, the ear is responding to energy that is delivered to the ear membranes by the sound wave. The range of sound intensity over which the human ear is sensitive is incredibly large. From the loudest sound that can be tolerated without pain to the softest sound that can be detected, the energy ratio is about 10^{12} — *twelve* orders of magnitude! If the sound energy reaching the ear membranes exceeds about 1 watt/m², we experience a painful effect. But at the other end of the range, the threshold of hearing corre-

TABLE 13-1 SPEED OF SOUND IN VARIOUS MATERIALS

MATERIAL	SPEED (m/s) AT 0 °C
Air	330
Lead	1210
Sea water	1450
Iron	4480
Granite	up to 6000

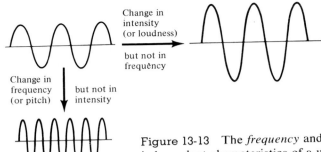

Figure 13-13 The *frequency* and the *intensity* are independent characteristics of a wave.

Change in frequency (or pitch) but not in intensity

Change in intensity (or loudness) but not in frequency

sponds to an energy input to the ear of only about 10^{-12} watt/m².

Because the range of audible energy values is so enormous, a special scale has been devised which compares sound intensities by powers of ten. The unit of sound intensity is the *bel,* named in honor of Alexander Graham Bell (1847–1922), inventor of the telephone. Usually, we refer to sound intensities in terms of the smaller unit, the *decibel* (dB): 1dB = 0.1 bel. If two sounds differ in intensity by a factor of 10, the louder sound has an intensity 1 bel (or 10 dB) greater than the softer sound. If the difference in intensity is 3 bel (or 30 dB), the energy ratio is 10^3 or 1000. A factor of 2 in intensity corresponds to approximately 3 dB. Table 13-2 lists some common situations and the associated sound intensities.

High sound intensities have been found to cause several noticeable physiological effects such as the constriction of blood vessels in the skin, dilation of the pupils of the eye, as well as fatigue and irritation. Prolonged exposure to high-level sound can permanently impair one's hearing. Rock music performers (as well as some dedicated fans) frequently have substantially reduced hearing capacity. Persons who work around jet aircraft are required to wear ear protectors in order to prevent them from rapidly becoming partially or even completely deaf. Because of the increased noise level in many occupations, a sizable fraction of the population has experienced a partial hearing loss. Noise pollution is exacting its toll.

The Doppler Effect

When you are standing on a sidewalk and a police car or ambulance races down the street with its siren blaring, you notice a definite difference in the pitch of the siren depending on whether the vehicle is coming toward you or going away. The frequency is high when the sound

TABLE 13-2 TYPICAL SOUND INTENSITIES IN VARIOUS SITUATIONS

	INTENSITY (watts/m²)	INTENSITY LEVEL (dB)
Physical damage	> 10	> 130
Painful	1	120
Rock music	0.3	115
Jetliner, 2000 ft overhead	3×10^{-2}	105
Power mower	10^{-2}	100
Riveter	3×10^{-3}	95
Heavy truck, 50 ft	10^{-3}	90
Busy street traffic	10^{-5}	70
Conversation in home	3×10^{-6}	65
Radio in home, turned "low"	10^{-8}	40
Whisper	10^{-10}	20
Rustle of leaves	10^{-11}	10
Threshold of hearing	10^{-12}	0

source moves toward you and it is low when the source moves away from you. Just as the vehicle passes, the pitch changes and you hear the familiar *whee-oo* sound. The dependence of the frequency of a wave on the motion of the source is called the *Doppler effect,* after the Austrian physicist Christian Johann Doppler (1803–1853), who extensively studied this phenomenon.

We can visualize the Doppler effect with water waves in the following manner. Suppose that we place the tip of a slender rod into a basin of water. If the rod is vibrated up and down, a pattern of circular waves will develop, as shown in Fig. 13-14a. All of the waves travel uniformly outward from the tip of the rod, and the wavelength of the waves can be measured at any position around the rod with identical results. Now, let the vibrating rod move through the water with constant velocity. Figure 13-14b shows that this motion causes the waves to bunch together in front of the source. In this direction, an observer would measure a wavelength that is *shorter* than the wavelength if the source were at rest. Behind the source, the waves are spread out and the wavelength is *longer* than normal. The photograph at the right clearly shows the Doppler effect for water waves.

A wavelength that is *shorter* than normal means that the frequency is *higher* than normal. (Remember, $v = \lambda \nu$; the velocity is constant, so that ν must increase if λ decreases.) Therefore, if an observer hears a sound from a source moving toward him, the frequency (or pitch) will be higher than if the source is at rest. Similarly, if the source moves away from him, the frequency will be lower than normal.

We can determine the wavelength and the frequency of a Doppler-shifted wave by the following simple analysis. Suppose that a listener L is stationary and that a source S is approaching L. The velocity of the source is v_S, and the velocity of sound in air is v. If the source were at rest, the listener would hear the normal source frequency ν_S. We can write the following relations:

number of waves emitted
  by the source during time $t = \nu_S t$

distance traveled
  by each wave during time $t = vt$

distance traveled
  by the source during time $t = v_S t$

distance in space occupied
  by the $\nu_S t$ waves $= vt - v_S t$

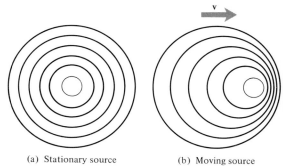

(a) Stationary source   (b) Moving source

Figure 13-14 (a) Circular waves spread out uniformly from a source at rest. (b) If the source moves through the medium, the waves are bunched together (shorter wavelength) in front of the source and the waves are spread out (longer wavelength) behind the source.

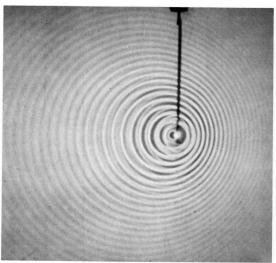

Photograph of water waves produced by a moving source. The vertical black line is the vibrating rod that is moved through the water toward the right. The wavelength in front of the source is shorter than normal and behind the source the wavelength is longer than normal.

That is, the waves emitted by the source are bunched up in space (see Fig. 13-14b) and occupy a smaller distance $(vt - v_St)$ than they would if the source were at rest (vt). The wavelength as determined by the listener is

$$\lambda_L = \frac{\text{distance occupied by waves}}{\text{number of waves}}$$

$$= \frac{vt - v_St}{v_St}$$

or

$$\lambda_L = \frac{v - v_S}{v_S} \qquad (13\text{-}6)$$

The frequency of the waves as determined by the listener is

$$\nu_L = \frac{v}{\lambda_L} = v \times \frac{\nu_S}{v - v_S}$$

or

$$\nu_L = \nu_S \left(\frac{v}{v - v_S} \right)$$

$$= \frac{\nu_S}{1 - v_S/v} \qquad (\text{S moving toward L}) \qquad (13\text{-}7)$$

(A slightly different expression is found for the case in which the listener moves toward a stationary source.)

Suppose that an airplane is traveling with half the speed of sound (Mach 0.5) and emits a sound signal with a frequency of 2000 Hz. What frequency does a listener hear if the stands in front of the oncoming airplane?

A speed of Mach 0.5 means that $v_S/v = 0.5$. Therefore,

$$\nu_L = \frac{\nu_S}{1 - v_S/v}$$

$$= \frac{2000 \text{ Hz}}{1 - 0.5} = 4000 \text{ Hz}$$

After the airplane has passed, the velocity of the source is *away* from the listener. We can still use Eq. 13-7 to obtain the frequency heard by the listener, but we must now change the sign of v_S. *Positive* v_S means that the source is approaching the listener; *negative* v_S means that the source is moving away from the listener. Thus,

$$\nu_L = \frac{\nu_S}{1 + v_S/v}$$

$$= \frac{2000 \text{ Hz}}{1 + 0.5} = 1333 \text{ Hz}$$

Shock Waves and Sonic Booms—How They Arise

Look again at Fig. 13-14. What will happen if we move the wave source through the medium with still higher speed? Clearly, the waves will bunch together even closer in front of the source. If the speed of the source is just equal to the speed of the waves in the medium, the waves will never be able to spread out in front of the source. And if the speed of the source *exceeds* the wave speed, the outgoing circular waves trail off behind the source as shown in Fig. 13-15. Notice how this diagram is the next step following from Figs. 13-14a and 13-14b. The traveling circular waves add together to produce the *wave front* indicated by the two solid lines in Fig. 13-15.

The diagonal wave fronts that are produced when a wave source moves through a medium with a speed greater than the wave speed in that medium are called *shock waves*. A simple example is the familiar V-shaped *bow wave* produced when a boat moves through water at a speed greater than that of surface waves. The photograph below clearly shows the conical shock waves that are produced at the tip and at the rear of a high-speed rifle bullet. The air along the wave front of a shock wave is highly compressed. When the shock waves from a rifle bullet reach the ear, they are heard (essentially simultaneously) as a sharp "crack" because of the rapid compressional action delivered to the ear membranes.

When an aircraft passes nearby at a speed greater than the speed of sound in air (about 750 mi/h), the shock waves that are produced are heard as a loud *sonic boom*. Figure 13-16 illustrates this situation. Two waves are

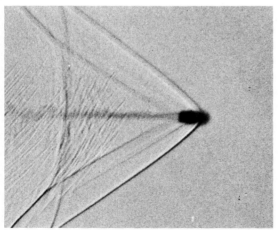

WINCHESTER-WESTERN

Shock waves produced by a high-speed rifle bullet. Notice that two waves are produced: one by the tip and one by the rear. The turbulent wake immediately behind the bullet is also evident. The shock waves and the turbulence can be seen because a special lighting technique produces shadows where the air has been highly compressed.

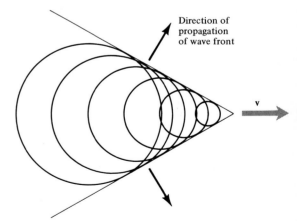

Figure 13-15 If the speed of a wave source is greater than the wave speed in the medium, a *shock wave* is produced, and the wave front moves away in a diagonal line from the direction of motion of the source.

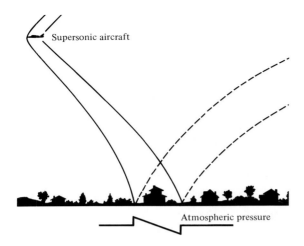

Figure 13-16 An aircraft flying at a speed greater than the speed of sound in air produces shock waves that are heard as a *sonic boom*. The solid lines are the shock waves and the dashed lines are the secondary, reflected waves. The diagram at the bottom shows that the air pressure rises sharply when the leading shock wave passes and again when the trailing wave passes. This produces a double sonic boom. The shock waves are curved slightly because of refraction in the varying density of the air.

produced: one from the front of the aircraft and one from the rear. As these shock waves pass over the ground, two loud booms are heard as the air pressure rises sharply for each wave. The intensity and the duration of the sonic boom increase with the size and the speed of the aircraft. For proposed operations of supersonic transports (SST), the area in which sonic booms will be heard will extend about 40 miles on each side of the ground track of the aircraft. Because these sonic booms would be annoying to persons and capable of producing some minor structural damage to homes, supersonic operations of transports must be confined to flights over water and over desolate land areas. Still, it might be unpleasant to be aboard an ocean liner when an SST passes overhead.

We find the Doppler effect in all kinds of wave motions—water waves, sound waves, even electromagnetic waves such as light. Indeed, by observing the Doppler shifts of the spectral features in the light emitted by stars and galaxies it is possible to determine how rapidly these objects are moving toward or away from the Earth.

13-5 REFRACTION, DIFFRACTION, AND INTERFERENCE

The Effect of Changes in Wave Speed

The speed at which a wave propagates—whether it is a water wave or a sound wave or an electromagnetic wave—depends upon the properties of the medium in which it travels. For example, the speed of sound in air is 330 m/s, but in sea water the speed is 1450 m/s (see

Table 13-1). What will be the effect on a wave when it moves from one medium into another medium where the wave speed is different?

When a water wave moves across the surface, the motion of the water molecules is greatest near the surface; the motion diminishes with increasing depth. In very deep water, the motion of the water molecules is not influenced at all by the presence of the bottom. Consequently, the speed of a wave in very deep water does not depend on the depth of the water. In shallow water, however, the motion of the water molecules can extend to the bottom. Therefore, as a wave runs in from deep water to shallow water, the wave "feels" the bottom, and because the motion of the water molecules is impeded by the presence of the bottom, the speed of the wave is decreased. Surface waves on water travel with a greater speed in deep water than in shallow water. If a water wave is incident obliquely on an abrupt change in depth, as indicated in Fig. 13-17, the speed change will cause the wavefront to bend, and the direction of propagation of the wave will be shifted. This phenomenon is called *refraction* and is exactly the same effect we will discuss in Section 15-1 for the case of light.

Notice in Fig. 13-17 that the wavelength λ_D of the wave in deep water is greater than the wavelength λ_S in shallow water. The reason is that $v = \lambda \nu$, and when the wave speed v decreases, the wavelength λ must decrease, because the frequency ν remains the same. (Can you see why ν must be the same on both sides of the depth change?)

Figure 13-17 shows that the left-hand portions of the waves are bent back because they run more slowly in the shallow water. In the reverse situation, in which the waves pass from shallow to deep water, the portions of the waves first entering the deep water will race ahead and the wave front will be bent in the opposite direction.

Diffraction

If we hold a piece of cardboard in bright sunlight, we notice that it casts a sharp shadow. And if we cut a small hole in the cardboard, the light waves pass through and we see a bright spot on the ground. What will happen if we try the same experiments with water waves? Will there be any difference in the two cases? Figure 13-18 shows the result of the first experiment with water waves. The cardboard barrier does *not* cast a shadow; the water waves move around the barrier al-

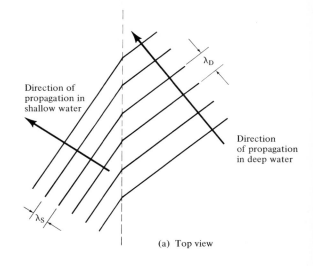

(a) Top view

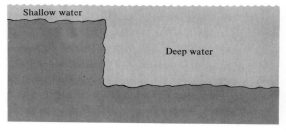

(b) Profile

Figure 13-17 When a water wave moves from a region of deep water (where the wave speed is high) into a region of shallow water (where the wave speed is low), the wave is *refracted* and the direction of propagation of the wave front changes.

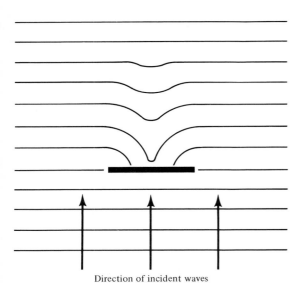

Figure 13-18 Water waves incident on a barrier are *diffracted* around the barrier. Several wavelengths beyond the barrier the wave shows no effects of having encountered any obstacle.

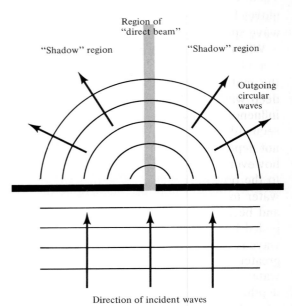

Figure 13-19 Water waves are incident on a barrier containing a narrow slot. The slot acts as a source of outgoing circular waves. As a result, the water waves beyond the barrier are not confined to the "direct beam" (as light waves would be), but penetrate into the "shadow" region.

most as if it did not exist! The water waves are bent around or *diffracted* by the barrier. This should not be too surprising a result because we know from experience that waves can bend around corners. After all, we can *hear* someone speak in another room even though we cannot *see* him. Sound waves, like water waves, exhibit diffraction.

Figure 13-19 shows the result of the second experiment: water waves are incident on a barrier containing a narrow slot. The water waves pass through the slot, but unlike light waves, they are not confined to the region of the "direct beam." The water waves spread out in all directions from the slot. Light waves leave a shadow region, but the water waves do not. The pattern of water waves "downstream" from the barrier is exactly the same as if the slot were the source of outgoing circular waves. The photograph in Fig. 13-20 shows an experimental verification of this statement.

Why is there such a great difference between the results of the experiments with light waves and those with water waves? The answer lies in the comparative sizes of the *wavelengths* of the two types of waves. A

Figure 13-20 Photograph of water waves producing the circular wave pattern shown in Fig. 13-19.

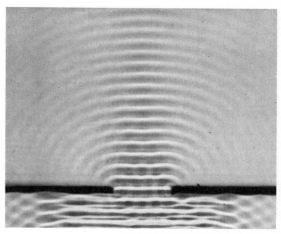

Figure 13-21 If the size of the slot is large compared to the wavelength of the incident waves, then water waves behave similar to light waves—a beam of waves is produced with shadow regions on either side.

typical wavelength for a light wave is 5000 Å or 5×10^{-7} m, whereas a water wave can have a wavelength of many centimeters or many meters. Thus, in Fig. 13-18 the size of the cardboard barrier is roughly comparable with the wavelength of the water waves, but it is enormously larger than the wavelength of light waves. The same is true of the slot in the barrier shown in Fig. 13-19. If we use slots that are sufficiently narrow (that is, comparable with the wavelength of light), the diffraction of light waves is easy to observe. Some examples of light diffraction effects are shown in the three photographs on page 326.

To see how the size of the wavelength compared to the size of the slot influences the wave patttern, let us next examine a case in which the width of the slot is increased to many times the wavelength. Figure 13-21 shows the result. The wave pattern beyond the slot now resembles that found for light waves—there is a direct beam with shadows on each side. We conclude that diffraction effects are large when the size of the slot is comparable with the wavelength of the wave and that the effects are small if the slot is much larger than the wavelength.

Notice that there is some diffraction around the corners of the wide slot in Fig. 13-21. In the case of the narrow slot (Fig. 13-19), the slot served as the source of outgoing circular waves. But in the case of the wide slot, *each point* along the slot acts as a source of circular

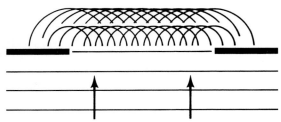

Figure 13-22 Each point of the slot acts as a source of outgoing circular waves. (Compare the case of the narrow slot in Fig. 13-19.) The outline of the individual waves duplicates the wave pattern shown in the photograph at the left. Notice that the wave front is straight over most of its length and that diffraction effects are evident only near the edges.

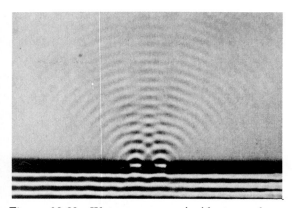

Figure 13-23 Water waves are incident on a barrier containing two slots. The circular waves originating at the slots combine (*interfere*) to produce a pattern of alternating regions of moving and still water.

waves. Figure 13-22 demonstrates how these individual circular waves add together to produce the wave pattern shown in Fig. 13-21.

Interference

We have already seen, in Fig. 13-8, that when two (or more) waves with different wavelengths are set up on a string, the waves combine and result in a complicated wave pattern. On a water surface we have more freedom to examine the ways in which waves combine because we can place wave sources at different locations. One way to study the pattern of waves from two nearby sources with identical wavelengths is to cut a second slot in a barrier such as that shown in Fig. 13-19. Each slot then acts as a separate source of outgoing circular waves. Figure 13-23 shows the pattern produced by the combination of the two sets of waves. Notice that there are regions of the surface which carry the crests and troughs of waves and that there are other regions which carry no wave motion at all. With only one slot open, waves cover the entire surface beyond the barrier (Fig. 13-20); but with two slots open, there is an absence of wave motion on certain parts of the surface. Why does this happen?

Look at Fig. 13-24 where several wave forms have been drawn, each with the same wavelength and same amplitude. If we combine two of these waves which have their crests and troughs together, as in Fig. 13-24a, the result is a wave with the same wavelength but with *twice* the amplitude. When the crests and troughs of two waves occur together, we say that the waves are *in phase*. The combining of in-phase waves is called *constructive interference*.

Figure 13-24b shows two waves with crests matching troughs. These waves are *out of phase* and they interfere *destructively,* completely canceling one another.

The wave pattern that results from two identical wave sources separated by a certain distance is due to the fact that the waves interfere constructively in some regions and destructively in others. Figure 13-25 illustrates the way to predict where constructive interference will occur. Circular waves originate at each slot, and the wavelength of these waves is equal to the wavelength of the incident waves. Therefore, using each slot as a center we draw a series of circular wave fronts, spaced one wavelength apart. These circles represent the wave

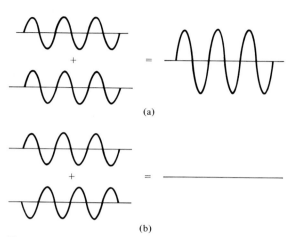

Figure 13-24 (a) In-phase waves of the same wavelength and amplitude interfere constructively to produce a wave with twice the amplitude. (b) Out-of-phase waves interfere destructively and completely cancel one another.

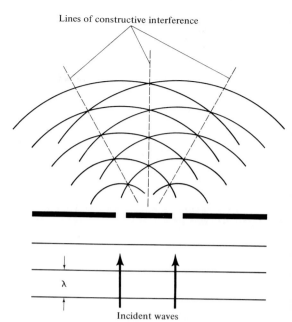

Figure 13-25 Geometrical construction for predicting the lines of constructive interference for circular waves originating in two slots.

crests at some particular instant of time. Where the circles intersect, both waves have crests and therefore interfere *constructively*. These points occur in a series of lines: one line is in the same direction as the incident waves and the others are at equal angles to either side. In between the lines of constructive interference, the waves are out of phase and interfere destructively. Consequently, the wave pattern consists of alternating lines of enhanced waves and regions of still water. (The angles that the diagonal lines of interference make with the central line depend on the distance between the slots compared with the wavelength of the waves.) This analysis agrees exactly with the photographic results shown in Fig. 13-23.

If the slots in a screen are made sufficiently narrow, then interference patterns similar to those we have found for water waves will also result for light waves. Because we cannot see the lines of interference for light waves as we can for water waves, we examine the interference by placing a screen some distance away from the pair of slots (which are now *slits*). This situation is

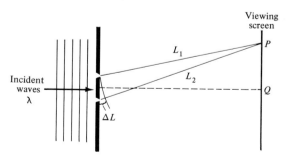

Figure 13-26 The interference at P will be constructive (a *bright* line) is the path difference ΔL corresponds to an integer number of wavelengths λ.

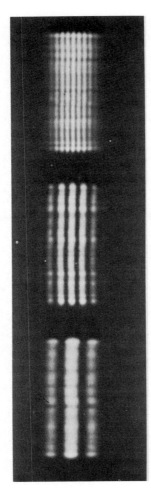

Double-slit interference patterns for light. The three cases correspond to different separations of the slits. In each case the wavelength of the light is the same.

shown in Fig. 13-26. At any point on the viewing screen where the light waves from the two slits are *in phase,* there will be constructive interference and a *bright line* will result. If the waves arrive at a point on the screen *out of phase,* the interference will be destructive and a *dark line* will result. The point Q on the viewing screen is equidistant from the two slits. Because the light originates at the two slits in phase, and because each light wave travels through the same distance to reach Q, the interference at this point is always constructive. That is, the central line in the interference pattern is a bright line. At point P, however, the light from the two slits has traveled through different distances. The distance to P from the upper slit is L_1 and the distance to P from the lower slit is L_2; the difference is $\Delta L = L_2 - L_1$. The interference at P will be constructive if the path difference ΔL corresponds to an integer number of wavelengths λ, for then the waves at P will be in phase. If ΔL differs from an integer number of wavelengths by one half wavelength, the waves will arrive out of phase and the interference will be destructive. We can summarize these statements in the following way:

constructive interference:
$$\Delta L = n\lambda \qquad \text{(bright line)}$$
destructive interference:
$$\Delta L = (n + \tfrac{1}{2})\lambda \qquad \text{(dark line)} \qquad (13\text{-}8)$$

Because ΔL changes as the viewing point is moved along the screen, the pattern observed consists of alternating bright and dark lines. The photograph at the left shows three such double-slit interference patterns, obtained with different separations of the slits.

In the next two chapters we will continue our examination of light as a wave phenomenon.

SUGGESTED READINGS

W. Bascom, *Waves and Beaches* (Doubleday, Garden City, New York, 1964).

D. R. Griffin, *Echoes of Bats and Men* (Doubleday, Garden City, N.Y., 1959).

Scientific American articles:

L. L. Beranek, "Noise," December 1966.

H. A. Wilson, Jr., "Sonic Boom," January 1962.

QUESTIONS AND EXERCISES

1. Sound waves do not propagate for great distances through air—eventually they "die out." What happens to the sound energy?

2. A traveling wave is set up in a long string by vibrating one end back and forth. If there is no friction, the wave amplitude will be the same all along the string. But if a traveling circular wave is set up in a large basin of water by vibrating a rod up and down, the amplitude of the wave will decrease with distance from the source even if there is no friction. Why? (You may want to use energy conservation in your explanation.)

3. A wave pulse travels with a speed of 24 m/s on a string that is connected between two posts, 1 m apart. What is the lowest frequency standing wave that can be set up on this string? Does this represent an audible sound?

4. The lowest frequency of standing waves that can be set up on a certain string suspended between supports 2 m apart is 5 Hz. What is the velocity of wave pulses on the string?

5. List some of the frequencies (starting with the lowest) of the standing waves that can be set up on a taut string 4 m long both ends of which are fixed and on which wave pulses propagate with a speed of 40 m/s.

6. A violin has only four strings, but these can be made to produce a large number of musical tones. How is this done? What is the effect of fingering?

7. If you hear a clap of thunder 8 seconds after you see a lightning flash, how far away did the flash occur?

8. What is the wavelength of a 220-Hz sound wave in air? What would be the wavelength in a bar of iron?

9. An organ note ($\lambda = 22$ ft) is sustained for 3 seconds. How many full vibrations of the wave have been emitted?

10. At the outer edge of an LP record, the record moves past the pick-up needle at a speed of approximately 0.5 m/s. If an 8000-Hz note has been cut into the record, what is the distance along the groove between successive peaks?

11. A certain tone has a wavelength in air of 10 cm. What is the frequency of the wave?

12. An *octave* is a range of frequencies that spans a factor of 2. That is, a musical note that has a frequency of 440 Hz is one octave higher than a note that has a frequency of 220 Hz. The range of normal human hearing spans how many octaves?

13. In an old western movie you may have seen the hero (or the villain) put his ear on a railroad track to listen for the approach of a train. Why is this better than simply listening for the normal sound?

14. A small explosive charge is set off on the surface of the ocean. At a listening station the sound is first picked up by a hydrophone (a device sensitive to sound waves in water) and 12 seconds later the sound is picked up by a microphone (in air). How far from the listening station was the explosion?

15. If you have ever watched a road construction crew or a strip-mining operation from a distance, you may have noticed that when the workers set off an explosion, you feel a ground tremor before you hear the explosion. Why?

16. At the same temperature, helium atoms have a greater speed than do air molecules. If a person inhales helium gas and then speaks, his voice is high-pitched and squeaky. Explain why. (*Hint:* The dimensions of the voice box are unchanged, so the standing wave patterns have the same wavelengths. How does the speed of sound change and how does this affect the frequency?)

17. A sound-intensity meter registers a level of 80 dB for a passing freight train at a certain point. What is the sound intensity in watt/m²?

18. Two sounds differ in intensity by 20 dB. What is the ratio of the *amplitudes* of the two sounds?

19. An ambulance has a siren that produces a 1500-Hz tone. What frequency sound do you hear if the ambulance approaches you with a speed of 60 mi/h? What will be the frequency of the sound after the ambulance has passed?

20. A sound source is moving toward a listener. As the source passes the listener, he hears the frequency of the sound drop to one half the previous value. What is the velocity of the source?

21. Suppose that you are looking directly north and see a train in the distance that is proceeding directly east at a speed of 20 m/s. The train's whistle has a frequency of 2 kHz. Do you expect to hear a whistle tone different from 2 kHz? (Assume that the train moves only a short distance while you are listening.)

22. An aircraft is traveling at Mach 2. Make a sketch of the sound waves from the aircraft. Make the sketch sufficiently accurate so that the proper angle of the shock wave is shown.

23. Refer to Fig. 13-17. Suppose that the change from deep to shallow water is not abrupt but, instead, takes place gradually. Sketch the way in which the wave fronts will behave as they proceed into the more shallow region. Use your result to argue that all waves tend to move inshore with wave fronts that are parallel to the beach.

24. In a poorly designed concert hall there may be certain locations in the hall where the sound intensity is particularly low or particularly high ("dead" spots and "live" spots). What is the reason for this effect? (Remember, sound can be reflected from walls and ceiling.)

25. Refer to the photograph on page 326 of the double-slit interference patterns for light. Which case corresponds to widest separation of the slits and which to the narrowest? Explain.

26. Make a sketch similar to Fig. 13-25 in which the wavelength of the waves is twice the separation of the slots. Compare the lines of constructive interferences with those in Fig. 13-25.

14

ELECTROMAGNETIC RADIATION

We are accustomed to thinking about many different types of radiations. We are all familiar with light, radio, and television waves, infrared (heat) radiation, and X rays. And, less frequently perhaps, we hear about ultraviolet radiation, microwaves, and gamma (γ) rays. We use these various types of radiations for many different purposes: we *see* with light; radio and television waves drive our receivers; and X rays are used to check our bones and teeth. Although these radiations perform a variety of tasks, they are all *electromagnetic radiations* and they all have the same basic physical features.

The only difference among the various types of electromagnetic radiations is that of *frequency*. The frequency of radio waves is *low* and the frequency of X rays and γ rays is *high,* but otherwise the nature of the radiations is the same.

The importance of a particular type of electromagnetic radiation depends upon the way in which this radiation interacts with matter, and this interaction depends, in turn, upon the frequency of the wave. High-frequency X rays will pass through fleshy tissue but will be absorbed by bony matter; therefore, X rays can be used effectively to examine the structure of bones and teeth. Visible light, on the other hand, will not penetrate the skin but it will pass through the jellylike parts of the eye and will register a signal on the eye's retinal surface. And radio waves, because they have very long wavelengths, will diffract around almost any object so that radio signals can be broadcast over long distances.

In this chapter we will discuss the production and the

properties of electromagnetic radiation. In some instances we will deal with the special cases of light or radio waves, but we should keep in mind that the basic features of all electromagnetic radiations are the same. We will emphasize here the *wave* character of electromagnetic radiation, but in Chapter 17 we will reexamine the situation and we will find that electromagnetic radiation has *particle* properties as well.

14-1 PRODUCING ELECTROMAGNETIC WAVES

Accelerating Charges and Changing Fields

Let us think for a moment about what we mean by *communication*. How can we transmit a piece of information from one place to another? A simple solution would be to write the message on a sheet of paper and then to deliver the letter to the desired location. Or, if the distance between the two locations is not great, *speaking* would serve the purpose. For longer distances, we could use a telephone. What is common about these various ways of transmitting information? In each case, some amount of *matter* was moved: the letter, the air molecules, or the electrons in the telephone wires. These forms of communication involve the transmittal of information by means of *moving matter*.

But what about light signals or radio waves? We know that information can be transmitted by these radiations. And we know that light signals and radio waves will propagate through empty space where there is no matter to move. After all, we can *see* the Moon, for example, and we can receive radio signals from instruments on the Moon. Light, radio signals, and other electromagnetic radiations are not carried by material particles. Instead, they depend upon electric and magnetic fields which can be set up in and which can propagate through empty space.

Evidently, information can be carried by electric and magnetic fields. But how does this happen? First, electric and magnetic fields act only on electrically charged particles. Suppose that we have two charged particles—electrons, for example—located at two positions, A and B, as in Fig. 14-1. How can we transmit a signal from A to B? If both electrons are at rest, we know that the only effect one electron has on the other is a steady

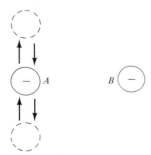

Figure 14-1   By moving the charge at *A* back and forth, a changing electromagnetic field is produced which can transmit a signal to the charge at *B*.

repulsive force. But a steady force is not a *signal*—it does not carry any information (other than to indicate the presence of a charge). The electric field around the stationary electron is *static;* we need a *changing* field to convey information.

Let us move the electron at *A* back and forth along a straight line, as indicated in Fig. 14-1. What will be the effect of this motion? The electric field at *B* will now be changing, both in magnitude and in direction. Furthermore, the motion of electron *A* is equivalent to a current flowing along the line of motion. This current sets up a magnetic field and this field changes as the direction of current flow is reversed. Therefore, at *B* there is a *changing electromagnetic field* to which electron *B* reacts. By altering the frequency or the amplitude of the back-and-forth motion of electron *A*, we can cause similar changes in the motion of electron *B*. Jiggling electron *A* (and, hence, also electron *B*) back and forth with a certain frequency could stand for the letter *H*; a different frequency could stand for the letter *E*; another frequency for the letter *L*; and a fourth frequency could stand for the letter *P*. That is, we can transmit a *signal* (or information) from *A* to *B* by means of a changing electromagnetic field generated by the motion of electron *A*. Electron *A* is the *transmitter* and electron *B* is the *receiver*.

Notice that the movement of electron *A* back and forth along the line involves changes in the direction of its motion. Electron *A* undergoes *acceleration*. An accelerating charge always produces a changing electromagnetic field, and only a changing field can transmit a signal. *All* electromagnetic signals are generated by accelerating charges.

Radiation from an Antenna

Let us now look at a more practical method for producing electromagnetic signals. Instead of isolated charges, we will consider electrons in wires that are attached to a source of alternating current. Figure 14-2 shows a simple antenna consisting of a straight wire that is cut at the midpoint and connected to a power source. Within the power source there is a means to reverse the polarity of the voltage on the wires that lead to the antenna. Notice that the sides of the antenna are not connected together to form a complete circuit. When the power source is connected to the antenna, current will flow only briefly, until the sides of the antenna become

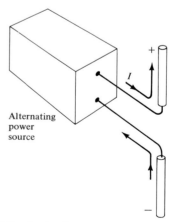

Figure 14-2 A simple antenna. By reversing the polarity of the power source attached to the lead-in wires, a current can be made to flow back and forth along the antenna. The changing electromagnetic field can transmit a signal to a receiver.

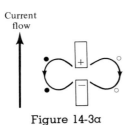

Current
flow

Figure 14-3a

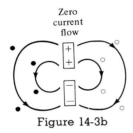

Zero
current
flow

Figure 14-3b

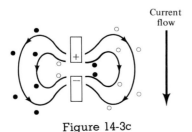

Current
flow

Figure 14-3c

charged. Reversing the polarity causes a similar brief flow of current in the opposite direction. By repeating the reversal of polarity, current flows back and forth in the antenna. Thus, we have a system of accelerating charges and the conditions are met for the transmission of electromagnetic signals.

How do the electric and magnetic fields change around the antenna? Let us follow the development of the fields with a series of diagrams. In Fig. 14-3a the current is beginning to flow *out* of the lower side of the antenna and *into* the upper side. The upper side therefore acquires a small positive charge and the lower side of the antenna acquires a corresponding negative charge. The electric field lines extend from the positive charge to the negative charge, just as they do in Fig. 12-3. Because a current is flowing, there is a magnetic field encircling the antenna. The direction of the magnetic field lines are shown in the diagram: a *solid* circle means a field line coming *out* of the page toward you and an *open* circle means a field line going *into* the page. Remember, each solid circle–open circle pair represents a single field line around the antenna.

Figure 14-3a, and all of the following diagrams, are meant to be schematic only. Figure 12-3, for example, is a much better illustration of the field lines connecting a positive charge and a negative charge; and Fig. 12-16 shows the magnetic field lines around a current-carrying wire. Remember, the complete field pattern is obtained by rotating each diagram around the axis of the charge pair or antenna.

In Fig. 14-3b we see the situation a short time after that illustrated in Fig. 14-3a. The current has continued to flow upward. More positive charge has accumulated on the upper section and more negative charge has accumulated on the lower section. A greater amount of charge means a more intense electric field, and we indicate this by an additional field line compared to Fig. 14-3a. This diagram represents the maximum accumulation of charge on the antenna. The current has ceased to flow and will next begin to flow in the opposite direction.

In Fig. 14-3c we see that the amount of charge on each section of the antenna has decreased because the current is now flowing in the opposite direction. The electric field lines follow the charges and begin to collapse on the gap between the two sides of the antenna. Notice also that the magnetic field lines near the antenna have reversed direction because of the change of direction of the current flow.

Finally, in Fig. 14-3d, the reversed flow of current has removed all of the charge from both sections of the antenna. The current is still flowing downward, however, and there are still magnetic field lines near the antenna. But notice what has happened to the electric field lines. In Section 12-1 we learned that electric field lines originate and terminate on electric charges. Indeed, this is the case in Figs. 14-3a, b, c. But when an alternating current flows in an antenna, positive charge accumulates first on one side of the antenna and then on the other side. For every complete cycle of charge transfer, there are two instants at which the antenna carries *zero* charge. At these instants there is no charge on which the field lines can originate or terminate. Consequently, the field lines pinch down toward the antenna gap, and, finally, at the instant when the charge is zero, the field lines are pinched off and form closed loops.

Notice in Fig. 14-3d that the magnetic field lines have opposite directions on the outer and inner parts of the loops of electric field lines. (Try to picture mentally the interlaced electric and magnetic field lines in three-dimensional space.)

As the current continues to flow, the lower side of the antenna becomes positively charged and new field lines are formed (but with directions opposite to those shown in the diagrams above). The pinched-off bundles move away from the antenna as the space near the antenna fills with new field lines.

A crucial feature of the electromagnetic field around the antenna is the fact that the field does not change instantaneously with a change in the current flow. At a certain distance from the antenna, the field is the result of the condition of the antenna at an earlier instant. That is, the field propagates outward from the antenna at a certain speed: the *speed of light*. Therefore, as the field bundles are pinched off from the antenna, as in Fig. 14-3d, they continue to propagate outward with the speed $c = 3 \times 10^8$ m/s.

After the current has oscillated back and forth through the antenna many times, a series of pinched-off bundles of field lines are propagating through space. Figure 14-4 shows three of these bundles. Notice how the directions of the electric and magnetic field lines alternate from one bundle to the next; this is due to the changes in the direction of current flow. Notice also that

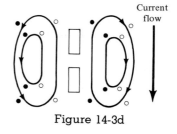

Figure 14-3d

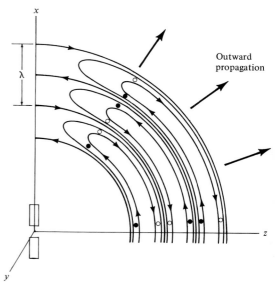

Figure 14-4 The outward propagation of the bundles of field lines from an antenna. Notice the indication of the wavelength of the radiation on the x-axis. The field pattern is duplicated below the z-axis and all around the x-axis. Each bundle of field lines, for example, constitutes a spherical shell.

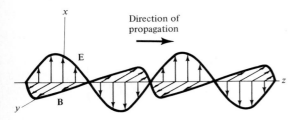

Figure 14-5 The electric and magnetic field vectors oscillate back and forth as the electromagnetic wave propagates through space.

the distance through which the fields propagate during one complete cycle of field changes constitutes the *wavelength* of the outgoing radiation. This wavelength is indicated on the x-axis in Fig. 14-4.

Waves in the Field

Let us now examine the electromagnetic field around the antenna from a different standpoint. We choose a set of coordinate axes as shown in Fig. 14-4. Now, we start at the antenna $(z = 0)$ and move outward along the z-axis, looking at the way the electric and magnetic fields change. The first electric field lines that we encounter point in the $+x$-direction and the next set of lines points in the $-x$-direction. As we continue moving outward, we see that the electric field lines change in a regular way, first pointing in the direction of $+x$ and then in the direction of $-x$. The magnetic field lines change in the same regular way, except that they point first in the direction of $+y$ and then in the direction of $-y$.

Figure 14-5 shows the variation of the field vectors, **E** and **B**, along the z-axis at a particular instant of time. Notice carefully how the directions of the field vectors correspond to the directions of the field lines shown in Fig. 14-4. The electric vector **E** always points in the $+x$ or $-x$ direction, and the magnetic vector **B** always points in the $+y$ or $-y$ direction. That is, both sets of field lines are always at *right angles* to the line drawn outward from the antenna (in this case, the z-axis).

TABLE 14-1 COMPARISON OF MECHANICAL AND ELECTROMAGNETIC WAVES

MECHANICAL WAVES	ELECTROMAGNETIC WAVES
Can be either *transverse* (for example, waves on a string) or *longitudinal* (for example, sound waves).	Always transverse.
Propagate by means of interactions among material particles.	Can propagate through vacuum.
Propagate with various speeds depending on the type of wave and the medium.	Always propagate with the speed of light. (In different materials, the speed of light is less than the speed in vacuum, 3×10^8 m/s.)
Characterized by the regular variation of a single quantity (for example, the density of air for sound waves or the amplitude of the vibrating particles for waves on a string).	Characterized by the regular variation of two quantities, **E** and **B**.
Carry energy and momentum.	Carry energy and momentum.

The regular variation of the field vectors with distance is exactly the same as the variation of the displacement of the particles in a string that carries a transverse wave (see Fig. 13-4). In fact, we have here an *electromagnetic wave*. The wave propagates outward from the antenna with the speed of light c and is characterized by the transverse oscillations of the field vectors, **E** and **B**. Electromagnetic waves are *transverse* waves. (The electromagnetic field cannot support longitudinal waves.)

A comparison of mechanical and electromagnetic waves is given in Table 14-1.

Receiving an Electromagnetic Signal

A radio transmitter impresses a signal on an electromagnetic wave by varying the frequency and the magnitude of the current flow in the antenna. The signal is then carried through space by the wave. How do we extract the signal from the wave at the receiving end? The procedure is essentially the reverse of that used at the transmitter. Figure 14-6 shows a receiving antenna which is identical to the transmitting antenna pictured in Fig. 14-2. When the electromagnetic wave strikes the antenna, the antenna is suddenly immersed in an electric field and current begins to flow in the antenna wire in the direction of the electric field lines. This current flow is detected in the receiver and the electrical signal is converted into an audible (radio) or visual (television) responses. In most radio and television applications, the magnitude of the current flowing in the receiving antenna is very small. Amplifying circuits are therefore necessary to increase the signals to the levels required to drive speakers and television tubes.

Information Carried by Electromagnetic Waves

We have seen how electromagnetic waves are generated, how they are transmitted and received by antennas. But how is information carried by such waves? An electromagnetic wave that consists of oscillations with uniform amplitude and constant frequency carries no information. Waves of this type are called *carrier waves* (Fig. 14-7a), and they must be altered in either amplitude or frequency (or both) in order to transmit information. Figures 14-7a and 14-7b show how this is accomplished.

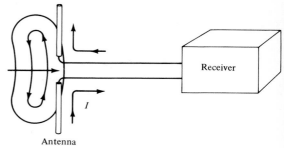

Figure 14-6　An electromagnetic wave incident on a receiving antenna causes a current to flow in the antenna wire. The receiver detects this current flow. (The wave bundle is only schematic.)

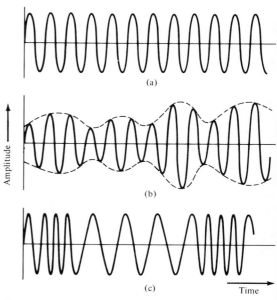

Figure 14-7　(a) A *carrier wave* with uniform amplitude and constant frequency. (b) An *amplitude modulated* (AM) wave in which the information signal alters the amplitude of the carrier wave but not its frequency. (c) A *frequency modulated* (FM) wave in which the information signal alters the frequency of the carrier wave but not its amplitude.

Radar—How It Works

A *radar* system involves a transmitter that produces a short burst of radiation in a particular direction. When this burst strikes a reflecting object, such as an airplane or a storm cloud, a portion of the electromagnetic wave is reflected back toward the transmitting site where it is detected (Fig. 14-8). The range to the reflecting object is determined by measuring the time interval between transmission and reception of the signal. (The signal travels with the speed of light.) A radar antenna is used only intermittently as a *transmitting* antenna. When not transmitting a signal, the circuitry is switched to act as a receiver and during this time the antenna acts as a *receiving* antenna. Thus, a radar antenna serves a dual purpose as a transmitting antenna and a receiving antenna.

SPERRY RAND—UNIVAC

Radar display of the type used in modern air traffic control systems. The controller communicates with the computer in this automated system through the keyboard at the right.

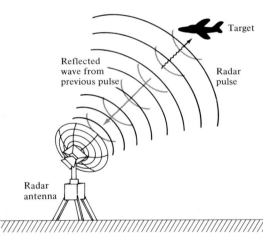

Figure 14-8 A short burst of electromagnetic radiation is emitted by a radar antenna. Part of the wave is reflected by the target and this returning signal is detected by the antenna, now acting as a receiving antenna.

Audible sound waves have frequencies much smaller than radio waves. In order to transmit sound frequencies via radio waves, the sound signals are first picked up by a microphone and converted into electrical signals with the same frequencies. These signals are then combined with the radio carrier wave. If the sound signals are made to alter the *amplitude* of the carrier wave, the result is that shown in Fig. 14-7b; this is an *amplitude modulated* (or AM) wave. If the sound signals are made to alter the *frequency* of the carrier wave, the result is that shown in Fig. 14-7c; this is a *frequency modulated* (or FM) wave.

When a modulated carrier wave is detected by a receiver, a special circuit eliminates the carrier wave. The modulation signal which remains is amplified and the signal is routed to a speaker which converts the electrical signal to a sound wave. Broadcasts made with FM waves are generally much less influenced by noise and static than are those made with AM waves. The reason is that the interference produced by electrical motors or thunderstorms affects the amplitude of the radio waves but not the frequency.

14-2 PROPERTIES OF ELECTROMAGNETIC WAVES

Reflection and Refraction

Because electromagnetic radiations are *waves,* they exhibit all of the properties we have discussed for mechanical waves. Thus, light waves and radio waves can be reflected and refracted, and they exhibit diffraction and interference. For example, when radio waves transmitted from the Earth's surface enter the atmospheric layer known as the *ionosphere,* the waves interact with the electrically charged ions and are refracted and reflected. Consequently, radio signals can be propagated over large distances by successive reflections between the ionosphere and the surface of the Earth, as shown in Fig. 14-9. High-frequency radiation will penetrate the ionosphere, whereas low-frequency radiation is reflected. Thus, radio waves are bounced back toward the Earth, but higher frequency waves are transmitted through the ionic layer. Television signals and light (which have frequencies much higher than radio waves) can be transmitted only through "line of sight" distances.

The Electromagnetic Spectrum

Although all electromagnetic radiations have the same basic properties, we assign different names to radiations with different frequencies (or different wavelengths). For example, the standard radio broadcast band extends

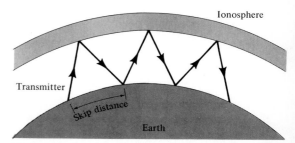

Figure 14-9 Radio waves are reflected by the layer of gas ions in the ionosphere. Radio waves can therefore "skip" large distances around the Earth.

from a frequency of 550 kHz to 1600 kHz. The wavelength of the radio wave transmitted by a station operating at a frequency of 600 kHz (6×10^5 s^{-1}) can be found by using Eq. 13-4:

$$\lambda = \frac{c}{\nu} = \frac{3 \times 10^8 \text{ m/s}}{6 \times 10^5 \text{ s}^{-1}} = 500 \text{ m}$$

Visible light is electromagnetic radiation in the frequency range from 4×10^{14} to 7.5×10^{14} Hz, corresponding to a wavelength range from 7500 to 4000 Å. (Recall that 1 Å $= 10^{-10}$ m.) In the high-frequency part of the spectrum we find X rays and gamma (γ) rays. Other parts of the spectrum are called ultraviolet (UV) radiation, infrared (IR) radiation, microwaves, and so forth. Figure 14-10 shows the major categories of electromagnetic radiations. Even though we use these convenient names for radiations with different frequencies, it is important to keep in mind that *all* of these electromagnetic radiations have the same fundamental characteristics.

By using electrical and electronic circuits of various types, it is possible to generate electromagnetic waves with frequencies ranging from a few Hz to about 10^{12} Hz. From 10^{12} Hz to about 10^{20} Hz, our only radiation sources are atomic systems, and for frequencies above 10^{20} Hz, nuclei are our primary sources. (The shorter the wavelength of the radiation, the smaller the size of the radiation source.) In a receiver, special circuits

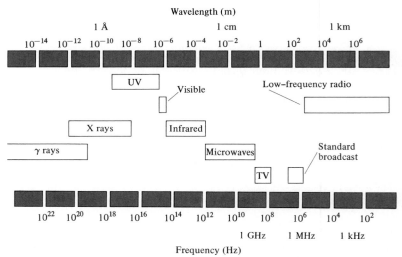

Figure 14-10   The electromagnetic spectrum. The frequency ranges for the various types of radiations do not have sharp limits and there is actually considerable overlapping.

During the middle of the 19th century, the subject of electromagnetism was placed on a firm foundation due largely to the efforts of only two men—Michael Faraday (1791–1867), the brilliant English experimentalist, and James Clerk Maxwell (1831–1879), the great Scottish theorist.

Faraday's accomplishments as an experimentalist were extraordinary. He began his scientific work as a chemist, an assistant to Sir Humphrey Davy (1778–1829). (Davy was a renowned chemist, justly famous for his many discoveries; but perhaps Davy's greatest discovery was Michael Faraday, whom he took into his laboratory from a bookbinder's shop.) Faraday's chemical research led to the discoveries of a method to liquefy gases, the laws of electrolysis, and the organic substance *benzene*. Turning his attention to electricity and magnetism, Faraday found a way to make a current-carrying wire revolve around a fixed magnet. This was the first demonstration of an *electromotive* effect and led to the development of the electric motor. In 1834, he discovered *electromagnetic induction*, the generation of an electric current from magnetism. (This is the reverse of the effect discovered by Oersted, the generation of magnetism from electric current.) Faraday constructed the first crude electrical generator which transformed mechanical motion into electrical current and set the stage for the development of the gigantic electrical industry. Out of these experiments on induction grew the idea of *lines of force*—Faraday had conceived the basic idea of modern *field theory*.

Michael Faraday

James Clerk Maxwell's contributions to theoretical science were as far-ranging as Faraday's were on the experimental side. Maxwell explained the composition of Saturn's rings; he studied color perception by the eye and he made the first color photograph; and even if he had done nothing else, his work on the kinetic theory of gases would have earned him a permanent place of honor in science. Ever since Maxwell had read Faraday's paper on lines of force, he was determined to incorporate this idea into a theory of electromagnetism. Finally, during the time of the American Civil War, he succeeded in producing a set of four equations that are a complete description of electromagnetic field phenomena. Combining these equations (now called *Maxwell's equations*) he was able to derive an equation that represented the *wave* motion of the field. The speed of propagation of the electromagnetic wave could easily be calculated from Maxwell's wave equation and from two experimental numbers. The result was that electromagnetic waves travel with the speed of light! Maxwell had identified light as an electromagnetic wave phenomenon, a connection that had not previously been seriously considered.

PHYSICS TODAY
James Clerk Maxwell

Although Maxwell's theory of electromagnetic waves was not widely accepted during his lifetime, numerous experiments have demonstrated that the theory is a correct description of the behavior of the electromagnetic field. The basic theory has not been altered since Maxwell's time and we use his equations today to describe all types of electromagnetic effects.

TABLE 14-2 THE RADIO SPECTRUM

APPLICATION	FREQUENCY RANGE
Marine radio, aircraft navigation beacons, and weather broadcasts	17.6–550 kHz
Standard AM radio broadcast	550–1600 kHz
Amateur bands, navigation	1.6–6 MHz
International short-wave, amateur bands	6–54 MHz
Television (VHF channels 1–6)	54–88 MHz
FM radio broadcast	88–108 MHz
Aircraft navigation and control	108–132 MHz
Police, taxi	150–160 MHz
Television (VHF channels 7–13)	174–216 MHz
Television (UHF channels 14–83)	470–890 MHz
Meteorological telemetry	1660–1700 MHz
Radio telephone, television relay	2000–11 700 MHz
Radar	300–30 000 MHz

allow you to "tune in" to a station by selecting from the range of frequencies the particular frequency of the desired station. Only the signals on this frequency are then amplified.

In order that different transmitters will not broadcast signals that interfere with one another, a definite frequency is assigned to each commercial or government transmitting facility. Certain types of transmitters (for example, those belonging to amateur radio operators) are assigned a *band* of frequencies in which they are permitted to broadcast. The entire range of frequencies below about 10^{12} Hz has been divided into various bands that are allocated by international agreement for various purposes. For example, the frequency range from 108 to 132 MHz is used by aircraft navigation and control transmitters. No other types of broadcasting are allowed in this frequency band so that there will be no interference with the control of flight operations. Within this band, different facilities are assigned frequencies at intervals of 50 kHz (0.05 MHz), and soon the spacing will be decreased to 25 kHz in order to accommodate the growing number of transmitters that are required to handle commercial and private air traffic. (Military aircraft usually use a different frequency band unless they are flying under civilian air traffic control.) Table 14-2 shows some of the band assignments within the radio part of the spectrum.

14-3 POLARIZATION

Transverse Waves

Because electromagnetic waves are *transverse,* they exhibit a number of interesting properties. Consider for a moment the case of transverse mechanical waves on a string. Suppose that we attempt to pass such waves through a slot in a barrier, as shown in Fig. 14-11. If the slot is at right angles to the direction of the transverse wave motion (Fig. 14-11a), the wave is blocked and is not transmitted through the slot. On the other hand, if the slot is in line with the wave motion (Fig. 14-11b), the wave is freely transmitted.

Can we produce a similar effect with transverse electromagnetic waves? Indeed we can. Consider again the wave shown in Fig. 14-5. In this case we have chosen to picture the **E** and **B** vectors exactly along the *x*- and

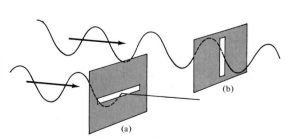

Figure 14-11 (a) A vertical wave on a string cannot pass through a horizontal slot in a barrier. (b) A vertical wave can, however, pass through a vertical slot.

y-axes, respectively, but this orientation is not necessary or required. All possible directions for the field vectors are allowed as long as they are mutually perpendicular and are perpendicular to the direction of propagation of the wave. In fact, electromagnetic radiation from a real source, such as light from the Sun or from an ordinary lamp bulb, consists of many individual waves with the **E** and **B** vectors pointing in all possible directions.

For the remainder of this chapter, we will focus on the electric field vector because **E** is of primary importance in all of the phenomena we will be discussing. Thus, we can represent the light from an ordinary source as a collection of **E** vectors pointing in all directions that are in the plane perpendicular to the direction of propagation, as in Fig. 14-12a. Light that is characterized in this way is said to be *unpolarized light.*

We know that we can represent any vector in terms of components that lie along mutually perpendicular axes. Therefore, we can take all of the individual **E** vectors in Fig. 14-12a and express them in terms of *x*- and *y*-components; the net result is shown in Fig. 14-12b. We still have unpolarized light, but now we have a simpler way to show the component **E** vectors. Unpolarized light can be considered to consist of a combination of electric field vectors with equal magnitudes that oscillate back and forth along the perpendicular *x*- and *y*-axes.

Now, if we have a source that produces waves with the **E** vector oscillating only along the *y*-axis, for example, or if we pass unpolarized light through a filter that absorbs the oscillations in the *x*-direction, we will have the situation illustrated in Fig. 14-12c. Here, all of the oscillations take place along the *y*-axis and we say that the light is *polarized* (or *linearly* polarized) in this direction. Electromagnetic radiation with any frequency can be polarized, but we will confine our discussion here to visible light because the phenomena are more familiar and are easier to appreciate.

Optical Polarizers

How can we extract polarized light from an unpolarized beam? Consider again the case of waves on a string. If the part of the string to the left of the barrier in Fig. 14-11b carries a complicated wave with vibrations in many different directions, the slot will filter out all of the wave components except the one that consists of up-and-down vibrations. The transmitted wave will then be

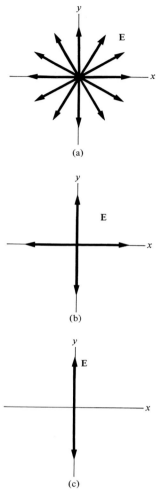

Figure 14-12   Electric field vectors as they appear looking along the *z*-axis toward the wave. (a) Unpolarized light consists of **E** vectors that point in all possible directions. (b) The individual **E** vectors in (a) have been resolved along the *x*- and *y*-axes; this is a simpler representation of unpolarized light. (c) Polarized light consists of **E** vectors that oscillate back and forth in only one direction; here, the light is polarized in the *y*-direction.

polarized. Therefore, if we pass a beam of unpolarized light through a suitable filter, we can expect that the transmitted light will be polarized.

We might consider using a grid of wires to polarize a light beam, but the spacing would have to be comparable to or smaller than the wavelength of the radiation. Otherwise, there will be no selective absorption and the transmitted wave will still be unpolarized. (If we enlarge the slot in Fig. 14-11a, the wave on the string will no longer be blocked.) A wire grid with a spacing fine enough to polarize light is extremely difficult to produce. But an effective optical polarizer can be constructed from a grid of aligned *molecules*.

In 1938, E. H. Land (who later developed the popular "picture-in-a-minute" photographic system) produced a molecular polarizer. This material, called *H-sheet,* is not only an effective polarizer but it is inexpensive and easy to manufacture. The polarizing agents in H-sheet are long molecules containing many iodine atoms. These molecules can be made to lie close together in almost perfect parallel alignment. The result is a fine grid which will pass only light with the electric vector oscillating in a particular direction.

When unpolarized light is passed through H-sheet, the transmitted light is polarized in the direction of the polarization axis of the material (Fig. 14-13). If this polarized light is incident on another piece of H-sheet with the same orientation, it will be passed freely with

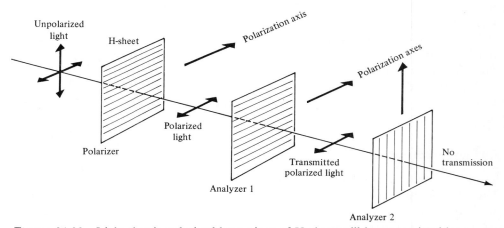

Figure 14-13 Light that is polarized by a piece of H-sheet will be transmitted by a second sheet with the same orientation. But if the orientation of the analyzer is perpendicular to that of the polarizer, there will be no transmission. (The polarization axis is actually *perpendicular* to the direction of alignment of the molecules in the polarizers and analyzers; see Exercise 13.)

very little attenuation. But if the beam is incident on a sheet with a perpendicular orientation, there will be no transmitted light. The polarizing sheet is called a *polarizer* and the second and subsequent sheets are called *analyzers*. No polarizer or analyzer can be made absolutely perfect, and so the beam is never 100 percent polarized nor is the transmission through the perpendicular analyzer ever actually zero. If you place two pieces of H-sheet together and look through them at a light source while slowly rotating one piece relative to the other, you will see the source change from bright to dark and back to bright with every 180° of rotation.

An unpolarized beam of light can be considered to consist of equal intensities of waves with the electric field vectors at right angles to one another (Figs. 14-12b and 14-13). A perfect polarizer would completely remove one of these components and would transmit the other component without attenuation. A perfect polarizer would therefore transmit exactly 50 percent of the incident unpolarized light. Because there is some absorption in any medium through which light passes, the transmission of any real polarizer is never 50 percent. H-sheet can be made to transmit 38 percent of the incident light, and by adding absorptive substances the transmission can be lowered to any desired amount (as is done, for example, in the manufacture of Polaroid sunglasses).

Polarization by Reflection and Scattering

A beam of unpolarized light can be polarized by passing it through a material such as H-sheet. But polarization occurs in certain natural situations as well. When unpolarized light is reflected from a transparent material such as glass or water, the reflected light is partially polarized. Or when sunlight is scattered by air molecules in the atmosphere, the scattered light is partially polarized. The polarizing effect is particularly strong when sunlight is scattered through 90° (Fig. 14-14). On a clear day, when there is little water vapor or dust in the air, and if the Sun is near the horizon, the light from overhead can be polarized to the extent of 70 or 80 percent.

On a sunny day, one often sees the *glare* of reflected light from surfaces of water or glass. Have you ever looked at such surfaces through polarizing sunglasses? (These sunglasses consist of H-sheet sandwiched between two pieces of glass and darkened to absorb

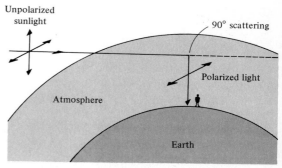

Figure 14-14 Unpolarized light from the Sun becomes almost completely polarized when scattered through an angle of 90° by molecules in the atmosphere.

Photographs of reflected glare with (top) and without H-sheet.

R. MARK, *SCIENTIFIC AMERICAN*

Polarized light is used to analyze the stresses in the arches of the famous Gothic cathedral at Chartres, France. A 1:180 model of a typical buttress section was constructed from plastic and then loaded in the way that would result from the prevailing wind conditions at Chartres. This photograph was taken with the polarized light that passed through the plastic model. The regions of stress are clearly shown by the interference patterns.

even more of the light.) The glare is noticeably reduced, sometimes eliminated completely. The reason is that the reflected light consists primarily of overhead light that is reflected horizontally to reach your eyes. The electric vectors in such light are polarized in the horizontal direction and if you use polarizing glasses designed to transmit vertically polarized light, much of the glare is suppressed.

Polarizing glasses are also beneficial to motorists in reducing highway glare and the glare of reflected light from the windshields of approaching automobiles. These glasses are often used by boating and water-sports enthusiasts to reduce water glare. Because the light reflected from water is highly polarized, fishermen sometimes use polarizing glasses to eliminate the reflection, thus enabling them to see into the water more clearly and to locate fish.

Polarized light is sometimes used in engineering design problems to analyze the stresses in structural members. If polarized light is passed through an unstressed transparent material, the analyzed light will show no particular features. However, if the material has internal stresses due to external forces or to the particular method of manufacture, the light analyzed by a piece of H-sheet will show interference patterns that indicate the regions of stress. In this way, areas of possible structural failure can be identified and corrective measures taken. An example of this kind of analysis is shown in the photograph at the left.

Polarization and the Eye

The human eye is not a very good detector of polarized light. Most persons see absolutely no difference between polarized and unpolarized light unless they use some material such as H-sheet as an analyzer. A few persons, however, can perceive polarization effects (such as the partial polarization of scattered sunlight) without any artificial aids.

Certain insects have eyes that are much more polarization-sensitive than the human eye. It has been demonstrated that some types of ants and beetles, and particularly the honeybee, can sense polarization. Indeed, the honeybee uses the polarization of scattered sunlight to navigate! Apparently, the primary navigation reference point for the honeybee is the position of the Sun. But if the Sun is obscured by clouds, the honeybee can continue to navigate properly if he can see any piece of blue sky. The direction of polarization of the scattered blue light allows the bee to deduce (if that is the proper word) the position of the Sun even though he cannot see it.

14-4 PHOTONS

Bundles of Electromagnetic Energy

What effects will we find if we allow the intensity of an electromagnetic wave to become weaker and weaker? Will the wave just gradually fade into nothing? We are asking here the same kinds of questions that we have previously asked about the structure and composition of matter. We see matter in the bulk form, but matter is ultimately composed of *atoms*. Is there any similarity between the atomic character of matter and the composition of electromagnetic radiation?

Suppose that we have a light source with a controllable intensity. Some distance away we assemble an array of light detectors, as in Fig. 14-15. We equip each detector with a device that produces an audible *click* when light is detected. With the light source turned up to full intensity, each detector emits a continuous series of clicks. As we decrease the intensity, the rate of clicking also decreases, but all of the detectors still respond in the same way. When the intensity has been decreased to an extremely low value, we notice a definite difference in the response of the detectors. No longer do the detectors click together; we hear first one detector produce its click, then another, then another. It appears as though the source is shooting out discrete bundles of light, first in one direction, then another, instead of emitting continuous waves.

Our conclusion is exactly right! Light that we ordinarily see has such a high intensity that the description of its behavior in terms of continuous waves is entirely correct. We expect that light waves should exhibit all of the normal properties of waves—reflection, refraction, diffraction, and interference—and experiments show this to be true. But ultimately the light waves are not continuous: they are composed of tiny bundles of radiation that we call *photons* or *quanta*.

We can think of a photon as a bundle of oscillating electromagnetic radiation that travels through space with the speed of light. The oscillations of any photon

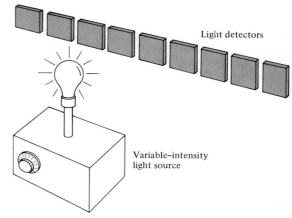

Figure 14-15 A light source with variable intensity and an array of light detectors to test the continuous wave character of light.

take place with a definite frequency and so when the number of identical photons is very large, they act as a continuous wave with the same definite frequency and propagation speed. The photons merge into an electromagnetic wave.

An electromagnetic wave carries energy and momentum, and so does a photon. The energy of a photon is directly proportional to its *frequency*. Thus, an X-ray photon has considerably more energy than a photon of visible light or infrared radiation. On the other hand, the energy of a photon that oscillates with a radio frequency is so low that we do not ordinarily discuss radio waves in terms of photons.

One of the most important properties of light is its photon character. In Chapters 17 and 18 we will turn to the fascinating story of the photon and trace the developments that led from the discovery of the discrete character of electromagnetic radiation to our modern theory of the structure of atoms.

SUGGESTED READINGS

R. M. Page, *The Origin of Radar* (Doubleday, Garden City, New York, 1962).

J. R. Pierce, *Waves and Messages* (Doubleday, Garden City, New York, 1967).

Scientific American articles:

J. R. Heirtzler, "The Longest Electromagnetic Waves," March 1962.

I. I. Shapiro, "Radar Observations of the Planets," July 1968.

QUESTIONS AND EXERCISES

1. Is *acceleration* always required to produce a wave? (Think about the ways that water waves, sound waves, and electromagnetic waves are generated.)

2. In describing the production of electromagnetic waves by the antenna in Fig. 14-2, it was stated that connecting the power source to the antenna would cause current to flow "only briefly." Why would the current flow only briefly and not for as long as the source is connected to the antenna?

3. A certain radar antenna sends out bursts of radiation at intervals of 10^{-4} s. What is the maximum distance at which a burst can be reflected from an aircraft and return to the antenna before the next burst is radiated?

4. A radar system records a difference of 2×10^{-5} s between the time of transmission of a burst of radiation and the time that a reflected signal is received. How far away is the object that reflected the radiation?

5. Why must long-range radar systems have powerful transmitters and sensitive receivers? (How does the strength of the signal received vary with the distance from the radar station to the target?)

6. By international agreement, the frequency 121.5 MHz is reserved for emergency radio transmissions from aircraft in distress. What is the wavelength of emergency signals?

7. In what wavelength range do radar signals lie? (Use Table 14-2.)

8. A certain sound wave in air has a wavelength of 0.1 m, and a certain electromagnetic wave in air has the same wavelength. Classify these waves (audible, inaudible; light, radio wave, etc.). Explain why two waves with the same wavelength can have such different properties.

9. An amateur radio operator has a transmitter that produces 30-m waves. At what frequency does the transmitter operate?

10. It is generally true that the wavelength of a radiation is as large as or larger than the size of the emitting system. Is this statement consistent with your observations (or knowledge) of the sizes of radio broadcast antennas, police radio antennas, and atoms?

11. What is the wavelength of the signal from an AM broadcast station that operates at "the top of your dial" (1600 kHz)?

12. How would you classify an electromagnetic wave with a wavelength of 10^{-9} m? With a frequency of 10^9 Hz?

13. Suppose that an unpolarized electromagnetic wave is incident on a wire grid, as shown in the diagram below. The distance between the wires in the grid is small compared to the wavelength of the wave. Which component of the wave—vertical or horizontal—will be transmitted? (Which component will be absorbed by inducing a current to flow in the wires? Why will the other component *not* induce a current? Refer to Fig. 14-6.)

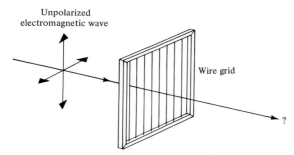

14. At noon on a clear day you examine the polarization of the light coming from various parts of the sky. From which regions will the light have the greatest polarization?

15. Is a polarizer that will polarize microwaves likely to work with light waves? Explain.

16. Explain how you could use two pieces of H-sheet to vary continuously the intensity of a light beam.

15

LIGHT

Of all the human senses, probably the most important is *sight*. Most of the information we receive concerning the world around us is obtained visually, by direct observation, by reading, and by referring to pictures. Cameras, binoculars, and television are all instruments that have been devised to improve our utilization of information that is carried by light. In scientific terms, too, light is the most important tool we have for investigating the macroscopic world of the solar system, stars, and galaxies, as well as much of the microscopic domain down to molecular sizes. We will examine in this chapter those characteristics of light and optical instruments that permit us to explore the Universe. We reserve for later chapters the discussions of the atomic origin of light and its electromagnetic properties.

15-1 BASIC FEATURES OF LIGHT AND LIGHT SOURCES

The Speed of Light

When a burst of light is emitted from a source, will it be seen immediately by an observer some distance away, or will there be a delay between the instant of emission and the instant of arrival at the position of the observer? The question whether or not light travels with an infinite speed was not answered until the late 1600's. Galileo (1564–1642) had attempted to measure the speed of light between two hilltops separated by about 2 miles. He and a companion took up positions on the two hills one night. Each was equipped with a lantern and a cover. Galileo removed the cover from his lantern and, upon seeing this light, his companion on the other hilltop removed the cover from his lantern. Galileo thought that by measuring the time between the moment he uncovered his lantern and the moment he saw the light from his companion's lantern, he would be able to determine the speed of light. It did not take long for him to realize that he was measuring not the speed of light, but the combined reaction times of himself and his companion. The speed of light is far too great to measure in this crude way. (The time required for light to travel the 4-mile round trip between the hills used in Galileo's experiment is about 0.00002 s, but the time required for the human reflexes to respond is a few *tenths* of a second—0.2 to 0.3 s.)

Probably the first reliable determination of the speed of light was made by the Danish astronomer Olaus Roemer (1644–1710), although there is some disagreement among historians as to whom should be credited this "first." Roemer made many precise measurements of the orbits of Jupiter's moons (discovered by Galileo in 1610). The innermost moon, Io, revolves around Jupiter once in 42.5 hours. The instant when Io passes behind the planet is easy to observe because the moon's light is suddenly cut off as it is eclipsed by Jupiter. The time of rotation (the *period* of Io) is the time between successive eclipses by Jupiter. Roemer found systematic discrepancies in the times of the onset of the eclipses. When the Earth was moving *away* from Jupiter in its orbit around the Sun (from A to B in Fig. 15-1), the onset of the eclipse came later and later than expected. And when the Earth was moving *toward* Jupiter (from C to D in Fig. 15-1), the onset gradually came earlier and earlier. Roemer correctly concluded that the discrepancies were due to the time required for light to travel the changing distance between Jupiter and the Earth. All of his measurements of the times of Io's eclipses could be brought into agreement if he assumed that it required approximately 1000 seconds for light to travel the distance of the diameter of the Earth's orbit. We therefore have a simple method for computing the speed of light:

$$\text{speed of light} = \frac{\text{diameter of Earth's orbit}}{1000 \text{ s}}$$

The distance from the Earth to the Sun is approximately

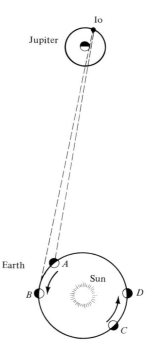

Figure 15-1 Roemer's method for determining the speed of light. The time at which the eclipse of Io by Jupiter occurs becomes later and later than expected as the Earth moves away from Jupiter (*A* to *B*) and becomes earlier and earlier as the Earth moves toward Jupiter (*C* to *D*).

1.5×10^{11} m; therefore,

$$\text{speed of light} = \frac{2 \times (1.5 \times 10^{11} \text{ m})}{10^3 \text{ s}}$$
$$= 3 \times 10^8 \text{ m/s}$$

Roemer did not have a good value for the Earth-Sun distance and consequently his value for the speed of light was about one-third smaller than the modern result – not bad for a first try.

Modern methods for determining the speed of light use Earth-based instead of astronomical measurements. Most of these techniques utilize a pulsing light source, and a measurement is made of the time for a round trip of the light pulses from the source to a fixed mirror and back to the starting point – a sophisticated version of Galileo's original method. Such techniques were used by A. A. Michelson in a series of precise measurements of the speed of light.

Albert A. Michelson (1852–1931) was one of the most celebrated American scientists of the late 19th and early 20th centuries. His measurements of the speed of light began in 1878 and continued off-and-on for the remainder of his life. The most famous of these experiments was carried out during 1923–1927. The base line Michelson chose for his measurements was between the peaks of Mount Wilson and Mount San Antonio in southern California. He surveyed this 22-mile line and obtained the separation between his two stations with an accuracy of better than one inch. At the Mount Wilson station he placed an 8-sided mirror which could be ro-

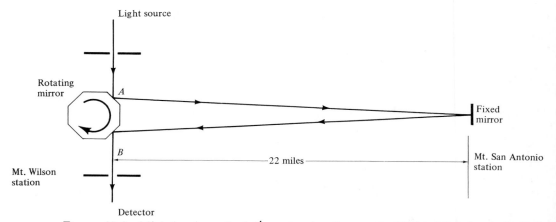

Figure 15-2   Michelson's method for measuring the speed of light. Michelson's 8-sided mirror was made for him by E. A. Sperry (1860–1930) inventor of the gyroscopic compass and founder of the successful Sperry Gyroscope Company.

tated at high speed. At the Mount San Antonio station he placed a fixed mirror (see Fig. 15-2). With the 8-sided mirror stationary, a light ray from the source would be reflected from side A, would travel to the Mt. San Antonio mirror and back to side B where it would be reflected into the detector. With the 8-sided mirror rotating, however, a flash of light is directed toward the Mt. San Antonio mirror only when one of the flat sides (such as A) is precisely in the position shown in the diagram. When the light pulse returns, no light will reach the detector unless the mirrors are lined up in exactly the same way that they were when the pulse of light was flashed toward the mirror on the mountain station. The light pulse will be able to make the entire journey and enter the detector if the 8-sided mirror moves through $\frac{1}{8}$ or $\frac{2}{8}$ or $\frac{3}{8}$ of a revolution between the start of the pulse and its return to the mirror. In practice, Michelson adjusted the speed of rotation so that the mirror rotated through exactly $\frac{1}{8}$ of a revolution during the round trip of the light pulse. The measurement therefore depends on the accurate determination of the distance between the two stations and the rotation speed of the 8-sided mirror.

Michelson's measurement in 1927 produced a result for the speed of light of 2.99798×10^8 m/s, with an uncertainty of a few parts per million. A few years later, he refined his technique still further and constructed a mile-long evacuated tube through which the light passed. Today, the most precise method for measuring the speed of light involves the use of laser beams. The currently accepted value is 2.99792458×10^8 m/s, with an uncertainty of only one digit in the last decimal, a precision of three parts per *billion*. In this book we will denote the speed of light by the symbol c and we will use the approximate value,

$$\boxed{c = 3 \times 10^8 \text{ m/s}} \qquad (15\text{-}1)$$

Light Rays

In the previous chapters we have learned that light is a wave phenomenon. The wavelength of visible light varies between about 4000 Å (4×10^{-7} m) and 7500 Å (7.5×10^{-7} m). These lengths are much smaller than the sizes of anything in our everyday experience. In Section 13-5 we discovered that diffraction effects in waves are apparent only when the wavelength is comparable with

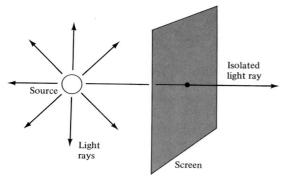

Figure 15-3 A screen can be used to isolate a single light ray from a source.

the size of the diffracting object or slit. Because the wavelength of light is so small, we do not ordinarily perceive diffractive effects. Our experience tells us that light travels in straight lines and does not bend (diffract) around corners. When we are dealing with objects of everyday size, this straight-line propagation picture is the most convenient way to describe most situations.

If we punch a pinhole in a screen and then place the screen in front of a light source, the light that passes through the opening represents a small portion of the total light output of the source. By following this *light ray* (Fig. 15-3) as it interacts with various objects and materials we can investigate the basic behavior of light. It is not really necessary to isolate a light ray in this way before we can make such studies. We can always consider the light from a source to consist of many light rays, with each ray acting independently of the others.

We will find it useful in a variety of situations to consider a bundle of *parallel* light rays. At a distance from a light source, a small bundle of rays will be *approximately* parallel (Fig. 15-4). By choosing only a narrow region around the central ray, the parallelism of the bundle can be made more exact. Or if the source is at a great distance, the rays in any small region will be almost perfectly parallel. The rays from the Sun, for example, can usually be considered to be perfectly parallel, and the rays from any star are always parallel. By using this incident light in the form of a single ray or a bundle of parallel rays, the analysis of the behavior of light in many situations is considerably simplified.

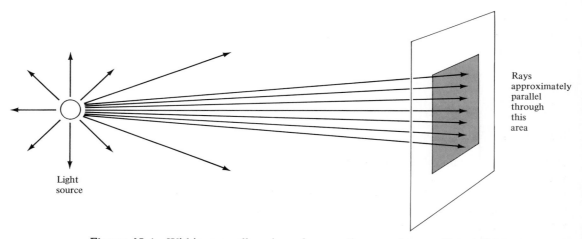

Figure 15-4 Within a small region of space the rays from a distant light source are approximately parallel.

Reflection

It was known even in ancient times that when a ray of light is incident on a flat polished surface, the ray is *reflected* with the angle of reflection equal to the angle of incidence (Fig. 15-5a). Notice that these angles, ϕ_r and ϕ_i, respectively, are measured from a line that is *perpendicular* to the reflecting surface. To an observer at position B in Fig. 15-5a, the light appears to originate *behind* the mirror and to come to his eye along the path through A' instead of along the actual path through A.

In Fig. 15-5b we have an object located in front of a mirror. Various rays from the tip of the object strike the mirror and are reflected. The relation $\phi_r = \phi_i$ is valid for each ray. By tracing each reflected ray backward, it is found that they all diverge from a single point behind the mirror. Moreover, this point is as far behind the mirror surface as the tip of the object is in front of the mirror. This same result is true for every other point on the object. Therefore, an observer at any position to the left of the mirror will see the object located apparently behind the mirror.

Refraction

If a light ray is incident on a piece of *transparent* material, two effects are evident (Fig. 15-6). First, a portion of the light is *reflected* in accordance with the reflection rule just discussed. Second, the nonreflected portion of the ray is *transmitted* through the material, but in a direction that is not the same as the incident direction. This bending of the light ray as it passes from one medium to another is called *refraction*.

When a light ray is incident from a medium such as air onto a more-dense medium such as glass or water, the refracted ray always lies *closer* to the perpendicular than does the incident ray. That is, in Fig. 15-6a, the angle θ is *less* than the angle ϕ. If the ray is incident from the denser medium, as in Fig. 15-6b, the ray is refracted *away from* the perpendicular. In this case, $\theta > \phi$. Because of the refraction of light when passing from one medium into another with different density, an object under water, viewed from outside the water, will appear to be displaced (Fig. 15-7).

Why does a light ray bend when it passes from one medium to another? The answer involves examining the *wave* properties of light. Consider a bundle of light rays incident on an air-glass surface, as shown in Fig. 15-8.

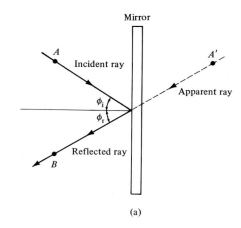

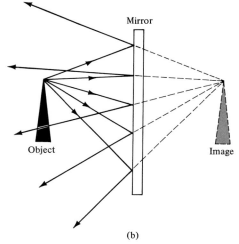

Figure 15-5   (a) The reflection of a light ray takes place with $\phi_r = \phi_i$. (b) An object viewed in a mirror appears to be located behind the mirror.

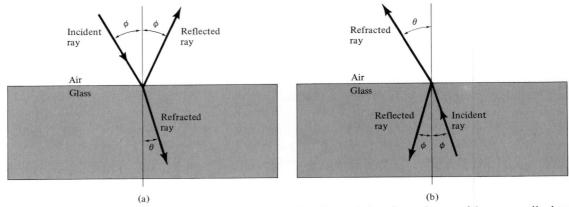

Figure 15-6 (a) A light ray entering glass from air is refracted *toward* the perpendicular, so that $\theta < \phi$. (b) A light ray passing from glass to air is refracted *away from* the perpendicular, so that $\theta > \phi$.

The front of the bundle is perpendicular to the rays and is indicated first by the line AB, and at a later time by the line CD. At C, one end of the line has reached the glass while the other end of the segment is still a distance DF from the surface. The speed of light within a dense medium, such as glass, is *less* than the speed in air. Therefore, the portion of the wave in the glass moves a shorter distance in a given period of time than does the portion in air. During the time that the right-hand end of the segment moves from D to F, the left-hand end (within the glass) moves through the shorter distance CE. Consequently, the direction of the original wave (AB and CD) is shifted upon passing into the glass (EF and GH).

The ratio of the speed of light in a medium to the speed in vacuum (or air) is called the *index of refraction* of the medium:

index of refraction $= n$

$$= \frac{\text{speed of light in vacuum}}{\text{speed of light in medium}} = \frac{c}{v} \quad (15\text{-}2)$$

Different types of glass have different values of n, but most are near $n = 1.5$. Water has $n = 1.33$. The values of n for two different materials in contact determine how much a light ray will bend in passing from one material to the other.

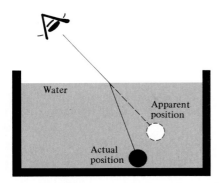

Figure 15-7 When an object under water is viewed obliquely, the refraction of light causes the apparent position of the object to be displaced from the actual position.

Snell's Law

We can derive the expression that relates the angles of incidence and refraction to the indexes of refraction by referring to Fig. 15-9, which is just Fig. 15-8 with more detail. We assume that a light beam originates in Medium 1 (which has index of refraction $n_1 = c/v_1$). This light beam is incident with an angle θ_1 on the boundary between Medium 1 and Medium 2 (which has index of refraction $n_2 = c/v_2$). During the time that the incident wave travels from B to D, the refracted wave travels from A to C. Thus, we can write

$$BD = v_1 t = \frac{ct}{n_1}$$

$$AC = v_2 t = \frac{ct}{n_2}$$

Next, look at the triangles $\triangle ABD$ and $\triangle ACD$. We can write

$$\sin \theta_1 = \frac{BD}{AD} \quad \text{or} \quad BD = AD \sin \theta_1$$

$$\sin \theta_2 = \frac{AC}{AD} \quad \text{or} \quad AC = AD \sin \theta_2$$

Dividing the first equation by the second,

$$\frac{BD}{AC} = \frac{\sin \theta_1}{\sin \theta_2}$$

Now, the ratio BD/AC is

$$\frac{BD}{AC} = \frac{ct/n_1}{ct/n_2} = \frac{n_2}{n_1}$$

Therefore,

$$\frac{n_2}{n_1} = \frac{\sin \theta_1}{\sin \theta_2}$$

or,

$$\boxed{n_1 \sin \theta_1 = n_2 \sin \theta_2} \qquad (15\text{-}3)$$

This expression is known as *Snell's law,* in honor of the Dutch mathematician, Willebrord Snell (1591–1626), who discovered the result experimentally in about 1621.

Suppose that a light beam is incident from air at an angle of 45° on a piece of glass that has $n_2 = 1.5$. What will be the angle of refraction?

The index of refraction for air is almost the same as that for vacuum, so we use $n_1 = 1$. Then, we have

$$\sin 45° = 1.5 \sin \theta_2$$

That is, θ_2 is the angle whose sine is $(\sin 45°)/1.5$. We

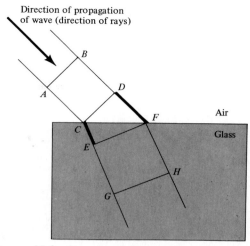

Figure 15-8 Because light travels with a lower speed in glass than in air, the advancing wave (AB and CD) is shifted in direction (EF and GH) upon passing from air into glass. The wave travels from C to E in the glass during the same time that it travels from D to F in air.

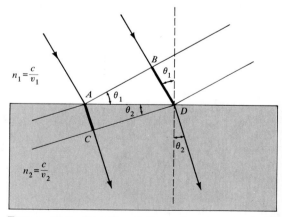

Figure 15-9 Details of the refraction of a light beam at the boundary between two media.

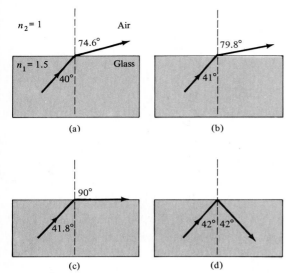

(a)

(b)

(c)

(d)

Figure 15-10   Light rays incident from glass on an air–glass boundary at various angles near the critical angle of 41.8° (c). For incident angles greater than the critical angle, the light is internally reflected (d).

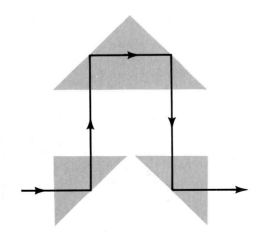

Figure 15-11   Internal reflection of a light ray in a series of 45° prisms.

express this statement as

$$\theta_2 = \sin^{-1}[(\sin 45°)/1.5]$$
$$= \sin^{-1}[0.707/1.5]$$
$$= \sin^{-1}0.471$$
$$= 28.1°$$

If Medium 2 in this case were water instead of glass, we would use $n_2 = 1.33$, and we would find $\theta_2 = 32.1°$.

Internal Reflection

If a light beam is incident from air on a piece of glass, there will be a transmitted beam for any angle of incidence. However, the reverse is not true. If the beam is incident on the air–glass boundary from the *glass* side, then there will be no transmitted beam for an angle of incidence greater than a certain angle called the *critical angle* θ_c. To see why this is true, we need only to use Snell's law.

In Fig. 15-10 we have a number of light rays incident on an air–glass boundary with various angles near 40°. (Notice that the glass is now the incident medium, Medium 1.) When the incident angle is 40° (Fig. 15-10a), Snell's law tells us that the refracted ray emerges into the air at an angle of 74.6°. Increasing the incident angle to 41° (Fig. 15-10b) causes the angle of refraction to increase to 79.8°. In Fig. 15-10c the incident angle is 41.8°, and the combination $n_1 \sin \theta_1$ equals 1. Because $n_2 = 1$, the combination $n_2 \sin \theta_2$ is simply $\sin \theta_2$. When $\sin \theta_2 = 1$, we have $\theta_2 = 90°$, and the refracted ray lies exactly along the surface. Any further increase in θ_1 causes the combination $n_1 \sin \theta_1$ to exceed 1. Because $\sin \theta_2$ can never exceed 1, there is no refracted ray for any incident angle greater than 41.8°. The angle 41.8° is the *critical angle* for glass with index of refraction equal to 1.5.

What happens when the incident angle is greater than the critical angle? No refraction can occur, so the light is totally reflected from the boundary and is confined within the glass. We call this effect *internal reflection* (or *total* internal reflection).

Because the critical angle for glass ($n = 1.5$) is a few degrees less than 45°, a glass prism cut with angles of 45° can be used as a mirror. A combination of 45° prisms is shown in Fig. 15-11. Here, the light ray passes through three prisms and is internally reflected four times. Such combinations of prisms are used in bin-

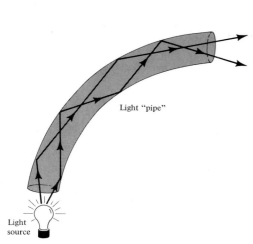

Figure 15-12 Because of internal reflection, light can be transmitted through curved cylinders of plastic or glass (light pipes).

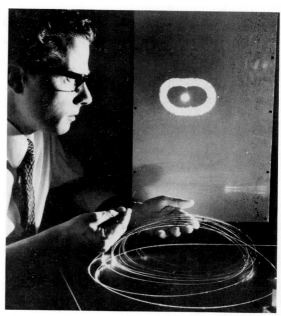

A bundle of plastic light pipes (called *optical fibers*) can be used to transmit light along tortuous paths. These devices are used to transmit light to or to receive light from inaccessible places. In this photograph the light is piped through the coiled fiber and is projected onto the screen.

oculars to increase the light path without increasing the length of the instrument.

Another application of internal reflection is in *light pipes.* When light is projected into the end of a small-diameter glass or plastic rod (Fig. 15-12), the angles with which the various rays strike the surface are greater than the critical angle. The light is therefore "piped" through the rod, even along a curved path, and emerges from the end. Bundles of tiny light pipes can be used in a variety of situations that do not permit the normal straight-line propagation of light. (For example, a physician can use a light pipe to see inside a patient's stomach.) Many other uses of light pipes (in the form of fiber optics) have been found in research areas and in communications systems.

15-2 LENSES

Focusing by Refraction

Thus far we have considered refraction only at *flat* boundaries between two media. If we allow the boundary surface to be *curved,* additional interesting effects occur. Figure 15–13 shows a glass rod whose end has been shaped into a hemispherical surface. A bundle of

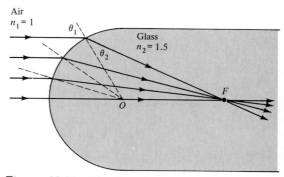

Figure 15-13 The focusing of a bundle of light rays by the refraction at a spherical surface between air and glass.

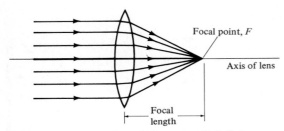

Figure 15-14 A lens brings parallel light rays together into a focus at the focal point F.

light rays is incident on the curved surface along paths that are parallel to the axis of the rod. The central ray is incident perpendicular to the surface and so passes along the axis of the rod, showing no refraction. The other rays, however, are incident on the surface at various angles. The ray farthest from the axis has the largest angle of incidence. Because the amount of refraction in this case depends on the distance of the ray from the axis, the various rays are made to converge toward a point. That is, the rays are *focused,* and the point F of convergence is called the *focal point.*

If we cut a thin slice of the rod's curved surface and shape the new surface in the same way as the original surface, we have a *lens* (Fig. 15-14). Refraction that depends on the distance from the axis now takes place at both surfaces and a parallel bundle of rays is brought to a focus outside the glass, as shown in Fig. 15-14.

In Fig. 15-14 the parallel rays are incident from the left and are brought to a focus at the right. Similarly, if rays are incident from the right, they will be brought to a focus at the left. A lens, therefore, has *two* focal points. If the curved surfaces of a lens have the same radius of curvature, the two focal points are equidistant from the lens. (We will consider only such symmetric lenses.)

Any ray that is traced in one direction through a lens can also be traced through the lens in the *opposite* direction. In Fig. 15-14, for example, we have a number of parallel rays that converge to the focal point F. If we place a light source at F, the rays will follow the paths indicated and will emerge from the lens as a parallel bundle at the left.

The formation of an *image* by a lens is shown in Fig. 15-15. Here, only the rays emanating from the tip of the object are shown; the rays that originate at any other point are also brought to a focus in a similar way. Notice that the position of the tip of the image is defined by the intersection of any two rays. There are three rays that are easy to draw. (1) The ray that enters the lens parallel to the axis passes through the right-hand focal point. (2) The ray that passes through the center of the lens encounters two surfaces that are essentially flat and parallel; this ray is therefore not refracted and can be drawn as a straight line. (Can you see why?) (3) The ray that passes through the left-hand focal point emerges from the lens along a path parallel to the axis. By using any two of these simple rays (or all three as a check), we can readily determine the position of the image for any object and any lens.

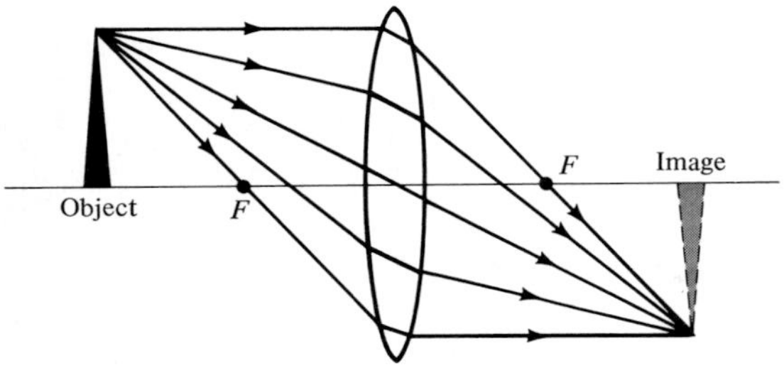

Figure 15-15 The light rays from an object are brought together by a lens to form a *real image* of the object.

The lens in Fig. 15-15 is a *convex* lens (so-called because the shape of the spherical surface is convex, or bowed outward), and the light rays from the object converge to form the image. Such an image is called a *real* image because it can actually be projected onto a screen or onto a film in a camera. A single convex lens will form a real image of any object that is placed farther away from the lens than the focal point. This is the case in Fig. 15-15.

Figure 15-16 shows what happens when an object is placed *closer* to the lens than the focal point. In this case the rays *diverge* after passing through the lens and an *apparent* or *virtual* (instead of a *real*) image is formed. That is, to an observer the rays *appear* to originate in the virtual image but, in fact, they do not. A virtual image can be observed with the eye but it cannot be projected onto a screen.

The observer in Fig. 15-16 sees an image that is larger than the object. That is, when an object is placed within its focal point, a simple convex lens acts as a *magnifier*.

In Fig. 15-17 we have a lens whose surfaces are *concave* instead of convex. A lens of this type will always cause light rays to *diverge*, never to converge. If a ray is incident on a concave lens along a path parallel to the lens axis, the ray will be refracted away from the axis and will appear, to an observer at the right of the lens, to have originated in a point that we call the focal point of the lens (even though the lens is incapable of actually focusing a beam of light).

Parallel rays incident on a convex lens from the left are brought to a focus on the right of the lens (Fig. 15-14). We say that a convex lens has a *positive* focal length. On the other hand, parallel rays incident on a concave lens from the left appear to diverge from a focal point on the left of the lens (Fig. 15-17a). We say that a concave lens has a *negative* focal length.

Figure 15-17b shows how a virtual image is formed by a concave lens. The rays originating in the object will be seen by an observer at the right of the lens to have come from a virtual image of reduced size. A concave lens is a *demagnifier*.

Locating the Position of an Image

We have already seen how it is possible to construct a diagram that locates the image for a particular lens and object position. In addition, we can easily derive an expression that relates the same quantities. Figure 15-18

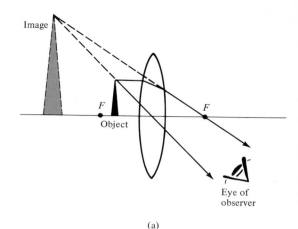

(a)

Figure 15-16 A *virtual* image is formed if the object is closer to the lens than the focal point.

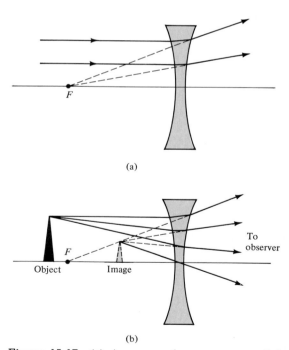

(a)

(b)

Figure 15-17 (a) A concave lens causes parallel rays to appear as though they originated at the focal point *F*. (b) The image formed by a concave lens is always *virtual*.

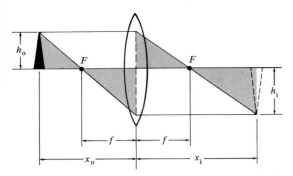

Figure 15-18 Geometry for deriving the simple lens equation.

shows the geometry of the lens, the object, and the image. The various symbols have the following meaning:

f is focal length of lens,
x_o is distance from lens to object,
x_i is distance from lens to image,
h_o is height of object, and
h_i is height of image.

Look at the two pairs of shaded triangles in Fig. 15-18. The two triangles in each pair (one on the object side of the lens and one on the image side) are *similar*. That is, the corresponding angles in each triangle of the pair are equal. Then, we can write the following proportionalities. For the triangles on the object side of the lens:

$$\frac{h_i}{h_o} = \frac{f}{x_o - f}$$

And for the triangles on the image side of the lens:

$$\frac{h_i}{h_o} = \frac{x_i - f}{f}$$

Equating these two expressions for h_i/h_o,

$$\frac{f}{x_o - f} = \frac{x_i - f}{f}$$

Then,

$$(x_o - f)(x_i - f) = f^2$$
$$x_o x_i - f x_i - f x_o + f^2 = f^2$$
$$x_o x_i - f x_i - f x_o = 0$$

Dividing each term by $x_o x_i f$, we have

$$\frac{1}{f} - \frac{1}{x_o} - \frac{1}{x_i} = 0$$

Finally, upon rearranging, we obtain

$$\boxed{\frac{1}{x_o} + \frac{1}{x_i} = \frac{1}{f}} \qquad (15\text{-}4)$$

This result is called the *simple lens equation*.

Suppose that an object is placed 6 cm from a lens that has a focal length of 4 cm; where will the object be located? Inserting $x_o = 6$ cm and $f = 4$ cm into Eq. 15-4, we have

$$\frac{1}{6} + \frac{1}{x_i} = \frac{1}{4}$$

$$\frac{1}{x_i} = \frac{1}{4} - \frac{1}{6} = \frac{3}{12} - \frac{2}{12} = \frac{1}{12}$$

Therefore,

$$x_i = 12 \text{ cm}$$

Next, suppose we move the object closer to the lens so that $x_o = 3$ cm. Then,

$$\frac{1}{x_i} = \frac{1}{4} - \frac{1}{3} = \frac{3}{12} - \frac{4}{12} = -\frac{1}{12}$$

Therefore,

$$x_i = -12 \text{ cm}$$

The object is again 12 cm from the lens. But now our result carries a negative sign. *Positive* distances for the image position are measured to the *right* of the lens. Therefore, $x_i = -12$ cm means that the image in this case is located to the *left* of the lens. This corresponds to the case illustrated in Fig. 15-16. Notice that the image is *virtual*.

To see how the simple lens equation works for a diverging lens, refer to the situation shown in Fig. 15-17b. Recall that the focal length of such a lens is *negative*. If $x_o = 6$ cm and $f = -5$ cm, we have

$$\frac{1}{x_i} = \frac{1}{f} - \frac{1}{x_o} = -\frac{1}{5} - \frac{1}{6} = -\frac{11}{30}$$

so that

$$x_i = -2.7 \text{ cm}$$

which means that the image is located to the *left* of the lens, as indicated in Fig. 15-17b (and is a *virtual* image).

Magnification and Magnifying Power

How do we determine the size of an image formed by a simple lens? Because the image size increases or decreases as the size of the object is increased or decreased, the useful quantity is the ratio h_i/h_o; we call this ratio the *magnification M* of the lens:

$$\text{magnification} = M = \frac{h_i}{h_o}$$

In Fig. 15-19 we have the same lens diagram shown

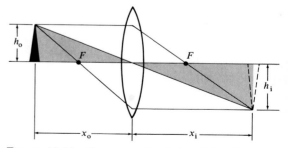

Figure 15-19 Geometry for determining the magnification of a lens.

in Fig. 15-18, except that we now use the shading to emphasize two different triangles. These triangles are similar, so we can write the ratio

$$\frac{x_i}{x_o} = \frac{h_i}{h_o}$$

Thus, the magnification is

$$\boxed{M = \frac{x_i}{x_o}} \qquad (15\text{-}5)$$

In one of the previous examples, we considered a case in which $f = 4$ cm and $x_o = 6$ cm. By using the simple lens equation, we then found $x_i = 12$ cm. The magnification in this case is, therefore,

$$M = \frac{x_i}{x_o} = \frac{12 \text{ cm}}{6 \text{ cm}} = 2$$

and the image is twice as large as the object.

This expression for the magnification is most useful when the image is *real;* it is less useful when the image is *virtual*. To see why, look at Fig. 15-20. Here, we have a simple magnifier consisting of a convex lens with an object located inside the focal point. (The rays that determine the location of the virtual image are not shown in this diagram; refer to Fig. 15-16.) Equation 15-5 for the magnification is valid for this case. (But notice that M is *negative;* this means that the image is located on the side of the axis opposite from that of a real image.) However, the object here is located near the focal point. A small change in the object position toward the focal point will cause the magnification to increase and, at the same time, will cause the image to move farther away. From the standpoint of the appearance of the image as viewed by the observer, the apparent magnification is determined, not by the ratio h_i/h_o, but by the ratio of the angles from the eye to the tip of the image and to the tip of the object. That is, the useful quantity is the ratio θ_i/θ_o (see Fig. 15-20). This ratio is called the *angular magnification* or the *magnifying power:*

$$\text{magnifying power} = \text{M.P.} = \frac{\theta_i}{\theta_o}$$

We will obtain an approximate expression for the magnifying power in the following way. First, notice that the observer's eye in Fig. 15-20 is located 25 cm from the lens. This is a typical distance when using a magnifying glass and corresponds to the minimum dis-

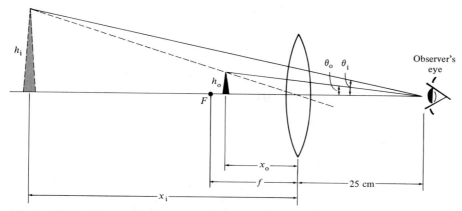

Figure 15-20 Geometry for determining the magnifying power of a lens. (This diagram is schematic only; for the focal length and object distance shown, the image will actually be much farther to the left and much larger.)

tance at which most persons' eyes can focus adequately. Next, we need a way to express the ratio of angles, θ_i/θ_o. Figure 15-21 shows two triangles which contain the angles θ_1 and θ_2. As long as these angles are not very large (about 20° or less), their ratio is equal to the ratio of the sides of the triangles:

$$\frac{\theta_1}{\theta_2} = \frac{y_1/l_1}{y_2/l_2}$$

Using this expression, we can write the magnifying power of the lens in Fig. 15-20 as

$$\text{M.P.} = \frac{\theta_i}{\theta_o} = \frac{h_i/(x_i + 25)}{h_o/(x_o + 25)}$$

Now, the image distance x_i is much larger than 25 cm and so we can neglect the term 25 cm in the numerator. Also, the object is located very near the focal point, so we replace x_o by f in the denominator. We then have

$$\text{M.P.} \cong \frac{h_i/x_i}{h_o/(f + 25)} = \frac{h_i}{h_o x_i} (f + 25)$$

In Fig. 15-20 look at the two triangles formed by h_i and h_o with the dashed line as hypotenuse. These triangles are similar, and we can write

$$\frac{h_i}{h_o} = \frac{x_i}{x_o}$$

Or, using the approximation that $x_o \cong f$,

$$\frac{h_i}{h_o x_i} \cong \frac{1}{f}$$

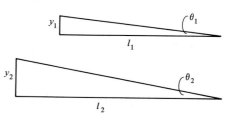

Figure 15-21 Geometry for determining the ratio of the angles θ_1 and θ_2.

Substituting this expression for $h_i/h_o x_i$ into the equation for M.P., we obtain

$$\text{M.P.} \cong \frac{f + 25}{f}$$

or,

$$\text{M.P.} \cong 1 + \frac{25}{f} \qquad (15\text{-}6)$$

where the focal length f is to be measured in cm.

If a magnifying glass has a focal length of 2.5 cm, the magnifying power will be $1 + 10 = 11$ for an object near the focal point. That is, the image will be seen 11 times larger than would the object with the lens removed.

15-3 OPTICAL INSTRUMENTS

Microscopes

Combinations of lenses can be used to produce a variety of optical effects. For example, Fig. 15-22 shows a two-lens system that is a simple *microscope*. The lens nearer the object to be magnified (called the *objective* lens) produces a real image between the two lenses. This first image acts as the object for the second lens (called the *eyepiece*), which produces an enlarged virtual image of the object that can be viewed by the eye.

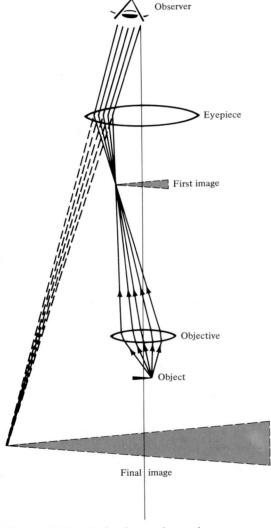

Figure 15-22   A simple two-lens microscope.

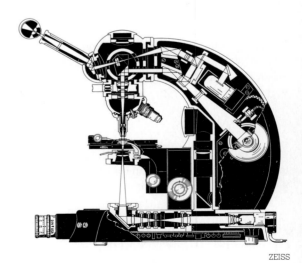

ZEISS

Cut-away drawing of a modern microscope, showing the lens system.

Modern research microscopes incorporate a large number of lenses and are designed to provide clear, sharp images for visual observation or for microphotography. By changing the objective and eyepiece lens combinations, magnifications from 10 or 20 up to about 1000 are possible. Because of the severity of distortion effects, special techniques are usually necessary to produce useful images at the highest magnifications.

Telescopes

Whereas a microscope serves to produce an enlarged virtual image of small objects placed close to the objective lens, a *telescope* serves to magnify the angular separation of distant objects or to increase the amount of light received by the eye from a distant point of light. One of the primary functions of a telescope is to collect the light from a weak source and to concentrate the bundle of rays so that the eye (or a photographic film) can register an image of the object. This is particularly true in observing faint stars, the light from which is made extremely weak because of their great distances.

The original inventor of the telescope is unknown, but a crude instrument similar to a telescope was described in the latter part of the 16th century. The telescope was reinvented in Holland in 1608, and during the early part of the following year a report of the instrument reached

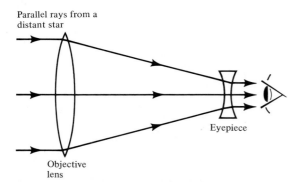

Figure 15-23 Diagram of an early Galilean astronomical telescope. The incident parallel rays from a star are brought toward a focus by the convex objective lens, but before they converge to a point, the rays are diverged into a parallel beam by a concave lens. The net effect is to increase the amount of light received by the eye from the star. The star is made to appear brighter and therefore nearer.

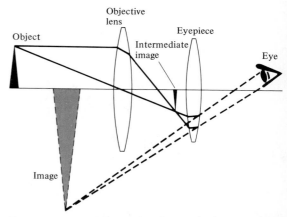

Figure 15-24 Schematic diagram of a two-lens telescope that produces an inverted image. Simple telescopes of this general design are used for viewing astronomical objects. For viewing Earth-objects, an additional lens or a prism must be used to produce an erect image.

Galileo in Italy. Because the report contained no information as to the details of construction, Galileo drew upon his knowledge of refraction and lenses to design and construct his own version. Galileo's first telescope, shown schematically in Fig. 15-23, had a magnifying power of three. But he soon constructed an instrument that magnified 8 times and he finally became sufficiently accomplished at grinding lenses that he was able to increase the magnification to more than 30.

Modern telescopes of this general type are constructed differently in that diverging lenses are not used. A simple design is shown in Fig. 15-24, where two converging lenses are used. Notice that the image is *inverted,* that is, the tip of the image points in the direction opposite to that of the object. An *erect* image is essential for observing objects on the Earth, but it is really not necessary for astronomical observations because there is no meaning to "up" or "down" for a star. In order to produce an erect image, a third lens or a prism must be added. (Ordinary binoculars use prisms to produce an erect image.)

Telescopes that make use of the refractive property of lenses are called *refracting telescopes* or *refractors*. If a

Replica of the reflecting telescope constructed by Newton in 1667 and demonstrated before the Royal Society. Focusing is accomplished by slight adjustments of the main mirror with the thumb screw in the base.

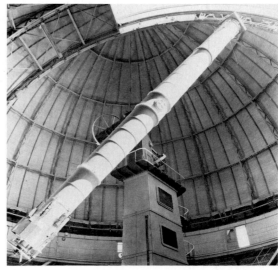

The 40-inch refracting telescope at the Yerkes Observatory of the University of Chicago. This is one of the few refracting telescopes still in use for research purposes.

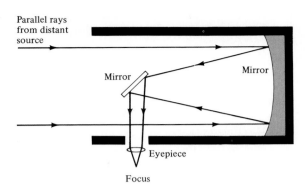

Figure 15-25 A Newtonian reflecting telescope. In order that the rays be brought to a proper focus in a reflecting telescope, the main mirror must be in the shape of a *paraboloid,* a surface generated by rotating a parabola around its axis. A *spherical* surface is adequate if the mirror is relatively small and is not intended for the most precise work.

refractor is to have a large light-gathering power (a necessity for astronomical observations), the lenses must be quite large. The instrument at the Yerkes Observatory (see the photograph on page 366) has an objective lens with a diameter of 40 inches. Lenses of such size are difficult to manufacture, have great weight, and are subject to cracking due to temperature changes. For these reasons, large-diameter refractors are not practical instruments, and very few are still in use for astronomical research.

In 1667 Isaac Newton devised a new kind of telescope that depends upon the *reflective* properties of a curved surface. A diagram of the Newtonian *reflecting telescope* (or *reflector*) is shown in Fig. 15-25. Parallel light rays from a distant source are incident on the mirror at the base of the instrument. Because the mirror surface is curved, the rays are converged toward a focus. Before the focal point is reached, however, the rays are intercepted by a small flat mirror which diverts the converging rays to an eyepiece external to the telescope. (The flat mirror is sufficiently small that it does not appreciably reduce the amount of light reaching the main mirror.)

Almost all telescopes now used in astronomical observing programs are *reflecting* telescopes. Several instruments with mirror diameters greater than 100 inches are in service. The largest American reflector is the 200-inch telescope on Mount Palomar. An even larger instrument (a 236-inch giant) has just been put into use by Soviet astronomers in the mountains of the Caucasus. All of these telescopes are equipped for photographic work and a number of different schemes, in addition to the Newtonian mirror deflector, are utilized for directing the focused beam to various positions for visual or photographic observations.

HALE OBSERVATORIES, R. W. PORTER

Cut-away drawing of the 200-inch Mount Palomar telescope. Notice that the Newtonian viewing system is not used; instead, the deflecting mirror directs the beam toward the base of the telescope and it emerges through a central hole in the main mirror.

Cameras—How They Work

One of the simplest of all optical instruments is the camera. Some cameras have complicated optical systems, but good quality photographs can be taken with a simple "box" camera that contains only a single lens. The first diagram in Fig. 15-26 shows the essential features of such a camera.

The image to be photographed is focused on the film by the lens. The movable bellows allows the lens–film distance to be adjusted for a sharp focus; different object distances require different lens–film distances. Behind the lens is a shutter than can be opened for a small fraction of a second in order to permit sufficient light from the object to produce a developable image on the film. The amount of light that reaches the film depends on the diameter and on the focal length of the lens. For a particular focal length, the amount of light admitted depends on the area of the lens, that is, on the square of the lens diameter. The ratio of the focal length to the lens diameter is called the *speed* or the *f-number* of the lens:

$$f\text{-number} = \frac{\text{focal length}}{\text{lens diameter}} = \frac{f}{d} \tag{15-7}$$

If two lenses with different focal lengths and different diameters have the *same* f-number, the degree of darkening of the film will be the same for exposures of the same period of time.

Many cameras have, in addition to a shutter, an *aperture* of variable diameter located behind the lens (see Fig. 15-27). The amount of light reaching the film can be controlled by adjusting either the aperture diameter or the shutter speed (or both). For a brightly lit object, one uses a small aperture and a fast shutter speed.

Aperture settings on a camera are usually indicated by markings that represent changes by a factor of $\sqrt{2}$ in f-number (that is, a factor of 2 in amount of light admitted). A typical lens for a 35-mm camera will have a focal length of 50 mm and diameter of 25 mm. The f-number for the aperture "wide open" is, therefore, 50 mm/25 mm = f2. By "stopping down" the aperture, a range of f-numbers can be obtained: f2, f2.8, f4, f5.6, f8, f11, f16, and f22. Changing the f-number by one position (by "one stop"), corresponds to changing the amount of light admitted by a factor of 2. Shutter speeds are

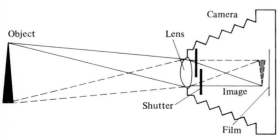

Figure 15-26

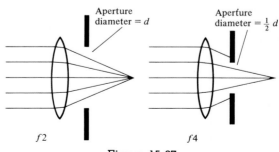

Figure 15-27

also changed in steps of a factor of 2. A typical 35-mm camera will have shutter speeds of 1/30, 1/60, 1/120, 1/250, 1/500, 1/1000, and 1/2000 of a second.

By adjusting the aperture and shutter speed, the proper exposure can be obtained for a variety of lighting conditions. But these adjustments are made for other reasons as well. For example, if a sharp photograph of a rapidly moving object is to be taken, then a fast shutter speed is necessary. The loss of light due to the fast shutter speed is compensated by using a large aperture (small f-number). When the aperture is small (for example, f 11 or f 16), light is admitted only very close to the axis of the lens. These near-axial rays will be in sharp focus for a wider range of object distances than will rays that enter near the edge of the lens. Therefore, small apertures are used when great "depth of focus" is required. Large apertures are used to decrease the depth of focus and to "fuzz out" any undesired background behind the object.

Because the index of refraction of a lens depends to some extent on the color of the light, the focal point for blue light and the focal point for red light will be slightly different. (This is called chromatic aberration and is one of the difficulties with lenses that Newton sought to overcome when he invented the reflecting telescope.) In order to overcome this defect, modern camera lenses are not single pieces of glass, but consist of two or three components with different indices of refraction to cancel the chromatic effects. See Fig. 15-28.

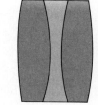

Figure 15-28 A *Cook triplet* is made from three different kinds of glass.

The Soviet 236-inch telescope will probably remain the largest astronomical instrument ever constructed. More information can be gathered with smaller telescopes that are placed into orbit on satellites or on permanent stations on the Moon than can be obtained by building ever larger Earth-based instruments that must contend with light filtered through and disturbed by the atmosphere. A 120-inch instrument is planned for installation on an orbiting station and will be capable of recording the light from galaxies that are 100 times fainter than those observable by the most powerful ground-based telescopes.

15-4 COLOR AND SPECTRA

Color

The light that we receive from the Sun or from an incandescent lamp is usually considered to be *white* light. But this light is actually composed of many colors. We can observe the breakup of white light into its component colors by passing the light through a triangular piece of glass called a *prism* (Fig. 15-29). A light ray is refracted upon entering and upon leaving the prism so that there

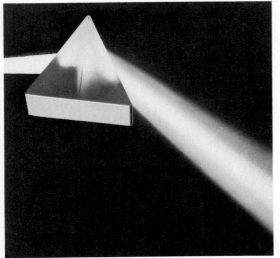

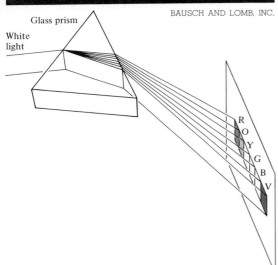

Figure 15-29 When white light, consisting of all colors, passes through a triangular glass prism, the light is dispersed into a spectrum of colors. Red light is refracted least and appears on the left-hand side of the spectrum in this diagram, followed by orange, yellow, green, blue, and violet.

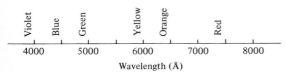

Figure 15-30 Wavelengths of colors in the visible spectrum of light. 1 angstrom (Å) = 10^{-10} m.

is a net deviation of the ray as it passes through the prism (just as there is when a noncentral ray passes through a lens). The *amount* of refraction, and therefore the angle of deviation, depends on the color of the light component. Red light is deviated least and violet light is deviated most. Between these two extremes we find the colors orange, yellow, green, and blue. Thus, a prism disperses white light into a *spectrum* of colors. When sunlight passes through raindrops, the spheres of water act as tiny prisms refracting the light and we observe the full spectrum of sunlight colors in a *rainbow*.

The reason that light with different colors is affected differently by a prism is that light is a *wave* phenomenon and the amount of refraction that light undergoes depends on its *wavelength*. That is, the index of refraction is wavelength dependent. In the visible spectrum of colors, red light has the longest wavelength (about 7.5×10^{-7} m) and violet light has the shortest wavelength (about 4.0×10^{-7} m). Thus, every spectral color can be specified quantitatively by the value of its wavelength (see Fig. 15-30), and with each wavelength there is associated a slightly different value of n for the prism glass.

For purposes of specifying light wavelengths, we often use a unit called the *angstrom* (Å): 1 Å = 10^{-10} m. Thus, red light has a wavelength of 7.5×10^{-7} m = 7500 Å.

Spectra

In 1802 the English scientist William Wollaston (1766–1828) was examining sunlight with a prism, in much the same way that Newton had done. But Wollaston observed that superimposed on the continuous spectrum of colors there were a number of sharp dark lines. Several years later these mysterious lines were studied in detail by the German optician Joseph von Fraunhofer (1787–1826) who cataloged 576 lines in the solar spectrum and measured their wavelengths. These lines are now known as *Fraunhofer lines*. Fraunhofer determined that the lines originate in the Sun (and are not due to some terrestrial effect) but he was unable to explain why they appear.

Because the Fraunhofer lines are *dark*, they represent the *absence* of light at particular wavelengths. That is, the light that would otherwise be present is *absorbed* in some way before it reaches the Earth. (The absorption must take place in the Sun because there can be no ab-

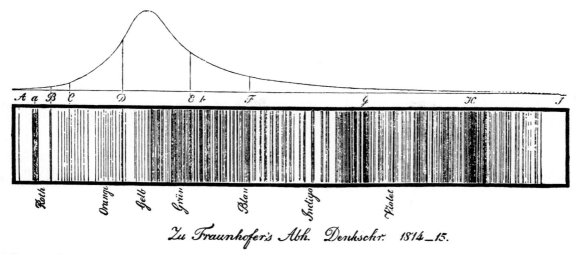

Fraunhofer's sketch of the dark lines in the solar spectrum.

sorption in the vacuum of space between the Sun and the Earth and the lines are not of terrestrial origin, as Fraunhofer demonstrated.) The interior of the Sun is extremely hot—so hot, in fact, that the atoms are completely stripped of their electrons due to the frequent, violent collisions between the rapidly moving particles. The light that originates in this region of the Sun is true *white* light, with no distinguishable features in the spectrum. The outer layers of the Sun, however, consist of cooler gases, and atoms (with at least some of their electrons) can exist in this region. When the white light from the interior passes through the relatively cool outer gases, the atoms in this layer absorb some of the light at the particular wavelengths characteristic of the gas atoms.

The absorption of light by gases can easily be demonstrated in the laboratory. Figure 15-31 shows a beam of white light projected through a cell containing vaporized sodium. The light is then analyzed by means of a prism. Examination of the resulting spectrum shows a pair of dark lines located in the yellow region of the spectrum. That is, the sodium atoms have absorbed some of the light at these particular wavelengths. After absorbing this light, the sodium atoms *re-emit* the light at the *same* wavelength, but the re-emitted light is radiated in all directions and so only a small amount finds its way into the original beam. Thus, the pair of sodium lines in the spectrum are *dark* lines. If we were to examine the light *emitted* by the vapor cell (instead of the light *transmitted* by the cell), we would find in the spectrum a pair

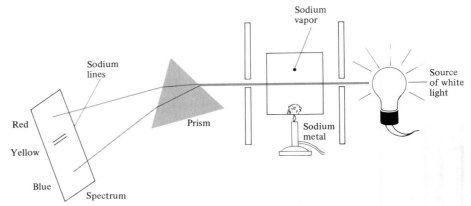

Figure 15-31   When white light passes through vaporized sodium, the sodium atoms absorb some of the light at particular wavelengths. This absorption is revealed by dark lines that appear in the spectrum. A pair of lines in the yellow region of the spectrum is characteristic of sodium.

of *bright* yellow lines representing the light emitted by the sodium atoms at these wavelengths. *Emission* lines are bright; *absorption* lines are dark.

Each chemical element emits (and absorbs) light at characteristic wavelengths. Figure 15-32 shows portions of the line spectra of several elements recorded photographically with a *spectrograph* (but one more sophisticated than a simple prism). These lines constitute a kind of "fingerprint" of the element emitting the light. Spectral measurements can therefore be used to identify

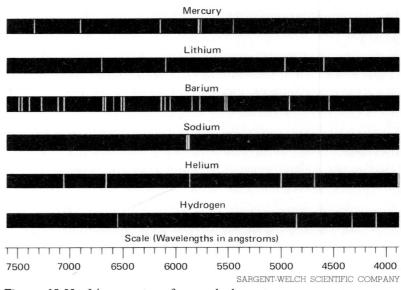

Figure 15-32   Line spectra of several elements.

A portion of the solar dark-line spectrum (center) and a comparison spectrum of iron (top and bottom) photographed with the same spectrograph in the laboratory. Many of the dark lines in the solar spectrum occur at the same positions as the bright iron lines, thus showing that iron exists in the outer layers of the Sun's atmosphere. (This photograph is from a *negative* so that the bright lines appear dark and vice versa.)

elements in laboratory samples of materials or, by attaching a spectrograph to a telescope, element analyses of stars can be carried out. This technique is, in fact, one of the most powerful methods that we have for investigating the details of stellar composition.

In Chapter 18 we will return to the subject of spectra and will discuss the relationship between spectra and atomic structure.

15-5 LIGHT AND VISION

Structure of the Eye

The human eye is a marvelous piece of light-detecting equipment. It is sensitive to a wide range of light intensities; it can render sharp images for both distant and nearby objects; it can sense subtle variations in color; and it requires very little maintenance. Modern technology has been unable to develop an optical instrument with comparable sensitivity, flexibility, and reliability. The eye and its associated electrical network that delivers optic signals to the brain is the most sophisticated system in the human body and is the channel through which we derive most of our information.

The adult human eye is a globular structure about one inch in diameter. The delicate inner parts are encased in a tough coat of elastic connective tissue called the *sclera* (Fig. 15-33). When light is incident on the eye, it passes first through the outer protective window called the *cornea*. Next, the light proceeds through the jellylike *aqueous humor,* through the *pupil* and *lens,* and into the central region of the eye which is filled with a transparent liquid called the *vitreous humor.* Finally, at the rear of the eye, the focused light is detected by the *rods* and *cones* in the *retinal* surface and a signal is sent to the brain through the optic nerve.

The cornea and the lens act in combination to pro-

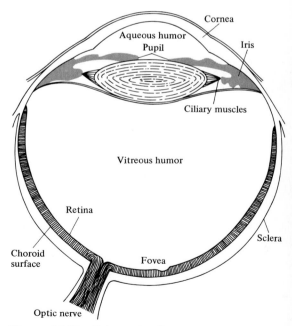

Figure 15-33   Schematic diagram of the human eye.

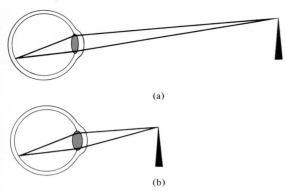

Figure 15-34 The ciliary muscles control the shape of the lens and adjust it for focusing the light from distant or nearby objects. (a) For a distant object, the lens shape must be relatively flat, and (b) for a nearby object the shape must be more rounded.

duce on the retina an image of the scene viewed. Light rays are refracted at the curved interface between the air and the cornea; further refraction takes place in the lens to focus the light on the retinal surface. The result is exactly the same as we have already discussed for a glass lens. But the lens of the eye has an additional capability. The degree of refraction in the lens (that is, its focal length) is controlled by the *ciliary muscles* which can cause the lens to become flatter or more rounded. If the object viewed is far away, the lens must be relatively flat in order to focus the light on the retina (Fig. 15-34a). On the other hand, the light from a nearby object will be properly focused only if the lens is made more rounded (Fig. 15-34b). These actions of the ciliary muscles are triggered by impulses from the brain when we concentrate our attention on objects at various distances. The focusing action of the eye is not completely automatic. After all, it is possible to consciously *defocus* the eyes so that no object in the line of sight is seen sharply.

Sometimes the ciliary muscles are incapable of properly adjusting the curvature of the lens surface. Moreover, as a person ages, the cells in the lens are continually dying, thus hampering proper focusing. (The lens contains no blood vessels and so supplying the lens cells with nutrients is very inefficient; consequently, the cells eventually die.) In these cases of inadequate natural focusing, the eye can be provided with artificial equipment—namely, glasses or contact lenses—to aid in the focusing process and to restore sharp vision.

The human eye can respond to light intensities that vary by a factor of 10^{10} (10 billion). At the upper end of the intensity range, the eye experiences some discomfort, and for even higher intensities (such as looking directly at the Sun), permanent damage can result.

Depending on the light conditions, the muscles in the *iris* (the pigmented part of the eye) control the size of the opening (the *pupil*) which admits light to the lens. When the light is dim, the pupil is large (7 or 8 mm in diameter) to admit as much light as possible. When the light is bright, the pupil is small (2 or 3 mm in diameter) to protect the inner parts of the eye from excess radiation. The iris therefore operates in the same way as the adjustable lens aperture in a camera. But the changes in the size of the pupil regulate the light entering the eye only in a ratio of about 10 to 1, a small range compared to the enormous sensitivity range of the eye. Thus, the size of the pupil plays only a minor role in adjusting the intensity of light that falls on the retina. The primary function of the adjustable pupil size seems to be to confine the rays, in a bright-light situation, to the central region of the retina where the focusing is most precise.

Rods and Cones

The light-sensitive part of the eye is a dense collection of sensors, called *rods* and *cones,* located in the retinal surface. Each human eye contains about 120 million rods and about 6 million cones. The electrical signals from these light-sensing elements are sent to the brain through a network of almost a million nerve fibers in the *optic nerve.* Rods and cones (Fig. 15-35) perform different functions and have different distributions over the retinal surface. A rod is about 500 times more sensitive to light than is a cone. But a cone senses color, whereas a rod does not. Thus, in a weak-light situation, the rods provide almost all of the visual information. Because the rods have no color sensitivity, we see only shades of gray in weak light. When the light is bright, the cones are active and colors can be perceived.

The light received from the central part of a scene that is viewed is concentrated on the central part of the retina. The packing of light receptors is much more dense in this part of the retina. Therefore, forward vision is much more acute than peripheral vision which is sensed by a smaller number of receptors. Furthermore, the cones are concentrated in the central region; consequently, peripheral vision, because it is sensed primarily by rods, is lacking in color perception. (Try to determine the color of an object that is seen only out of the "corner" of your eye.)

At the center of the retina there is a small depressed region, called the *fovea,* which contains only cones. The density of receptors in the fovea is extremely high—about 150 000 cones per square millimeter. Most of the light-sensitive elements in the eye do not connect directly to the brain: instead, some signal processing is accomplished in the neural cells attached to the bases of the rods and cones. The foveal receptors, however, have a direct line to the brain, making this region the most acute part of the eye.

How do the rods and cones sense light? We know that chemical reactions induced by light (*photochemical* reactions) take place on the tips of the rods and cones, but we do not know exactly how these reactions produce the neutral impulses that carry signals to the brain.

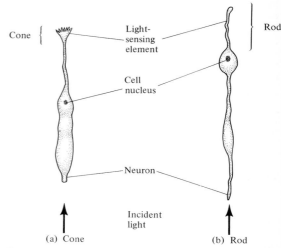

Figure 15-35 The differing structures of the (a) cones and (b) rods located in the retinal surface of the eye. Notice that the light-sensing element is at the end away from the incident light.

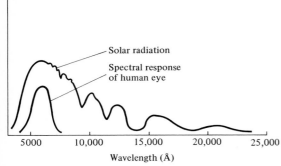

Figure 15-36 The spectrum of solar radiation reaching the surface of the Earth compared to the spectral response of the human eye. The solar spectrum is not smooth due to the selective absorption of radiation by molecules in the atmosphere.

TABLE 15-1 COLORS IN THE VISUAL SPECTRUM

COLOR	WAVELENGTH (Å)
Red	7500
Orange	6100
Yellow	5900
Green	5400
Blue	4600
Violet	4000

In the absence of light, the retina has a reddish-purple color which is due to the presence of a chemical substance called *rhodopsin* (or *visual purple*). When light is incident on the retina, photochemical reactions change the rhodopsin into various compounds, finally forming *vitamin A* (which is colorless). Under ordinary lighting conditions, no more than about 2 percent of the rhodopsin is reduced at any instant. Regeneration processes act to restore the rhodopsin (and restore the light-sensing ability of the receptor) by converting the vitamin A. As few as 5 photons received by the rods can trigger the photochemical reactions with rhodopsin and produce a visual sensation.

Because of the production of vitamin A in the photochemical process that senses light, the myth has grown up that by taking increased amounts of vitamin A into the system (over and above that normally ingested), a person can improve his vision, particularly night vision. This notion is entirely false.

Color Vision

The human eye is sensitive to a highly restricted range of wavelengths in the electromagnetic spectrum. The eye will not respond to radiation unless the wavelength is between about 4000 Å and about 7500 Å. Why has the evolutionary process developed visual acuity in this particular wavelength range? The reason is to be found by looking at the spectrum of solar radiation. Figure 15-36 shows the spectrum of radiation from the Sun as measured at the Earth's surface. The spectrum contains very little ultraviolet radiation (wavelengths shorter than about 3500 Å) because of the strong absorption of these radiations in the atmosphere. Furthermore, the infrared part of the spectrum (wavelengths longer than about 8000 Å) contains many irregularities due to the selective absorption in certain wavelength regions by air molecules. The resulting spectrum shows a maximum near 6000 Å, and this is precisely the wavelength at which the eye is most sensitive. But why is the eye's response cut off on the long wavelength side? One reason is that the longer wavelengths undergo relatively little refraction at the surface of the cornea and in the lens, making it difficult to focus these radiations. Another reason is probably that there is simply too much long wavelength radiation in the solar spectrum for the optic system to handle. The eye is sensitive to only about 14 percent of the Sun's radiation that reaches the surface of the Earth.

How does the eye perceive color? We know that every wavelength in the visual part of the spectrum corresponds to a particular color (see Table 15-1). But not every color corresponds to a particular wavelength. For example, there is no brown, no purple, and no pink in the visual spectrum, and yet these colors are readily perceived by the eye. Somehow the eye is capable of responding to *combinations* of wavelengths and issuing a signal that reports shades and hues of color.

There cannot be a separate receptor for every conceivable shade of color—there are simply too many possibilities. (One can purchase paint in at least a thousand different shades!) Instead, we distinguish colors by summing the signals from three different kinds of cones which have different spectral responses (Fig. 15-37). If the eye receives light that is a mixture of green (5400 Å) and orange (6100 Å), visual signals will be transmitted by the *green*-sensitive cones (curve *B* in Fig. 15-37) and by the *red*-sensitive cones (curve *A* in Fig. 15-37). The brain interprets these signals as the color *yellow* and we have the same visual sensation that would be produced by light with a wavelength of 5900 Å. Other combinations produce the effect of different shades and hues of color. For example, a mixture of red (7000 Å) and violet (4000 Å) will produce an intense purple. Purple is not a spectral color and can be produced only through a mixture.

The three different types of cones have response peaks corresponding to the colors red (*A*), green (*B*), and blue (*C*). These are the three *primary colors* from which all other colors can be obtained by various combinations.

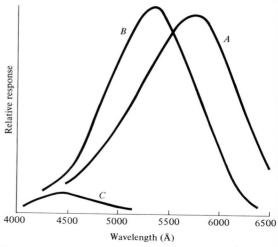

Figure 15-37   Relative spectral responses of the three different types of cones. Curve *A* is for the *red*-sensitive cones (even though the peak is in the *yellow* part of the spectrum); Curve *B* is for the *green* sensitive cones; and curve *C* is for the *blue*-sensitive cones, which are much less sensitive than the other two types.

SUGGESTED READINGS

Sir William Bragg, *The Universe of Light* (Dover, New York, 1959).

B. Jaffe, *Michelson and the Speed of Light* (Doubleday, Garden City, New York, 1960).

Scientific American articles:

P. Connes, "How Light is Analyzed," September 1968.

E. H. Land, "Experiments in Color Vision," May 1959.

QUESTIONS AND EXERCISES

1. In Michelson's measurement of the speed of light (see Fig. 15-2), the round-trip distance traveled by the light was 44 miles. How fast did the 8-sided mirror rotate (in rpm,

revolutions per minute) when the light entered the detector? (The mirror rotated $\frac{1}{8}$ revolution or 45° during the time required for the round trip of the light pulse.)

2. In Michelson's experiment (Fig. 15-2), he found that light entered the detector when the mirror rotated at a certain speed. If the rotation speed were then *doubled* or *halved,* what would be the results? How did Michelson know which rotation speed to use in calculating the speed of light?

3. How long is required for light to travel from the Sun to the Earth?

4. Accurate measurements of the distance from the Earth to the Moon are being made by pulsing the light from a laser beam that is directed toward the Moon. These pulses are reflected by a special mirror left on the Moon's surface by the Apollo 11 astronauts. The measurements show that the time required for the round trip of the laser pulses is approximately 2.48 seconds. What is the Earth–Moon distance? (The result is actually the distance from the *surface* of the Earth to the *surface* of the Moon.)

5. In the 2nd century B.C., Hero of Alexandria explained the equality of the angles of incidence and reflection as a result of the axiom that a light ray, in traveling from one point to another by means of reflection, takes the *shortest* possible path between the points. Make a sketch and, by measuring the lengths of several different light paths, show that the path for which $\phi_r = \phi_i$ has the shortest length.

6. If you lie on the bottom of a swimming pool looking upward and away from the vertical, you will see some objects that are outside the pool and other objects that are on the bottom of the pool. Make a sketch to show why this is so.

7. If you look at a swimmer who is standing in waist-deep water, he appears to have stubby legs. Make a sketch to explain this effect.

8. Suppose that a transparent rod is made from plastic which has $n = 1.33$. If such a rod is placed in water, will it be visible? Explain.

9. What is the speed of light in water (index of refraction = 1.33)?

10. A light ray passing through water enters a flat piece of glass ($n = 1.50$) at an angle of 35° (with respect to the perpendicular). What is the direction of the ray in the glass?

11. A light ray in air enters a piece of glass at an angle of 45°. Within the glass the direction of the ray makes an angle of 26° with respect to the perpendicular. What is the index of refraction for this piece of glass?

12. A flat piece of glass ($n = 1.50$) is partially immersed in water with the flat surfaces parallel to the water surface. A light ray in air enters the exposed side of the glass at an angle of 35°. What will be the direction of the ray in the water after it has passed through the glass?

13. What is the critical angle for a water–air surface?

14. What is the critical angle for a surface between water and glass ($n = 1.50$)? From which side must a light ray be incident in order to be totally reflected?

15. Make a sketch similar to Fig. 15-17b except with the object located closer to the lens than the focal point. Describe the image. (Is it real, virtual, erect, reversed, enlarged, reduced?)

16. An object with a height of 2 cm is located at a distance of 8 cm from a lens with a focal length of 6 cm. Where is the image located? What is the magnification?

17. Suppose that the object in Exercise 16 above is moved 4 cm closer to the lens. What is the new position of the image? (Your answer will now contain a negative sign. What does this mean?) Sketch the situation. On which side of the lens is the image? Is the image real or virtual?

18. An object is placed at a distance of 2 cm in front of a convex lens with a focal length of 3 cm. What is the magnification of the image? What is the magnifying power of the lens?

19. What is the magnification of an object that is placed at the focal point of a convex lens? What is the physical reason for your result?

20. Is the idea of magnifying power appropriate for a lens that produces a real image? Explain.

21. Use the simple lens equation (twice) to determine the position of the image in the situation illustrated in the diagram below. Characterize the image. (Is it real, virtual, erect, reversed, enlarged, reduced?)

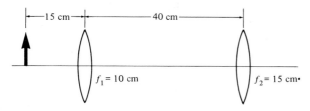

22. If appropriate camera settings to photograph a particular scene are $f5.6$ at $1/120$ s, what must be the shutter speed if the aperture setting is changed to $f2.8$?

23. Explain the difference between *dark*-line spectra and *bright*-line spectra. Can both be used to identify the composition of materials? Explain.

24. The light from a source of unknown composition is examined with a spectograph. Strong bright lines are found at wavelengths of 6678 Å, 5875 Å, 5461 Å, and 4358 Å. Use the spectra in Fig. 15-32 to identify the elements that emit light with these wavelengths and thereby determine the composition of the source.

25. The claim is made that the eye's lens projects an inverted image on the retina. But this would mean that we would see everything "upside down." Comment.

26. Light from a point on which you concentrate your attention (called the *fixation* point) enters the eye and is concentrated at the center of the retina—that is, on the fovea. The fovea contains only cones. Therefore, if you want to observe a weak source of light (for example, a faint star), should you look *directly at* the source? Explain. (Try a few observations.)

16

RELATIVITY

Most of the physical ideas that we have discussed so far are easy to accept. The notions that *force* produces *acceleration* and that *gravity* acts between the Sun and the Earth just as it acts between a baseball and the Earth do not offend our sensibilities. Even the statement that an electromagnetic wave can carry *energy* through a vacuum becomes reasonable when we realize that this is exactly how we receive energy from the Sun. These ideas do not seem strange to us because we have often encountered them in our everyday activities even though we may not have been aware of the precise scientific statements of the physical principles involved.

The upbringing of an individual in today's world automatically involves an introduction (whether consciously or subconsciously) to a number of scientific concepts. These ideas become a part of one's intuition. (Einstein once said that *intuition* is the layer of prejudices that is built up in one's mind before the age of eighteen.) Statements or observations that agree with one's intuition are easy to accept. But those that are at variance seem unreal, even impossible. When Man first attempted the construction of a flying machine, the idea seemed preposterous — there was nothing in one's experience suggesting that such a feat was possible. But today we accept airplanes as readily as we accept automobiles. Intuition changes because experiences change.

We now turn our attention to some ideas that do not fall within the realm of our intuition. The failure of our intuition in the areas of relativity and quantum theory is due to very obvious reasons. Relativistic effects are important only when extremely high speeds are involved — speeds near the speed of light, $c = 3 \times 10^8$ m/s. And quantum effects are important only when we consider objects in the submicroscopic domain of molecules, atoms, and nuclei. We do not have direct contact with either of these areas in our everyday experience; the world we see consists of objects composed of bulk matter and which move with relatively low speeds. Thus, when we first hear of the ways in which Nature behaves in the relativistic and quantum domains, the ideas seem particularly strange. But we must realize that our intuition is not infallible. If our intuition does not agree with the facts, we must learn to accept the facts instead of intuition. We must not force our intuitive ideas into areas where they do not apply. After all, intuition once dictated that the Earth is flat.

16-1 THE BASIS OF RELATIVITY

The Ether Concept Fails

Following the triumph of Newtonian dynamics as an explanation of the motion of all kinds of objects, there grew up during the 18th century a general mechanistic view of Nature. The basic theme of this outlook was that Newtonian reasoning could explain all natural phenomena. Many of these early ideas persisted until the 20th century.

An object can be moved by a push or a pull exerted by another object — this is a very real and understandable kind of force between two objects in contact. But how does one explain the gravitational and electromagnetic forces? These forces can act between objects *not* in contact, even through empty space. It seemed so unreal that a force could act through a complete vacuum that a substance was invented to fill this void. The mysterious substance that was considered to fill all space was called the *ether*. It was the ether that transmitted forces between objects not actually in contact, and it was the vibrations of the ether that carried electromagnetic waves. But what was the ether? No one could really answer this question because the ether had no directly measurable properties!

Although the ether satisfied the intuitive need for something to fill the vacuum of space, it was more trouble than it was worth. When the ether concept was used to interpret various experiments, serious contradictions were discovered. Many efforts were made to modify the ether idea to bring it into agreement with experiment, but none was successful.

Einstein's Postulates

When Albert Einstein (1879–1955) saw the enormous difficulties surrounding the concept of the ether, he decided that it would be unproductive to attempt once more to patch up the old theory. Instead, he adopted a new and radical view. A light wave had previously been considered to be a vibration set up in the ether, propagating in the same way that a sound wave travels through air or some other material medium. The ether was regarded as a *real* substance, and a light wave was considered to be only a distortion of the medium. But Einstein regarded the electromagnetic field as the real entity. A light wave, to Einstein, was a disturbance of the field; the disturbance propagates freely through empty space, requiring no material medium for its existence.

By discarding the ether concept, Einstein simultaneously abolished the idea of "absolute motion." If there is no ether to serve as the basic reference frame against which all motion is measured "absolutely," then it becomes possible to describe motion only in terms of an object moving *relative* to another object. To describe the relative motion of two objects, it does not matter whether we select a reference frame attached to one object or to the other, or whether we use some other frame in uniform motion with respect to both objects. The basic laws of physics must be the same no matter what reference frame we use for describing them (as long as the reference frame is not accelerating).

This idea is not really new. In Section 4-2 it was pointed out that Newton appreciated the fact that if the laws of dynamics are valid in one reference frame, they are also valid in any other frame in uniform (nonaccelerated) motion relative to the first frame. Einstein extended this idea to include electromagnetic as well as mechanical phenomena. Because there is no place for "absolute motion" in Einstein's formulation of the physical laws, his theory is called the theory of relativity.

So far, we have discussed only the first postulate of Einstein's theory, which can be simply stated as:

I. All physical laws are the same in all inertial (nonaccelerated) reference frames.

The second of Einstein's postulates is more subtle. Think about a water wave moving across a pool. Any observer, no matter what may be his motion with respect to the pool, will agree that the pool represents the identifiable medium through which the wave propagates. This conclusion is possible because the pool is a *material* substance and any motion of an observer relative to the pool can be measured. But what about a light wave that propagates through empty space? There is no material medium, such as the pool of water, to serve as the obvious frame of reference. Every observer of the light wave can describe the wave in terms of his own reference frame. None of these possible reference frames has any preferred status compared to the others. If an observer measures the speed of the light wave, he makes the measurement with respect to his own frame. (There is no "ether frame" with respect to which an "absolute" speed can be measured.) What do the various observers find for the speed of the light wave in their own frames? Einstein gave a surprising answer: all of the observers measure the same speed! Stated more completely, this answer becomes Einstein's second postulate:

> II. The velocity of light (in vacuum) is the same for any observer in an inertial reference frame regardless of any relative motion between the light source and the observer.

This postulate is in striking contradiction to our intuitive expectation. Accordingly, we will devote the next sections to an examination of this new idea and some of its consequences.

16-2 THE VELOCITY OF LIGHT

The Velocity of Light Is Constant

A good baseball pitcher can throw a ball with a speed of 100 mi/h. If the pitcher makes his throw inside a railway car moving with a velocity of 100 mi/h in the direction of the pitch, what will an observer *on the ground* measure for the baseball's speed? The answer is simple enough; as shown in Fig. 16-1, the ground speed of the baseball is just the velocity of the ball relative to the railway car plus the velocity of the railway car relative to the ground:

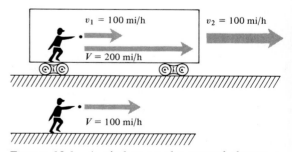

Figure 16-1 A pitcher on the ground throws a baseball with a speed relative to the ground of $V = 100$ mi/h. The same pitcher in a railway car moving with a velocity of 100 mi/h throws a ball with a velocity relative to the ground of $V = v_1 + v_2 = 200$ mi/h.

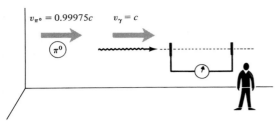

Figure 16-2 Experiment to test Einstein's second postulate. A beam of neutral pions (π^0) enters the laboratory with a velocity $v_{\pi^0} = 0.99975c$. The pions decay in flight into γ rays, and the speed of the γ rays is determined by measuring the flight time between two fixed detectors. The result is that $v_\gamma = c$.

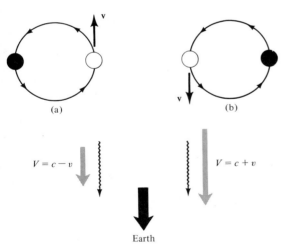

Figure 16-3 (a) The shining star of a binary pair moves away from the Earth and emits "slow" light toward the Earth. (b) The star moves toward the Earth and emits "fast" light. No evidence has ever been found that substantiates this variation in the speed of light.

$$V = v_1 + v_2$$
$$= 100 \text{ mi/h} + 100 \text{ mi/h}$$
$$= 200 \text{ mi/h}$$

Suppose that we now increase the speed of the railway car to one-half the speed of light, $v = 0.5\,c$, and we replace the baseball pitcher with a source that emits pulses of light. To an observer inside the railway car and moving with it, the speed of each light pulse is, quite understandably, equal to $c = 3 \times 10^8$ m/s. What will be the speed of the light pulses as measured by an observer stationed on the ground? According to our previous analysis, the speed should be $V = v + c = 0.5\,c + c = 1.5\,c$. But Einstein's second postulate states that the speed of light measured by *each of the observers* is the same, namely, c.

How can we justify this curious postulate which asserts that the speed of light is always the same, regardless of relative motion between the source and the observer? We can only appeal to experiment. The most convincing laboratory demonstration of the validity of Einstein's second postulate was carried out in 1964 by a group in Geneva, Switzerland. The experimenters prepared a beam of neutral pions, π^0 (see Table 2-3). These pions decay in a short time into electromagnetic quanta (γ rays) which are the same as light photons except that their energy is considerably higher. The velocity of the pion beam in the laboratory was determined to be $v = 0.99975\,c$. When the pions decayed in flight, a measurement was made of the speed of the emitted γ rays that were moving in the same direction as the original pions (Fig. 16-2). According to the old view, we would expect the γ rays to be moving with a speed almost equal to $2\,c$. But the result was that the γ rays traveled with a speed in the laboratory equal to c to within 1 part in 10^4.

Further evidence comes from astronomical observations. Many of the stars in the sky are actually not single stars but are *binary* stars, two stars orbiting around one another. Sometimes it happens that only one star of the pair shines brightly while the other is dim or even dark. Figure 16-3 shows such a pair, with only one star emitting any significant amount of light. At certain times in its orbit, the shining star will be traveling *away from* an observer on Earth (Fig. 16-3a) and at other times the star will be traveling *toward* the observer (Fig. 16-3b). If the speed of the emitted light depends on the motion of the source, then in Fig. 16-3a the speed of the light moving toward the Earth is $V = c - v$, whereas in Fig. 16-3b the speed is $V = c + v$. We therefore have a situation in which the "fast" light emitted at a certain time can overtake the "slow" light emitted at an earlier time. Thus, on Earth we would see the star in two positions at once! No observation of a binary star system has ever revealed any such curious behavior. All observations are consistent with the postulate that the speed of light does not depend on the relative motion between the source and the observer.

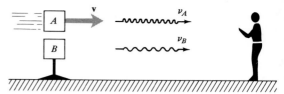

Figure 16-4 If the observer measures the speed of the light emitted by the stationary and the moving sources, he will obtain the same value. But because of the Doppler effect, the frequency of the light from the moving source will be found to be greater than that from the stationary source: $\nu_A > \nu_B$.

What Distinguishes a Moving Source from a Stationary Source?

Figure 16-4 shows an observer viewing the light from two identical sources, one of which is stationary and one of which is moving in his reference frame. We have already concluded that if the observer measures the speed of the light from each of these sources he will obtain exactly the same result. But there is a *physical* difference in the two cases — one source is moving relative to the observer and the other is not — and so there should be some *physical* difference in the light from the two sources. The speed of the light from each source is the same, but because of the Doppler effect (Section 13-4), the *observed frequency* is not. The frequency of the light from the source moving toward the observer is greater than the frequency of the light from the stationary source: $\nu_A > \nu_B$. Therefore, if the observer deals with identical sources, he can always determine which are stationary and which are in motion toward or away from him. A measurement of the speed of the light from the two sources will, however, always yield the same value.

Simultaneity

We are accustomed to thinking of space and time as separate and distinct concepts. But Einstein's postulates force upon us another interpretation. We can see this most clearly if we examine the way we determine the sequence of events taking place at different locations. Because of the constant velocity of light, we find that observers who are in relative motion can give *different* answers to questions of "before" and "after."

Consider the situation pictured in Fig. 16-5. Again, we have a moving railway car with one observer.

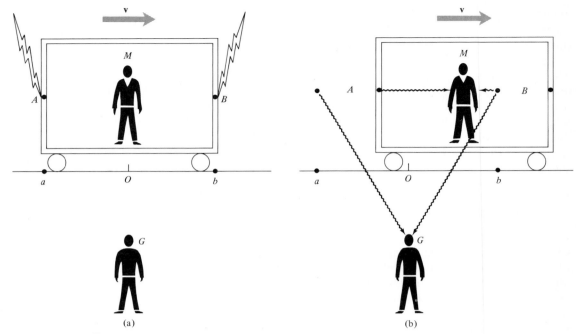

Figure 16-5 The ground observer *G* sees two lightning bolts strike the railway car *simultaneously,* but the moving observer *M* sees the flash from *B before* he sees the flash from *A*. *G* and *M* do not agree on the time sequence of the events.

Melvin, moving with the car and one observer, George, on the ground. (The initials of our observers will remind us who is Moving and who is on the Ground.) Melvin measures the length of the car and positions himself in the exact center. George positions himself opposite a point *O* on the tracks. The car moves with a velocity **v** and as it passes George, two lightning bolts strike the ends of the car at *A* and *B* and leave burn marks on the track at *a* and *b* indicating the position of each end of the car at the moment of strike (Fig. 16-5a). Light flashes emanate from *A* and *B* as the burn marks are produced and arrive at George's position at the *same instant* (Fig. 16-5b). George then measures the distance from *O* to each of the burn marks *aO* and *Ob*, and he finds these distances to be equal. Because the light flashes reached George at the same instant and because they traveled equal distances, George concludes that each lightning bolt struck the car at the *same instant.* That is, to George the strikes were *simultaneous.*

The lightning strokes can be seen by Melvin inside the car as flashes of light emanating from *A* and *B*. George knows that Melvin is moving *toward* the flash from B and *away from* the flash from A. Therefore, George concludes, Melvin must see the B flash prior to the A flash. Melvin confirms that this is the case.

So far we have concluded nothing out of the ordinary. (We have not, for example, invoked Einstein's second postulate.) But now let us inquire how the moving observer, Melvin, interprets the situation. Melvin knows that the velocity of light in the railway car does not depend on any relative motion between the car and the ground. (Now we *have* introduced the second postulate.) He also knows that the distance from his position to each end of the car is the same. Therefore, when Melvin sees the B flash arrive *before* the A flash, he concludes that the lightning strike at B must have occurred at a time earlier than the strike at A. Melvin does *not* conclude that the strikes were simultaneous.

Albert Einstein was born in 1879 in the German city of Ulm, near München. After studying in Germany, Italy, and Switzerland, he entered the Swiss Federal Polytechnic School in Zürich. Einstein showed little interest in higher mathematics at this time because he believed that physical ideas could be best expressed in terms of simple mathematics. He graduated in 1900 and for a time held no regular job and showed almost no interest in scientific matters. He soon changed, however, and in 1902 he was glad to take a position in the Swiss Patent Office; this provided a measure of security and allowed him time to devote to his reawakened interest in physics and mathematics. In 1905 he published several important papers—in *three* different fields of physics. His explanation of the photoelectric effect so profoundly influenced the thinking in this field that this work earned for Einstein the 1921 Nobel Prize in physics. Next, he wrote a paper that provided a detailed explanation of Brownian motion. Finally, he published the famous paper, "On the Electrodynamics of Moving Bodies," which announced the new views of relativity theory. By 1909, his work was sufficiently highly regarded to earn him a professorial position at the University of Zürich. In 1911, he moved to the University of Prague where he published the first paper on the general theory of relativity, the theory that deals with gravitation. Einstein's fame had grown to such proportions that a new Chair of Mathematical Physics was created for him at Zürich. But in 1914, he moved once more, to head his own research institute at the Kaiser Wilhelm Institute in Berlin. He held this post until 1933 when the anti-Semitism of Hitler's Germany compelled him to move to the United States where he established himself at the Institute for Advanced Studies at Princeton. With the discovery of nuclear fission in 1939, Einstein wrote the famous letter to President Roosevelt that stimulated the formation of the Manhattan Project which resulted in the successful construction of an atomic weapon in 1945. Always searching for a geometrical description of physical phenomena, Einstein remained at the Institute until his death in 1955.

The key point in this argument is that both observers agree that the flash of light from B reaches Melvin before the flash from A. It is the *interpretation* of this fact in terms of the second postulate that leads to a difference in the sequence of events as seen by George and Melvin. Observers who are in relative motion will reach different conclusions regarding the time ordering of events that take place *at different locations*. But even moving observers will agree on the time sequence of events that take place *at the same location*. We can understand this in the following way. Suppose that the flash from A is a lethal laser beam, and suppose that the flash from B contains the message, "Get out of the way!" If the two flashes reach Melvin at the same instant, he has no time to react to the warning message and is killed by the laser pulse. However, if the message (the B flash) reaches him first, he will move and be saved from the fatal A flash. Now, after both flashes have reached Melvin, he will either be dead or alive. If the B flash arrives first, he will be alive, and no observer—whatever his state of motion—will conclude that Melvin is dead. If George concludes on the basis of his observations that the B flash reaches Melvin *before* the A flash, *all* other observers (including Melvin) must reach the same conclusion.

16-3 RELATIVISTIC EFFECTS ON TIME AND LENGTH

Time and Moving Clocks

We now turn to a closer examination of the way in which the ideas of relativity theory influence the interpretation of time and length measurements. In order to measure time, we require a clock. But this does not need to be the customary kind of timepiece. All that is necessary is a regular sequence of events which can be counted. We can imagine a clock constructed in the following way. At the origin O there is located a light source that emits short pulses of radiation. A distance L away there is a mirror M that reflects each light pulse back to the origin where it is detected (Fig. 16-6a). Each round trip of a light pulse through the distance $2L$ represents one "tick" of the clock. We call this time interval t.

Now, suppose that we have two of these "light clocks," each attached to a particular reference frame.

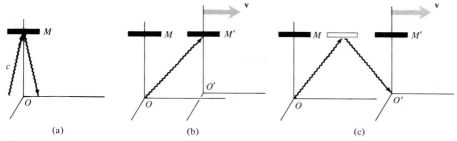

Figure 16-6 (a) A light clock. One "tick" of the clock corresponds to the time interval for a light pulse to travel from the source at the origin to the mirror and back to the origin. (b, c) Any observer will conclude that a moving light clock (or any other kind of clock) will run more slowly than an identical clock at rest in his reference frame.

Let us imagine that we locate ourselves in the frame with origin O and mirror M; this frame therefore becomes the "stationary" frame. The second clock is in a reference frame with origin O' and mirror M'. The $O'M'$ system moves with a uniform velocity $\mathbf{v}$ with respect to the stationary system OM (Fig. 16-6b, c). At the instant that the two origins coincide, each light source emits a pulse. We observe our stationary clock to register one "tick." The duration of the "tick" for the OM clock is the distance traveled $2L$ divided by the speed of the light pulse:

$$t = \frac{2L}{c} \qquad (16\text{-}1)$$

What about the moving clock? As we observe the ticking of the moving clock (Fig. 16-6b, c), we see that the light pulse must travel along a slanted path to reach the mirror M' and return to the origin O'. We know that all light signals propagate with the same speed. Therefore, we conclude that the time interval t' of a "tick" in $O'M'$ is *longer* than the interval t required for a "tick" in our stationary frame OM. That is, the moving clock runs more slowly than our stationary clock.

What does an observer in $O'M'$ conclude? He sees OM moving past his frame with a velocity $-\mathbf{v}$ and he sees the light pulse travel a longer distance than in his own frame. He therefore reaches the conclusion that the OM clock runs more slowly than his own. Both of these conclusions can be summarized by the statement:

Any observer will find that a moving clock runs· more slowly than an identical clock that is stationary in his reference frame.

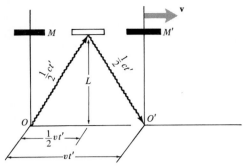

Figure 16-7 The geometry of Fig. 16-6c. After one-half of a "tick" in the $O'M'$ system, the light pulse has reached the mirror and has traveled a distance $\frac{1}{2}ct'$. During the same interval the origin O' has moved a distance $\frac{1}{2}vt'$.

We can derive the relationship between the time intervals t and t' by referring to the geometry of Fig. 16-6c. This situation is shown again in Fig. 16-7 where the important distances are indicated. Look at the right triangle that has sides with lengths $\frac{1}{2}ct'$, L, and $\frac{1}{2}vt'$. Using the Pythagorean theorem, we can write

$$(\tfrac{1}{2}ct')^2 = (\tfrac{1}{2}vt')^2 + L^2$$

and substituting the value of L from Eq. 16-1, we have

$$(\tfrac{1}{2}ct')^2 = (\tfrac{1}{2}vt')^2 + (\tfrac{1}{2}ct)^2$$

Canceling the factor $(\tfrac{1}{2})^2$ and transposing the term $(vt')^2$, we find

$$(c^2 - v^2)t'^2 = c^2 t^2$$

Solving for t'^2,

$$t'^2 = \frac{c^2 t^2}{c^2 - v^2} = \frac{t^2}{1 - \dfrac{v^2}{c^2}}$$

Finally, taking the square root, we obtain

$$t' = \frac{t}{\sqrt{1 - \dfrac{v^2}{c^2}}} \qquad (16\text{-}2)$$

In this expression, remember that t represents the duration of a "tick" in the observer's rest frame, and that t' represents the duration of a "tick" of an identical clock in the moving reference frame.

The factor $\sqrt{1 - (v^2/c^2)}$ is always less than 1, so t' is always greater than t. For example, if $v = 0.8\,c$, then

$$t' = \frac{t}{\sqrt{1 - (0.8)^2}} = \frac{t}{\sqrt{0.36}} = \frac{t}{0.6} = \frac{5}{3}\,t$$

The lengthening of the duration of a "tick" of a moving clock is termed *time dilation*. (To *dilate* means to *enlarge*.)

It must be remembered that here (as always) we are dealing with *relative* motion. The observer in O sees the O' clock in motion relative to his own frame. But the observer in O' sees the O clock in motion relative to *his* frame. That is, to *each* observer the other clock is in motion and ticks more slowly than his own clock.

Testing Time Intervals
in Moving Systems

We can again use the properties of pions to test the prediction concerning the dilation of time. An observer who is at rest with respect to a collection of charged pions will find that half of the pions will decay into muons in a time of approximately 2×10^{-8} s (see Table 2-3). That is, the half-life of pions measured in a reference frame in which they are at rest is 2×10^{-8} s. The internal characteristic of pions that causes decay constitutes a kind of elementary clock; this is the clock we will use in the reference frame at rest with respect to the pions. Suppose that we prepare a beam of charged pions in which every pion moves with a velocity $v = 0.8\ c$ relative to an observer in the laboratory. The distance from the point at which they are formed to the point at which half of the pions will have decayed is expected to be

$$l = vt = (0.8 \times 3 \times 10^8 \text{ m/s}) \times (2 \times 10^{-8} \text{ s})$$
$$= 4.8 \text{ m} \qquad (16\text{-}3)$$

But if we perform the experiment, we find that the pions actually travel a *greater* distance in the laboratory. According to relativity theory, the laboratory observer will see the pion clock run more slowly than if the pions were at rest in the laboratory. The half-life of 2×10^{-8} s is valid only if it is measured in the reference frame of the pion; the laboratory observer must use the time-dilated value. Therefore, in calculating l, we should use t' (from Eq. 16-2) instead of t. Then, the predicted distance of travel becomes

$$l = vt' = v \times \frac{t}{\sqrt{1 - \dfrac{v^2}{c^2}}} = \frac{vt}{\sqrt{1 - (0.8)^2}}$$

Using $vt = 4.8$ m from Eq. 16-3, we find

$$l = \frac{4.8 \text{ m}}{\sqrt{0.36}} = \frac{4.8 \text{ m}}{0.6} = 8.0 \text{ m} \qquad (16\text{-}4)$$

This experiment has been performed using a pion beam from the Columbia University cyclotron. Instead of the 4.8-m distance predicted by the ordinary nonrelativistic calculation, the pions were found to move a distance of 8 m before half had decayed, thus confirming the relativistic prediction.

Length Contraction

Suppose that we now view the decaying pions from a frame of reference *moving with the pions*. That is, we imagine that we are observers in a frame in which the pions are at rest. In this frame the pion half-life is 2×10^{-8} s. Furthermore, in this frame the laboratory appears to be moving past the pions with a velocity $v = 0.8\ c$. The distance that the laboratory moves during the pion half-life is the same as that calculated in Eq. 16-3, namely, 4.8 m. But the *physical position* in the laboratory where decay takes place — the spot on the floor above which the decay occurs — must be the same regardless of the frame from which it is viewed. The observer in the laboratory will say that this spot is 8 m from the source, but the observer moving with the pion will say that the distance is only 4.8 m. This is an example of another consequence of relativity theory:

> Lengths and distances in a moving system are contracted in comparison with the equivalent lengths and distances in an observer's rest frame.

The two different ways of viewing the pion decay — by time dilation or by length contraction — must be equivalent. In order to compare *time* in the two moving frames, we used Eq. 16-2. In order to compare *length* in the two frames, we must use an expression that also involves the relativistic factor $\sqrt{1 - (v^2/c^2)}$. If the length $l = 4.8$ m in the pion frame is to become $l' = 8$ m in the laboratory frame, then l and l' must be related by

$$ l' = l\ \sqrt{1 - \frac{v^2}{c^2}} \qquad (16\text{-}5) $$

Then,

$$
\begin{aligned}
l' &= (8\ \text{m}) \times \sqrt{1 - (0.8)^2} \\
&= (8\ \text{m}) \times 0.6 \\
&= 4.8\ \text{m}
\end{aligned}
$$

This *length contraction* effect has the following implication. If a meter stick moves past an observer with a velocity of $0.8\ c$, the observer will measure the length of this meter stick to be

$$
\begin{aligned}
l' &= (1\ \text{m}) \times \sqrt{1 - (0.8)^2} \\
&= (1\ \text{m}) \times (0.6) \\
&= 0.6\ \text{m}
\end{aligned}
$$

Length contraction takes place only in the direction *parallel* to the direction of relative motion of the two reference frames. All distances *perpendicular* to this direction are unaffected by the relativistic contraction. We anticipated this result when we constructed the light clocks (Fig. 16-6) by placing the mirrors in positions so that OM and $O'M'$ were perpendicular to the direction of relative motion.

Notice that time dilation and length contraction are not distinct and independent effects. As we have seen in the example of pion decay, the phenomenon can be explained in equivalent ways by using either time dilation or length contraction. The results are simply two different manifestations of the same relativistic effect (Fig. 16-8).

"Think of two witches on identical broomsticks. As they glide past each other, each notes with pride that her own status symbol is the longer!" (L. Marder, in *Time and the Space-Traveler*.)

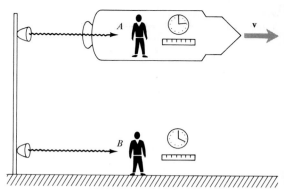

Figure 16-8 Observers *A* and *B* measure the same velocity for the light signal even though they are in relative motion. However, each observer measures the other's meter stick and finds it to be contracted, and each determines that the other's clock runs more slowly than his own.

Are Relativistic Effects *Real?*

Are the relativistic results for time and length *real* effects? Or are they somehow merely optical illusions? In a physical science the only way in which we can answer such questions is in terms of *measurements* that we can make. To speculate or philosophize on such matters is fruitless—we must perform *experiments*. Only in terms of the results of actual measurements can we make clear and meaningful statements. Many different kinds of experiments have been performed that involve length and time in moving systems. *All* such measurements have demonstrated the correctness of the relativistic predictions. We must therefore conclude that time dilation and length contraction are *real* effects. Nature behaves in such a way that the characteristics of our measuring instruments—clocks and meter sticks—are different when viewed in motion than when viewed at rest.

If relativistic effects are *real,* why do we not see them at work around us? Why are lengths and times not all jumbled up as we move around? If a meter stick moves past us at a velocity of 0.8 *c*, then it appears to have a length of only 0.6 m. But we do not ordinarily see meter sticks (or any piece of bulk matter) moving with such tremendous speeds. More commonly we see objects moving with much lower speeds. Suppose that a meter stick moves past us with a velocity $v = 30$ m/s (about 67 mi/h). If we measure the length of this meter stick, what result do we obtain? Using Eq. 16-5,

$$l = (1 \text{ m}) \times \sqrt{1 - \left(\frac{30 \text{ m/s}}{3 \times 10^8 \text{ m/s}}\right)^2}$$

$$= (1 \text{ m}) \times \sqrt{1 - 10^{-14}}$$

$$= 0.999\ 999\ 999\ 999\ 995 \text{ m}$$

TABLE 16-1 LENGTH CONTRACTION AND TIME DILATION FACTORS

v/c	LENGTH CONTRACTION FACTOR (length of a moving meter stick, m), $\sqrt{1-(v^2/c^2)}$	TIME DILATION FACTOR[a] (duration of a 1-s "tick" of a moving clock, s), $1/\sqrt{1-(v^2/c^2)}$
0	1.000	1.000
0.2	0.980	1.021
0.4	0.917	1.091
0.6	0.800	1.250
0.8	0.600	1.667
0.9	0.436	2.294
0.95	0.312	3.203
0.98	0.199	5.025
0.99	0.141	7.089
0.999	0.0447	22.37
0.9999	0.0141	70.71
0.99999	0.00447	223.6

[a] This column is also the *mass increase factor* and represents the mass in kg of an object with a rest mass of 1 kg (see Section 16-4).

Small wonder, then, that relativistic effects do not manifest themselves in everyday matters!

Useful Formulas

In calculating relativistic effects, the following approximate expressions are useful when the velocity v is small compared to the velocity of light ($v \cong 0.3\ c$ or smaller):

$$\sqrt{1 - \frac{v^2}{c^2}} \cong 1 - \frac{1}{2}\frac{v^2}{c^2} \qquad (16\text{-}6)$$

$$\frac{1}{\sqrt{1 - \frac{v^2}{c^2}}} \cong 1 + \frac{1}{2}\frac{v^2}{c^2} \qquad (16\text{-}7)$$

To check the validity of Eq. 16-6, we compare the exact result for $v = 0.2\ c$ to the approximate value:

$$\sqrt{1 - (0.2)^2} = \sqrt{1 - 0.04} = \sqrt{0.96}$$
$$= 0.979796\ \text{(exact)}$$

Using Eq. 16-6,

$$\sqrt{1 - (0.2)^2} \cong 1 - \tfrac{1}{2}(0.04)$$
$$= 0.98\ \text{(approximate)}$$

For smaller values of v/c, the approximate result is even closer to the exact result.

Table 16-1 gives the length contraction and time dilation factors for various values of v/c.

Relativity and Space Travel

The dilation of time at high speeds prompts some exciting speculations concerning the possibility of long-distance space travel. Suppose we are considering a trip to a star that is at a distance of 10 light-years (L.Y.). Let us consider what would happen if we set out on such a journey with a velocity only slightly less than the velocity of light: $v = 0.995\ c$. At this speed the trip would require a time just slightly greater than 10 years according to an observer who monitored the voyage from the Earth. However, the space traveler's clock (according to the Earth observer) will run more slowly than the Earth clock. In fact, the duration t' of a "tick" of the space traveler's clock compared to the duration t of a "tick" of the Earth clock will be

$$t' = \frac{t}{\sqrt{1 - (0.995)^2}} = \frac{t}{\sqrt{1 - 0.99}}$$

$$= \frac{t}{\sqrt{0.01}} = \frac{t}{0.1} = 10t$$

That is, the Earth observer sees his own clock "tick" 10 times for every "tick" of the clock in the space vehicle. According to the Earth observer's measurement with his own clock, the trip will require 10 years. But according to the clock in the space vehicle as viewed by either observer, the trip is completed in only 1 year.

How does the space traveler analyze the trip? With respect to his own reference frame in the space vehicle, the space traveler sees the Earth speeding away from him with $v = 0.995\ c$ and he sees the star approaching him with the same velocity. If he measures the Earth–star distance, he will find the distance contracted to

$$l' = (10\ \text{L.Y.}) \times \sqrt{1 - (0.995)^2}$$
$$= (10\ \text{L.Y.}) \times \tfrac{1}{10} = 1\ \text{L.Y.}$$

Therefore, because the star moves toward him with a speed almost equal to the speed of light, the star will arrive in only 1 year. In terms of time dilation or length contraction, the same result is obtained.

Suppose that the two observers are twins: Sam is the space traveler and Ernest is the Earth-bound twin. (Again, the initials remind us who is the Space traveler and who remains on Earth.) Sam makes the trip to the star and returns. According to Sam's clock, the round trip required 2 years, but according to Ernest's clock, the elapsed time was 20 years. Therefore, when Sam returns, he is 18 years *younger* than his twin brother.

Is this analysis really correct? If we take the viewpoint of Sam's frame of reference, it is *Ernest* (and the Earth) who makes the trip. Then, according to Sam, it is Ernest's clock that is running slowly and when Ernest (and the Earth) returns from the trip, *Ernest* should be younger by 18 years. That is, it should not matter which twin makes the trip. Each sees the other moving away from him and then returning; each sees the other's clock running slowly. Thus, we seem to have a paradox (usually called the "twin paradox").

In reality there is no paradox. The seemingly impossible situation has been created by assuming that it does not matter which twin we consider to have made the space voyage. But it *does* matter. All of the relativistic effects we have discussed make use of the fact that every observer is in an inertial reference frame. Ernest, who remains on Earth, is in an inertial frame throughout the entire episode. Sam, on the other hand, must be *accelerated* to a velocity of 0.995 c with respect to the Earth; he must be *accelerated* when he turns around at the distant star; and he must undergo *acceleration* once more when he slows down to land on Earth. Therefore, Sam does *not* remain in an inertial frame throughout the trip. Consequently, the analysis of the situation must be made very carefully. A proper calculation does, in fact, show that Sam, the space-traveler, ages less than his Earth-bound twin.

The effect of motion on clocks has recently been demonstrated directly by flying a set of atomic clocks around the world on airliners. When these clocks were compared with control clocks that remained on the Earth, it was found that the discrepancy in the times was just that predicted by relativity theory. Time dilation had previously been observed in measurements of the lifetimes of moving and stationary elementary particles, but this experiment is the only one that has demonstrated the effect with man-made clocks.

Because of time dilation effects it is therefore conceivable that journeys to distant stars can be made within the lifetime of the space travelers. But they would return home to find quite a different Earth, an Earth that had advanced by hundreds or thousands of years. An exciting prospect indeed! Of course, such trips are possible only if the space vehicle can be accelerated to a speed that is only fractionally smaller than the speed of light. In order to accelerate a 50 000-ton space vehicle to a velocity of 0.995 c, the energy requirement would be 100 000 times the annual world-wide energy consumption! Hence, there is little prospect that we will ever be able to take advantage of time dilation effects for space travel.

We must also note that relativistic time travel can be made only into the future. There is no way to move backward in time. Time always moves forward, but it can move more rapidly for one observer than for another.

16-4 MASS AND ENERGY

The Variation of Mass with Velocity

Relativity theory has shown us that two observers who are in motion relative to one another will obtain different results for measurements of length and time. The measurement of the third fundamental physical quantity — mass — is also affected by relative motion.

Suppose that we accelerate an electron to a high

velocity **v** and project it into a magnetic field **B,** as shown in Fig. 16-9a. We know from the discussion in Section 12-4 that the electron will execute a circular orbit within the field. According to Eq. 12-13, the radius of the orbit will be $R = mv/eB$. Next, let us increase the accelerating voltage so that the electron enters the field with a velocity **2v,** as in Fig. 16-9b. We would now expect the orbit radius to increase to $2R$. But if we perform the experiment and measure the orbit radius, we find that the path actually followed by the electron has a radius *greater* than $2R$: $R' > 2R$. What can account for this effect? The magnetic field **B** is the same as in the first case, and the charge e of the electron cannot change. Because the velocity is measured in each case, we must conclude that the electron *mass* has changed. In fact, one of the results of relativity theory can be stated in the following way:

> The mass of an object in motion with respect to an observer is greater than the mass of an identical object at rest with respect to the observer.

The equation that expresses this statement is

$$m = \frac{m_0}{\sqrt{1 - \dfrac{v^2}{c^2}}} \qquad (16\text{-}8)$$

The quantity m_0 is the mass of the particle when measured at rest, and is therefore an intrinsic property of the particle. The quantity m, on the other hand, is not really a property of the particle alone; instead, m depends on the relative motion of the particle and the observer. We call m_0 the *intrinsic mass* or the *rest mass* of a particle.

In order to analyze correctly the motion of the electron projected into the magnetic field, we must use Eq. 16-8 in the expression for the orbit radius:

$$R = \frac{m_0 v}{eB\sqrt{1 - \dfrac{v^2}{c^2}}} \qquad (16\text{-}9)$$

From this equation we can see that R will more than double if v is doubled.

Figure 16-10 shows the way in which m varies with velocity. The quantity plotted on the vertical axis is the mass m in units of the rest mass m_0 (that is, m/m_0), and the horizontal axis gives the velocity in units of c. Notice that the deviation of m/m_0 from unity is very slight

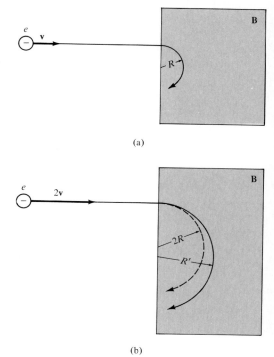

(a)

(b)

Figure 16-9 (a) An electron moving with a velocity **v** enters a magnetic field **B** and executes a circular orbit with a radius R. (b) If the velocity is increased to **2v**, it is found that the orbit radius increases to R' which is greater than $2R$. The reason is the increase in *mass* of the moving electron.

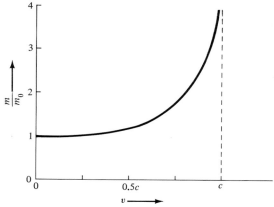

Figure 16-10 The relativistic increase of mass with velocity.

when v is small (the expected nonrelativistic result), but that m/m_0 becomes extremely large when v is close to c. When v is approximately one-half the velocity of light, the mass has increased by about 20 percent. As v is increased beyond $0.5\,c$, the mass increases rapidly and dramatically. The electrons that emerge from the 2-mile-long Stanford Linear Accelerator (SLAC) have $v = 0.999999999\,c$ and a mass approximately equal to that of an iron atom!

The Einstein Mass–Energy Relation

When v is small compared to c, we can use Eq. 16-7 for the relativistic term in the expression for the mass. Then, Eq. 16-8 for the mass of a particle becomes

$$m = m_0 \left(1 + \frac{1}{2} \frac{v^2}{c^2} \right)$$

To convert this into an *energy* equation, we multiply both sides by c^2.

$$mc^2 = m_0 c^2 + \tfrac{1}{2} m_0 v^2 \qquad (16\text{-}10)$$

In this equation we note that the term $\tfrac{1}{2} m_0 v^2$ is the ordinary expression for kinetic energy. The other term on the right-hand side represents an energy that does not depend on the particle's velocity—this is the *rest energy* of the particle. The sum of the *rest* energy and the *kinetic* energy is the *total* energy of the particle:

$$
\boxed{
\begin{array}{ccc}
mc^2 & = & m_0 c^2 & + & \text{K.E.} \\
\text{(total energy)} & & \text{(rest energy)} & & \text{(kinetic energy)}
\end{array}
}
\qquad (16\text{-}11)
$$

(In this derivation, we have used the approximate expression for the mass when v is small compared to c. A rigorous derivation would actually yield the result expressed by Eq. 16-11 even when v is comparable with c.)

If we denote the total energy by $\mathscr{E}$, Eq. 16-10 becomes

$$\mathscr{E} = mc^2 \qquad (16\text{-}12)$$

This is the famous Einstein equation that expresses the relationship between mass and energy. The equation does *not* imply (as is sometimes incorrectly stated) that "mass and energy are the same thing." Mass and energy are distinct physical concepts—they are related by the Einstein equation but they are definitely *not* the "same thing."

Mass can be converted into energy and energy can be converted into mass, the conversions always obeying Eq. 16-12. For example, when a nucleus of ^{235}U undergoes *fission,* the combined mass of the fission products is *less* than the mass of the original uranium nucleus. Some mass has *disappeared,* and in its place is an equivalent amount of energy in the form of kinetic energy of the moving fission products. No protons or neutrons are destroyed in the fission process—the total number is the same before fission as afterward. But the *arrangement* of the protons and neutrons is different, and the two arrangements have different mass. It is this mass difference that appears as energy.

When a bulk sample of uranium undergoes fission, each kilogram of uranium is converted into 0.999 kg of fission products. That is, for each kilogram of uranium, one gram of mass disappears and the energy release is

$$\mathcal{E} = (10^{-3} \text{ kg}) \times (3 \times 10^8 \text{ m/s})^2$$

$$= 9 \times 10^{13} \text{ J}$$

$$= (9 \times 10^{13} \text{ J}) \times \left(\frac{1 \text{ kWh}}{3.6 \times 10^6 \text{ J}}\right)$$

$$= 2.5 \times 10^7 \text{ kWh}$$

This amount of energy, which is obtained by the conversion of only 1 gram of matter into energy, represents one day's output of a modern power generating facility.

Because mass and energy are connected by the Einstein relation, we really do not have *separate* conservation principles for mass and energy. Properly, we have only *one* conservation principle, that of mass–energy. In all everyday situations, however, conversions between mass and energy are a small part of the total picture and, for all practical purposes, mass and energy are separately conserved.

The Ultimate Velocity

In the relativistic expressions for length, time, and mass, a common factor is $\sqrt{1 - (v^2/c^2)}$. If v were equal to c, this factor would become *zero.* Then, lengths would contract to zero, clocks would cease to run, and masses would become infinite. These are impossible, nonphysical situations. We must conclude that a material object (that is, any object that possesses an intrinsic mass) can never travel with a velocity equal to or greater than the velocity of light.

If we attempt to accelerate a material object to higher

and higher velocities, we find that this becomes more and more difficult as v approaches c. Each increase in velocity requires an expenditure of more energy than the last, and no amount of energy can ever produce $v = c$. The velocity of light is therefore the *ultimate* velocity.

Light photons and neutrinos have no intrinsic mass and these objects always travel with the velocity of light. No observer can ever place himself in a reference frame in which he can view a photon or a neutrino at rest.

One sometimes hears questions of the following type: "I know that relativity theory says that we must always have $v < c$, but what would happen *if* an object or a particle could be accelerated from some small velocity to a velocity greater than c?" There is no answer to such a question within the realm of physical science because there is no conceivable way in which we can put the question to an experimental test. Physical science cannot address itself to any question that does not permit a *measurement* to settle the issue.

16-5 THE GENERAL THEORY

The Principle of Equivalence

In 1915 Einstein enlarged upon the scope of the *special theory* of relativity (which is restricted to considerations of effects in nonaccelerated reference frames) to include a treatment of gravitational fields and accelerated reference frames. This aspect of relativity is called the *general theory*. In fact, the general theory is a theory of *gravitation*.

Suppose that an observer is in a closed laboratory on the surface of the Earth (Fig. 16-11a). By making measurements on the behavior of a ball that is dropped, the observer can determine that the ball experiences an acceleration g. Next, we place the observer and his closed laboratory on a rocket that is located far from any massive body and is therefore free of any gravitational effects (Fig. 16-11b). If the rocket is moving with constant velocity, when the ball is released it will continue to move with the same velocity as the rocket. That is, the ball will remain motionless with respect to the laboratory. On the other hand, if the rocket is accelerating when the ball is released, the ball will continue to move with *constant velocity* whereas the floor of the

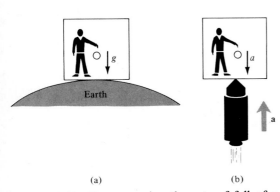

(a)   (b)

Figure 16-11 By measuring the rate of fall of a dropped ball in his closed laboratory, an observer can determine the acceleration of the ball, but he cannot determine whether that acceleration is due to a gravitational force or to the acceleration of his reference frame.

laboratory will accelerate toward the ball. Thus, the observer sees the ball accelerating toward the floor. If $a = g$, the ball will behave in exactly the same way in the two situations. If the observer cannot view or communicate with the outside world, he cannot determine by measurements made within his laboratory whether his laboratory is in a gravitational field or whether it is accelerating. This idea that gravitational and acceleration effects cannot be distinguished is called the *principle of equivalence,* and is a basic postulate of the general theory.

The bold step forward that Einstein took in formulating the general theory was the merger of space and time into a single entity. In this theory, distance and time are treated on an equal basis. The view of the Universe in the general theory is not in terms of three-dimensional space plus the separate concept of time. Instead, the description is one in which we have a *four-*dimensional space-time world.

Mass plays a unique role in the general theory. Not only does mass possess inertia, thereby giving rise to apparent forces in accelerated reference frames, but it is also the source of gravitation. In Einstein's space-time world, mass is given a geometrical interpretation. The presence of mass results in the curvature or warping of the space-time fabric; and, by the same token, a curvature of space-time reveals itself as mass. Thus, in the general theory, space and time are woven together and mass is associated with distortions of the space-time surface. The general theory is a geometrical theory of gravitation.

Tests of the General Theory

The various predictions of the special theory of relativity—time dilation, length contraction, the increase of mass with velocity, and the equivalence of mass and energy—have been verified to high precision in a large number of experiments. The general theory, however, makes fewer predictions and the experimental tests are all extremely difficult to perform. But because the general theory is a theory of gravitation, it has implications for the large-scale distribution of matter in the Universe and is intimately connected with our theories concerning the structure and the evolution of the Universe. It is therefore, of considerable importance to test the theory as severely as possible even though the experiments are formidable.

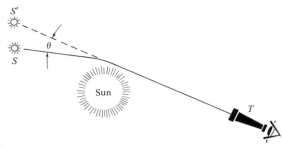

Figure 16-12 Starlight that passes near the Sun is slightly bent, resulting in an apparent shift in the position of the star as viewed by the observer using the telescope T.

The gravitational wave detection apparatus used by Professor Joseph Weber at the University of Maryland. The key part of the system is a 1400-kg aluminum cylinder which responds to gravitational waves by vibrating ever so slightly.

According to the Einstein mass–energy relation, any object that possesses mass has associated with it an equivalent amount of energy, and vice versa. A light photon possesses energy and so we can consider an amount of mass $m = \mathscr{E}/c^2$ to be associated with every photon. One of the first predictions of the general theory to be tested involves an effect that depends upon this equivalent "mass" of a light photon. If the light from a distant star passes close to the Sun, the Sun's gravitational field will act on the light and the light will be deviated from its straight-line path (Fig. 16-12). The effect is exceedingly small, however, and amounts to an angular deflection of only 0.0005 of a degree for starlight that just grazes the Sun's surface. Measurements of the deflection were first carried out during the solar eclipse of 1919 when stars lying near the Sun's disk could be observed. The experiment has been repeated several times during subsequent eclipses. The results are in agreement with the prediction of the general theory although the precision is only about 10 percent. It is not yet possible to state that a definitive test has been made, but new techniques are now being employed, and it appears that the precision of the experiment can be significantly improved.

According to Newton's theory of gravitation, the planets should orbit the Sun in elliptical paths. In the general theory, however, the orbits are predicted to deviate slightly from ellipses. This effect is due, in part, to the fact that the mass of a planet is not constant: the mass is greatest when the planet is nearest the Sun and its speed is high (see Fig. 6-2). The magnitude of the predicted deviation from an elliptical shape is again small, but precise astronomical measurements have confirmed this result of the general theory.

Black Holes and Gravitational Waves

When an electrical charge is accelerated, it radiates electromagnetic waves (Section 14-1). By the same reasoning, if a *mass* is accelerated, should it not radiate *gravitational* waves? According to the general theory, this should in fact be the case. Because gravitational radiation is so weak, it has only been within the last few years that equipment sufficiently sensitive to detect gravitational waves has been constructed and put into use. The data accumulated by Professor Joseph Weber using the equipment shown in the photograph on this page may have revealed gravitational radiation emanating predominantly from the center of our Milky Way Galaxy. It is too early to state whether these experiments have conclusively demonstrated the detection of gravitational waves.

Where do gravitational waves come from? What physical phenomenon in the Universe can produce sufficient accelerations of large masses to radiate detectable amounts of gravitational waves? If a light photon moves outward from a massive star, the star's gravity will act on the equivalent mass of the photon. If the gravitational field of the star is sufficiently strong (this means a very small, very massive star), photons will be unable to escape and the star appears dark. Such an object is called a *black hole*. Because of its huge gravitational attraction, mass is accelerated toward and "falls into" the black hole. When a large chunk of matter disappears into a black hole, a burst of gravitational radiation will be emitted. It seems possible that Weber's instruments have detected such catastrophic events.

General relativistic effects are associated with the large-scale properties of the Universe. Theoretical and experimental studies of these effects have already shown us startling new aspects of our Universe, and further studies are certain to reveal even more intriguing cosmological phenomena.

Einstein in his later years at Princeton.

SUGGESTED READINGS

J. Bernstein. *Einstein* (Viking, New York, 1973).

D. W. Sciama, *The Physical Foundations of General Relativity* (Doubleday, Garden City, New York, 1969).

Scientific American articles:

R. H. Dicke, "The Etvos Experiment," December 1961.

R. S. Shankland, "The Michelson-Morley Experiment," November 1964.

QUESTIONS AND EXERCISES

1. The value of the velocity of light is extremely large. How would the relativistic effects described in this chapter be altered if the velocity of light were truly *infinite?*

2. Describe some of the effects that would be evident if the velocity of light were 60 mi/h.

3. A manufacturer of a certain cathode ray tube (similar to a television picture tube) claims that the spot made on the face of the tube by the electron beam can be made to move with a speed of 5×10^8 m/s (this is called the *writing speed* of the tube). Do you believe his claim? Explain. (Consider this question carefully.)

4. According to an observer on Earth, the clocks on board a space vehicle moving directly away from the Earth run at exactly one-half speed. What is the velocity of the space vehicle with respect to the Earth?

5. How long would it require (according to the space traveler) to make a voyage to Alpha Centauri (4.3 L.Y. away) if the velocity of his spacecraft relative to the Earth is 0.99 c?

6. An Earth observer O views a space vehicle traveling with a velocity $v = 0.8\ c$, as shown in the diagram below. In the space vehicle are two 6-foot astronauts, one standing "up" and one lying "down." What does O measure for the heights of astronauts A and B?

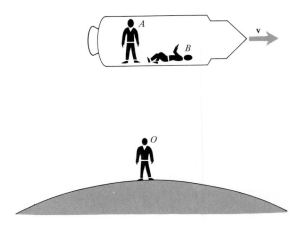

7. Three observers move at high speeds relative to one another along the same straight line. Each measures the length of the same meter stick and each obtains a value of 1 m. What can be said about the *orientation* of the meter stick? Each observer also measures the length of another meter stick. One observer obtains a value of 1 m, but the other two observers find values less than 1 m. What can be said about the *motion* of this meter stick?

8. A billboard is 3 m high and 5 m long and is located parallel to a roadway. If you travel along the road with a velocity of 0.8 c, what do you observe for the dimensions of the billboard?

9. An observer measures the length of a moving meter stick and finds a value of 0.5 m. How fast is the meter stick moving?

10. What is the observed fractional difference in length between a meter stick at rest and one that is moving relative to an observer with a speed of 10 km/s?

11. What is the observed fractional increase in mass of an object that moves with a speed of 0.8c with respect to an observer?

12. A meter stick has a rest mass of 1 kg and moves past an observer. The observer finds the mass to be 2 kg. What does the observer find for the *length* of the meter stick?

13. Nuclear reactions in the Sun convert approximately 4×10^9 kg of matter into energy each second. What is the power output of the Sun in kW? What fraction of the mass is converted into energy each year? Even though the Sun's mass is changing, is it therefore reasonable to refer to *the* mass of the Sun?

14. An observer A sees a space vehicle P moving directly away from him with a velocity $v_1 = 0.8\ c$. An observer B on P sees another space vehicle Q moving directly away from him (and away from A) with a velocity $v_2 = 0.8\ c$. With what velocity does A see Q

moving away from him? (A, P, and Q are on the same straight line.) Einstein analyzed this problem and found that the velocity V of Q with respect to A is

$$V = \frac{v_1 + v_2}{1 + \dfrac{v_1 v_2}{c^2}}$$

Calculate V for this case and compare with the value that would result from the neglect of relativistic considerations.

15. A space vehicle moves directly away from the Earth with a velocity $v_1 = 0.9\ c$. A rocket is fired from this space vehicle with a velocity $v_2 = 0.6\ c$. What does an observer on Earth measure for the rocket velocity if (a) the rocket moves directly *away from* the Earth and (b) the rocket moves directly *toward* the Earth? (Use the velocity expression in the previous exercise and remember the *signs* of the velocities.)

16. Will binary stars radiate gravitational waves? Explain. (A binary star system is illustrated in Fig. 16-3.)

17

ELECTRONS
AND PHOTONS

At about the time that Einstein was formulating his ideas concerning space and time which were to lead to the development of relativity theory, other scientists were investigating the nature of light and electrons. Electrons were known to be *particles* and light was acknowledged to be a *wave* phenomenon.

We all have rather clear intuitive ideas about waves and particles. We know that a wave is an extended propagating disturbance in a medium. We know that a particle is an object that can be located at a particular point in space whereas a wave cannot. And we have come to accept the existence of atomic particles — electrons, protons, and neutrons. What could be simpler? A wave is a wave, and a particle is a particle; the distinction is clear.

17-1 THE PHOTOELECTRIC EFFECT

The Ejection of Electrons from Metals

But it is not all this simple. As the 20th century began, scientists were confronted with new questions concerning waves and particles. Consider the results that Hertz and others had obtained. It was found, for example, that if a piece of clean zinc is exposed to ultraviolet (UV) radiation, the zinc acquires a positive charge. The radiation can carry no charge to the zinc, so this result must mean that electrons (the carriers of negative

406

charge) are literally knocked off the zinc by the action of the UV radiation (Fig. 17-1). The removal of electrons causes the zinc to become charged positively. This phenomenon is called the *photoelectric effect*.

There was nothing really controversial or even unexpected about this result. All electromagnetic radiation (including UV) was known to carry energy and momentum; the transferral of this energy and momentum to the zinc can account for the ejection of the electrons. But in addition to the charging action of UV radiation it was found that visible light could *not* eject electrons from zinc and this *was* unexpected. If a piece of zinc is exposed to red light, there is no buildup of positive charge. If the frequency of the light is increased, and yellow and then blue light is incident on the zinc, there is no change in the charge on the metal, *regardless of how intense the light is made*. Only when the frequency is increased into the UV region are photoelectrons ejected. Thus, the photoelectric effect exhibits a frequency limit, and if the frequency is too low, the effect cannot be produced even if the light intensity is made very large. The traditional view of the wave nature of electromagnetic radiation was unable to provide a satisfactory explanation of this curious behavior.

After the qualitative aspects of the photoelectric effect had been discovered, detailed measurements were made with apparatus similar to that shown in Fig. 17-2. Light with a definite frequency from an external source is incident on a metal electrode A which is sealed inside an evacuated glass tube. The light from the source can be varied in frequency allowing the choice of infrared, visible, or ultraviolet light. The electrons that are emitted from A travel along the tube and are collected by electrode B. These electrons are called *photoelectrons*. The two electrodes are connected by an external circuit and the current of photoelectrons flowing in the circuit is measured by the ammeter.

When measurements are carried out with this sort of apparatus, several important results are obtained:

(1) For low frequencies of the incident light (that is, for infrared and red light), there is no photoelectric current, even if the light intensity is very high.

(2) As the frequency is increased, photoelectrons begin to be emitted at a *threshold frequency* ν_0 (which is in the blue or ultraviolet part of the spectrum for most materials).

(3) For frequencies above ν_0, the kinetic energy of the

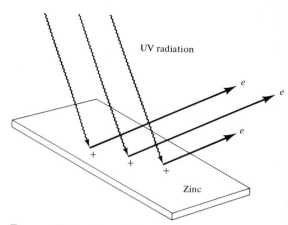

Figure 17-1 Ultraviolet radiation knocks electrons off a piece of zinc by means of the *photoelectric effect*, leaving the zinc positively charged.

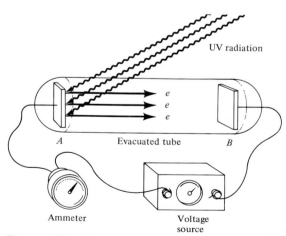

Figure 17-2 Apparatus for studying the photoelectric effect. The current of photoelectrons moving from electrode A to electrode B is measured by the ammeter.

photoelectrons is directly proportional to the frequency of the incident light. The kinetic energy does *not* depend on the intensity of the light.

(4) If different materials are used for electrode *A* (the *photoemissive surface*), exactly the same behavior is found, except that each material has its own characteristic value of the threshold frequency ν_0.

According to classical electromagnetic theory, the energy transferred by a wave is proportional to its intensity and does not depend on the frequency. But the experiments clearly show that the frequency of the light is crucial in the photoelectric effect. The traditional explanation is therefore completely inadequate to interpret the photoelectric experiments. A new idea is needed before any progress can be made.

Einstein's Photoelectric Theory

In the same year (1905) that he published his first paper on relativity, Einstein proposed an explanation for the results of the experiments on the photoelectric effect. Einstein's photoelectric theory is characterized by the same clarity and simplicity that are exhibited in his relativity theory. In formulating this theory, Einstein used and enlarged upon an idea that had been proposed by the German physicist Max Planck (1858–1947). According to Planck, when electromagnetic radiation interacts with matter, energy is exchanged, not in arbitrary amounts, but only in discrete bundles which are called *quanta*. We now know of other examples in Nature that involve discreteness—for example, electric charge can be transferred only in definite units, namely, integer numbers of electron charges. But these cases were not known in Planck's day and his idea was therefore new and unprecedented. In fact, Planck's suggestion was so radical that it was generally ignored by the scientists of that time. It remained for Einstein to give respectability to the quantum concept by incorporating this idea into his theory of the photoelectric effect.

Einstein expanded Planck's quantum hypothesis by asserting that all electromagnetic radiation (not just the exchange of energy between radiation and matter) is quantized and that the energy of a quantum is directly proportional to its frequency:

$$\mathscr{E} = h\nu \tag{17-1}$$

A commercial photoelectric light-sensing device. Phototubes of this general type have been manufactured since 1935.

where the proportionality constant h is called *Planck's constant* and has the value

$$h = 6.625 \times 10^{-34} \text{ J-s} \qquad (17\text{-}2)$$

A quantum of electromagnetic radiation is called a *photon*.

If all electromagnetic radiation occurs in discrete bundles, how could this fact have been overlooked for so many years of intensive study of electromagnetic phenomena? In order to see the reason, let us compute the energy of a photon of visible light. Yellow light, in the middle of the visible spectrum, has a wavelength of approximately 6000 Å or 6×10^{-7} m. Using the relation $\nu = c/\lambda$ (Eq. 13-4), we can express Eq. 17-1 for the energy of a photon in terms of λ:

$$\mathscr{E} = h\nu = \frac{hc}{\lambda} \qquad (17\text{-}3)$$

Then, for yellow light we find

$$\mathscr{E} = \frac{(6.6 \times 10^{-34} \text{ J-s}) \times (3 \times 10^8 \text{ m/s})}{6 \times 10^{-7} \text{ m}}$$

$$= 3.3 \times 10^{-19} \text{ J}$$

or, using Eq. 12-6 to convert this result to the electron-volts,

$$\mathscr{E} = (3.3 \times 10^{-19} \text{ J}) \times \left(\frac{1 \text{ eV}}{1.6 \times 10^{-19} \text{ J}} \right)$$

$$= 2.1 \text{ eV}$$

That is, the energy of a photon of yellow light is only slightly more than the energy of a single electron having accelerated through the voltage of an ordinary 1.5-volt flashlight battery! All everyday events require much larger energy exchanges and therefore involve such tremendous numbers of photons that the quantum characteristic is never evident. It is only when we investigate phenomena in the atomic domain that the discrete nature of light becomes observable and important.

Returning now to the situation shown in Fig. 17-2, when a photon of UV radiation strikes the photoemissive surface A, the photon interacts, not with the electrode as a whole nor even with an entire atom, but only with a *single* electron (one of the *conduction* electrons). Some of the photon's energy is expended in removing the electron from the surface of the electrode material; the remainder appears in the form of the electron's kinetic energy. The photon disappears, leaving behind an energetic electron.

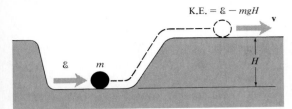

K.E. $= \mathscr{E} - mgH$

v

$\mathscr{E}$ m

H

Figure 17-3   Mechanical analog of the photoelectric effect. If an energy $\mathscr{E}$ is supplied to a ball in a trough, the kinetic energy after emerging from the trough will be K.E. $= \mathscr{E} - mgH$.

Actually, not all of the electrons emerging from a photoemissive surface will have the same kinetic energy. There are two reasons for this. First, the conduction electrons in a material have slightly different energies. Those with the highest energy will be the easiest to remove and will therefore have the greatest kinetic energy. Second, if a photoelectron originates beneath the surface of the material, it will lose some of its kinetic energy through collisions before emerging. We wish to deal here only with the photons that have the greatest possible kinetic energy. These are the conduction electrons that have the highest energy within the material and which are located on the surface of the material.

The ejection of a photoelectron from a substance is quite similar to the mechanical situation pictured in Fig. 17-3. A ball of mass m is in a trough of depth H. In order to remove the ball from the trough, an amount of energy mgH must be expended. If more than this amount of energy is supplied to the ball, the excess will appear as kinetic energy. Thus, if an amount of energy $\mathscr{E}$ is given to the ball, the kinetic energy after emerging from the trough will be

$$\text{K.E.} = \mathscr{E} - mgH \qquad (17\text{-}4)$$

In the photoelectric case, the input energy is the photon energy $h\nu$ and the energy by which the electron is bound in the material is the *work function* $\phi = h\nu_0$. Thus, the energy equation equivalent to Eq. 17-4 is

$$\boxed{\text{K.E.} = \mathscr{E} - \phi = h\nu - h\nu_0} \qquad (17\text{-}5)$$

That is,

$$\boxed{\begin{aligned}\text{(Electron kinetic energy)} = \\ \text{(Photon energy)} - \text{(Work function)}\end{aligned}}$$

$$(17\text{-}6)$$

TABLE 17-1   PHOTOELECTRIC PROPERTIES OF SOME METALLIC ELEMENTS

METAL	WORK FUNCTION, $\phi = h\nu_0$ (eV)	THRESHOLD FREQUENCY (Hz)	THRESHOLD WAVELENGTH (Å)	
Cesium (Cs)	1.9	4.6×10^{14}	6500	
Potassium (K)	2.2	5.3	5600	Visible light
Sodium (Na)	2.3	5.6	5400	
Calcium (Ca)	2.7	6.5	4600	
Zinc (Zn)	3.8	8.9	3400	UV
Platinum (Pt)	5.3	12.9	2300	

The photoelectric properties of some metallic elements are listed in Table 17-1 and some photon energies are given in Table 17-2.

Suppose that blue light ($\lambda = 4600$ Å) is incident on a piece of potassium. What will be the kinetic energy of the ejected photoelectrons? From Table 17-2 we see that $\mathscr{E}$ (blue light) = 2.7 eV, and from Table 17-1 we have ϕ (potassium) = 2.2 eV. Therefore,

$$\text{K.E.} = \mathscr{E} - \phi = 2.7 \text{ eV} - 2.2 \text{ eV} = 0.5 \text{ eV}$$

Light Is a Particle?

What is the significance of Einstein's explanation of the photoelectric effect? The quantum hypothesis and its incorporation into the successful photoelectric theory represented a new departure in physical ideas. No longer could electromagnetic radiation be considered exclusively a wave phenomenon. There are circum-

TABLE 17-2 SOME PHOTON ENERGIES

COLOR	FREQUENCY (Hz)	WAVELENGTH (Å)	PHOTON ENERGY (eV)
Red[a]	4.0×10^{14}	7500	1.6
Orange	4.9	6100	2.0
Yellow	5.1	5900	2.1
Green	5.6	5400	2.3
Blue	6.5	4600	2.7
Violet[a]	7.5	4000	3.1

[a] Extremes of the visible spectrum.

Motion Picture Sound Tracks—How They Work

Photoelectric tubes (or *phototubes*) are used in the motion picture industry to sense the sound "images" that are placed alongside the visual images on movie film. As shown in Fig. 17-4, light from a special lamp is focused by a pair of lenses on the *sound track*. The transmitted light is focused by another lens onto the photoemissive surface of a phototube. The sound track is a strip of the film that is darkened in varying degrees. The amount of transmitted light therefore changes as the film progresses through the projector. The varying light intensity on the phototube produces a varying output of photoelectric current. This electrical signal contains the sound message, and it is amplified and radiated to the audience through loudspeakers. Two or more separate sound tracks are used for stereophonic effects.

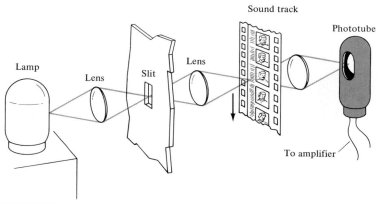

Figure 17-4

stances, such as those of the photoelectric effect, in which light exhibits a highly localized behavior by interacting with a single electron. That is, a light photon behaves in some respects as a *particle* and is found at a particular point in space, not spread out as is a wave. After a hundred years during which light had been treated exclusively in terms of a wave theory, it was suddenly apparent that light was a more complex physical phenomenon than had been supposed. New experiments and new ideas were necessary to understand the fundamental principles governing the interaction of radiation and matter at the atomic level.

17-2 THE WAVE NATURE OF PARTICLES

Waves and Particles

We are all accustomed to the fact that an object with a well-defined edge casts a sharp shadow when placed in a beam of light (Fig. 17-5a). But we have also learned that light is a wave phenomenon, and if we closely examine the edge of the shadow we find that it is really not sharp. Instead, we see a series of fringes—bright and dark lines—in the region that should be fully illuminated (Fig.

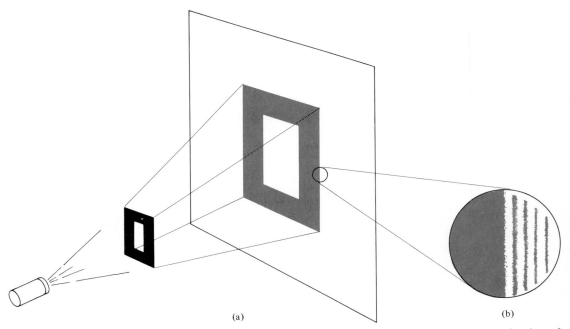

(a)   (b)

Figure 17-5   The block "O" appears to cast a sharp shadow (a), but close examination of the edge of the shadow reveals a series of interference fringes (b).

17-5b). These interference fringes are evidence of the wave character of light. The photographs on this page show the fringes in the region near the shadow of a straight edge and the well-developed interference pattern surrounding the shadow of a razor blade.

Although light produces interference fringes surrounding a shadow, surely particles do not. Suppose that we project small particles (for example, tiny paint droplets from a spray can) at a post that stands on a piece of cardboard, as in Fig. 17-6. Those paint droplets that do not strike the post are collected on the cardboard. The region behind the post is free of paint and a sharp line of demarcation separates the painted and unpainted regions. There are no interference fringes here — the paint droplets are *particles*.

We have seen that light can exhibit properties that are associated with both waves and particles. The paint droplets, however, exhibit only particlelike properties. Will this "particle-only" behavior persist if we examine particles considerably smaller in size than paint droplets? What will we find, for example, with electrons?

Figure 17-6   When sprayed with tiny paint droplets, the post casts a sharp "shadow." The paint droplets act as particles and there is no evidence of any wave interference effects.

de Broglie's Hypothesis

In 1924 a young Frenchman, Louis de Broglie (1892–), proposed an answer to the above questions in his doctoral thesis. De Broglie argued that if light can exhibit both wavelike and particlelike properties, then perhaps particles should behave in a similar way. In order to put his hypothesis in equation form, de Broglie first expressed the wavelength of light in terms of *momentum*. We know that

$$(\text{momentum}) = (\text{mass}) \times (\text{velocity})$$

We also know that the equivalent mass associated with a photon of energy $\mathscr{E}$ is $m = \mathscr{E}/c^2$ (see Eq. 16-12 and Section 16-5). The photon velocity is c and so the momentum p becomes

$$p = m \times c = \frac{\mathscr{E}}{c^2} \times c = \frac{\mathscr{E}}{c} \qquad (17\text{-}7)$$

Next, we use the fact that $\mathscr{E} = h\nu$ and that $\lambda = c/\nu$. Therefore, we can write

$$p = \frac{h\nu}{c} = \frac{h}{\lambda} \quad \text{or} \quad \lambda = \frac{h}{p} \qquad (17\text{-}8)$$

De Broglie argued that this equation, which is valid for

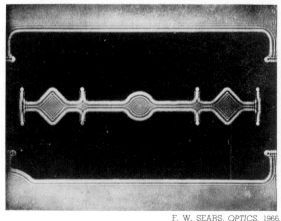

F. W. SEARS, *OPTICS*, 1966.
ADDISON-WESLEY, READING, MASS.

Interference fringes in the region near the shadow of a straightedge (above) and the fringe pattern surrounding the shadow of a razor blade (below).

NIELS BOHR LIBRARY, AIP

Louis Victor Pierre Raymond de Broglie, winner of the 1929 Nobel Prize in physics for his prediction of the wavelike properties of matter.

light photons, should also be true for particles:

> de Broglie wavelength for particles: $\lambda = \dfrac{h}{p}$ (17-9)

Thus, de Broglie combined the mass–energy relation, $\mathscr{E} = mc^2$, and the energy-frequency equation, $\mathscr{E} = h\nu$, to produce an expression for the *wavelength* of a *particle*.

If particles do indeed exhibit wavelike properties with wavelength $\lambda = h/p$, why did the paint droplets fail to produce any observable interference fringes (Fig. 17-6)? Let us calculate the particle wavelength for this case. The diameter of a typical droplet is about 0.2 mm and its mass is about 10^{-8} kg. The velocity is about 20 m/s, so the wavelength is

$$\lambda = \frac{h}{p} = \frac{h}{mv}$$
$$= \frac{6.6 \times 10^{-34} \text{ J-s}}{(10^{-8} \text{ kg}) \times (20 \text{ m/s})}$$
$$= 3.3 \times 10^{-27} \text{ m}$$

which is far smaller than the size of the droplet and, in fact, is far smaller than the size of nuclear particles (10^{-15} m). That is, the wavelength is so small that the interference fringes are completely unobservable even in the most favorable of circumstances.

For comparison, the wavelength of an electron that has been accelerated through a voltage of 20 V is computed as follows. First, we must find the momentum. We write the kinetic energy in terms of the momentum, $p = mv$:

$$\text{K.E.} = \frac{1}{2} mv^2 = \frac{1}{2m} (mv)^2 = \frac{p^2}{2m}$$

Solving for p,

$$p = \sqrt{2m \text{ K.E.}} \qquad (17\text{-}10)$$

Hence, the wavelength is

$$\lambda = \frac{h}{p} = \frac{h}{\sqrt{2m \text{ K.E.}}} \qquad (17\text{-}11)$$

Substituting the appropriate values into this equation, we find for the 20-eV electron,

$$\lambda = 2.73 \times 10^{-10} \text{ m} = 2.73 \text{ Å}$$

The wavelength of a 20-eV electron is much shorter than the wavelength of visible light (4000 Å–7000 Å), but is about the same as an X-ray wavelength.

Electron Interference

Within three years after de Broglie made his ingenious proposal, diffraction experiments had been performed which directly and conclusively demonstrated that electrons have wave properties and exhibit interference effects. These investigations were made independently by George P. Thomson (1892– , son of Sir J. J. Thomson) in England and by C. J. Davisson (1881–1958) and L. H. Germer (1896–) in the United States.

In order to observe diffraction effects with light, it is necessary to have a slit with a width that is comparable with the wavelength of the light (see Section 13-5). As we have seen, electron wavelengths tend to be considerably smaller than the wavelengths of visible light. How, then, is it possible to construct a slit that is sufficiently narrow to permit the observation of electron diffraction effects? Such slits cannot be prepared by machine methods (as can optical slits), but, fortunately, suitable slits occur naturally in the form of the planes of atoms in crystals (see, for example, Fig. 19-3). In Thomson's experiments and in those of Davisson and Germer, crystals were used as the diffraction "slits." A typical result is shown in the photograph on this page. Observations of electron interference effects in many different situations have demonstrated unequivocally that particles can behave as *waves*.

The interference pattern produced by electrons passing through a single crystal of sodium chloride. Because a crystal is a three-dimensional structure, there are only a limited number of directions in space in which the interference effects will be constructive. Therefore, this interference pattern consists of a set of bright *spots* (instead of *lines,* as shown on page 326). The bright central spot is due to the direct beam.

The Wave–Particle Duality

If photons can appear as waves and as particles, and if electrons can also exhibit both particlelike and wavelike properties, what meaning can we attach to the concepts of "waves" and "particles"? When is a wave a *wave* and when is it a *particle?* The problem is that our intuition has led us astray (as was the case in relativity theory). We have learned to think of waves and particles in terms of *classical or Newtonian* ideas. A wave is a propagating disturbance in a medium—a water wave or a sound wave or a wave on a string. A particle is a localizable material object—a paint droplet or a BB or a grain of sand. But photons and electrons are *not* classical quantities—they are *quantum* entities—and we must not force classical ideas upon nonclassical objects. A photon is not *either* a wave *or* a particle—it is a *photon.*

We cannot perform experiments without apparatus of some sort, and any apparatus is necessarily of macroscopic size and is subject to *classical* interpretation. Any experiment with photons does not consist of photons alone—it consists of photons *plus* apparatus. If we choose to perform a diffraction experiment with photons, then by the very nature of the equipment we use and the measurements we make, we have ensured

The Electron Microscope—How It Works

The usefulness of an optical microscope in examining small objects is limited by diffraction effects. The image produced by a microscope for any object will always exhibit interference fringes around the edges; close examination will reveal these fringes regardless of the size of the object (note the fringes surrounding the razor blade in the photograph on page 413). As long as we are interested in observing features of the object that are large compared to the wavelength of light, the interference effects present no serious problem. But if we wish to examine some object or some feature of an object the size of which is comparable with or smaller than the wavelength of light, the diffraction of light around the object will result in a blurred image or no image at all. (In Fig. 13-18, note how the water waves diffract around the barrier; some distance away the wave form shows no trace of the influence of the barrier.) Consequently, optical microscopes lose their utility for objects smaller than about $10\ 000\ \text{Å} = 10^{-6}$ m.

In order to overcome the diffraction limit associated with visible light, we could use radiation with shorter wavelengths—ultraviolet radiation. There are, however, severe problems connected with the absorption and the detection of UV radiation, and UV microscopes operating in the wavelength region below 2000 Å are not practical.

The fact that electrons have a wavelike behavior provides a means for extending the useful range of microscopic observations down to a few angstroms. An electron that is accelerated through a potential difference of 1000 V will have a de Broglie wavelength of only 0.4 Å. But what kind of lenses can we use to take advantage of the short wavelength of energetic electrons? Electrons cannot be focused by ordinary glass lenses; in fact, because they interact electrically with the atomic electrons in matter, electrons cannot penetrate a piece of glass that has any appreciable thickness. The answer to the problem is to use *magnetic* lenses to focus the electrons. Devices that use this principle are called *electron microscopes*. As shown in Fig. 17-7, the arrangement of magnetic focusing coils in an electron microscope is similar to the arrangement of glass lenses in an optical microscope. The electrons that pass through the specimen form a first image and a section of this image is further expanded and projected on the viewing screen as a final image.

Electron microscopes have proved particularly useful in the examination of the details of biological material and in the study of solids. Recent improvements in electron microscopy include the development of a method for tracing the electron beam back and forth over the sample in much the same way that the electron beam scans across the face of a television picture tube. These *scanning electron microscopes* provide a three-dimensional quality to the image, as shown in the photograph below.

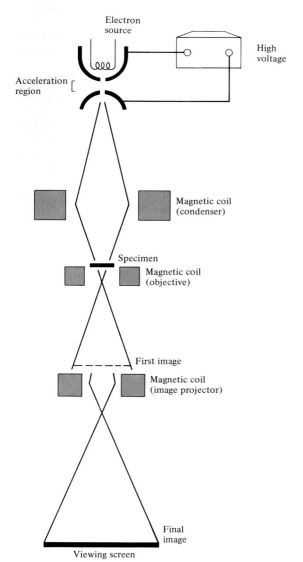

Figure 17-7 Schematic diagram of the arrangement of magnetic focusing lenses in an electron microscope. The entire apparatus is contained within a vacuum cell in order to prevent the deflection of the electrons by colliding with air molecules.

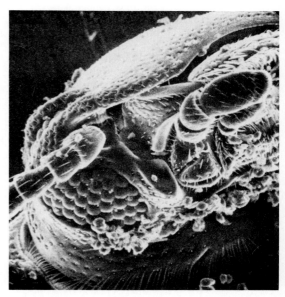

Micrograph taken with a modern scanning electron microscope at the Lawrence Berkeley Laboratory, University of California. Subject is the head of a living flour beetle, enlarged here 230 times.

that the interpretation will be in terms of *waves*. As soon as we allow photons to interact with classical apparatus, we have forced a result that is classical in character. If our apparatus is designed to examine wave properties, we will obtain a *wave* result. If our apparatus is designed to examine particle properties (for example, a photoelectric tube), we will obtain a *particle* result. It is the way in which we make our measurements that determines whether a wave or a particle answer will be obtained.

According to Max Born, the originator of the probabilistic interpretation of quantum theory, "Every process can be interpreted either in terms of particles or in terms of waves, but . . . it is beyond our power to produce proof that it is actually particles or waves with which we are dealing, for we cannot simultaneously determine all the other properties which are distinctive of a particle or of a wave, as the case may be. We can, therefore, say that the wave and particle descriptions are only to be regarded as complementary ways of viewing one and the same objective process. . . ."

17-3 QUANTUM THEORY

Probability

When light is incident on a narrow slit in a panel, we know that the light is spread out (or *diffracted*) and produces a series of interference fringes on a screen placed some distance away (Fig. 17-8). This is the result when we have a *beam* of light, as we found in Section 13-5. But what will happen if we reduce the intensity of the beam—reduce it, in fact, to such a low level that at any instant there is only a *single* photon in the vicinity of the apparatus? Will we obtain the same interference pattern as for the beam of light?

In order to answer these questions, we must remember two points. First, the fact that we are performing a diffraction experiment, even though it is with only one photon at a time, means that the *wavelike* property of the light will be evident. (The argument is the same whether we consider *photons* or *electrons*.) Moreover, the second part of the experiment involves the *detection* of the photon. For this purpose we can use a photographic film or a series of electronic counters. Each photon will be registered at a definite point on the

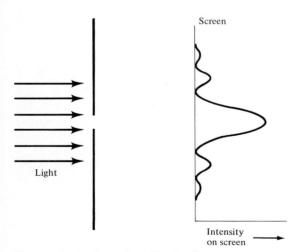

Figure 17-8 Intensity distribution on a screen due to light diffracted by a narrow slit.

screen by rendering developable one of the grains in the film or by triggering one of the counters. That is, in the detection process a photon exhibits its *particlelike* property.

Because a photon interacts with the screen only at one point, a single photon cannot produce a complete interference pattern. An extremely large number of photons must pass through the apparatus before the total pattern is built up. The *same* final intensity distribution is produced by a number of photons whether they strike the screen one at a time or essentially all together.

What is the meaning of the intensity pattern for the case of a single photon? Where will a single photon be detected on the screen? Before the event occurs, *we have absolutely no way to predict where the photon will interact with the screen.* Only after a large number of photons have been detected will the intensity pattern have the appearance shown in Fig. 17-8. If we know the wavelength and the direction of the incident electrons, we can calculate the *probability* or likelihood that an individual electron will interact at a particular point. This probability is exactly the same as the final intensity curve. That is, for *each and every photon* the probability is greatest that it will be detected in line with the slit, at the peak of the intensity curve. Of course, not every photon will be detected at this position, but more will interact here than at any other point along the screen. The probability of an interaction at a particular point is always proportional to the magnitude of the intensity curve at that point. As the photons strike the screen one at a time, the impact points are scattered about, as shown in Fig. 17-9a. Eventually, the individual photon impacts build up to correspond to the wave-theory intensity curve (compare Fig. 17-9e and Fig. 17-8).

Quantum theory deals with *probabilities*. But the theory *can* make precise predictions in terms of the *average value* of a large number of identical measurements. As far as an individual photon is concerned, it is possible to give only the *probability* that an interaction with the screen will occur, for example, at the position corresponding to the secondary maximum to the right (or the left) of the central maximum in Fig. 17-8 or 17-9. However, if the results for a large number of photons are analyzed, the theory can predict with high precision the positions on the screen where, for example, the secondary maxima will occur. If we wish precision in the realm of photons, electrons, atoms, and nuclei, we must

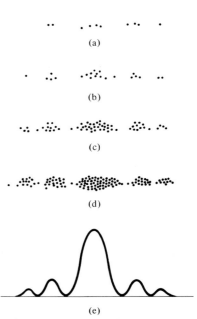

Figure 17-9 The accumulation of single-photon events on a diffraction screen. (a) After only a few photons have passed through the apparatus, there is only a scattering of detection sites on the screen. As more and more photons are detected (b, c, d), the pattern grows more pronounced until the final distribution corresponds exactly to the wave-theory intensity curve (e).

be content with predicting *average values,* not individual events.

Wave Packets

A photon can exhibit both wavelike and particlelike properties. How can we describe such an entity in a way that gives emphasis to these dissimilar aspects of its existence? A light wave, according to the classical description, can be represented in terms of the variation of the electric and magnetic field vectors in space and in time (see Section 14-1). On the other hand, the quantum description of a photon must be made in terms of *de Broglie waves,* waves that represent the *probability* of locating the photon at a particular position at a particular time. The *probability* amplitude is usually given the symbol ψ (psi). The amplitude ψ varies with space and time (just as **E** and **B** do) and can take on positive and negative values. Negative values for **E** or **B** have physical meaning—opposite signs for two electric vectors simply means that the vectors point in opposite directions. But negative probability amplitude has no meaning, and ψ has no physical (or *measurable*) significance. The probability of locating a photon at a particular position is proportional to the *square* of the amplitude at that position. The quantity ψ^2 is always positive (or zero) and represents a *measurable* probability. ψ^2 is physically meaningful; ψ is not.

The de Broglie wave or ψ wave for a photon (or an electron) has the appearance shown in Fig. 17-10. The oscillatory character of ψ is evident, and it is this aspect that gives to the photon (or electron) its wavelike property. It is also apparent that the ψ oscillations are localized in space and it is this feature that gives to the photon (or electron) its particlelike property. This group of ψ waves is called a *wave packet* and the variation of ψ within the packet carries all of the information that specifies the probability of locating the photon (or electron) at any position.

The Uncertainty Principle

Suppose that we attempt to perform what appears to be a very simple experiment, namely, the measurement of the position of an electron. How can we accomplish this and what accuracy can we hope to attain? Any real physical measurement will always be subject to limita-

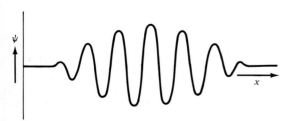

Figure 17-10 The variation of ψ with position x for a photon (or an electron) takes the form of a *wave packet.*

tions imposed by imperfections in our equipment. But we now wish to examine the fundamental physical limitations on experiments, not inaccuracies that are introduced by the measuring apparatus. Therefore, we can imagine that we have ideal equipment at our disposal, so that the ultimate precision in the result will be limited by Nature and not by imperfect apparatus.

The most gentle way to probe for an electron is with a photon. That is, the "touch" of a photon should be less than that of another electron or an atom or some more massive object. In order to use a photon, we must know *where* the photon is. But a photon is an oscillating bundle of radiation and its location can be known only to an accuracy approximately equal to its wavelength. Therefore, when a photon is used to probe for an electron, the position of the electron can be determined only to this same accuracy. Thus, we say that the uncertainty in the position of the electron is $\Delta x \cong \lambda$. The electron can be located only if the photon is scattered by the electron, and in this process some momentum must be transferred to the electron. The momentum transferred will be approximately equal to the photon momentum which, according to Eq. 17-8, is $p = h/\lambda$. The uncertainty in the electron momentum is therefore approximately equal to this value, $\Delta p \cong h/\lambda$. The product of the position uncertainty and the momentum uncertainty is $\Delta x \, \Delta p \cong \lambda \times (h/\lambda)$. Thus,

$$\boxed{\Delta x \, \Delta p \cong h} \qquad (17\text{-}12)$$

This relationship, which was first derived (in a more rigorous manner) by the German theorist, Werner Heisenberg (1901–   ), is the expression of the *Heisenberg uncertainty principle*. Because the product $\Delta x \, \Delta p$ is equal to a constant, this means that if we attempt to locate the electron more precisely (that is, *reduce* Δx), then we lose information regarding the electron momentum (that is, Δp *increases*). On the other hand, if we use lower frequency radiation (longer wavelength) in an effort to disturb the electron less and to reduce Δp, then we are unable to determine the position as precisely. Thus, if we desire increased precision in the determination of either position or momentum, we must pay for this additional information by sacrificing accuracy in the other quantity.

(Actually, the uncertainty relation is an *inequality*, and should properly be expressed as $\Delta x \, \Delta p > h$. Thus, the uncertainty in a particular case will always be *greater* than that calculated using Eq. 17-12. However, we will not concern ourselves with this distinction here.)

Suppose that we use optical radiation to determine the position of a free electron to within an uncertainty equal to the wavelength of the light, 5000 Å $= 5 \times 10^{-7}$ m. What will be the resulting uncertainty in the electron's velocity? How precisely will we know the position of the electron one minute after the measurement? Nonrelativistically, the uncertainty relation can be expressed in the following way:

$$\Delta p = m \times \Delta v \cong \frac{h}{\Delta x}$$

so that the velocity uncertainty is

$$\Delta v \cong \frac{h}{m \, \Delta x}$$
$$\cong \frac{6.6 \times 10^{-34} \text{ J-s}}{(9.1 \times 10^{-31} \text{ kg}) \times (5 \times 10^{-7} \text{ m})}$$
$$\cong 1.5 \times 10^{3} \text{ m/s} = 1.5 \text{ km/s}$$

Therefore, after one minute the electron could be anywhere within a distance of (60 s) $\times$ (1.5 km/s) $=$ 90 km! By locating the electron to as small an interval as 5000 Å, we severely limit our knowledge of the position of the electron at future times.

The Meaning of the Uncertainty Principle

Why are quantum objects restricted by the uncertainty principle? Is this an attempt by Nature to prevent us from looking too deeply into the way things *really* behave in the atomic domain? There is no reason to view the uncertainty principle in this way. A quantum object possesses a dual character—it has particlelike and wavelike properties—and the uncertainty principle simply expresses the limitations that are inherent in dealing with any wavelike object.

Because we cannot know with precision both the position and the momentum of an electron or a photon, we cannot predict where the electron or photon will be in the future. Thus, when we project an individual photon through a slit, as in Fig. 17-8, we cannot predict the point on the screen where it will interact. We can only give the probability for an interaction at a particular position. Thus, the uncertainty principle underlies the probabilistic nature of events that take place at the most elementary level.

There seems to be no way to escape the consequences of the uncertainty principle. One might say, "Well, there appears to be some limitation on the measurements we make, but an electron is always at some *precise* location and is moving with some *precise* velocity—the only problem is that we cannot measure the precise location and the precise velocity at the same time." Such a statement really is outside the realm of physical science because the only quantities that are physically meaningful are those that we *can* measure. To state that an object possesses some property that we cannot measure falls in the realm of metaphysics, not physics.

The application of quantum theory to the domain of molecules, atoms, and particles has been magnificently successful. Many elementary phenomena can be predicted with remarkable precision. (That is, the *average value* of a large number of identical measurements can be predicted with high precision.) But we also know that the classical theories of mechanics and electrodynamics are extremely successful in dealing with large-scale phenomena. Does this mean that we have one set of physical principles that is correct in the macroscopic domain and a completely different set in the microscopic world? Some measure of satisfaction is afforded by the fact that this is not the case. If we begin by applying quantum theory to elementary systems—atoms and molecules—and then increase the size and complexity of the systems, we find that as they grow larger and larger the systems are described in terms that approach closer and closer to classical theory. When the systems have been increased to macroscopic size, the last vestiges of quantum effects have disappeared and the description is entirely classical. Thus, there is a unity between the microscopic and macroscopic worlds: both are described correctly by quantum theory, but for large-scale phenomena, the quantum description is indistinguishable from the results of classical theory.

SUGGESTED READINGS

B. L. Cline, *Men Who Made a New Physics* (New American Library, New York, 1969).

J. R. Pierce, *Electrons and Waves* (Doubleday, Garden City, New York, 1964).

Scientific American articles:

G. Gamow, "The Principle of Uncertainty," January 1958.

E. Schrödinger, "What Is Matter?" September 1953.

1. What is the kinetic energy of the photoelectrons emitted from sodium when light with $\lambda = 4000$ Å is incident?

2. Light with wavelength 5800 Å will not produce photoelectrons when incident on a certain material but light with wavelength 5760 Å will produce photoelectrons. What is the approximate value (in eV) of the work function for the material?

3. What is the energy (in eV) of a 5000-Å photon?

4. Electrons with a maximum kinetic energy of 0.5 eV are ejected from potassium when irradiated with light of a certain color. What is the color?

5. The human eye will respond to as few as 5 photons of green light. How much energy absorption (in joules) does this represent?

6. Electrons and photons both have wavelike and particlelike properties. Why were electrons originally considered to be *particles* whereas light was considered to be a *wave* phenomenon?

7. What is the kinetic energy (in eV) of an electron with a wavelength of 1 Å?

8. What is the wavelength of a *proton* that is accelerated through a potential difference of 100 V?

9. What is the energy (in eV) of a quantum of electromagnetic energy with $\lambda = 300$ m (which is near the middle of the standard AM broadcast band)? Does it make sense to discuss radio waves in terms of *quanta?* Explain.

10. A photon and an electron each have an energy of 5 eV. What are their wavelengths?

11. What kinds of measurements would distinguish between a photon with $\lambda = 1$ Å and an electron with $\lambda = 1$ Å? What kinds of measurements would *not* be suitable?

12. Consider two photons, one with $\lambda = 10^{-4}$ m and one with $\lambda = 10^{-12}$ m. For which photon will its particle aspect be more important? Explain.

13. A 0.1-kg rock is thrown with a velocity of 30 m/s. What is the de Broglie wavelength of the rock? On the basis of your result, do you expect the wave properties of the rock to be observable? (Compare the wavelength with the size of the rock.)

14. What is the wavelength of an electron that is "at rest?" Would it be possible to measure the position of such an electron? Explain.

15. Why is the uncertainty principle of no significance in everyday (that is, large-scale) phenomena?

16. It is desired to measure the velocity of a moving electron to a precision of 1 m/s. If this measurement is made, how precisely can we state the position of the electron at the time of measurement?

17. An electron is confined to a box that is 1 mm in length. How accurately can the electron's velocity be determined?

18. Just as position and momentum are related by an uncertainty relation, so are *energy* and *time*. In fact, a way to express the uncertainty principle that is entirely equivalent to Eq. 17-12 is $\Delta E \, \Delta t \cong h$. That is, the uncertainty in the energy of a process multiplied by the uncertainty in the duration of the event that yields that energy is approximately equal to Planck's constant. When an atom of a substance absorbs energy and is raised to an excited state, it will remain in this condition for an average time that is typically 10^{-8} s. If the nominal wavelength of the radiation that is emitted when the atom returns to its normal condition is 6000 Å, what is uncertainty in this wavelength? (This uncertainty will manifest itself in a range of wavelengths that will be observed from a large number of identical atoms excited in the same way.)

18

THE MODERN VIEW OF ATOMS

By 1912 new and important discoveries concerning atoms and radiation had been accumulating for about 15 years. The electron had been identified by Thomson in 1897. Planck had made his quantum hypothesis in 1900, and Einstein had adopted this idea in 1905 to explain the photoelectric effect. Rutherford's nuclear model of the atom was proposed in 1911. And in that year, a young Dane, Niels Bohr (1885–1962), came to work in Thomson's laboratory at Cambridge.

Bohr wondered what connection there could be between the quantized nature of radiation and the structure of atoms. Spectrographic studies had shown that atoms emit radiation only with certain definite wavelengths and that each atomic species has its own characteristic spectrum of emitted radiation. Bohr concluded that if an atom could emit radiation only with definite wavelengths (that is, with discrete energies), then the internal energetics of the atom must also be quantized. Thomson, Bohr's host at the Cavendish Laboratory, would not accept the idea that atoms possess a quantized structure—he much preferred a classical atomic model. Several sharp arguments over the matter took place and this unpleasantness caused Bohr to decide to leave Cambridge and spend the remainder of his fellowship in a more forward-looking atmosphere. Bohr chose Manchester, where Rutherford and his colleagues were investigating atomic structures with radioactivity methods.

Thus begins one of the most exciting chapters in the history of scientific discovery—the development of the ideas concerning the inner working of atoms that has culminated in our modern quantum theory of atoms and molecules and has produced a unification of the fields of chemistry and physics. But before we can continue with the story of Bohr and the quantum, we must look at one important result that was obtained 25 years earlier.

18-1 THE HYDROGEN ATOM

The Balmer Formula

The science of spectroscopy began in the middle 1800's and methods were soon developed for the precise measurement of the wavelengths of atomic radiations. The Swedish spectroscopist Anders Ångström (1814–1874) carefully measured many wavelengths, including those of the hydrogen atom. The hydrogen spectrum (Fig. 18-1) showed a curious regularity, but the reason for the progression of lines toward the violet with decreasing spacing was unknown.

In 1885, Johann Balmer (1825–1898), an obscure teacher at a Swiss girls' school, published an article in which he presented a simple formula that reproduced Ångström's values for the wavelengths of the hydrogen spectral lines with remarkable accuracy (see Table 18-1). If we convert Balmer's original wavelength formula into an expression for frequency (which is the more useful form), we have

$$\nu = cR\left(\frac{1}{2^2} - \frac{1}{n^2}\right), \qquad n = 3, 4, 5, 6, \ldots \quad (18\text{-}1)$$

where R is a constant that Balmer adjusted to give best agreement with the data and where n can be any integer number starting with 3. Each value of n that is substituted into the formula gives the frequency of a different spectral line in the series.

Balmer had no basis for his extraordinary formula—it was simply an empirical representation of the ex-

TABLE 18-1 HYDROGEN SPECTRAL LINES IN THE BALMER SERIES

n	λ (observed by Ångström) (Å)	λ (calculated by Balmer) (Å)
3	6562.10	6562.08
4	4860.74	4860.8
5	4340.1	4340.0
6	4101.2	4101.3

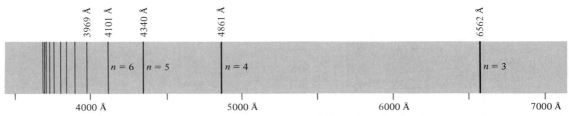

Figure 18-1   The Balmer series of spectral lines from hydrogen.

perimental data. Any formula as accurate as Balmer's is very likely to have some fundamental significance. But *what?* When Bohr arrived in Manchester in 1912, he realized that understanding the significance of the Balmer formula would be a crucial step in solving the puzzle of atomic structure.

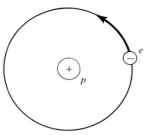

Figure 18-2 In Bohr's model of the hydrogen atom, the single atomic electron is considered to move around the nuclear proton in a circular orbit.

The Bohr Model of the Hydrogen Atom

Rutherford's newly developed picture of the atom intrigued Bohr. But there were unanswered questions. According to the Rutherford model, most of the mass and all of the positive charge of an atom resides in the tiny central nucleus which is surrounded by the atomic electrons. What prevents the attractive electrical forces from pulling the electrons into the nucleus and collapsing the atom? How does the atom maintain its stability? Bohr reasoned that the only way in which a nuclear atom could resist collapse would be for the electrons to move around the nucleus just as the planets move around the Sun. The solar system has a dynamical stability and so should an atom. In the simplest case—the hydrogen atom—a single electron orbits around the nuclear proton (Fig. 18-2).

It was now necessary to impose upon the planetary atomic model of hydrogen some restriction that would reproduce the discrete wavelength spectrum and would account for the Balmer formula. Bohr needed a procedure to *quantize* the hydrogen atom. We can see the situation more clearly if we convert the Balmer *frequency* equation into an *energy* equation. The energy of a photon is $\mathscr{E} = h\nu$, and so multiplying both sides of Eq. 18-1 by h, we can write

$$\mathscr{E} = h\nu = \frac{hcR}{2^2} - \frac{hcR}{n^2} \qquad (18\text{-}2)$$

That is, the photon energy $\mathscr{E} = h\nu$, is given by the difference of two energy terms. (In energy units, the combination of constants hcR is equal to 13.6 eV.) Bohr saw in this result the evidence that the structure of the atom must be restricted to certain definite configurations, each with a definite amount of energy. He interpreted each term on the right-hand side of Eq. 18-2 as representing a discrete energy state of the atom. He assumed that an atom must be able to exist in one of a number of discrete energy states and *only in these states*. When an atom makes a transition from a higher energy state to a lower energy state, the energy difference is radiated as a photon. As the individual atoms in a collection make transitions between the allowed energy states, the entire set of spectral lines is emitted and the spectrum has the appearance shown in Fig. 18-1.

During his stay in Manchester, Bohr pondered the situation and learned more about Rutherford's nuclear atom. Upon his return to Denmark in 1913, Bohr brought his ideas together. In doing so, he was confronted with two important questions. First, what is the reason for the quantization of atomic energies? Second, it was known from electromagnetic theory that accelerated electric charges radiate energy (see Section 14-1): since an orbiting atomic electron is continually acceler-

Niels Bohr was born in 1885 in Copenhagen, the son of a physiology professor at the University of Copenhagen. In 1903 he entered the University where he studied physics (and became a first-rate soccer player). Bohr obtained his doctorate in 1911 and then received a fellowship to study abroad. He spent his year at Cambridge and at Manchester, where he absorbed Rutherford's ideas concerning atomic structure. In 1913 he published his now-famous paper on the hydrogen atom, but at the time, his mixed classical and quantum ideas did not generate much of a following. Only when de Broglie put forward his matter–wave hypothesis did Bohr's ideas finally appear reasonable. In 1917, at the age of 31, Bohr became professor of physics at the University of Copenhagen. In 1922 he received the Nobel Prize in physics, and with the sponsorship of the Carlsberg Brewery, he founded the Institute for Theoretical Physics. The Institute rapidly became (and still is) a gathering place for physicists from all over the world to meet and discuss the current problems and theories. In 1939 Bohr visited the United States, bringing with him the news that the fission of the uranium nucleus had been discovered. He soon developed (along with John Wheeler of Princeton) a theory of the fission process based on the similarity of the nucleus to a liquid drop. Bohr returned to Copenhagen in 1940, but in 1943 he was forced to flee from the Nazis. He and his family were smuggled to Sweden in a fishing boat. Subsequently, he was flown to England in the bomb bay of a bomber where he nearly died from lack of oxygen because no mask could be found to fit him; he was unconscious upon landing but survived. Bohr went on to the United States and worked on the atomic bomb project at Los Alamos, New Mexico, until 1945. He had grave misgivings about the use of nuclear energy in warfare (a situation which nearly caused Winston Churchill to issue an order for his arrest). Bohr labored for the rest of his life in the cause of peaceful uses of nuclear energy. In 1957 he received the first Atoms for Peace award. He died in Copenhagen in 1962.

ated, why does it not radiate away its energy and fall toward the nucleus? A calculation using classical electromagnetic theory shows that the electron in a hydrogen atom will radiate all of its energy within a small fraction of a second. But, of course, this does not happen. Why should an atom radiate energy *only* when it makes a transition between two allowed states?

Bohr's approach to this question was unorthodox, to say the least. He simply hypothesized that an atomic electron defies classical theory and does not radiate when moving in an allowed orbit; radiation takes place *only* when the electron moves from one orbit to another—that is, when the atom makes a transition from one energy state to another. Thus, Bohr did not *answer* the question; he *abolished* it! And Bohr's answer to the question regarding the reason for energy quantization required an equally bold step.

Bohr's Angular Momentum Hypothesis

Bohr was not satisfied with merely hypothesizing that the energy states of the hydrogen atom are quantized—he sought some more fundamental idea that would lead to the result in a straightforward way. His solution was to propose that *angular momentum* is quantized. Bohr found that he could derive an expression for the energies of the photons emitted by hydrogen atoms if he assumed that the angular momentum of the orbiting electron is limited to integer multiples of Planck's constant divided by 2π. Using Eq. 5-5 for the angular momentum, we can write Bohr's quantization condition as

$$L = mvr = n\frac{h}{2\pi}, \qquad n = 1, 2, 3, \ldots \quad (18\text{-}3)$$

Starting with this quantization rule, Bohr proceeded to calculate the electron orbits in the hydrogen atoms as if the electron were a planet revolving around the nuclear "Sun." In this calculation, Bohr substituted the electrical force between the electron and the nuclear proton for the gravitational force that holds a planet in orbit around the Sun. As shown in the section beginning on page 430, the result of Bohr's calculation is that the electron can occupy only discrete orbits with definite radii and can move only with definite velocities. Each orbit corresponds to a different number n in the angular momentum quantization condition (Eq. 18-3).

Bohr's Calculation

Bohr began his calculation by noting that the force between the orbiting electron and the nuclear proton is one of electrical attraction between the negative electron charge and the positive proton charge. The proton and the electron each carry a charge of magnitude e. Inserting this value into the expression for the electrical force. $F_E = Kq_1q_2/r^2$ (Eq. 6-9), we find

$$F_E = K \frac{e^2}{r^2} \tag{18-4}$$

This is the mutual force of attraction between the two particles, but the proton is much more massive than the electron and therefore remains essentially stationary. Accordingly, we will focus attention on the motion of the electron. Newton's law, $F = ma$, tells us that the force on the electron, F_E, must equal the mass of the electron multiplied by its acceleration (which is the centripetal acceleration, Eq. 3-19):

$$F_E = ma_c = \frac{mv^2}{r} \tag{18-5}$$

Equating these two expressions for the force, we have

$$\frac{mv^2}{r} = K \frac{e^2}{r^2}$$

and solving for the velocity,

$$v = \sqrt{\frac{Ke^2}{mr}} \tag{18-6}$$

There is nothing extraordinary about this result—we have made only a straightforward application of classical ideas. Indeed, we used exactly this procedure in Section 6-4 when we calculated satellite orbits (except that the force was gravitational instead of electrical). Now, we introduce the quantum suggestion of Bohr. We use the result for v in Eq. 18-3 for the angular momentum:

$$L = mvr = m \times \sqrt{\frac{Ke^2}{mr}} \times r = n \frac{h}{2\pi}$$

To facilitate solving for the radius r, we square the last equality, obtaining

$$Kme^2r = n^2 \frac{h^2}{4\pi^2}$$

and, finally, solving for r, we have

$$\boxed{r_n = \frac{n^2h^2}{4\pi^2Kme^2} \qquad n = 1, 2, 3, \ldots} \tag{18-7}$$

where we have attached a subscript n to the radius r to designate that r_n takes on different values depending on n.

Equation 18-7 shows that the angular momentum quantization condition leads to a series of discrete radii for the electron orbits in the hydrogen atom. By substituting $n = 1$ into this equation we can compute the size of the hydrogen atom in the most compact state predicted by the Bohr analysis:

$$r_1 = \frac{h^2}{4\pi^2 Kme^2}$$

$$= \frac{(6.6 \times 10^{-34} \text{ J-s})^2}{4\pi^2 \times (9 \times 10^9 \text{ N-m}^2/\text{C}^2) \times (9.1 \times 10^{-31} \text{ kg}) \times (1.6 \times 10^{-19} \text{ C})^2}$$

$$= 0.53 \times 10^{-10} \text{ m} = 0.53 \text{ Å}$$

The state with this orbit radius is the lowest possible energy state of the hydrogen atom and is called the *ground state*. The diameter of the normal hydrogen atom (that is, the atom in its ground state) is therefore approximately 1 Å.

Using the result for $n = 1$, we can express the radius of the hydrogen atom for any value of n as

$$r_n = n^2 \frac{h^2}{4\pi^2 Kme^2} = n^2 r_1 = n^2 \times (0.53 \text{ Å}) \tag{18-8}$$

What is the significance of the discrete orbits in Bohr's model of the hydrogen atom? We can answer this question by examining the energies associated with atoms in the various orbital states.

The Energy States of the Hydrogen Atom

Bohr had found a way to quantize the *structure* of the hydrogen atom. He was now in a position to complete the solution to the problem by calculating the *energies* of the various atomic states, thereby explaining the origin of the hydrogen spectral lines. The total energy of an orbiting electron is equal to the sum of the electrical potential energy and the kinetic energy. The potential energy depends on the separation of the electron and the nucleus (that is, on r_n) and the kinetic energy depends on v (which, according to Eq. 18-6, can also be expressed in terms of the radius of the orbit). Bohr made this calculation and then substituted for r_n from Eq. 18-7. He found the energy of an electron in the orbit labeled by n to be

$$\mathcal{E}_n = -\frac{2\pi^2 K^2 me^4}{h^2} \cdot \frac{1}{n^2}, \qquad n = 1, 2, 3, \ldots \tag{18-9}$$

The total energy $\mathscr{E}_n$ is negative because we choose the zero energy condition to correspond to infinite separation between the electron and the nucleus—that is, $r = \infty$ or $n = \infty$. All other energies are lower than this energy and the lowest possible energy, that for the normal or ground state, is found for $n = 1$. If we substitute into Eq. 18-9 the values of the various constants, this energy is

$$\mathscr{E}_1 = -13.6 \text{ eV}$$

Thus, Eq. 18-9 can be expressed as

$$\mathscr{E}_n = -\frac{13.6 \text{ eV}}{n^2} \qquad (18\text{-}10)$$

If 13.6 eV of energy (or more) is supplied to a hydrogen atom in its lowest energy state, the electron will be removed from the atom and the atom will become an *ion*. The *binding energy* or *ionization* energy of a hydrogen atom is therefore equal to 13.6 eV.

If a hydrogen atom is in an energy state specified by n and makes a transition to a lower energy state specified by n', the difference in energy between the two states is carried off in the form of a photon. That is, the photon energy is

$$\mathscr{E} = \mathscr{E}_n - \mathscr{E}_{n'} \qquad (18\text{-}11)$$

when Eq. 18-10 is used for the energies $\mathscr{E}_n$ and $\mathscr{E}_{n'}$ of the atom,

$$\boxed{\mathscr{E} = (13.6 \text{ eV}) \times \left(\frac{1}{n'^2} - \frac{1}{n^2}\right), \ n > n'} \qquad (18\text{-}12)$$

Bohr had now succeeded in deriving the Balmer formula. (When $n' = 2$, we have exactly the result expressed by Eq. 18-2 because $hcR = 13.6$ eV.) Bohr's mixture of classical physics and quantum ideas (still incompletely understood) had made a significant breakthrough in solving the mystery of atomic spectra.

Transitions in the Hydrogen Atom

Figure 18-3 shows the quantized orbits of the hydrogen atom for $n = 1$ through 5. Each orbit corresponds to a definite energy state of the atom. The straight lines originating on the $n = 3$, 4, and 5 orbits and terminating on the $n = 2$ orbit represent transitions in the Balmer series. If the transitions terminate instead on the $n = 1$ orbit, the energy differences are greater and the radiations fall in the ultraviolet part of the spectrum. This set

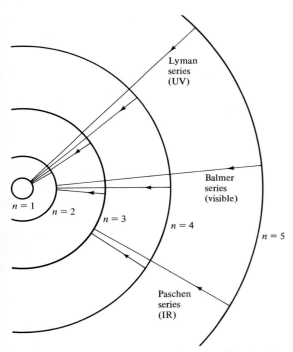

Figure 18-3 The discrete allowed orbits in the Bohr model of the hydrogen atom. Some of the transitions in three of the spectral series are shown.

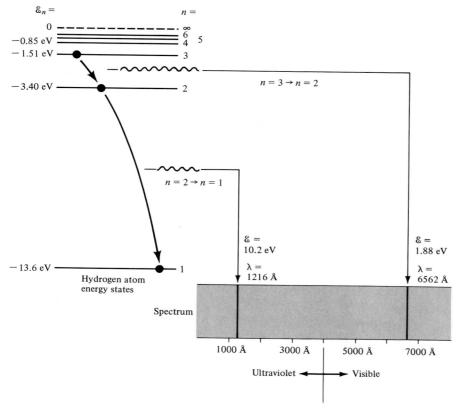

$\mathcal{E}_n =$ $n =$

0 - - - - - - - - - - - - - - - ∞
                                              6
-0.85 eV   4   5
-1.51 eV ●————————— 3

————〰〰〰〰〰————
                                    $n = 3 \rightarrow n = 2$
-3.40 eV ●————————— 2

                —〰〰〰—
                $n = 2 \rightarrow n = 1$

                                    $\mathcal{E} =$ $\mathcal{E} =$
                                    10.2 eV   1.88 eV

-13.6 eV ————————●———— 1 $\lambda =$ $\lambda =$
  Hydrogen atom   1216 Å   6562 Å
  energy states

Spectrum

        1000 Å   3000 Å   5000 Å   7000 Å

                Ultraviolet ◄———► Visible

Figure 18-4 Some of the energy states of the hydrogen atom are shown on the left. Two of the possible transitions and the resulting spectral lines are also shown.

of spectral lines is called the *Lyman series*. Also shown are the lower energy, infrared lines of the *Paschen series* which terminate on the $n = 3$ orbit.

Figure 18-4 shows (on the left) an energy diagram which indicates the relative energies for several of the hydrogen atom states. Also shown are two of the possible transitions that result in lines in the hydrogen spectrum. The transition from the $n = 3$ state to the $n = 2$ state produces a photon with $\lambda = 6562$ Å; this spectral line occurs in the Balmer series. In the transition from the $n = 2$ state to the $n = 1$ state, a photon with $\lambda = 1216$ Å is emitted; this line is part of the Lyman series.

Suppose that a beam of electrons is incident on a collection of hydrogen atoms all of which are in the lowest energy state ($n = 1$). What is the minimum energy that the electrons can have if they are to excite the hydrogen atoms into the $n = 2$ state? According to Eq.

18-10, the energy of the $n = 2$ state is

$$\mathscr{E}_2 = -\frac{13.6 \text{ eV}}{(2)^2} = -3.4 \text{ eV}$$

Therefore, the energy difference between the ground state and the $n = 2$ state is

$$\mathscr{E} = \mathscr{E}_2 - \mathscr{E}_1 = (-3.4 \text{ eV}) - (-13.6 \text{ eV}) = 10.2 \text{ eV}$$

This is the minimum energy required to excite the atom into the $n = 2$ state. If a 10.5-eV electron struck an atom, 10.2 eV would be transferred to the atom by way of excitation energy and the electron would retain 0.3 eV of kinetic energy.

A convenient expression for the wavelength λ in Å of a photon with energy $\mathscr{E}$ in eV is

$$\lambda = \frac{12\,400}{\mathscr{E}} \qquad (18\text{-}13)$$

Thus, the wavelength of the photon emitted in the transition between the $n = 2$ and $n = 1$ orbits of the hydrogen atom (for which $\mathscr{E} = 10.2 \text{ eV}$) is

$$\lambda = \frac{12\,400}{10.2} = 1216 \text{ Å}$$

which is the value indicated in Fig. 18-4.

The key step in Bohr's analysis of the structure of the hydrogen atom was the hypothesis of angular momentum quantization. This idea has a much wider validity than first realized by Bohr. In fact, *all* angular momenta satisfy exactly the requirements that Bohr had assumed: *all* angular momenta due to orbiting particles must equal an integer number multiplied by $h/2\pi$. Thus, Bohr's keen perception permitted him to discover a universal rule, not merely one that was limited to the case of the hydrogen atom.

The New Quantum Numbers

In Bohr's model of the hydrogen atom, if the principal quantum number n is specified, everything is known about the atomic state. In particular, the quantum number n gives the energy of the state according to Eq. 18-9 and the angular momentum of the state according to Eq. 18-3. But more detailed examinations of the spectra of hydrogen and other elements revealed small discrepancies that could not be explained in terms of Bohr's scheme. The idea of simple circular orbits had to be abandoned, and the description of atoms rapidly became much more complex than in the original Bohr model. By the time the theory was completely developed, it had been found necessary to add three new quantum numbers. Instead of specifying an atomic state in terms of the single number n, it is actually necessary to specify *four* quantum numbers.

The first of Bohr's original ideas to be modified was the angular momentum condition, $L = nh/2\pi$ (Eq. 18-3). Bohr had been perfectly correct in his hypothesis of angular momentum quantization; the problem was that for a particular value of n, several different angular momenta are allowed. A new quantum number labeled l, must be introduced to specify the angular momentum of a state:

$$L = l\frac{h}{2\pi}, \qquad l = 0, 1, 2, \ldots, n-1 \quad (18\text{-}14)$$

where values of the angular momentum quantum number l can range from zero to a maximum equal to $n - 1$. That is, the state with $n = 1$ can have only $l = 0$; the state with $n = 2$ has two possibilities, $l = 0$ and $l = 1$; and so forth.

For convenience in describing atomic systems, the states are labeled by a number-and-letter system that indicates the values of the quantum numbers, n and l: for example, 1S, 2P, and 3D. The number is the value of n and the letter stands for the value of l according to the scheme:

$l = 0$	S state
$l = 1$	P state
$l = 2$	D state
$l = 3$	F state
$l = 4$	G state

with higher values of l following in alphabetical order. The various possible angular momentum states and their designations for $n = 1$ through 5 are shown in Table 18-2. This scheme of labeling values of the angular momentum with letters is a holdover from the pre-quantum-theory days of spectroscopy when certain spectral lines were designated strong (S), principal (P), diffuse (D), and fundamental (F).

This is not the complete story concerning angular momentum. Angular momentum is a vector quantity—it has magnitude and direction. As we have seen, the *mag-*

TABLE 18-2 DESIGNATIONS FOR SOME ATOMIC STATES

n	S $l = 0$	P $l = 1$	D $l = 2$	F $l = 3$	G $l = 4$
1	1 S				
2	2 S	2 P			
3	3 S	3 P	3 D		
4	4 S	4 P	4 D	4 F	
5	5 S	5 P	5 D	5 F	5 G

nitude L is quantized in units of $h/2\pi$. In addition, the *direction* of the angular momentum vector **L** is also quantized. In particular, if an atom is in a magnetic field **B**, the vector **L** can point only in certain definite directions relative to **B**. This restriction can be expressed by stating that the component of **L** in the direction of **B** (we call this the z-direction) is limited to discrete multiples of $h/2\pi$; that is,

$$L_z = m_l \frac{h}{2\pi} \qquad (18\text{-}15)$$

The new quantum number m_l (which is called the *magnetic* quantum number) can have the following values:

$$l = 0: \qquad m_l = 0$$
$$l = 1: \qquad m_l = -1, 0, +1$$
$$l = 2: \qquad m_l = -2, -1, 0, +1, +2$$
$$l = 3: \qquad m_l = -3, -2, -1, 0, +1, +2, +3$$

and so forth. Negative values of m_l mean that the z-component of **L** is in the negative z-direction (opposite to the direction of **B**); $m_l = 0$ means that **L** is perpendicular to **B**. For any value of l, there are always $2l + 1$ possible values of m_l. Each value of m_l represents a *magnetic substate*.

Finally, there is a fourth quantum number, the *spin* quantum number. In 1925, Samuel Goudsmit (1902–   ) and George Uhlenbeck (1900–   ) showed that certain spectroscopic results can be explained only if it is assumed that an electron possesses some intrinsic angular momentum, quite independent of the angular momentum that it possesses because of its orbital motion. In a classical way, we can picture the electron as a tiny ball spinning around its own axis, and this spinning motion has associated with it a certain angular momentum. This picture is useful to gain an idea of the origin of spin angular momentum, but the notion of a spinning ball is contrary to the quantum view of an electron. All we can really state is that an electron possesses an intrinsic angular momentum just as it possesses an intrinsic mass and an intrinsic charge. Angular momentum is simply one of the basic properties of an electron.

The spin angular momentum of an electron is quantized. If we measure the component of the spin along a particular direction, the z-direction (for example, the direction of a magnetic field), we find only two allowed values: either $\frac{1}{2}(h/2\pi)$ *along* the z-direction or $\frac{1}{2}(h/2\pi)$ *opposite to* the z-direction. Stating this result in a way analogous to that for the z-component of **L**, we can write

TABLE 18-3 THE FOUR ELECTRON QUANTUM NUMBERS

QUANTUM NUMBER	PHYSICAL SIGNIFICANCE	EQUATION	ALLOWED VALUES
n	Energy	$E_n = -\dfrac{13.6 \text{ eV}^a}{n^2}$	$1, 2, 3, \ldots$
l	Angular momentum	$L = l\dfrac{h}{2\pi}$	$0, 1, 2, \ldots, n-1$
m_l	Component of orbital angular momentum	$L_z = m_l \dfrac{h}{2\pi}$	$-l, \ldots, 0, \ldots, +l$
m_s	Component of spin angular momentum	$S_z = m_s \dfrac{h}{2\pi}$	$+\frac{1}{2}, -\frac{1}{2}$

^{a} This result is valid only for the hydrogen atom; an expression for the energy in terms of n that is valid for all cases cannot be given.

the z-component of the spin angular momentum as

$$S_z = m_s \frac{h}{2\pi}, \qquad m_s = +\tfrac{1}{2} \text{ or } -\tfrac{1}{2} \qquad (18\text{-}16)$$

The fourth quantum number, m_s, is called the *spin* quantum number. We often refer to the two possible spin states simply as *spin up* and *spin down*.

The state of an electron in an atom is completely specified only when *all four* quantum numbers are given. These quantum numbers and the significance of each are summarized in Table 18-3.

18-2 QUANTUM THEORY OF THE HYDROGEN ATOM

Electron Waves

The new developments had significantly altered Bohr's original model. Instead of a single quantum number n, it had been discovered that four quantum numbers are necessary for the complete description of an atomic state. But the model was still a curious combination of classical theory and quantum ideas. The first indication that the model was consistent with the emerging concept of the wave nature of matter was demonstrated by de Broglie. The most conspicuous holdover from classical ideas that appeared in the modified Bohr model was the concept of a particlelike electron moving in planetlike orbits. But in the early 1920's, electron diffraction experiments showed that matter possessed wave properties as proposed by de Broglie. How were *electron waves* to be fitted into Bohr's scheme?

In each Bohr orbit the electron has a definite velocity (Eq. 18-6) and the de Broglie wavelength (for the state labeled n) is

$$\lambda_n = \frac{h}{p_n} = \frac{h}{mv_n}$$

The angular momentum of the electron in this state is

$$L_n = mv_n r_n = n \frac{h}{2\pi}$$

Solving for mv_n,

$$mv_n = \frac{nh}{2\pi r_n}$$

and substituting this into the equation for the de Broglie wavelength, we have

$$\lambda_n = \frac{h}{(nh/2\pi r_n)} = \frac{2\pi r_n}{n}$$

Thus, n wavelengths exactly equals the *circumference* of the orbit:

$$n\lambda_n = 2\pi r_n \qquad (18\text{-}17)$$

That is, n de Broglie wavelengths exactly fit into the nth orbit — the first orbit contains exactly one wavelength, the second orbit contains exactly two wavelengths, and so forth. If we plot the electron wave in the conventional manner (Fig. 18-5) and then deform the center line into a circle, this circle exactly matches the corresponding Bohr orbit.

Why is this result significant? Figure 18-6 shows the reason. When the de Broglie electron wave exactly fits into a Bohr orbit, the wave reinforces itself by constructive interference (Fig. 18-6a); therefore, the wave persists. On the other hand, if the de Broglie wave does not fit into the orbit, as in Fig. 18-6b, the wave interferes destructively with itself and rapidly cancels; these waves cannot persist. Thus, the existence of discrete orbits and discrete energy states follows directly from de Broglie's hypothesis concerning the wave properties of matter.

Despite these successes for the ideas of Bohr and de Broglie, the strange combination of a quantized classical theory with a dash of wave properties was too arbitrary to set well with most physicists. Moreover, there were still small but unreconcilable conflicts between the theory and experiment. There was something of fundamental importance in the behavior of atoms that re-

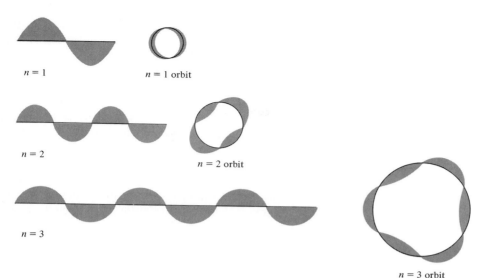

n = 1   n = 1 orbit

n = 2

n = 2 orbit

n = 3

n = 3 orbit

Figure 18-5   De Broglie electron waves and the Bohr orbits in the hydrogen atom for n = 1, 2, 3. The lengths of the waves shown are exactly equal to the circumferences of the orbits. (The n = 1 wave and orbit are shown for clarity on a scale $2\frac{1}{2}$ times larger than for the other cases.)

mained hidden. The search for the missing idea became the central issue in the efforts to discover an *entirely* quantum description of atomic spectra. The picture of electrons "jumping" from one orbit to another was abandoned and a completely wave-oriented theory was developed by Erwin Schrödinger, Werner Heisenberg, Max Born, Paul Dirac, Wolfgang Pauli, and others. The theory that emerged is the modern *quantum theory,* a complex but powerful theory which now permits us to give extremely precise descriptions for all types of atomic and molecular phenomena.

Probability in Quantum Theory

In proceeding from the original Bohr model of the hydrogen atom to the modern quantum description, we forego completely the idea of *orbits.* In its place we have the *probability* interpretation of the electron wave (see Section 17-3). It is important to understand what this idea means for atomic electrons. Suppose that we prepare a sample of hydrogen atoms in the 1S state. (This state is the normal or ground state.) Next, we make a series of measurements to determine the distance of the electron from the nucleus for particular atoms. We do not find a single well-defined distance. (The Bohr theory predicts the orbit size to be 0.53 Å.)

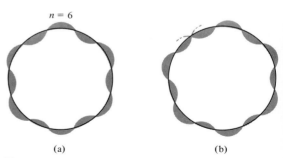

n = 6

(a)   (b)

Figure 18-6   (a) If an integer number of wavelengths of the de Broglie electron wave exactly fits into a Bohr orbit, the wave reinforces itself and persists. (b) If the wave does not exactly fit into the orbit, destructive interference causes the wave to cancel.

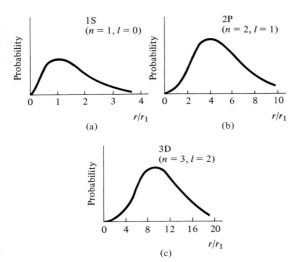

Figure 18-7 The probability of finding the electron in the hydrogen atom at various distances from the nucleus. The distances are given in units of the radius r_1 of the first Bohr orbit.

Instead, we find a range of values (Fig. 18-7a). The *most probable* result is 0.53 Å, but there is an appreciable probability for finding both larger and smaller values. The "orbits" of electrons in atoms are fuzzy things.

Figure 18-7 shows the probability curves for the 1S, 2P, and 3D states for the hydrogen atom. The distances are given in units of the radius r_1 of the first Bohr orbit (0.53 Å). According to the Bohr model, the radii for the $n = 1$, $n = 2$, and $n = 3$ orbits should be r_1, $4r_1$, and $9r_1$, respectively (see Eq. 18-8). As we have already noted, the maximum probability for the 1S state does occur at r_1. Moreover, the maxima for the 2P state occurs at $4r_1$ and that for the 3D state occurs at $9r_1$. But in each case the electron wave has no well-defined distance from the nucleus. There are no electron "orbits" of the type that exist for the motion of planets around the Sun.

The probabilistic interpretation of the behavior of atoms was the key idea that finally and completely divorced the quantum theory of matter from the classical concepts of Newton. At this point, only one additional refinement remained to be incorporated into the theory. When relativity theory was joined with quantum theory and the equations of electrodynamics were included, the last discrepancies between theory and experiment disappeared. So successful is today's relativistic quantum electrodynamics that the most precise experimental results (10 significant figures) can be duplicated by theoretical calculations. Indeed, relativistic electrodynamics is the most "perfect" scientific theory ever devised.

Although we now have at our disposal a powerful theory for describing atomic and molecular matter, precise calculations can be carried out only for the simpler systems. It is one thing to have a theory—to apply the theory in particular cases is an entirely different matter. The properties of elementary particles and the simpler atomic and molecular systems can be described with high precisions by modern quantum theory. But to make precise calculations for complicated molecules is beyond our present abilities. The theory is not at fault. The problem is our ability to make calculations—even our largest electronic computers are not sufficient. In these cases it is necessary to resort to approximate methods and the results of such calculations do not represent the full capabilities of the theory.

It must be emphasized again that quantum theory does not invalidate Newtonian theory. In the realm of atoms and molecules, quantum theory is necessary to interpret the behavior of matter. But as the system under study becomes larger, the effects of quantum phenomena become less apparent. When we reach the size of everyday things, quantum theory gives way to Newtonian theory. In the macroworld of large-scale objects, Newton's principles provide a correct description of the way Nature behaves.

18-3 COMPLEX ATOMS AND THE PERIODIC TABLE

The Need for a New Principle

Between 1869 and 1871 the Russian chemist, Dmitri Mendeléev (1834–1907) published a series of articles in which he advanced his views on the way to arrange the known chemical elements in a form that emphasizes their similarities and differences. Mendeléev's chart, in which he purposely left blanks for elements undiscovered at the time, we now call the *periodic table of the elements*. In this table, shown in Fig. 18-8, the elements are arranged according to *groups* and *periods*. Each *group* (or vertical column) contains elements with similar chemical properties; for example, the alkali metals—lithium, sodium, potassium, rubidium, and cesium—fall into Group I. Each *period* (or horizontal row) terminates with an inert, monatomic gas—these are the noble gases: helium, neon, argon, krypton, xenon, and radon.

The underlying reason for the obvious regularity in the periodic table was unknown in Mendeléev's day, and it remained unknown until the development of quantum theory in the 1920's. Even when the importance of the four quantum numbers was realized, there still was no fundamental understanding of the periodic behavior of the chemical elements.

The solution to the problem was provided by the German theorist Wolfgang Pauli (1900–1958). The key was a simple but profound point first stated by Pauli in 1925:

> No two electrons in an atom can have identical sets of quantum numbers.

Thus, if one of the electrons in an atom has a particular set of quantum numbers (for example, $n = 2$, $l = 1$, $m_l = 0$, $m_s = -\frac{1}{2}$), no other electron in that atom can have exactly the same set. (If another electron also has $n = 2$, $l = 1$, and $m_l = 0$, then it must have the other

Periodic table of the elements.

Period	Group I	II						Transition elements							III	IV	V	VI	VII	VIII
1	1 H 1.00797																			2 He 4.0026
2	3 Li 6.939	4 Be 9.0122													5 B 10.811	6 C 12.01115	7 N 14.0067	8 O 15.9994	9 F 18.9984	10 Ne 20.183
3	11 Na 22.9898	12 Mg 24.312													13 Al 26.9815	14 Si 28.086	15 P 30.9738	16 S 32.064	17 Cl 35.453	18 Ar 39.948
4	19 K 39.102	20 Ca 40.08	21 Sc 44.956	22 Ti 47.90	23 V 50.942	24 Cr 51.996	25 Mn 54.9380	26 Fe 55.847	27 Co 58.9332	28 Ni 58.71	29 Cu 63.54	30 Zn 65.37			31 Ga 69.72	32 Ge 72.59	33 As 74.9216	34 Se 78.96	35 Br 79.909	36 Kr 83.80
5	37 Rb 85.47	38 Sr 87.62	39 Y 88.905	40 Zr 91.22	41 Nb 92.906	42 Mo 95.94	43 Tc (99)	44 Ru 101.07	45 Rh 102.905	46 Pd 106.4	47 Ag 107.870	48 Cd 112.40			49 In 114.82	50 Sn 118.69	51 Sb 121.75	52 Te 127.60	53 I 126.9044	54 Xe 131.30
6	55 Cs 132.905	56 Ba 137.34	57–71 *	72 Hf 178.49	73 Ta 180.948	74 W 183.85	75 Re 186.2	76 Os 190.2	77 Ir 192.2	78 Pt 195.09	79 Au 196.967	80 Hg 200.59			81 Tl 204.37	82 Pb 207.19	83 Bi 208.980	84 Po (210)	85 At (210)	86 Rn (222)
7	87 Fr (223)	88 Ra (227)	89–103 †	(104)	(105)															

*Lanthanide elements

57 La 138.91	58 Ce 140.12	59 Pr 140.907	60 Nd 144.24	61 Pm (145)	62 Sm 150.35	63 Eu 151.96	64 Gd 157.25	65 Tb 158.924	66 Dy 162.50	67 Ho 164.930	68 Er 167.26	69 Tm 168.934	70 Yb 173.04	71 Lu 174.97

†Actinide elements

89 Ac (227)	90 Th 232.038	91 Pa (231)	92 U 238.03	93 Np (237)	94 Pu (242)	95 Am (243)	96 Cm (245)	97 Bk (249)	98 Cf (249)	99 Es (254)	100 Fm (252)	101 Md (256)	102 No (254)	103 Lw (257)

Key:

26 — Atomic number (Z)
Fe — Element symbol
55.847 — Atomic mass of the naturally occurring isotopic mixture; for the elements that are naturally radioactive, the numbers in parentheses are mass numbers of the most stable isotopes of these elements.

Figure 18-8 Periodic table of the elements. All elements with atomic number Z greater than 83 are radioactive and have no stable isotopes. The elements 104 and 105 have not yet been named. Element 101 (mendeleevium) is named for Dmitri Mendeléev.

allowed value of m_s, namely, $m_s = +\frac{1}{2}$.) Pauli's idea is known as the *exclusion principle*.

Atomic Shell Structure

How does the exclusion principle affect the structure of atoms? First, recall the previous statement that the principal quantum number n determines the energy of the atomic electron. Actually, the energy also depends to some extent on the value of the angular momentum quantum number l. (But the values of m_l and m_s have practically no influence on the energy.) Figure 18-9 shows a schematic energy diagram for the various atomic states. It is evident that the energy of a state depends largely on the value of n: the $n = 2$ states lie above the $n = 1$ state (that is, the $n = 2$ states have greater energy than the $n = 1$ state); the $n = 3$ states lie above the $n = 2$ states; and so forth. But the diagram also indicates that the states with a particular n have somewhat different energies depending on whether they are S, P, D, or F states. Each small square in the

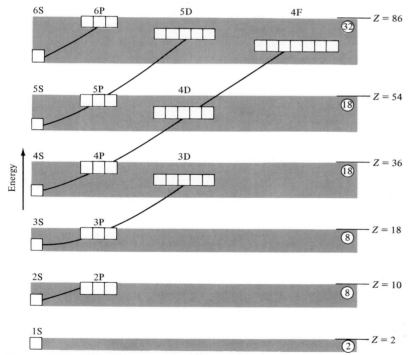

Figure 18-9 Schematic energy diagram for the various electron states. Each small square can accommodate two electrons (spin-up, spin-down). The total number of electrons in each shell is shown in the circle at the right, and the atomic number Z at shell closure is given at the right. The sloping lines connect states with the same value of n.

Dmitri Mendeléev (1834–1907), who first proposed the correct way to arrange the elements into a periodic table. Mendeléev was a Siberian and he received his first instruction in chemistry from a political prisoner. He finished his university training in 1855, at the top of his class. In 1866 he became professor of chemistry at St. Petersburg. Mendeléev was one of the most forward-looking chemists of the 19th century, and for his work on the systematization of the chemical elements the Royal Society of London awarded him the Davy medal in 1882.

Sculpture based on Bohr's model of the sodium atom, as modified by Arnold Sommerfeld to include elliptical orbits. The large orbit is that of the 3S valence electron.

diagram represents a magnetic substate. There is only one such substate for $l = 0$ $(m_l = 0)$; there are three substates for $l = 1 (m_l = -1, 0, +1)$; and so forth. Two electrons can occupy each substate (spin *up, spin down*).

Let us begin to "build" the elements by adding electrons one at a time, following the energy scheme represented by Fig. 18-9. (Of course, we also add positive charge to the nucleus so that every atom is electrically neutral.) Each new electron is always to be placed in the position of lowest available energy. In this way we ensure that we consider only the *ground states* of the various atomic species.

The first atomic electron occupies the $n = 1$, $l = 0$ state—that is, the 1S state. Because two electrons can be accommodated in this state, the second electron can also be placed in the 1S state. This accounts for the first two elements, hydrogen and helium, $Z = 1$ and $Z = 2$. Two electrons completely fill the 1S state—the exclusion principle prevents additional electrons from having the quantum numbers of this particular state. Therefore, the third electron must be placed in the next higher state, the 2S state. But notice that there is a substantial energy difference (an *energy gap*) between the 1S and 2S states. That is, two electrons fill the first electron *shell* and additional electrons are in a different and higher energy region.

The first electron shell (which is called the *K shell*) is completed with the element helium $(Z = 2)$ and the second shell (the *L shell*) begins with lithium $(Z = 3)$. Eight electrons can be accommodated in the L shell—two in the 2S state and two in each of the three 2P substates. Another energy gap separates the 2P states and the 3S states. Thus, the L shell closes with $Z = 10$, the inert gas neon.

If we were to continue this scheme, we would expect that the third shell contains all of the electrons with $n = 3$. But this is not the case. As shown in Fig. 18-9, the next energy gap occurs, not between the 3D and 4S states, but between the 3P and 4S states. Therefore, the third shell contains only eight electrons—two 3S electrons and six 3P electrons—and closes with $Z = 18$, corresponding to the inert gas argon. The electron configurations for the first three shells are shown schematically in Fig. 18-10.

The fourth shell contains two 4S electrons, ten 3D electrons, and six 4P electrons. This shell closes with $Z = 36$, corresponding to the inert gas, krypton. But beginning with the fourth shell, we find a new feature of the periodic table. Between Group II and Group III are ten elements—scandium ($Z = 21$) to zinc ($Z = 30$)— that are assigned to neither group (see Fig. 18-8). These elements are all metals with rather similar properties; they are grouped together and called *transition elements*. The reason for the occurrence of the transition elements can be seen in Fig. 18-9. The first two elements of Period 4 are formed by adding 4S electrons, and the last six elements are formed by adding 4P electrons. In between (in terms of energy) there are the ten electron positions corresponding to the 3D states. As

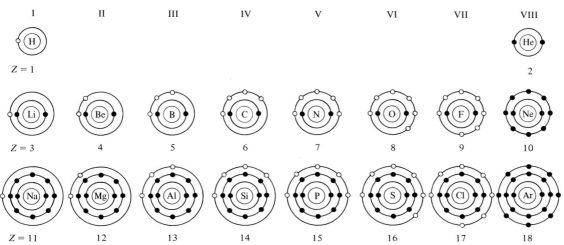

Figure 18-10 Electron configurations in the first three shells. Electrons in the shell that is being filled are indicated by open circles; those in the filled shells are indicated by black dots. The K shell can accommodate two electrons, whereas the next two shells can each accommodate eight electrons. Such a diagram is only schematic; electrons do not exist in well-defined "orbits."

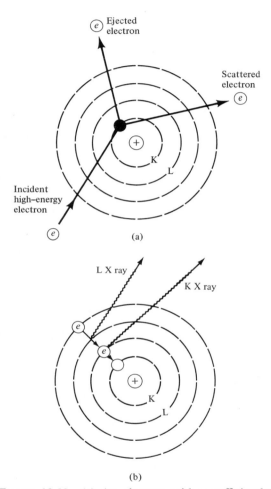

Figure 18-11 (a) An electron with a sufficiently high energy can eject an inner electron from an atom with a high atomic number. (b) An electron from the L shell makes a transition to fill the vacancy in the K shell and a K X ray is emitted. Subsequently, an electron from a higher shell makes a transition to fill the L-shell vacancy and an L X ray is emitted.

the 3D states are filled, the transition elements are formed. This same phenomenon occurs in the higher shells as well. In Periods 6 and 7 there occurs another type of "back-filling" of a passed-over set of states which crowds a series of similar elements into a single position at the beginning of the transition elements. These elements are known as the *lanthanide series* and the *actinide series,* respectively. The lanthanides are sometimes called the *rare-earth* elements, but they are neither rare nor earthlike. The actinides are all radioactive (as are all elements with atomic number greater than 83); some of these elements occur naturally in the Earth but others must be produced artificially.

18-4 X RAYS

Inner-Shell Transitions

The energy required to ionize a hydrogen atom is 13.6 eV. An even smaller energy (7.4 eV) is sufficient to remove the outermost electron from an atom of lead ($Z = 82$). But if we attempt to remove one of the inner-shell electrons from a lead atom, we find that a considerably higher energy is required. In fact, the removal of a K-shell electron from lead can only be accomplished by expending an energy of 88 000 eV (88 keV). The reason for this vast difference between the energies by which the two electrons in lead are bound is that the electrical force exerted by the positively charged nucleus on the inner electron is much greater than that exerted on the outer electron. This is due, in part, to the fact that the outer electron is at a much greater distance from the nucleus than is the inner electron (and the force varies as $1/r^2$). But in addition, all of the electrons that lie between the outermost electron and the nucleus tend to cancel the electrical effect of the nucleus on the outermost electron. That is, the outermost electron experiences a force due to a much reduced positive charge. (The nuclear charge is partially *shielded* by the inner electrons.)

The large amount of energy required to remove an inner electron from an atom with a high atomic number can be provided by the collision of a high-energy electron. Thus, if an electron is accelerated through a potential difference in excess of 88 000 volts, it will have sufficient energy to eject a K-shell electron from a lead atom (Fig. 18-11a). The atom then becomes an *ion*, and the process is called *K-shell ionization*.

The removal of a K-shell electron leaves a vacancy in the shell. This condition is energetically unfavorable—the ion is in an *excited state* and it will rapidly adjust itself to a lower energy situation. It is most likely that the ion will accomplish this by a process in which one of the L-shell electrons makes a transition to fill the K-shell vacancy (Fig. 18-11b). In this process energy is released in the form of an energetic photon or *X ray* (called a *K X ray* in this case). Now there is a vacancy in the L shell and this vacancy is filled by an electron making a transition from a higher shell. An additional X ray (an *L X ray*) is emitted in this process. Thus, following a K-shell ionization several X rays are emitted as electrons cascade down to fill the lower-shell vacancies. Eventually, the ion captures an electron from its surroundings and returns to an electrically neutral condition.

X-Ray Photography—How It Works

Ordinary optical radiation (visible light) is reflected or absorbed by quite thin layers of most materials and is transmitted by only a restricted class of substances—for example, glass, water, and certain plastics. X radiation, on the other hand, being of much higher energy than visible light, can penetrate all types of materials. Because X rays interact with atomic electrons, the depth of X-ray penetration depends upon the density of electrons in the material. If the electron density is high (as it is, for example, in lead), the penetration will be slight; a thin sheet of lead will stop most X rays. If the electron density is low (as it is, for example, in plastic materials or biological tissue), the penetration is considerably greater. For these reasons, X radiation has found important uses in examining the internal structures of many different types of objects in medical as well as industrial situations.

A schematic diagram of an X-ray tube is shown in Fig. 18-12. The source of electrons is a heated coil of wire that is attached to the negative terminal of a high voltage supply. (The coil is heated in order to facilitate the release of electrons.) The positive terminal of the supply is connected to the target (or anode). Therefore, electrons released at the cathode are accelerated toward the anode. When they strike the target, the electrons have an energy in eV numerically equal to the voltage between the cathode and the anode. If a 100 000-V supply is used, the electrons will have an energy of 100 keV when they strike the target.

The X rays that are produced when the energetic electrons strike the target are allowed to be incident on the object being studied—for example, an arm

Wilhelm Roentgen (1845–1923), the discoverer of X rays, showed an X-ray photograph of his own hand when he announced his discovery before the Physical Medical Society of Wurzburg on December 28, 1895.

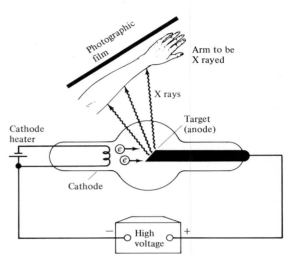

Figure 18-12 Schematic diagram of an X-ray tube. The cathode (the source of electrons) and the anode (the target) are located within an evacuated tube. Because large electron currents are frequently used, the anode must be a massive block of metal (often water cooled) in order to withstand the high temperatures.

in which a bone is broken, as indicated in Fig. 18-12. The X rays readily pass through the fleshy material of the arm and are registered on the photographic film placed behind the arm. Some of the X rays must pass through the bones of the arm before reaching the film. Because the density of atomic electrons is higher in bony material than in tissue and muscle, many of the X rays are absorbed in the bones. Consequently, fewer X rays reach the film immediately behind the bones and these areas appear sharply distinguished when the film is developed. X-ray photographs of this type are of great assistance to the physician in determining the proper method to use in setting broken bones or in probing for foreign matter in the body.

Although X-ray techniques are of great value in many medical applications, high energy X rays are capable of producing significant biological damage and must be used with great care. (See Section 21-4.)

X-ray techniques are also used for a variety of industrial purposes. Voids in metallic castings and imperfections in welds can be detected by energetic X rays. It is standard procedure, for example, to X ray various parts of aircraft structures after extended use in order to determine whether there are any areas that may have a tendency to fail because of excessive fatigue.

18-5 LASERS

Stimulated Emission

An atom in its ground state can be raised to a higher energy state by the absorption of a photon with an energy equal to the energy difference between the states. This process—called *excitation*—is illustrated in Fig. 18-13a. An atom that is in an excited energy state will spontaneously emit a photon and return to the ground state in a *deexcitation* process, as shown in Fig. 18-13b. In each case the photon energy is $h\nu = \mathscr{E}_1 - \mathscr{E}_0$.

The quantum description of these two processes is identical—the only important consideration is the fact that a transition between $\mathscr{E}_0$ and $\mathscr{E}_1$ takes places. It does not matter whether the transition is one of excitation or deexcitation. What will happen, then, if a photon of energy $h\nu = \mathscr{E}_1 - \mathscr{E}_0$ is incident on an atom that is in the energy state $\mathscr{E}_1$? The photon cannot *excite* the atom because it is already in the excited state. Therefore, the photon produces the equivalent effect, namely it *deexcites* the atom. This process is called *stimulated* emission, and is shown schematically in Fig. 18-14.

There is a significant and important difference between stimulated and spontaneous radiation. In the case of spontaneous radiation, the photons are emitted in random directions, as are the photons from all ordinary light sources. A stimulated photon, on the other hand, leaves the atom in the same direction as the incident photon. Furthermore, the incident photon forces the stimulated photon to oscillate in conformity with its own oscillations. That is, the stimulated photon is *in phase* with the incident photon (Fig. 18-14). Due to the addition of the amplitudes of the two in-phase photons, the light intensity in the direction of the incident photon is increased because of the stimulated emission. A device which makes use of this effect for optical radiation is called a *laser*. (This name is an acronym for *light amplification by the stimulated emission of radiation*.)

Optical Pumping

How can we produce a collection of atoms which are in the appropriate excited energy state so that laser action can be initiated by an incident photon? The atoms can be excited by irradiating them with photons of energy $h\nu = \mathscr{E}_1 - \mathscr{E}_0$. If we use a conventional source of *white* light (light of all frequencies) for this purpose, then only

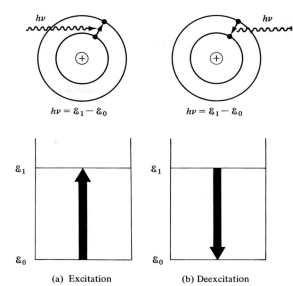

$$h\nu = \mathscr{E}_1 - \mathscr{E}_0 \qquad h\nu = \mathscr{E}_1 - \mathscr{E}_0$$

(a) Excitation   (b) Deexcitation

Figure 18-13   (a) Excitation of an atom from the ground state $\mathscr{E}_0$ to an excited state $\mathscr{E}_1$ by a photon of energy $h\nu = \mathscr{E}_1 - \mathscr{E}_0$. (b) Deexcitation of an atom in a state $\mathscr{E}_1$ by the emission of a photon of energy $h\nu = \mathscr{E}_1 - \mathscr{E}_0$.

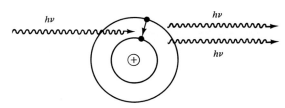

Figure 18-14   An atom in an excited state can be stimulated into radiating by the incidence of a photon of the proper frequency. The incident photon and the stimulated photon leave the atom in the same direction and in phase.

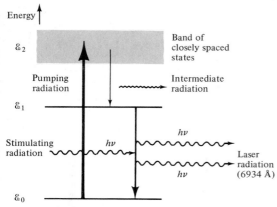

Figure 18-15 Simplified energy diagram for ruby. White light pumps energy into the band of states $\mathscr{E}_2$. A transition to the state $\mathscr{E}_1$ follows, and then laser action can be initiated by a photon with energy $h\nu = \mathscr{E}_1 - \mathscr{E}_0$. The emitted radiation ($\lambda = 6934$ Å) is visible red light. In other types of lasers (particularly those that operate continuously instead of in pulses as does the ruby laser), the laser radiation is emitted between states that are both excited states—that is, the laser transition does not connect to the ground state as in this diagram.

a tiny fraction of the photons will have the correct frequency to excite the atoms. For a practical laser, we need a much higher excitation efficiency.

Some materials have the property that they possess large numbers of excited energy states so closely spaced that the states form a continuous *band*. One such material is ruby, which consists of aluminum oxide with a small amount of chromium as an impurity. (Pure aluminum oxide is colorless; it is the presence of the chromium impurity that gives to ruby its characteristic red color.) Figure 18-15 shows a simplified energy diagram for ruby, in which the band of states is labeled $\mathscr{E}_2$. This band has sufficient width that when white light is incident on ruby, there is an efficient pumping of energy into the band by absorption. When excited into the band $\mathscr{E}_2$, ruby does not simply reradiate photons of the same energy and return to the ground state. Instead, a lower energy transition takes place which leaves the atoms in the state $\mathscr{E}_1$. This is the state which exhibits laser action. If a photon with energy $h\nu = \mathscr{E}_1 - \mathscr{E}_0$ is incident on an atom in the state $\mathscr{E}_1$, stimulated emission will occur and two photons with this energy will be emitted (Fig. 18-15).

Laser Construction

Finally, we need a method for channeling the stimulated radiation into a narrow beam, for in this way we can make maximum use of the radiation. This problem has been solved in the following way. As shown in Fig. 18-16, the laser material (in this case, ruby) is formed into a long cylinder which is encircled by a flash lamp that produces the pumping radiation. One end of the ruby cylinder is coated with a metallic film and acts as a mirror. The opposite end receives only a thin metallic coating and becomes a partially reflecting mirror. That is, some of the light incident on this end is reflected and the remainder is transmitted.

When the lamp is flashed by sending through it a sudden surge of current from the power supply, many of the atoms are pumped into the band $\mathscr{E}_2$ and then make the transition to the state $\mathscr{E}_1$, the state that exhibits laser action. The atoms will not remain indefinitely in this state; in fact, spontaneous radiation will deplete this state following the same kind of random emission that is found in radioactive decay (see Section 20-1). But as soon as a few atoms have radiated spontaneously, stimulated emission begins. The spontaneous photons can be

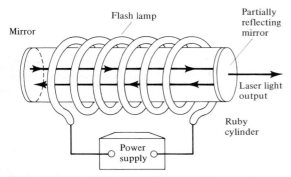

Figure 18-16 Schematic diagram of a ruby laser. The laser beam emerges from the end that has been coated to form a partially reflecting mirror.

emitted in any direction, and those that move toward the cylinder walls will be lost. But a few of the photons will be emitted along the cylinder axis and these are responsible for initiating the laser action. Each photon will stimulate the emission of other photons and these photons in turn will stimulate still more photons. Because of reflections at the ends of the cylinder, the multiplication process is further enhanced, with each photon having many opportunities to stimulate additional radiation. In this way the excited atoms, instead of radiating spontaneously, are stimulated to release their energy rapidly in the form of in-phase radiation that is directed along the cylinder axis. This radiation emerges in a narrow beam from the end that is partially transmitting and constitutes the *laser beam*.

All of the photons that are emitted in the laser process necessarily have the same energy. Thus, a laser beam is not only highly directional but it also has a pure frequency. In the case of the ruby laser, the light has the single wavelength, $\lambda = 6934$ Å, in the red part of the visible spectrum.

Laser Applications

Light from an ordinary source consists of spontaneous photons that are emitted in random directions. Laser photons, on the other hand, are emitted in a narrow beam that retains its small size even though it travels a substantial distance through space. For this reason, lasers are frequently used in situations that involve critical alignment problems, such as surveying over large distances. The problem of making certain that a tunnel dug from opposite sides of a mountain will actually meet in the middle can be much more easily solved by using surveying lasers than by conventional methods.

A spectacular application of lasers is in a continuing experiment to measure the distance from the Earth to the Moon. In this experiment a short burst of laser radiation is projected through a telescope that is directed at the Moon. The pulse travels to the Moon where it strikes a special reflector that was placed on the Moon's surface by the Apollo 11 astronauts in 1969. A portion of the initial pulse is reflected toward the Earth and is viewed by a highly sensitive photoelectric device at the focus of the telescope. The velocity of light is known with precision. Therefore, by accurately measuring the time for the laser pulse to make the round trip to the Moon (approximately 2.6 s), the Earth–Moon distance

HUGHES AIRCRAFT

A high-intensity laser beam strikes a metal plate and produces a shower of sparks.

The following warning appeared in *Notices to Airmen* beginning in August 1969 to alert pilots to the possible danger of flying through the lunar ranging laser beam when operating near the Earth station at McDonald Observatory in the mountains of western Texas.

SPECIAL NOTICE: Extensive Laser operns will be conducted for an indefinite period from the McDonald Observatory located at 30°40'17''N, 104°01'30''W near Marfa VOR in conjunction with a scientific moon project. Pilots should avoid flying from surface to FL 240 within a rectangular area bounded by lines 4NM N and 10NM S of an E/W line through the location of the McDonald Observatory and 13NM E and 13NM W of a N/S line through the location of the McDonald Observatory. Permanent eye damage may result if a person is exposed to the Laser beam. Hrs of opern may be obtained by contacting El Paso, Marfa, Salt Flat, Wink, Midland, Pecos, Ft Stockton, Cotulla or Rock Springs Rdo and Albuquerque ARTCC. The location of the Observatory is further described as being on the 340° rad 22.5NM NNW of Marfa VOR. (8–69)

Abbreviations:
VOR = VHF (very high frequency) omnidirectional range (radio navigation transmitter)
FL 240 = Flight level 240 = 24 000 ft
NM = nautical mile
ARTCC = Air Route Traffic Control Center
rad = radial (direction)

can be determined. The uncertainty in this result at present is about 0.15 m, and it is expected that in the near future the accuracy can be improved further. The measurement of the changes in the Earth–Moon distance will provide important clues regarding the structure of the Moon and the Earth.

In addition to basic research activities, such as studies of the interaction of radiation and matter and the investigation of the physical properties of the Moon, lasers have found wide application in a variety of technical situations. High-power lasers are used to weld materials that resist other methods, and microholes can be drilled into even the hardest substances by laser beams. It has been found that a laser can deposit just the right amount of energy to "weld" a detached retina in an eye onto the choroid surface that lies beneath it. This technique was first used in the treatment of human patients in 1964 and since then thousands of cases have been treated successfully. Lasers are also used in many other types of minor surgery, such as the removal of warts and cysts.

Laser light has a single, pure frequency. Some types of lasers can be *tuned* to a desired frequency—for example, the exact frequency at which a particular atomic or molecular species will absorb radiation. If such a beam is directed through air that does not contain the absorbing species, a detector placed some distance away will register very little attenuation of the beam. However, if the air does contain absorbing molecules, there will be a sharp drop in the intensity of light reaching the detector. Therefore, tunable lasers can be used as extremely sensitive detectors of pollutants in the atmosphere. Undesirable gases in automobile exhaust emissions or in smokestack effluents are detectable even though the concentrations of the offending gases may be only a few ppb (parts per billion). Experiments have shown that it is possible to detect the hydroxyl radical OH (which plays a crucial role in the production of smog) in concentrations down to 1 part in 10^{13} using laser techniques.

Another application of the unique in-phase characteristics of laser light is in the field of *holography*. It has been found possible to record on a single piece of film sufficient information to allow the reconstruction of a *three-dimensional* image of an object instead of the usual two-dimensional (or *flat*) image. By directing a laser beam through the special film, an image can be produced that stands lifelike in space. This technique, still in its infancy, is certain to be widely used in the future, not only in scientific fields but also in the entertainment industry.

A large number of different materials—solids, liquids, and gases—are now known to exhibit laser action. Some can produce continuous beams, whereas others (for example, ruby) must be flashed or pulsed because continuous operation would cause excessive heating. The highest power lasers can produce short bursts of energy at rates in excess of 10^{13} watts!

Obviously, the concentrated power in a laser beam can be dangerous. A laser of moderate power can cause skin burns and even a low-power laser can produce eye damage. Direct exposure of the eye to a 2-milliwatt (0.002 W) laser beam for 1 second is likely to cause a retinal burn. *Never look directly down a laser beam toward the laser.*

To replace the phrase "to exhibit laser action," another new word has entered the English language (see *The American Heritage Dictionary of the English Language,* 1969): "**lase** (lāz) *intr. v.* To function as a laser."

CARTIER

A hologram of a woman's hand holding a diamond necklace was projected from the front window of Cartier, the famous New York jewelry firm, in November 1972. This was one of the first commerical applications of holography. The holographic display was so striking that traffic was stopped on Fifth Avenue. One passerby attacked the image with her umbrella, declaring it to be the "devil's work."

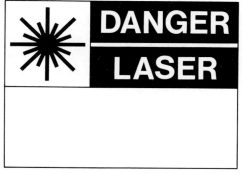

Placard that is posted near operating lasers to warn of possible biological damage, particularly to eyes, due to the concentrated power in a laser beam.

SUGGESTED READINGS

G. Gamow, *Thirty Years That Shook Physics* (Doubleday, Garden City, New York, 1966).

R. E. Moore, *Niels Bohr: The Man, His Science and the World They Changed* (A. A. Knopf, New York, 1966).

Scientific American articles:

G. Gamow, "The Exclusion Principle," July 1959.

A. L. Schawlow, "Laser Light," September 1968.

QUESTIONS AND EXERCISES

1. The hydrogen atom contains only a single electron and yet the hydrogen spectrum contains many lines. Why is this so?

2. What is the velocity of the electron in the $n = 1$ orbit of the hydrogen atom? (Use Eq. 18-6.) Compare your result with the velocity of light. Is it reasonable to use a nonrelativistic expression (such as Eq. 18-6) for this calculation?

3. According to the Bohr theory, the principal quantum number n determines what three physical properties of the atom?

4. White light is incident on a cell containing hydrogen gas. In the transmitted light certain wavelengths will be absent due to the absorption of some of the radiation by hydrogen atoms. These wavelengths will be the same as those in which series of hydrogen spectral lines? (Refer to Fig. 18-3.)

5. How much energy is required to ionize a hydrogen atom that is in the $n = 3$ state? (Refer to Fig. 18-4.)

6. A free electron (with essentially zero kinetic energy) is captured by a free proton into the $n = 2$ orbit. What is the wavelength of the photon that is emitted in this process?

7. Use Eq. 18-10 to compute the energies of the hydrogen atom states with $n = 1$ through 5.

8. Use Eq. 18-13 and compute the wavelengths of the lines of the Lyman series that are shown in Fig. 18-3. (Use the results of the preceding exercise for the energies of the various states.)

9. What is the equivalent energy difference between the mass of a hydrogen atom and the combined mass of a free proton and a free electron?

10. What is the *shortest* wavelength radiation in the Balmer series? (What value of n must be used?)

11. Write expressions analogous to Eq. 18-1 for the frequencies of radiations in the Lyman and Paschen series. (See Fig. 18-3.)

12. A beam of electrons with kinetic energy of 12.5 eV is incident on a sample of hydrogen. What radiations (express the energies in eV) do you expect to see coming from the sample? (Use Fig. 18-4.)

13. Consider the $n = 3$ state in the hydrogen atom. Use Eq. 18-8 and compute the radius and the circumference of the corresponding Bohr orbit. Next, use Eq. 18-6 and compute the velocity of the electron in the $n = 3$ orbit. Finally, compute the de Broglie wavelength of the electron and compare with the circumference of the orbit. What do you find?

14. Derive Eq. 18-9 by proceeding in the following way. The total energy $\mathscr{E}_n$ for the state with principal quantum number n is the sum of the kinetic energy $(\frac{1}{2}mv_n{}^2)$ and the electrical potential energy $(-Ke^2/r_n)$. Substitute for v_n from Eq. 18-6 (with r_n in place of r). Substitute for r_n from Eq. 18-7 to obtain the expression for $\mathscr{E}_n$. Finally, insert the values of the various constants and show that $\mathscr{E}_n = (-13.6 \text{ eV})/n^2$ (Eq. 18-10).

15. Sketch de Broglie wave pictures similar to that in Fig. 18-6a for $n = 4$ and 5.

16. How many electrons are there in the outermost shell of the Group II elements? How many electron vacancies are there in the outermost shell of the Group VI elements?

17. Because all of its isotopes are radioactive, technetium (Tc, $Z = 43$) was one of the last elements to be discovered. What elements are chemically similar to technetium?

18. What are the quantum numbers for the outermost (or *valence*) electron for (a) potassium, and (b) cesium?

19. Hydrogen, which is a Group I element (see Fig. 18-8), is sometimes also listed as a Group VII element. Why is this reasonable?

20. Suppose that electrons have no spin angular momentum and that the Pauli exclusion principle is still valid. What difference would this make in the structure of atoms? Shell closures would occur for which elements in this situation?

21. Some of the ideas and terminology of quantum theory have made their way into other fields. For example, in the August 14, 1972 issue of *Newsweek,* Paul D. Zimmerman, reviewing Woody Allen's movie, *Everything You Always Wanted to Know about Sex but Were Afraid to Ask,* makes the following comment: "Allen as director has deliberately sacrificed laughs for visual quality and, in this respect, he has succeeded all too well, achieving a quantum jump in cinematography at the expense of the breathless, ragtag comedy that distinguished his earlier film *Bananas.*" How do you interpret "quantum jump" in this context?

19

THE
STRUCTURE
OF MATTER

We now turn to a discussion of matter in the bulk state. We have seen the way in which electrons fill the atomic shells and how complex atoms are constructed. The matter that we ordinarily see and use, however, is in the form of large-scale collections of atoms or of atoms that are bound together as molecules. Does bulk matter just *occur?* Or is the structure related to the behavior of electrons, atoms, and molecules? As we will see, we live in a world of large-scale matter that is organized according to microscopic principles. Quantum theory does not apply exclusively to the atomic domain; directly or indirectly, quantum effects influence the behavior of many everyday things and much of everyday life.

In this chapter we will learn how atoms join together in various ways to produce molecules, and we will see how aggregates of atoms form crystalline bulk matter. In addition, we will discuss two of the special forms of bulk matter — semiconductors and superconductors — that are important in modern technology.

19-1 IONIC BONDS AND IONIC CRYSTALS

Types of Atomic Bonds

The binding together of atoms to form molecules is accomplished in two general ways. The first mechanism involves the transfer of an electron from one atom to another so that the resulting ions are electrically attracted to one another—this is called *ionic bonding*. The second type of bonding involves the sharing of electrons by two atoms. Because a pair of electrons is a part of the electron structure of each atom, the atoms are bound together—this is called *covalent bonding*.

Some types of molecules are formed predominantly by ionic bonds and others are formed mainly by covalent bonds. But, in reality, all molecules have a mixed ionic–covalent character. In general, *inorganic* compounds (minerals, salts) tend to be ionic, whereas *organic* compounds (compounds found in living matter) tend to exhibit covalent bonding. We shall discuss these two types of bonds in turn.

Electron Transfer

We know that two objects which carry opposite electrical charges will attract each other. Therefore, if we remove an electron from an atom A (thereby forming the ion A^+) and then attach this electron to an atom B (thereby forming the ion B^-), the two ions will experience an electrical attraction. If A^+ and B^- remain permanently ionized under their mutual attraction, they will be electrically bound together as an *ionic molecule*.

Under what conditions will this electron transfer process produce an ionic binding between atoms? To answer this question we must recall the structures of the outermost electrons in atoms. First, if an electron is removed from an atom (by whatever means), then the remaining electrons will always spontaneously readjust themselves into the configuration of lowest possible energy. (Compare the way in which X rays are produced—Section 18-4.) This means that a positive ion in its stable state will always have a vacancy in the position occupied by the outermost electron in the normal atom. For example, the outermost electron in a sodium atom is a 3S electron and it is this electron that is missing in the sodium ion, Na^+ (see Fig. 19-1). The removal of the 3S electron from a sodium atom requires the expenditure of 5.1 eV of energy. That is,

$$Na + 5.1 \text{ eV} \longrightarrow Na^+ + e^-$$

To remove an electron from an atom with a *closed* outer shell requires considerably more energy; for example, to produce the ion Ne^+ requires 21.5 eV. In fact, as we proceed from the Group I elements (of which sodium is a member) to the Group VIII elements (of which neon is a member), the ionization energy steadily

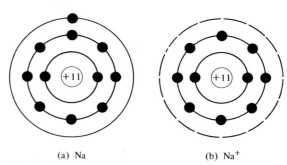

(a) Na   (b) Na^+

Figure 19-1 (a) Electron configuration of the normal sodium atom. (b) In the sodium ion the outermost (3S) electron is missing.

increases. The atoms that can easily be converted into positive ions and which will therefore participate in forming ionic molecules are located on the left-hand side of the periodic table, in Groups I and II.

What elements are likely candidates for the production of *negative* ions? If we attempt to attach an electron to a Group VIII element, the extra electron must be placed into a new shell outside the closed shell. In such a position the electron is far from the nucleus and experiences very little attractive force. Negative ions of the inert gases are therefore difficult to produce. But if we examine the neighboring Group VII elements we find one position in the outer electron shell that is vacant. These elements have an affinity for electrons because the addition of a single electron to an atom produces a closed outer shell and this is an energetically favorable situation. Energy is *released* in the formation of these negative ions; for example,

$$Cl + e^- \longrightarrow Cl^- + 3.7 \text{ eV}$$

Let us consider now the formation of Na^+ and Cl^- ions by removing an electron from a sodium atom and attaching it to a chlorine atom. We can express this as

$$Na + Cl \longrightarrow Na^+ + Cl^-$$

An energy of 5.1 eV has been expended in producing the Na^+ ion and 3.7 eV of energy has been recovered by using the electron to form the Cl^- ion. Therefore, there has been a net energy expenditure of 5.1 eV $-$ 3.7 eV $=$ 1.4 eV. Where does this energy come from? If we have two oppositely charged objects a certain distance apart, the system possesses electrical potential energy. If we allow the mutual attraction to pull the charged objects toward one another, some of the potential energy will be converted to another form of energy. For example, if the objects are simply released, they will begin to move, thereby converting potential energy into kinetic energy.

In the case of the Na^+ and Cl^- ions, the 1.4 eV necessary to effect the electron transfer is obtained from the electrical potential energy associated with the two ions. We can see this in the graph of Fig. 19-2 which shows the potential energy of the Na^+ and Cl^- ions as a function of their separation. An amount of work equal to 1.4 eV is performed in transferring the electron from the sodium atom to the chlorine atom. Therefore, the system possesses a potential energy of 1.4 eV when $r = \infty$. As the two ions are brought closer together, the potential energy decreases. When a separation of 10 Å

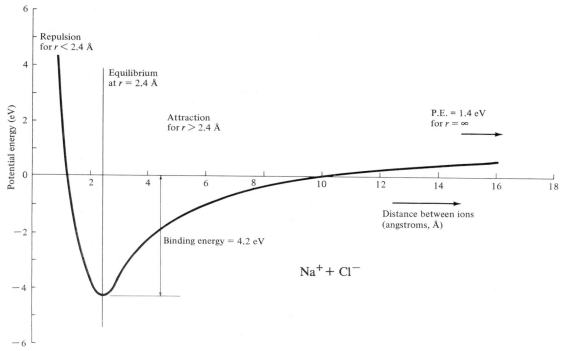

Figure 19-2 Electrical potential energy of a Na$^+$ ion and a Cl$^-$ ion as a function of the distance between the ions. At the equilibrium separation of 2.4 Å, the binding energy is 4.2 eV. (The *binding energy* is the energy that must be supplied in order to separate the system into neutral atoms.)

(10^{-9} m) is reached, the potential energy is zero; that is, at this separation the energy required to form the two ions has been recovered. But there is still an attractive force between the ions and they continue to approach one another. As the separation decreases further, the binding between the ions becomes greater. At a separation of 2.4 Å, the binding energy is 4.2 eV (see Fig. 19-2). However, if the separation is made less than 2.4 Å, the force between the ions actually becomes *repulsive*. Thus, $r = 2.4$ Å represents the equilibrium separation of the ions.

Why do the ions not approach closer than 2.4 Å? There are two reasons. First, when the separation becomes less than about 3 Å, the electron structures of the ions begin to overlap. The nuclei are then no longer electrically shielded from one another and they begin to exert a mutual repulsive force. In addition, for small separations, the electron structures coalesce and form one single electron system instead of two independent systems. The Pauli exclusion principle then applies to *all* of the electrons together and not merely to the two individual electron systems. This has the effect of forcing some of the electrons into new and higher energy states, thereby absorbing some of the electrical potential energy and decreasing the binding energy. If the separation distance is decreased to a value smaller than about 1 Å, the ions no longer represent a bound system (see Fig. 19-2).

A diagram such as that in Fig. 19-2 indicates what we mean by the *size* of an atom—or, in this case, an *ion*. At equilibrium, the nuclei of the sodium and chlorine ions are separated by 2.4 Å. This distance is the sum of the radii of the Na^+ and Cl^- ions. By studying other compounds, we can find values for different sums of radii and we can eventually work out the individual radii. The radius of the Cl^- ion is found to be 1.8 Å, and that of the Na^+ ion is 0.6 Å. Notice that these values are not far different from the radius of the hydrogen atom (0.53 Å) which we calculated from the crude Bohr model in Section 18-1. In fact, the radii of all ions and all atoms are in the range from 0.5 to 2.5 Å.

Ionic Solids

In our discussion thus far we have indicated that two isolated ions, Na^+ and Cl^-, will bind together to form the ionic molecule Na^+Cl^-. Actually, this is not strictly true. In Nature we do not find isolated Na^+Cl^- molecules. Instead, we find sodium chloride in solid form. In all solid matter, the atoms are packed closely together and are bound by electrical forces into more-or-less permanent positions with respect to one another. If we examine matter at the atomic level, we find that some substances, such as *glass,* consist of atoms in a random, disorderly array. From the atomic standpoint, glass is a highly disorganized substance. (In fact, glass is actually a supercooled liquid and possesses all of the random characteristics of a normal liquid.) Materials such as glass are said to be *amorphous;* there is no regular pattern for the arrangement of the atoms in these substances. On the other hand, the examination of sodium chloride (the mineral *halite* or common table salt) with only a magnifying glass will reveal the tiny cubic *crystals* that characterize this substance. The individual atoms that make up a sample of solid sodium chloride are arranged in a regular, repeating pattern so that the bulk material has a characteristic cubic shape.

Most solids have crystalline structures. Some materials, such as quartz and diamond, occur naturally as large single crystals. It is much more common, however, to find crystals of extremely small size. A bar of iron, for example, does not appear to be crystalline. But when the surface is cleaned with acid and viewed under a microscope, the iron is seen to consist of many *microcrystals* of various sizes.

A pure crystalline substance has a definite chemical composition—quartz, for example, is always SiO_2 and halite is always NaCl. But how is a crystal formed? In a very real sense, a crystal *grows* by solidifying from the molten state or from a solution. Starting with the tiniest bit of the material, the sample increases in size by attaching to itself layer after layer of additional atoms. In this process the atoms are arranged in an orderly geometrical way with respect to one another. Each successive layer that solidifies follows exactly the same pattern. Consequently, all crystals of the same substance have the same basic shape. The basic arrangement of atoms in sodium chloride is that of a cube. When many of these cubes are assembled to form a bulk sample of the material, the crystal has an overall cubic or rectangular appearance, as shown in the photograph.

The atomic reason for the crystalline structure of such substances as sodium chloride was first proposed in 1898 by William Barlow who visualized the NaCl crystal as a cubic arrangement of tightly packed ball-like atoms. Modern experiments have shown that Barlow's scheme is essentially correct. But we now know that the basic units in the crystal are *ions,* not *atoms.* The oppositely charged ions attract one another and bind the crystal together. The arrangement of Na^+ and Cl^- ions in a crystal of sodium chloride is shown in Fig. 19-3. Notice that each sodium ion is surrounded by 6 chlorine ions and that each chlorine ion is surrounded by 6 sodium ions. This is a particularly stable arrangement of the ions and accounts for the tightly bound cubic structure of the crystal.

There exist many types of ionic crystalline substances in addition to sodium chloride. Any of the Group I elements (of which sodium is a member) can combine with any of the Group VII elements (of which chlorine is a member) to form an ionic compound. Some of these compounds are sodium bromide (NaBr), potassium fluoride (KF), and cesium chloride (CsCl). The crystal structure of cesium chloride is shown in Fig. 19-4. Notice that although this crystal is cubic, it has a form that is different from that of sodium chloride (Fig. 19-3). The

Crystals of natural halite, formed on the basic cubic pattern of sodium chloride.

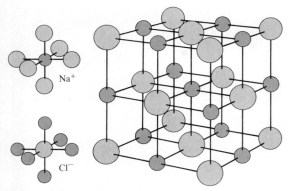

Figure 19-3   Schematic arrangement of Na⁺ and Cl⁻ ions in a crystal of sodium chloride. Each ion is surrounded by 6 ions of the other type. (The sizes of the ions relative to their separations have been reduced in order to show clearly the lattice structure of the crystal.)

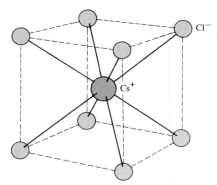

Figure 19-4   Structure of the cesium chloride crystal.

NaCl crystal is called a *face-centered* cubic crystal, whereas the CsCl crystal is called a *body-centered* cubic crystal.

19-2 COVALENT BONDS AND COVALENT CRYSTALS

The Hydrogen Molecule

When two hydrogen atoms combine to form a hydrogen molecule, H_2, they do so in a way quite different from the electron transfer process we have been discussing. Instead of transferring an electron to form H^+ and H^- ions, the two atoms *share* their two electrons. A closed atomic electron shell is an extremely stable configuration, and by sharing their electrons, each hydrogen atom behaves as though its K shell is complete. This situation is represented schematically in Fig. 19-5 and is called *covalent bonding*. A similar bonding takes place when electrons are shared to complete a higher shell; Fig. 19-6 shows the case of the fluorine molecule, F_2, in which the shared electrons fill the L shell. It is important to realize that covalent bonding is a quantum phenomenon and it cannot be pictured accurately in terms of classical or nonquantum ideas.

For an isolated hydrogen atom, the probability of finding the electron in any particular position around the nucleus is spherically symmetric; that is, at a particular distance from the nucleus, the probability of finding the electron is the same in all directions. The electron probability "cloud" is spherical and has the radial variation in density shown in Fig. 18-7a. The electron "clouds"

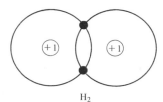

Figure 19-5   Two hydrogen atoms form a covalent bond by sharing their two electrons, thereby completing the K shell for each atom.

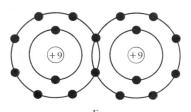

Figure 19-6   Two fluorine atoms form a covalent bond by sharing two electrons, thereby completing the L shell for each atom.

for two isolated hydrogen atoms are illustrated in Fig. 19-7a. In the covalent bonding of two hydrogen atoms, the electron "clouds" overlap and the electron probability density is concentrated *between* the atoms (Fig. 19-7b). Each of the nuclear protons is electrically attracted toward the concentration of negative charge, thereby producing a strong bond between the atoms.

A graph of binding energy versus separation for the hydrogen molecule is very similar to that for the Na^+Cl^- system (Fig. 19-2). At the equilibrium position for H_2, the separation is 1.06 Å and the binding energy is 2.65 eV. Notice that the separation of the nuclei in H_2 is just twice the Bohr-model radius for the hydrogen atom.

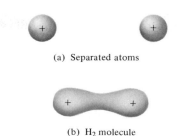

(a) Separated atoms

(b) H_2 molecule

Figure 19-7 (a) The electron probability "clouds" around isolated hydrogen atoms are spherically symmetric. (b) In the hydrogen molecule, the electron probability "cloud" is concentrated between the atoms.

Hybrid Bonding

The carbon atom is a vital participant in all molecules of living matter (*organic* molecules). Carbon is bound in these molecules by covalent bonds — but not in exactly the way that would be expected on the basis of the electron configuration of the isolated atom. The carbon atom has two 2S electrons and two 2P electrons in the L shell. We can represent this structure according to the system of boxes at varying energies that was used in Fig. 18-9. In this scheme, the occupation of the various states by electrons in the normal carbon atom is shown in Fig. 19-8a. There are two paired electrons in each of the S states and two unpaired electrons in the 2P state. Thus, we would expect that a carbon atom would join with other atoms through two covalent bonds. But carbon does not interact in this way. If we supply to a carbon atom only about 2 eV of energy, the 2S electron pair is broken apart and one of the electrons is promoted into the P state. Then, we have *four* unpaired electrons (one S electron and three P electrons) that are equally effective in covalent bonding. One does not distinguish between the S and P bonds in this case — all are equivalent in their bonding action. This type of bonding is called *hybrid* or SP^3 *bonding*. When a molecule is formed by this type of bonding, the two additional bonds provide extra binding energy significantly in excess of the 2 eV expended in adjusting the electron configuration to furnish the new bonding positions. (Compare the case of recovering the energy required to form ions from neutral atoms in ionic bonding.) The arrangement of the four SP^3 electron probability "clouds" around the nucleus of the carbon atom is shown in Fig. 19-9. Notice

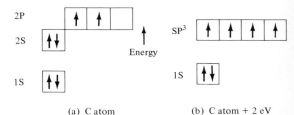

(a) C atom

(b) C atom + 2 eV

Figure 19-8 (a) Electron configuration of the normal carbon atom. (b) By supplying only about 2 eV to a carbon atom, the 2S pair is broken apart and four unpaired electrons are made available for bonding to other atoms.

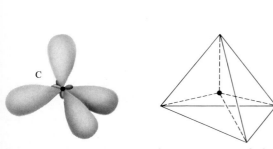

Figure 19-9 The tetragonal arrangement of the four electron probability "clouds" around the nucleus of the carbon atom.

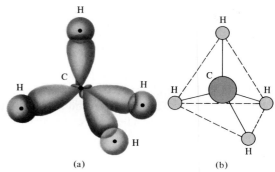

(a)   (b)

Figure 19-10 (a) The methane molecule, CH_4, according to the orbital picture. (b) Ball-and-stick model of the tetragonal methane molecule.

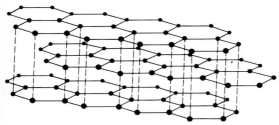

Figure 19-11 In the graphite form of carbon, the atoms are joined together to form planes.

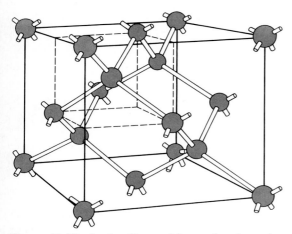

Figure 19-12 In the diamond form of carbon, the atoms are joined together in a rigid three-dimensional structure.

that the "clouds" lie at equal angles with respect to one another—that is, the configuration is tetragonal in shape.

The simplest molecule that carbon forms through hybrid bonding is methane, CH_4. Figure 19-10 shows the distribution of the electron probability "clouds" and the structure of the molecule.

Covalent Crystals

Atoms that bind together through covalent bonds also form crystals. Carbon is typical of this class of materials, and carbon forms two different types of crystal structures by utilizing its four covalent bonds in different ways. In the graphite form of carbon, the atoms are arranged in planes of interconnecting hexagons, as shown in Fig. 19-11. The binding between adjacent planes of atoms is very weak. Consequently, the planes easily slip over one another and graphite has an almost greasy feel. Indeed, powdered graphite is an excellent lubricant; one of its virtues is that it can be blown into inaccessible places, such as door locks. The various types of mica (for example, biotite) have crystal structures similar to graphite and are therefore easy to split into thin sheets.

In the diamond form of carbon, each atom is joined to four other atoms. Unlike graphite, however, the carbon atoms in diamond do not lie in planes. As a result, when the atoms are linked together they form an extremely stable three-dimensional structure (Fig. 19-12). This arrangement of the carbon atoms and the high strength of the covalent bond between carbon atoms makes diamond the hardest substance known.

19-3 HYDROGEN BONDS

Polar Molecules

Hydrogen is a unique substance because when the single electron in a hydrogen atom is pulled away from the nucleus to participate in a bond, the nuclear proton is left almost completely exposed. In molecules formed with atoms that are both small in size and highly electronegative, the bonding electron of hydrogen will be substantially displaced from the nucleus. Therefore, when hydrogen combines with atoms of fluorine, oxygen, or nitrogen, the hydrogen atoms in their interaction with neighboring molecules behave almost as bare protons.

Consider the water molecule, shown again in Fig. 19-13. The effect of the electron displacement is so great that the hydrogen portions of the molecule are positively charged and, in compensation, the oxygen portion carries a negative charge. Thus, there is a separation of charge in the molecule, even though the molecule as a whole remains electrically neutral. Such molecules are said to be *polar*. In an electric field, polar molecules are aligned with the field direction—that is, the substance as a whole becomes *polarized* (Fig. 19-14). Thus, the behavior of polar molecules in an electric field is similar to the behavior of the polarizable objects shown previously in Fig. 12-5.

Bonding of Polar Molecules

When a water molecule is in the presence of other molecules, it attracts to itself the oppositely charged portions of these other molecules (Fig. 19-15). Thus, even in the liquid state, water is not entirely a random collection of molecules—some degree of order exists because of the electrical attraction of the molecules for one another. When water freezes and becomes ice, the orderly arrangement of the molecules becomes particularly evident. As shown in Fig. 19-16, the H_2O molecules in ice form a hexagonal structure with a large void space in the middle of the ring. These rings are attached to one another in a regular crystalline array. The electrical forces that exist between adjacent molecules and that are due to the exposed nature of the hydrogen proton, are called *hydrogen bonds*.

In Fig. 19-16, notice that each molecule participates in two hydrogen bonds to molecules in the same ring.

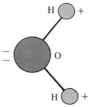

Figure 19-13 In a water molecule the hydrogen electrons are displaced toward the oxygen atom, resulting in a separation of charge and producing a polar molecule.

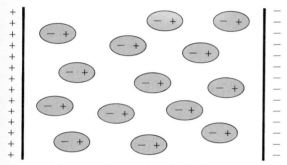

Figure 19-14 With no electric field, the orientation of polar molecules is random, but in a field the molecules become aligned along the field direction.

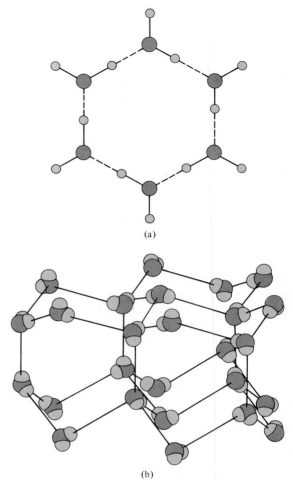

(a)

(b)

Figure 19-16 (a) Hydrogen bonds (dashed lines) join together H_2O molecules in ice crystals. (b) The hexagonal structure repeats throughout the crystal. (The sticks that represent the intramolecular bonds are eliminated in this model for the sake of clarity. The sticks here indicate the hydrogen bonds.)

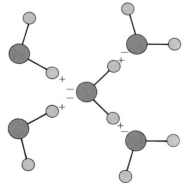

Figure 19-15 A water molecule, because it is polar, electrically attracts other water molecules to itself.

The regular patterns of snowflakes is a result of the repeating hexagonal structure of ice crystals.

There is an additional bond to a molecule in the neighboring ring. But, as shown in Fig. 19-15, each water molecule can bond to *four* other molecules. Therefore, in ice (Fig. 19-16), there is a fourth bond available to each molecule. This bond joins to a molecule in the ring above or below the ring shown. Consequently, an ice crystal is a three-dimensional solid bound together by intermolecular hydrogen bonds.

The large void space in each ring of molecules in ice makes the density of ice less than that of water. (Compare Figs. 19-15 and 19-16.) Thus, when water freezes, it *expands*.

19-4 SEMICONDUCTORS

Conduction by Electrons and Holes

In a material that is a good conductor of electricity, there are many free electrons (or *conduction* electrons) that can be pushed along by an electric field to produce a current. In an insulator, on the other hand, there are no (or very few) conduction electrons available to carry a current. A *semiconductor,* as the name implies, is a material that is neither a good conductor of electricity nor a good insulator. Semiconductor materials, such as silicon and germanium, have only small numbers of conduction electrons. When a voltage is placed across these materials, the amount of current flow is very small.

The semiconductor elements silicon and germanium belong to Group IV of the Periodic Table and therefore have four valence electrons. These elements form crystal structures similar to that of carbon in the diamond form (see Fig. 19-12). Because of this character-

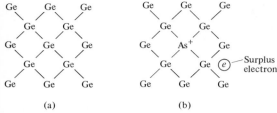

(a)   (b)

Figure 19-17 (a) The diamondlike crystal structure of pure germanium. (Compare Fig. 19-12 which shows the three-dimensional aspect of this type of crystal.) (b) When a germanium crystal is doped with arsenic, one of the five valence electrons of each arsenic atom becomes available to conduct current through the crystal.

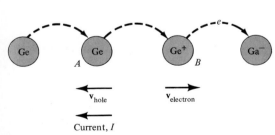

Figure 19-18 Electrons move from atom to atom in a gallium-doped germanium crystal. The motion of the electrons is in one direction and the "motion" of the electron holes is in the opposite direction. Notice that the direction of current flow is opposite to that of the electron flow and in the same direction as the "motion" of the holes.

istic crystalline form, the conductivities of these semiconductors can be increased to a useful level by adding to the material a carefully controlled amount of a particular impurity.

Figure 19-17a shows the diamondlike crystal structure of pure germanium. Arsenic is a Group V element and has 5 valence electrons compared to 4 for germanium. Therefore, if an arsenic atom is substituted for a germanium atom in a crystal, the arsenic atom uses 4 electrons to bond to the neighboring germanium atoms and has one electron "left over" (Fig. 19-17b). The introduction of an impurity into an otherwise pure crystalline substance is called *doping* and the impurity is called a *dopant*. Group V elements that are used as dopants for silicon and germanium include arsenic, phosphorus, and antimony.

n-Type and p-Type Semiconductors

If an electric field is applied to an arsenic-doped germanium crystal, current will flow, carried by the surplus electrons contributed by the arsenic atoms. Doped semiconductors which conduct current by means of *negative* charge carriers (namely, electrons) are called *n*-type semiconductors. When a Group III element is used as a dopant in silicon or germanium, a different effect occurs. Gallium has only 3 valence electrons and therefore a neutral gallium atom cannot bond to 4 germanium atoms. However, if a gallium atom is introduced into a germanium crystal, it will "steal" an electron from a germanium atom in order to provide itself with the fourth atomic bond. In the process the gallium atom becomes a negative ion and the germanium atom which lost the electron becomes a positive ion. The absence of an electron in the germanium ion is referred to as an electron *hole*.

If a germanium atom at position *A* contributes an electron in order to neutralize the germanium ion at site *B*, an electron moves from *A* to *B* and, consequently, a hole "moves" from *B* to *A* (Fig. 19-18). That is, electrons skip from one atom to another in one direction while holes "move" in the opposite direction. When an electric field is applied to a crystal of gallium-doped germanium, electrons (and holes) readily move and a current flows. Materials in which holes (that is, *positive* charge carriers) are involved in current flow are called *p*-type semiconductors. Some *p*-type dopants used in silicon and germanium are gallium, indium, and thallium, which are all Group III elements.

Diodes

Electrical energy is almost always transported from the generating plant to the user as alternating current (see Section 11-3). Through any section of wire carrying AC, the current flows first in one direction and then in the other (Fig. 19-19a). For many applications, however, it is direct current (DC), and not AC, that is required, AC-to-DC conversion is accomplished by inserting into the circuit a device that allows the current to flow freely in one direction but not in the other direction. Such a device is called a *diode*, and its effect in an AC circuit is shown in Fig. 19-19b. When the current flowing through the diode is passed through additional circuit elements, the bumps are smoothed out (*filtered*) and a steady flow of current in one direction results — this is DC (Fig. 19-19c).

Before the day of solid-state electronics, all AC-to-DC conversion was accomplished by means of electron tubes (vacuum tubes). But, beginning in about 1950, semiconductors have been increasingly used in all types of electronic circuits and now virtually all everyday circuits employ semiconductor diodes to produce direct current.

A semiconductor diode consists of n-type and p-type material in intimate contact. The basic operation of a p–n diode is as follows. When the two different types of material are in contact, some of the surplus electrons in the n-type material drift (or *diffuse*) across the boundary

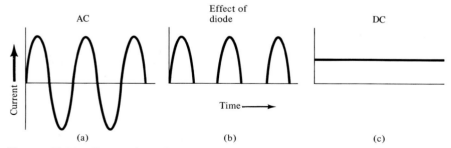

Figure 19-19 Conversion of AC to DC. (a) Alternating current flows first in one direction (indicated in the graph by positive values) and then in the opposite direction (indicated by negative values). (b) When a diode is introduced into the circuit, current flows only in one direction. (c) Passing the current through additional circuit elements smooths out (or filters) the bumps in the current and produces a steady current flow — this is DC.

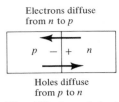

Holes diffuse
from *p* to *n*

Figure 19-20　The diffusion of electrons and holes across the boundary in a *p–n* diode builds up a potential difference with the *n* side positive and the *p* side negative.

into the *p*-type material. Similarly, some of the holes in the *p*-type material diffuse into the *n*-type material (Fig. 19-20). The net result is that a potential difference is developed across the diode: positive on the *n*-type side and negative on the *p*-type side. When the diode is placed in a circuit, this potential difference *aids* the flow of current from the *p*-type to the *n*-type materials, but it *suppresses* the flow in the opposite direction (Fig. 19-21). It is the unique ability of semiconductor materials to set up a potential difference by the migration of electrons and holes across a boundary that renders these materials so extraordinarily useful in electronic devices.

Photodiodes and Light-Emitting Diodes

Certain types of semiconductor materials have the interesting property that they can convert electromagnetic energy in the form of light directly into electrical energy. That is, when light is incident on such a material, an electrical current is caused to flow—the greater the light intensity, the greater the current flow. Devices that employ these materials are called *photodiodes,* and they are used in a variety of situations. A *solar cell* is a photodiode that is used to extract energy from sunlight. Solar cells are always used to generate power in satellites that must remain in orbit for long periods of time and transmit radio signals to Earth. No light-weight battery could be made with sufficient energy content to power the on-board electronics, so solar cells are used to keep a small set of batteries charged. (Batteries must still be used even though the satellite is equipped with solar cells because power is required while the vehicle is in the Earth's shadow.) A satellite with a panel of solar cells is shown in the photograph on page 173.

It has recently been discovered that photodiodes can be made to operate *backward.* That is, when an electrical current is passed through the diode, light is emitted. Such a device is called a *light-emitting diode* (LED). LEDs can be used in battery-operated devices because only a tiny amount of power is required for light to be emitted; conventional (incandescent) lighting methods represent a high power drain on any battery system. Several everyday products using LEDs—such as digital clocks and midget calculators—have already been developed.

Light-emitting diodes can be made in extremely small sizes, and by incorporating different materials into the diodes, they can be made to emit light with various

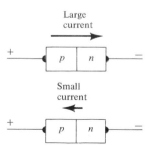

Large
current

Small
current

Figure 19-21　Because of the potential difference across the boundary (Fig. 19-20), current flows readily in only one direction through a *p–n* diode.

colors. We can probably look forward in the near future to the availability of color television sets that use panels of LEDs instead of bulky picture tubes. Such sets will be extremely compact, perhaps only an inch thick. We will then be able to hang our television sets on walls like pictures.

Transistors

A *transistor* is a device that consists of three semiconductor elements arranged in *p–n–p* or in *n–p–n* fashion (Fig. 19-22). Transistors are used to amplify and control electrical signals in an extremely wide variety of electronic circuits—from "transistor" radios to huge digital computers. We can illustrate the functioning of a transistor in analogy with fluid flow in the following way. Figure 19-23 shows a section of a pipe into which is inserted a tube with a movable cap. The extension of the cap determines how much fluid will flow through the pipe, and the cap extension is controlled by the pressure in the side tube. By changing the control pressure up and down, the flow rate through the pipe can be decreased or increased. Notice that the fluid in the side tube is isolated from the main pipe—it is the *pressure* in the side tube, and not flow through it, that regulates the flow in the pipe. The movable cap is analogous to the center element in a transistor and the pressure in the side tube is analogous to the control voltage applied to a transistor. The fluid flow through the pipe resembles the current flow through a transistor.

Semiconductor technology has revolutionized the electronics industry. Vacuum-tube circuits were always subject to frequent breakdowns, usually due to the heat generated by the tubes themselves. Transistor circuits generate very little heat and because they are solid materials, the reliability is very high. (Which fails more often, your television set—with vacuum tubes—or your transistor radio?) Communications devices, computers, and control circuitry of every type now rely almost exclusively on transistors to process electrical signals. The program of space exploration—in which extremely light-weight and reliable control circuits are required for on-board operation—would not be possible were it not for transistors. About the only places where vacuum tubes are still used in everyday devices are in television circuits. Even in these circuits, transistors are replacing vacuum tubes, except in the high-voltage circuits where transistors tend to break down.

The numerical readout of a midget calculator utilizes light-emitting diodes (LEDs). The internal computations are carried out by integrated circuit chips containing many diodes and transistors. The unit is powdered by a small rechargeable battery.

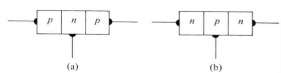

Figure 19-22 (a) A *p–n–p* transistor. (b) An *n–p–n* transistor.

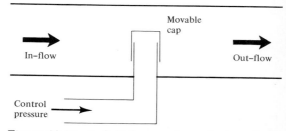

Figure 19-23 A fluid-flow analogy of a transistor. The movable cap, which controls the flow rate through the pipe performs the same function as the center element in a transistor.

Different types of encapsulated transistors.

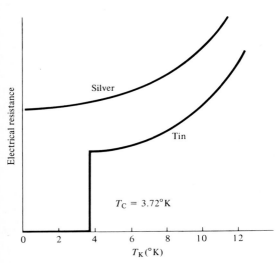

Figure 19-24 Electrical resistance as a function of temperature for silver (a nonsuperconductor) and for tin (a superconductor). For temperatures below the critical temperature T_C at 3.72 °K, the resistance of tin is *zero*.

Semiconductor technology has now progressed to the point that hundreds of transistors, together with resistors and capacitors, can be manufactured on a single wafer of material only a millimeter square. These integrated-circuit (or IC) assemblies are complete circuits and can be designed for all kinds of special purposes. By using IC wafers, digital computers (which once required thousands of vacuum tubes and covered thousands of square feet of floor space) can now be made no bigger than a breadbox.

The 1956 Nobel Prize in physics was shared by John Bardeen, Walter Brattain, and William Shockley who discovered and developed the transistor (1949).

19-5 SUPERCONDUCTORS

Zero Electrical Resistance

All ordinary, conductors resist the flow of electrical current to some extent. If there were no electrical resistance, electrons would flow freely through wires — there would be no electrical heating and no power losses in current-carrying wires. However, no material is known that is a perfect conductor (that is, has zero resistance) at room temperature. But in 1911 the Dutch physicist Kamerlingh Onnes (1853–1926) discovered that a column of mercury suddenly loses all electrical resistance when it is cooled below 3 °K. The resistance of mercury near absolute zero does not simply become very small — it is *zero*! Since Onnes' discovery, two dozen elements and thousands of alloys have been found to exhibit the phenomenon of *superconductivity*.

Figure 19-24 shows the electrical resistance as a function of temperature for an ordinary metal (silver, a nonsuperconductor) and for a superconductor (tin). The resistance of silver decreases as the temperature approaches absolute zero but even at the very lowest temperatures studied the resistance is nonzero. On the other hand, the resistance of tin suddenly drops to zero at the critical temperature $T_C = 3.72$ °K. This behavior of the resistance–temperature curve is typical of the entire class of superconducting elements and alloys. Each superconductor has a characteristic value of the critical temperature, most of which are below about 20 °K. Curiously enough, the elements that are the best conductors at room temperature (gold, silver, and copper) do *not* exhibit superconductivity. Evidently, the electron–atom interactions which cause normal metals to be *poor* conductors are essential for a metal to reach the superconducting state.

Although numerous experiments were performed to investigate the properties of superconductors, the phenomenon of superconductivity was essentially a mystery until a fundamental explanation was put forward in 1957 by John Bardeen, Leon Cooper, and Robert Schrieffer. Their explanation involves the application of the exclusion principle to matter in the bulk state and emphasizes the point that quantum phenomena are sometimes evident on a macroscopic scale.

The essence of the Bardeen–Cooper–Schrieffer theory is the following. In ordinary metals, the free electrons act as individual particles. In some metals, such as tin, the conduction electrons change their configuration when superconductivity sets in. At some very low temperature, the interactions of the electrons with one another and with the atoms in the crystal cause the conduction electrons to form into closely associated pairs (spin up–spin down). Thus, in a superconductor the fundamental units of charge are electron *pairs* and not single electrons.

The exclusion principle applies to all particles that have spin $\frac{1}{2}$—for example, single electrons. (See Section 18-1 for the discussion of spin.) But a spin up–spin down electron pair acting as a unit has zero net spin and the exclusion principle does not apply to zero-spin particles. (This is a new point; we did not make the distinction between spin-zero and spin-$\frac{1}{2}$ particles in discussing the exclusion principle in Section 18-3.) Consequently, there is no restriction on the energy state that an electron *pair* can occupy. In particular, at low temperatures thermal agitation is minimal, and *all* of the electron pairs can occupy the same energy state, namely, the lowest possible energy state. With all of the electron pairs in the lowest state, no energy exchanges can take place. (A pair has no energy to give to another pair or to an atom in the crystal lattice because it already has the least possible energy.) If no energy exchanges can take place, the normal resistive energy losses are not possible. The electron pairs move unimpeded through the metal: the metal has zero electrical resistance and exhibits superconductivity.

The 1972 Nobel Prize in physics was shared by Bardeen, Cooper, and Schrieffer for their work on the theory of conductivity. Bardeen thereby joined Marie Curie and Linus Pauling as the only scientists to win

TEXAS INSTRUMENTS

This integrated circuit assembly consists of a tiny silicon wafer approximately $\frac{1}{4}$ in. square (located in the center of the plastic support) containing hundreds of individual transistors and other circuit elements.

two Nobel Prizes, and only Bardeen's Prizes were in the same field. (Marie Curie, physics and chemistry; Pauling, chemistry and peace.)

Applications of Superconductors

During the last several years, the importance of superconductors in various areas of practical engineering have been realized. The transportation of electrical energy involves the use of ordinary wires carried on the high-voltage towers seen almost everywhere throughout the country. Substantial losses occur in these wires because of electrical resistance effects. These losses could be almost entirely eliminated by using superconducting wires laid underground in cooled, vacuum pipes. Perhaps we will someday be able to eliminate the unsightly high-voltage towers and at the same time transport our electrical energy more economically.

The production of large magnetic fields by conventional methods is expensive. An electromagnet requires a continuous current in order to function, and if the field desired necessitates a substantial current, a system must be provided to carry away the heat produced by the resistance of the magnet windings. Permanent magnets require no current, but they are bulky, inflexible, and, of course, cannot be turned off.

Once a current is set up in a loop of superconducting wire, the current will flow forever because there are no current losses due to resistance effects. The magnetic field produced by the current will likewise persist forever. In a practical situation, we usually extract some energy from the magnetic field and therefore an equivalent amount of energy must be supplied to the superconductor to maintain the current flow. But we need supply only the amount of energy that is used—no additional energy is required simply to maintain the field as is the case with ordinary electromagnets.

Superconducting magnets have already found uses in research areas that require large magnetic fields. And there have been several proposals to use superconducting magnets to suspend railway-type cars in special transportation systems. Such cars would run freely without appreciable frictional losses and would be capable of producing smooth, high-speed rides.

Even closer to realization is the construction of superconducting electrical generators. Westinghouse hopes to have a superconductor power plant in operation in the mid-1980's. The generator is expected to be only about

This superconducting magnet is constructed from windings of niobium–tin (Nb_3Sn) and at 3 °K generates a magnetic field of 165 kilogauss in a central cavity that has a diameter of 1 inch.

one-tenth the size of conventional generators for the same power output. Moreover, the losses due to electrical resistance effects are expected to be only about one-third as great. Smaller versions of these generators will find uses aboard ships and as auxiliary power plants (for example, as emergency units in hospitals).

It must be remembered that the superconducting state of a material is reached only at very low temperatures. Therefore, in any application, special low temperature (*cryogenic*) apparatus is necessary, along with a supply of liquid helium. This poses significant problems and expense, especially for proposed field applications such as superconducting power transmission lines. It would be much simpler (and much less expensive) to construct and maintain such systems if materials could be found that become superconductors at higher temperatures. Several special alloys (such as Nb_3Ge) have been observed to enter the superconducting state at temperatures near 23 °K. Because this temperature is above the liquefaction temperature of hydrogen (20 °K), liquid hydrogen can be used instead of liquid helium to cool these "high temperature" superconductors. It seems likely that other materials will be discovered to have superconducting transition temperatures that are even higher. It is conceivable that some exotic material may become a superconductor near room temperature. This would be an exciting prospect indeed!

SUGGESTED READINGS

L. Pauling and R. Hayward, *The Architecture of Molecules* (Freeman, San Francisco, 1964).

J. R. Pierce, *Quantum Electronics* (Doubleday, Garden City, New York, 1966).

Scientific American articles:

L. Halliday, "Early Views on Forces Between Atoms," May 1970.

D. P. Snowden, "Superconductors for Power Transmission," April 1972.

QUESTIONS AND EXERCISES

1. Two chlorine atoms combine to form Cl_2 and a chlorine atom combines with a sodium atom to form NaCl. Why does sodium not form the molecule Na_2?

2. As one proceeds down Group I of the periodic table toward elements with higher atomic numbers, the ionization energies become *smaller*. Explain why this is so.

3. Atoms participate in chemical reactions primarily through their outer or valence electrons. Neutral atoms of sodium have a greater chemical activity than Na^+ ions. Why?

4. Write down the chemical formulas of the compounds of lithium that are ionic substances similar to sodium chloride.

5. Explain why He_2 molecules do not exist. (Consider covalent bonding and remember the exclusion principle.)

6. The molecule H_3 does not exist, but the ion H_3^+ *does* exist. Why?

7. Write down the chemical symbols for the forms of the following gases as they occur in Nature: chlorine, bromine, and iodine. Explain.

8. In a molecule of carbon dioxide, CO_2, the oxygen molecules are on directly opposite sides of the central carbon atom. Do you expect the carbon dioxide molecule to be *polar*? Explain.

9. Suppose that, upon freezing, water molecules became more closely packed than in the liquid state. Explain how a lake would then freeze in winter.

10. What is the purpose of *doping* a semiconductor?

11. Selenium is a semiconductor element. What dopant could be used to convert selenium into a *p*-type material? Into an *n*-type material?

12. What are some of the problems that would be encountered in replacing underground electrical transmission lines with superconducting lines?

20

NUCLEI AND NUCLEAR POWER

Only a little more than 60 years ago was it first realized that the atoms which make up all things contain at their centers concentrated bits of matter—*nuclei*. Rutherford's analysis of the experiments in which rapidly moving α particles were deflected by thin sheets of matter had shown, in 1911, that every atom possesses a central nucleus of extremely small size. Although *atoms* have sizes that are only about 10^{-10} m, *nuclei* are approximately 10 000 times *smaller*. Thus, almost the entire mass of an atom is packed into a region with a typical dimension of 10^{-14} m.

In spite of their extremely small size and their shielded position within atoms, nuclei have been the subject of extensive investigation. These studies have provided a wealth of information concerning the behavior of fundamental matter and they have unlocked the doors to the enormous reservoirs of nuclear energy. Beginning in 1945, the world has witnessed the incredible destructive power of nuclear energy (in the form of weapons) as well as the great benefit that it can be to Mankind (in the form of radioisotopes and nuclear power plants). In this chapter we will review some of the properties of nuclei and the ways in which they interact. Then, we will discuss how energy can be liberated from nuclei and put to use.

20-1 RADIOACTIVITY

Early Discoveries

At about the time that Thomson was investigating the properties of electrons, another discovery of great importance was made by Henri Becquerel (1852–1908), a French physicist. In 1896 Becquerel found, quite by accident, that when he placed a sample of uranium salts (potassium uranyl) on a piece of unexposed photographic film, the developed film revealed an outline of the crystals. The same result was obtained even when the film was wrapped in heavy black paper, a sufficient shield to exclude all light from the film. Furthermore, the darkening of the film was observed when *any* substance containing uranium was placed on the film. Clearly, it was uranium, and not light, that had caused the film to show an outline of the crystals, and Becquerel reasoned that the uranium must be emitting some different kind of radiation, rays that had not been detected before. This new phenomenon was called *radioactivity*.

Before the end of the 19th century, the study of radioactivity had led to the discovery of two new elements. In 1897 Marie Curie selected as her doctoral research problem the investigation of the mysterious rays emitted by uranium. In order to determine whether elements other than uranium produced these rays, Madame Curie tested every known element. Only two were found to be radioactive—uranium and thorium. We now know that a large number of elements exhibit radioactivity in their natural forms, but these activities are weak and Madame Curie's methods were not sufficiently sensitive to detect their presence. She used various materials in her experiments, sometimes pure elements and sometimes minerals. One curious fact emerged: the mineral *pitchblende* (an ore of uranium) was a much more prolific source of radiation than was pure uranium metal. Because pitchblende contains no thorium, Madame Curie wondered whether there could be an undiscovered element, an impurity in the pitchblende, that could account for the exceptional radioactivity of this ore. She then began a series of tedious chemical procedures designed to isolate the source of the intense radioactivity in pitchblende. By the end of 1898, Marie Curie and her husband Pierre (neither of whom were chemists) had succeeded in preparing two tiny samples of highly radioactive substances which they had laboriously separated from pitchblende.

All tests showed that these substances were not compounds but new elements. The Curies named their elements *polonium* and *radium*. The name *polonium* was chosen to honor Marie Curie's native country, Poland, and *radium* was chosen because of the great intensity of radiation emitted by this substance.

Alpha, Beta, and Gamma Rays

Within a few years after the Curies' discoveries, three different types of emanations from radium and other radioactive substances had been identified. For lack of any better names for these new radiations, they were labeled by the first three letters of the Greek alphabet, designations that we still use:

(a) *Alpha rays:* positively charged particles with relatively large mass. (Symbol: α.)
(b) *Beta rays:* negatively charged particles with mass much less than that of alpha rays. (Symbol: β.)
(c) *Gamma rays:* neutral rays with no detectable mass. (Symbol: γ.)

All of these radiations were found to be emitted with high speeds from a variety of radioactive materials.

Alpha rays and beta rays (or α particles and β particles) were studied in much the same way that Thomson had investigated cathode rays and positive rays in discharge tubes, namely, by bending the particles in an electric field (Fig. 20-1) and by measuring the buildup of electrical charge on surfaces or wires that collected the radiation. These experiments showed that β particles are identical to *electrons* and that α particles are the same as *helium nuclei* (that is, helium atoms from which two electrons have been removed). Gamma (γ) rays proved to have properties similar to those of light, except that they lie outside the range of the visible spectrum. Gamma rays are photons of electromagnetic radiation just as are light quanta and X rays. The only difference between gamma rays and other forms of electromagnetic radiation is that gamma rays have much higher frequencies.

Nuclear Changes

Radioactivity is a *nuclear* phenomenon and it does not depend in any way on chemical or physical changes that the *atom* may undergo. The rate and the speed with which α particles are emitted from radium are the same

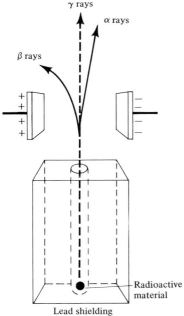

Figure 20-1 The three types of radiations emitted by a radioactive sample are affected in different ways by an electric field. The bending of the β rays toward the positive plate shows that these particles are *negatively* charged. Similarly, the bending of the α rays in the opposite direction shows that these particles are *positively* charged. (The less massive β particles are bent by a much greater amount than are the α particles.) Gamma rays are unaffected by the electric field; they are *neutral* rays.

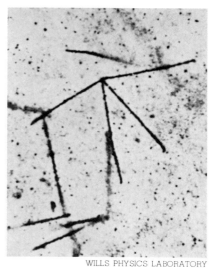

The path traveled by a *single* nuclear particle can be recorded by using a special photographic film (called a *nuclear emulsion*). This photomicrograph shows the tracks left by several α particles emitted from a single radioactive parent nucleus and its radioactive daughters. In this process, a thorium nucleus emits an α particle, leaving a radioactive daughter nucleus; this nucleus emits another α particle, again leaving a radioactive nucleus; and so on. The length of the longest track in this picture is approximately 3×10^{-5} m, or 0.03 mm.

whether the radium is in the form of the pure metal or whether it is in a chemical compound. Radioactivity is unaffected by temperature, pressure, or chemical form (except to a very small extent in special circumstances).

When an α particle, a β particle, or a γ ray is emitted by a radioactive substance, it emerges from the *nucleus* of the material. But because the electron structure of an atom depends on the amount of electrical charge in the nucleus, if there is a change in the nuclear charge, there will be a corresponding change in the number of atomic electrons. For example, the radium nucleus ($Z = 88$, $A = 226$) has 88 protons and 138 neutrons. When ^{226}Ra emits an α particle (^{4}He), two protons and two neutrons are carried away (Fig. 20-2). Therefore, the residual nucleus has 86 protons and 136 neutrons. The product of radium α decay (the *daughter*) is a different element: *radon* ($Z = 86$). The atomic electron structure changes, following the decay event, to accommodate the new nuclear charge by releasing two of its 88 electrons. These two electrons, or their equivalent, eventually attach themselves to the emitted α particle and form a neutral atom of ^{4}He. Thus, the original neutral radium atom decays and two neutral atoms are formed, one of radon and one of helium.

In the β-decay process, an electron is emitted from the nucleus. (But this electron does not preexist in the nucleus; the electron is formed in the β decay process and is immediately ejected.) The removal of a negative charge from the nucleus means that the (positive) nuclear charge *increases* by one unit (that is, by $+e$). Thus, a nucleus with atomic number Z that undergoes β decay becomes a nucleus with atomic number $Z + 1$. But no proton or neutron is emitted in a β radioactivity process and so the mass number A of the daughter nucleus is the same as the mass number of the parent nucleus. When radioactive ^{14}C (6 protons, 8 neutrons) emits a β particle, the new nucleus contains 7 protons and 7 neutrons—that is, ^{14}N is formed (Fig. 20-2).

The α decay of ^{226}Ra and the β decay of ^{14}C can be represented by the following schematic nuclear "equations":

$$\left. \begin{array}{l} {}^{226}_{88}\text{Ra} \xrightarrow{\alpha \text{ decay}} {}^{222}_{86}\text{Rn} + {}^{4}_{2}He \\ {}^{14}_{6}\text{C} \xrightarrow{\beta \text{ decay}} {}^{14}_{7}\text{N} + {}^{0}_{-1}e \end{array} \right\} \quad (20\text{-}1)$$

where we use the nuclear notation to show that the electron has $A = 0$ and $Z = -1$.

In *stable* nuclei, those that do not exhibit radioac-

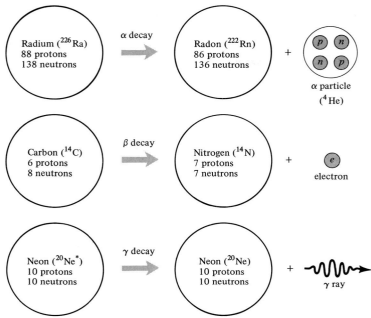

Figure 20-2 The three types of radioactive decay process. Alpha and beta decay are nuclear disintegration events in which the original nucleus changes into a different species. Gamma radiation usually follows α and β decay as the protons and neutrons of the daughter nucleus rearrange themselves; no disintegration process is involved in the emission of γ rays. (The excited nucleus that exists before γ-ray emission takes place is represented by an asterisk, *.)

tivity, the protons and neutrons exist together permanently with no changes. However, if a neutron is removed from a nucleus (by means of a nuclear reaction) and becomes a *free* neutron, it cannot exist permanently. In fact, a free neutron undergoes exactly the same kind of β decay as does a radioactive nucleus such as ^{14}C:

$$\ _{0}^{1}n \longrightarrow\ _{1}^{1}H +\ _{-1}^{0}e \qquad (20\text{-}2)$$

Indeed, we can view radioactive β decay as a process in which one *nuclear* neutron changes into a proton (with the accompanying emission of an electron). This is exactly the process by which ^{14}C is converted into ^{14}N (see Fig. 20-2).

Conservation Laws in Radioactive Decay

Two important facts about radioactive decay processes should be noted:

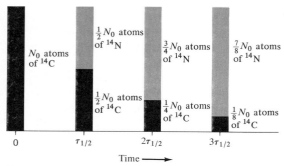

Figure 20-3 The radioactive decay law for the case of the ^{14}C $\rightarrow$ ^{14}N decay. In each interval of time $\tau_{1/2}$, the number of atoms of ^{14}C surviving is equal to one-half of the number that existed at the beginning of that interval.

(a) The total number of protons and neutrons present before the decay takes place is exactly equal to the number after the decay. For example, the mass number of ^{226}Ra (226) equals the sum of the mass numbers of ^{222}Rn and ^{4}He (222 + 4); and similarly for the β decay of ^{14}C.

(b) The total electrical charge is the same before and after the decay takes place. For example, in the α decay of ^{226}Ra, there are 88 protons present before decay and 86 + 2 after decay. In the β decay of ^{14}C, there are 6 protons present before decay and 7 protons afterward; but an electron is also present after decay, so there is a balance of electrical charge [$6e = 7e + (-e)$].

These two facts actually represent important *conservation laws* of Nature that apply to *all* types of processes, not just radioactive decay:

(a) The total number of protons and neutrons in any system remains constant.

(b) The total electrical charge in any system remains constant.

The Half-Life

An atom of radioactive carbon (^{14}C) can undergo β decay and become an atom of nitrogen (^{14}N). But what happens to a sample of ^{14}C, consisting of a large number of atoms, as time goes on? The sample does not suddenly become ^{14}N. Nor does the amount of ^{14}C decrease uniformly to zero after some period of time. Instead, the process of radioactive decay obeys a different kind of law. Every radioactive species has associated with it a characteristic time, which is called the *half-life* and is denoted by the symbol $\tau_{1/2}$. The half-life has the following significance. Suppose that we begin with a sample of ^{14}C consisting of N_0 atoms. After a time $\tau_{1/2}$ (which for ^{14}C is 5730 years) one-half of the ^{14}C atoms will have decayed and the sample will consist of $\frac{1}{2}N_0$ atoms of ^{14}C and an equal number of ^{14}N atoms (Fig. 20-3).

What happens during the time from $\tau_{1/2}$ to $2\tau_{1/2}$? We can apply the same reasoning as before. We start with $\frac{1}{2}N_0$ atoms of ^{14}C at time $\tau_{1/2}$, so after an interval of one half-life (that is, at the time $2\tau_{1/2}$), one-half of the sample with which we started will have decayed. Therefore, at time $2\tau_{1/2}$, we will have remaining only $\frac{1}{4}N_0$ atoms of ^{14}C and there will be $\frac{3}{4}N_0$ atoms of ^{14}N. Similarly at time $3\tau_{1/2}$, we will have $\frac{1}{8}N_0$ atoms of ^{14}C. In every interval of time $\tau_{1/2}$, the amount of ^{14}C will decrease by one-half.

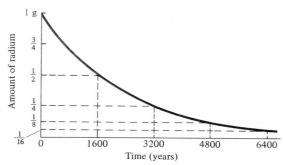

Figure 20-4 Radioactive decay curve for ^{226}Ra. The half-life is $\tau_{1/2} = 1600$ years.

Figure 20-4 shows the way in which a sample of radium (^{226}Ra) decreases with time. The half-life of radium is approximately 1600 years. Therefore, if we start with 1 gram of ^{226}Ra, after 1600 years $\frac{1}{2}$ gram of radium will remain, after 3200 years (that is, an additional half-life) $\frac{1}{4}$ gram will remain, after 4800 years $\frac{1}{8}$ gram will remain, and so on.

The range of known half-lives for α and β decay extends from a small fraction of a second to many billions of years. Some typical values are listed in Table 20-1. We will discuss some of the many uses of radioactivity in the next chapter.

TABLE 20-1   SOME RADIOACTIVE HALF-LIVES

NUCLEUS	TYPE OF DECAY	HALF-LIFE
Thorium (^{232}Th)	α	1.41×10^{10} y
Radium (^{226}Ra)	α	1602 y
Plutonium (^{238}Pu)	α	87.4 y
Polonium (^{214}Po)	α	1.64×10^{-4} s
Potassium (^{40}K)	β	1.28×10^{9} y
Carbon (^{14}C)	β	5730 y
Cobalt (^{60}Co)	β	5.26 y
Neutron (n)	β	760 s
Krypton (^{93}Kr)	β	1.29 s

20-2 NUCLEAR MASSES

Binding Energy

When Chadwick discovered the neutron in 1932, it became clear that atomic cores consist of two different types of particles—protons and neutrons—that are tightly bound together into tiny nuclei. These two particles have very similar properties—they have approximately the same mass (1 AMU) and the same size; the primary difference is that the proton carries one unit of positive electrical charge whereas the neutron carries no charge at all. A nucleus that contains a total of A protons and neutrons has a *mass number A* and a mass that is *approximately A* AMU (see Section 2-3). The nucleus of a helium atom, for example, contains two protons and two neutrons and therefore has a mass of approximately 4 AMU.

But nuclei do not have masses that are *exactly* equal to the sum of the masses of the constituent protons and neutrons, and the discrepancy is of vital importance for the existence of nuclei. Let us consider the simplest nucleus that contains a neutron, namely, the nucleus of deuterium (or heavy hydrogen, ^{2}H). This nucleus consists of one proton and one neutron; that is, a deuterium atom is the same as an ordinary hydrogen atom except that it contains in addition a nuclear neutron. If we sum the masses of a proton and a neutron, we find (see Eq. 2-2)

$$m_{\text{proton}} = 1.0073 \text{ AMU}$$

$$m_{\text{neutron}} = 1.0087 \text{ AMU}$$

$$\overline{m_{\text{proton}} + m_{\text{neutron}} = 2.0160 \text{ AMU}}$$

NUCLIDE CORPORATION

Nuclear masses can be measured by comparison with the mass of the standard ^{12}C. (The mass of ^{12}C is defined to be exactly 12 AMU; see Section 2-3.) One method for determining nuclear masses involves the use of a *mass spectrograph,* as shown here. Ions of carbon and the isotope or isotopes to be studied are accelerated through the same voltage and are projected into a magnetic field. The radii of curvature of the orbits in the field depend on the masses of the isotopes. By comparing the radius of the orbit of ^{14}N, for example, with that for ^{12}C, the mass of ^{14}N in AMU can be determined. Such methods are capable of high precision (1 part in 10^{7}).

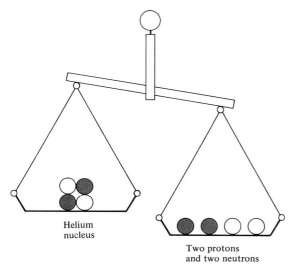

Figure 20-5   The mass of any nucleus (for example, the helium nucleus shown here) is *smaller* than the combined mass of the constituent protons and neutrons in the free state. The mass difference corresponds to the *binding energy* of the nucleus.

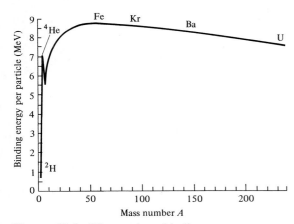

Figure 20-6   The nuclear binding energy curve. The binding energy per particle in the nucleus reaches a maximum in the vicinity of iron and gradually decreases toward heavier elements.

The mass of deuterium nucleus, however, is 2.0136 AMU. That is, the mass of deuterium is *smaller* than the combined mass of a proton and a free neutron by an amount $2.0160 - 2.0136 = 0.0024$ AMU, or about 0.1 percent of the mass of deuterium. This may appear to be a very small, almost trivial difference but, nonetheless, it is an extremely important difference.

According to the Einstein mass–energy relation (Eq. 16-12), a *mass* difference is entirely equivalent to an *energy* difference. The equation $\mathscr{E} = mc^2$ expresses this relationship. If we supply the value of the velocity of light c and the factor that converts joules to electron volts, we find (see Exercise 8) that a mass of 1 AMU is equivalent to an energy of 9.31×10^8 eV or 931 MeV (million electron volts). Therefore, the mass difference found for deuterium (0.0024 AMU) amounts to an energy of 2.2 MeV.

What is the significance of this energy difference? if we wish to convert a deuterium nucleus into a free proton and a free neutron, we must *increase* the mass of the system. That is, we must supply energy to a deuterium nucleus in order to split it into its component parts. If this amount of energy (or more) is not supplied, the deuterium nucleus can never break apart—it is *bound* by 2.2 MeV and this energy value is called the *binding energy* of the nucleus. The *smaller* the mass of a nucleus (compared to the mass of the same number of free protons and neutrons), the *greater* is the binding energy of the nucleus.

All nuclei have this property possessed by the deuterium nucleus. *All* nuclei have masses that are smaller than the combined masses of the constituent protons and neutrons (Fig. 20-5). Indeed, independent and precise measurements of nuclear masses and binding energies have been used to verify the correctness of the Einstein mass–energy relation.

The Binding Energy Curve

One of the most useful ways to summarize the information that has been accumulated regarding nuclear masses is to plot the data in the way shown in Fig. 20-6. The binding energy of deuterium is 2.2 MeV, but the binding energy of ^{235}U is 1760 MeV. Therefore, in order to show the vast range of binding energies on a convenient scale, we divide the binding energy of a nucleus by its mass number A. That is, the quantity plotted is the binding energy *per particle* in the nucleus. As seen in Fig. 20-6, this quantity is approximately the same for most nuclei, varying only between 7.5 and 8.7 MeV per particle for all A greater than about 16. The lighter nuclei have somewhat smaller binding energies. But notice that the binding energy of ^{4}He is considerably greater than that of any of its neighbors—that is, the α particle is an exceptionally tightly bound nucleus.

The binding energy curve reaches a maximum for nuclei in the vicinity of iron (Fe) and then gradually decreases toward the heavier elements. This behavior is responsible for the fact that the *fusion* and *fission* processes release energy. We shall return to these interesting topics after first discussing the basic ideas of nuclear reactions.

20-3 NUCLEAR REACTIONS

Modern Alchemy

One of the most ancient dreams of Man was to be able to transform some cheap and plentiful material, such as lead, into gold. Alchemists devised many fantastic recipes for such processes and were able to extract a great deal of gold from their unwary sponsors, but they obtained none from lead. No chemical or ordinary physical process can change one element into another. Radioactive decay processes alter the nuclear charge and therefore do transform the atoms of one element into atoms of a different element. It is actually possible to produce gold in this way, but in order to do so it is first necessary to prepare a radioactive isotope of platinum—certainly not a practical way of achieving the alchemists' dream!

Ernest Rutherford's experience with radioactivity led him to wonder whether there might be ways, other than radioactive decay, to transmute one element into another. He reasoned that if a nuclear projectile, such as an α particle, could be fired at another nucleus with sufficient speed, the projectile might disrupt the target nucleus and form a nucleus of a different element. Rutherford chose for his projectiles the high-speed α particles emitted in radioactive α decay. When these particles were projected through nitrogen gas, he found that some of the α particles struck and reacted with nitrogen nuclei, thereby producing two different elements—oxygen and hydrogen. The nuclear reaction first

observed by Rutherford was

$$^{14}\text{N} + {}^{4}\text{He} \longrightarrow {}^{17}\text{O} + {}^{1}\text{H} \qquad (20\text{-}3)$$

nitrogen helium   oxygen hydrogen

Figure 20-7 shows that the net result of this reaction is to transfer two neutrons and one proton from the helium nucleus to the nitrogen nucleus, thereby forming ^{17}O.

Since the time of Rutherford's first observation of a nuclear reaction, thousands of reactions have been studied in the laboratory. Some of these reactions involve the simple capture of an incident particle by a target nucleus and others involve complex disintegrations. The techniques of reaction investigations have been refined to the extent that nuclear transmutations are now a routine laboratory practice.

Features of Nuclear Reactions

Protons and neutrons cannot be destroyed (or created) in nuclear reactions—they are only rearranged in such processes. Therefore, in any nuclear reaction we must have a balance of protons and neutrons before and after the reaction. For the case of Rutherford's $^{14}\text{N} + \alpha$-particle reaction, we have

$$\left.\begin{array}{l} {}^{14}\text{N} + {}^{4}\text{He} \longrightarrow {}^{17}\text{O} + {}^{1}\text{H} \\ \text{Number of protons:}\quad 7 + 2 = 8 + 1 \\ \text{Number of neutrons:}\quad 7 + 2 = 9 + 0 \end{array}\right\} \;(20\text{-}4)$$

One of the interesting features of a nuclear reaction is that it can be *reversed*. If we use some device, such as a *cyclotron,* to produce high-speed protons (radioactivity will not do because individual protons are never emitted in radioactive decay) and if we project these protons at a gas consisting of the isotope ^{17}O, then disintegration events will occur that produce nitrogen and helium nuclei. That is, the reaction is the reverse of Eq. 20-3:

$$^{17}\text{O} + {}^{1}\text{H} \longrightarrow {}^{14}\text{N} + {}^{4}\text{He} \qquad (20\text{-}5)$$

The other isotopes of oxygen, ^{16}O and ^{18}O, will undergo nuclear reactions when bombarded with fast protons, but these reactions will have features quite different from the $^{17}\text{O} + {}^{1}\text{H}$ reaction. The *atomic* structures of the oxygen isotopes are all the same, and, consequently, these isotopes participate in *chemical* reactions in exactly the same way. But the *nuclear* structures are different—each isotope contains a different number of neutrons—and so the *nuclear* reactions

P.M.S. BLACKETT

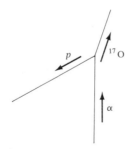

A cloud chamber is a device for rendering visible the paths of nuclear particles by the condensation of water droplets on the ions that the particles leave in their wakes as they pass through the gas in the chamber. This photograph shows the disintegration of a nitrogen nucleus by a fast α particle in a cloud chamber. This picture, taken by P. M. S. Blackett in 1925, is the first photograph of a nuclear reaction. Only one reaction event is seen amidst the tracks of many α particles that do not induce reactions.

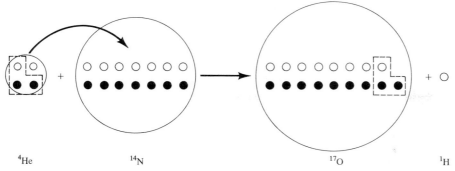

^{4}He ^{14}N ^{17}O ^{1}H

Figure 20-7   The net result of the reaction $^{14}N + {}^4He \rightarrow {}^{17}O + {}^1H$ is the transfer of one proton and two neutrons from the helium nucleus to the nitrogen nucleus, thereby forming ^{17}O and leaving a single proton (a hydrogen nucleus).

that are produced by the proton bombardment of ^{16}O, ^{17}O, and ^{18}O all have distinctive features.

Nuclear Energetics

Energy (or more correctly, mass–energy) must be conserved in any nuclear process. When protons bombard ^{7}Li, for example, the reaction produces two helium nuclei:

$$^7Li + {}^1H \longrightarrow {}^4He + {}^4He \qquad (20\text{-}6)$$

The mass of the various nuclei are

^{7}Li: 7.0160 AMU	^{4}He: 4.0026 AMU
^{1}H: 1.0078 AMU	^{4}He: 4.0026 AMU
8.0238 AMU	8.0052 AMU

The original nuclei are seen to have a combined mass that is greater than that of the product nuclei. Therefore, energy is *released* in this reaction and the mass difference ($8.0238 - 8.0052 = 0.0186$ AMU) is converted into 17 MeV of kinetic energy that is shared by the helium nuclei.

The $^7Li + {}^1H$ reaction was one of the first reactions studied when accelerators capable of producing high-speed nuclear projectiles were developed in the 1930's. The large energy release in this reaction prompted speculation that nuclei constituted an enormous reservoir of energy which could be tapped for useful purposes. Lord Rutherford, and other leading nuclear scientists of the day, scoffed at the idea (see the 1933 newspaper clipping on page 488). As late as 1937, shortly before his death, Rutherford stated that "the outlook for gaining

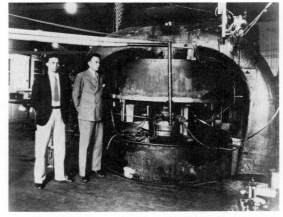

An early model of the cyclotron, invented by Ernest O. Lawrence (at the right) for the acceleration of nuclear particles to high speeds, and constructed at the University of California in the 1930's.

SEPTEMBER 12, 1933

Atom-Powered World Absurd, Scientists Told

Lord Rutherford Scoffs at Theory of Harnessing Energy in Laboratories

By The Associated Press

LEICESTER, England, Sept. 11.— Lord Rutherford, at whose Cambridge laboratories atoms have been bombarded and split into fragments, told an audience of scientists today that the idea of releasing tremendous power from within the atom was absurd.

He addressed the British Association for the Advancement of Science in the same hall where the late Lord Kelvin asserted twenty-six years ago that the atom was indestructible.

Describing the shattering of atoms by use of 5,000,000 volts of electricity, Lord Rutherford discounted hopes advanced by some scientists that profitable power could be thus extracted.

"The energy produced by the breaking down of the atom is a very poor kind of thing," he said. "Any one who expects a source of power from the transformation of these atoms is talking moonshine. . . . We hope in the next few years to get some idea of what these atoms are, how they are made and the way they are worked."

Lord Rutherford

Ernest Rutherford's keen insight had enabled him to make enormous progress in unraveling the mysteries of the nucleus, but his prophecy concerning the future of atomic power proved to be completely in error. The views expressed more than a decade earlier by the British scientist Sir Oliver Lodge were more accurate. In 1920 Lodge wrote, "The time will come when atomic energy will take the place of coal as a source of power. . . . I hope that the human race will not discover how to use this energy until it has brains enough to use it properly."

useful energy from the atoms by artificial processes of transformation does not look very promising." Although Rutherford's intuition in scientific matters had guided him to many important discoveries, his views on the prospects for nuclear energy were, within a few years, shown to be far too cautious.

20-4 NUCLEAR FISSION

The Splitting of Heavy Nuclei

In 1938, just before the outbreak of the Second World War in Europe, the German radiochemist Otto Hahn (1879–1968), working with Fritz Strassman (1902–   ), bombarded uranium with neutrons and studied the radioactive material that resulted from the interaction. Hahn and Strassman found that the products of the uranium-plus-neutron reaction included radioactive barium ($Z = 56$), an element with a mass much less than that of the original uranium ($Z = 92$). What kind of reaction could produce a nucleus so much lighter than the bombarded nucleus? The mystery was soon solved by Lise Meitner and Otto Frisch, refugees from Nazi Germany who were then working in Sweden. Meitner and Frisch suggested that the absorption of neutrons by uranium produced a breakup (or *fission*) of the nucleus into two fragments, each with a mass approximately one-half the mass of the original uranium nucleus:

$$U(Z = 92) + n \longrightarrow Ba(Z = 56) + Kr(Z = 36) \quad (20\text{-}7)$$

This type of nuclear reaction was quite unlike any that had been studied before 1938. All other reactions had involved the transfer of only a few protons and neutrons from one nucleus to another, as in the $^{14}N + {}^4He$ reaction. Fission, on the other hand, is the splitting of a heavy nucleus into two approximately equal fragments with the release of a substantial amount of energy. It was promptly recognized that the unique features of the

---

Otto Hahn was trained as a chemist but he showed little potential until he began graduate work toward a doctoral degree. In 1901 he received his Ph.D. from the University of Marburg. During 1905–6 he worked in Rutherford's laboratory, learning the techniques of radiochemistry. In 1918 he and Lise Meitner discovered the new radioactive element protactinium (Pa, $Z = 91$). Hahn and Strassman isolated barium from the products of neutron bombardment of uranium in 1938. Although the notion of nuclear fission occurred to Hahn, he was reluctant to put forward such an unheard-of suggestion. It remained for Meitner and Frisch to make the proposal. Hahn did not participate in the German atomic bomb effort and he remained a firm opponent of nuclear weapons.

fission process offered the possibility for the release of nuclear energy on a gigantic scale.

The Energy Release in Fission

The graph in Fig. 20-6 shows that the binding energy of uranium is approximately 7.5 MeV per particle, whereas the binding energies for barium and krypton are each about 8.5 MeV per particle. That is, the combined mass of barium and krypton is approximately 1 MeV per particle *less* than the mass of uranium. Thus, when a uranium nucleus splits into nuclei of barium and krypton, there is an energy release of about 1 MeV for each proton and neutron involved. The fission of each uranium nucleus therefore releases just over 200 MeV of energy.

Because the binding energy curve exhibits a smooth decrease from iron to uranium, there is nothing unique about the particular fission process, $U + n \rightarrow Ba + Kr$. Essentially the same amount of energy would be released in the fission of uranium into two other nuclei, for example,

$$U(Z = 92) + n \longrightarrow$$
$$Ce(Z = 58) + Se(Z = 34) \quad (20\text{-}8a)$$

or,

$$U(Z = 92) + n \longrightarrow$$
$$Xe(Z = 54) + Sr(Z = 38) \quad (20\text{-}8b)$$

Indeed, both of these fission processes, as well as many others, have been observed. Moreover, any heavy nucleus can undergo fission and many have been studied, but only two—uranium and plutonium—have been utilized in large-scale applications.

An energy release of 200 MeV per nucleus represents a staggering amount of energy that is available in a bulk sample of a heavy element. The fission energy that can be released from 1 kg of uranium is sufficient to raise the temperature of 200 000 000 gallons of water from room temperature to the boiling point (approximately 8×10^{13} joules).

Chain Reactions

When a heavy nucleus undergoes fission, not only are two lighter nuclear fragments formed, but two or three neutrons are released as well. Therefore, Eq. 20-7 expressed in more detail is

$$^{235}\text{U} + n \longrightarrow {}^{139}\text{Ba} + {}^{94}\text{Kr} + 3n \qquad (20\text{-}9a)$$

or,

$$^{235}\text{U} + n \longrightarrow {}^{139}\text{Ba} + {}^{95}\text{Kr} + 2n \qquad (20\text{-}9b)$$

Most of the isotopes produced in fission processes (for example, ^{139}Ba, ^{94}Kr, ^{95}Kr, as well as many others) are *highly radioactive.*

The fact that a fission event is induced by *one* neutron and the event releases *two* or *three* neutrons means that it is possible to construct a system in which the fission process is *self-sustaining.* If each of the neutrons released in a primary fission event is absorbed by another uranium nucleus, thereby producing additional events, the process multiplies rapidly and can consume all of the available uranium in a small fraction of a second. Figure 20-8 shows schematically the cascading of fission events (a *chain reaction*) that leads to the rapid release of the fission energy — a nuclear explosion. This is the principle of the *atomic bomb* (properly, a *nuclear* bomb).

In order for a fission device to explode, the cascading of the fission events is essential: the neutrons must be prevented from leaving the sample so that they are

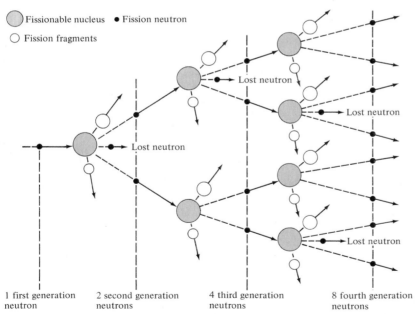

Figure 20-8   An uncontrolled chain reaction of fission events. Each event releases two or three neutrons; in each case two neutrons are shown initiating new fission events and the third neutron (if released) is assumed to leave the sample. The uncontrolled multiplication of fission events leads to a nuclear explosion.

The world's first detonation of a nuclear device occurred at the Trinity site near Alamagordo, New Mexico, on July 16, 1945. The sky was brightly illuminated by the rising cloud of incandescent gases.

available to induce additional fission events. If the sample is too small, neutrons will escape and an insufficient number of fission events will take place in a short time to constitute an explosion. (The sample will merely become hot.) But if the sample is large enough, the neutrons will be contained and an explosion will result. The minimum size is called the *critical mass* of the material. One of the major problems in constructing a nuclear bomb was to devise a method for bringing together two subcritical masses (which cannot explode) into a single mass that is greater than the critical mass (and which will immediately explode). This problem was solved by the scientists and engineers of the Manhattan Project in 1945. The details are still classified information but it is known that the critical mass of ^{235}U is several kilograms.

The first explosive atomic device was detonated on July 16, 1945, in the desert near Alamagordo, New Mexico. The device had been prepared by a huge scientific team from the Allied countries working in the Manhattan Project laboratory at Los Alamos, New Mexico. After the successful Alamagordo test, two weapons of different design were constructed and made available to the military. These weapons were dropped on the Japanese cities of Hiroshima and Nagasaki in August 1945. The explosions caused more than 100 000 casualties and forced the Imperial Japanese government to capitulate, thus ending the Second World War.

If the fission events in a sample of uranium are allowed to multiply in an uncontrolled way, an explosion results. But if the system is designed so that, on the average, exactly *one* neutron from each fission event triggers another event (Fig. 20-9), the fission energy can be released in a slow and controlled manner. This is the basic operating principle of the nuclear *reactor*. The construction and operation of reactors is discussed in the following section.

A self-sustaining chain reaction is analogous to population growth. An uncontrolled chain reaction, in which the number of neutrons continues to grow, corresponds to "population explosion." A controlled chain reaction, in which the number of neutrons remains constant, corresponds to "zero population growth."

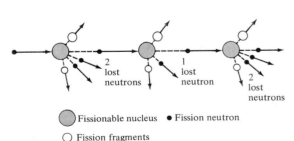

Fissionable nucleus • Fission neutron

○ Fission fragments

Figure 20-9 The rate at which energy is released from nuclear fission can be controlled by arranging a system in which exactly one neutron from each fission event initiates another event. In this way, the cascading process characteristic of an explosive device (Fig. 20-8) is avoided.

Plutonium

Naturally occurring uranium consists of the isotopes ^{238}U (99.3 percent) and ^{235}U (0.7 percent). The isotope that undergoes fission when it absorbs a slowly moving neutron is ^{235}U. When ^{238}U absorbs a slow neutron, ^{239}U is formed and fission does not take place. Consequently, natural uranium cannot be used in a conventional chain-reacting device because the abundant isotope ^{238}U absorbs too many neutrons for the reaction to be self-sustaining. One of the major problems faced by the Manhattan Project scientists was to devise a method to separate ^{235}U from natural uranium so that the fissioning isotope would be free from the difficulties produced by its isotopic partner. The separation techniques developed during the war years are still used to process the large quantities of uranium required by the nuclear power industry.

The isotope ^{238}U, although it does not undergo fission in the presence of slow neutrons, is nevertheless useful in preparing fission fuel. When ^{238}U absorbs a neutron, it becomes ^{239}U, a radioactive isotope. The β decay of ^{239}U produces the element *neptunium* (Np, $Z = 93$):

$$^{239}\text{U} \xrightarrow{\beta \text{ decay}} {}^{239}\text{Np} \qquad (^{239}\text{U half-life} = 23.5 \text{ min})$$

The new isotope ^{239}Np is also radioactive and decays to *plutonium* (Pu, $Z = 94$):

$$^{239}\text{Np} \xrightarrow{\beta \text{ decay}} {}^{239}\text{Pu} \qquad (^{239}\text{Np half-life} = 2.35 \text{ days})$$

^{239}Pu is also radioactive, but the half-life for decay is sufficiently long (24 360 years) that substantial quantities of the isotope can be accumulated. The importance of ^{239}Pu lies in the fact that it undergoes fission as readily as does ^{235}U. Therefore, ^{238}U, which serves only to prevent a self-sustaining chain reaction in natural ura-

---

The Manhattan Project was begun in 1941 in an effort to develop a nuclear weapon for possible use by the Allies in the Second World War. There was no guarantee that a bomb could actually be constructed, but it was suspected that scientists in Nazi Germany were already working on such a project and it seemed essential to start a counterproject. (The German team, working under the direction of the Nobel Prize winner Werner Heisenberg, never succeeded in producing a weapon.) The Manhattan Project was the largest-scale scientific effort ever undertaken (until the spaceflight program), involving thousands of scientists and technicians and costing more than a billion dollars. In December 1942, the first controlled self-sustaining chain reaction was achieved in a laboratory beneath the West Stands of Stagg Field (now demolished) at the University of Chicago. It required two and a half more years of intensive work before a nuclear weapon was actually built.

An underground weapons test at the U.S. Atomic Energy Commission's Nevada Test Site. The explosion carves out a huge cavern and the overburden collapses the crust, as seen in this photograph. The crust does not fracture catastrophically, and the radioactivity is confined to the cavern. Small breakthroughs have occurred, however, and on at least one occasion enough radioactivity was released to require the evacuation of the test area.

nium, can be converted into a useful fission fuel. Many of the fission devices now available, including low-yield military weapons, utilize plutonium as the fission material.

All of the *transuranic* elements (that is, elements with atomic numbers Z greater than that for uranium, $Z = 92$) are radioactive and have half-lives that are short compared with geologic times. Therefore, these elements, even though they may have been present when the Earth was formed 4.5 billion years ago, have since decayed to stable elements with lower Z. Only recently have trace amounts of plutonium (^{244}Pu, half-life = 80 million years) been found in natural uranium ores. This is the only transuranic isotope to have been found in Nature.

The Testing of Nuclear Devices

Although no nuclear weapons have been detonated in anger since the Hiroshima and Nagasaki bombs of 1945, the great powers of the world continue to expand and to improve their nuclear arsenals. In the absence of an international agreement to halt the construction of nuclear weapons, this practice will presumably continue indefinitely.

Weapons improvements require testing. Prior to 1963, such tests were usually carried out in the atmosphere where the appropriate measurements could be made with relative ease. As a result, large amounts of radioactive materials were released in the air and were carried by winds around the world. The situation became potentially hazardous—for example, radioactive strontium, ^{90}Sr, a fission product, appeared in alarming concentrations in milk—and in 1963 the Nuclear Test Ban Treaty was signed. According to this agreement, the atmospheric testing of nuclear devices is prohibited. Among the nuclear powers, the United States, Great Britain, and the Soviet Union signed the treaty. France and China, however, are not signatories, and these countries have continued their programs of atmospheric testing.

The United States and the Soviet Union have shifted their tests to underground sites where the radioactivity resulting from the blasts is confined and only tiny amounts from occasional tests have entered the atmosphere. The U.S. program is carried out primarily at the Nevada Test Site, a 1350-square-mile facility near Las Vegas, Nevada; the Soviet test stations are located in remote areas in Siberia.

One of the drawbacks to the extension of the Nuclear Test Ban Treaty to include underground testing has been the difficulty in policing the ban without on-site inspections. Recent advances in seismological techniques now make it possible to detect all but the very smallest underground nuclear tests anywhere in the world. With the new detection techniques it would not be necessary for a country to allow foreign inspectors into its territory to check for possible treaty violations. Perhaps the treaty will one day be extended and all nuclear testing will then cease.

20-5 NUCLEAR REACTORS

General Features

When a heavy nucleus undergoes fission, most of the 200 MeV of energy that is released appears in the form of kinetic energy of the fission fragments. The rapidly moving fragments collide with the atoms in the sample and quickly dissipate their energy. As a result, the energy that represents the mass difference between the heavy nucleus and the fission fragments eventually appears as *heat*.

In the generation of electrical power from fossil fuels, chemical energy is extracted by burning the fuels in order to heat water and convert it into steam. The steam is then used to turn a turbine which operates an electrical generator. Many of the nuclear power plants in operation today are similar in design. The main difference is that a fission reactor is used to produce the high pressure steam — the subsequent steps in generating electricity are the same as those in a conventional power plant.

A schematic diagram of a pressurized water reactor is shown in Fig. 20-10. Water is pumped through the core of the reactor which can be at a temperature of 1200 °C. (A reactor could operate more efficiently if the temperature were even higher, but the temperature must be maintained safely below the melting point of the core materials.) High-pressure water emerges at a temperature of 300 °C and converts the water in a second loop into steam. This steam passes into a turbine where it turns the blades at high speeds. The turbine shaft is connected to an electrical generator that produces electrical power which is fed into the power network over conventional transmission lines.

COMBUSTION ENGINEERING, INC.

Cutaway view of a nuclear power plant. The reactor is in the center. Notice the control rods that enter the reactor through the top and the air-tight door that leads to the interior of the large outer containment vessel. The large pipes carry steam to the turbines on either side of the reactor. The man in the doorway at the lower right has been included to give an indication of the scale.

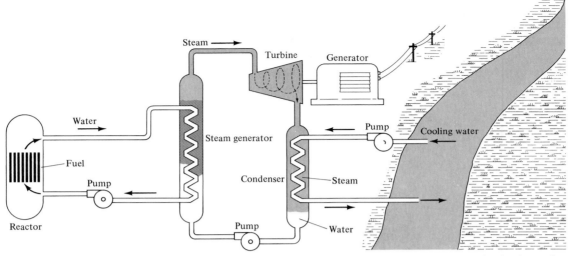

Figure 20-10 Schematic diagram of a nuclear power plant. The water in the loop that passes through the core of the reactor is at high pressure. This type of system is called a *pressurized water reactor* (PWR). Here, the cooling water is shown being drawn from a river, but many of the newer plants use cooling towers so that excess heat is exhausted into the atmosphere instead of bodies of water.

TABLE 20-2 SIGNIFICANT EVENTS IN THE DEVELOPMENT OF NUCLEAR POWER

1939	Discovery of nuclear fission.
1940	Discovery of plutonium.
1942	First self-sustaining fission chain reaction.
1945	First successful test of an explosive fission device; first (and only) use of nuclear weapons in warfare.
1946	U.S. Atomic Energy Commission (AEC) established.
1951	First significant amount of electric power (100 kW) produced from a test reactor.
1954	Commissioning of first nuclear-powered submarine, *Nautilus*.
1957	First reactor designed exclusively for the production of commercial electric power becomes operational.
1972	First breeder reactor becomes operational (USSR).
1975	AEC split into Nuclear Regulatory Commission (NRC) and Energy Research and Development Administration (ERDA).

Figure 20-10 also shows that the steam exhausted from the turbine (now at a lower temperature) is condensed back to water by cooling coils supplied with water from some sort of reservoir. The source of this cooling water can be a river or bay, or it can be water that is circulated through a cooling tower. In the former case, the reactor's surplus heat is exhausted into the water system, whereas in the latter case it is released into the atmosphere. We will return to this problem of *thermal pollution* later in this section.

The Moderation and Control of Fission Neutrons

The neutrons that are emitted in the fission process have an average energy of several MeV and therefore move with very high speeds. Fission neutrons are *fast* neutrons. The fission of ^{235}U or ^{239}Pu, however, is considerably more efficient for *slow* neutrons (neutrons with energies of an eV or less) than for fast neutrons. Consequently, the design of an efficient reactor must include a provision for slowing down (or *moderating*) the fast fission neutrons.

If a billiard ball makes a head-on collision with an identical stationary ball, the laws of energy and momentum conservation demand that the moving ball stop and that the struck ball move off with the same velocity as the original incoming ball. The same principles hold when a neutron collides with the nucleus of an atom. In the head-on collision between a neutron and a stationary proton (which has a mass essentially equal to the neutron mass), the proton is set into motion and the neutron comes to rest. Because the proton (unlike the neutron) is a *charged* particle, it rapidly loses its energy through electrical interactions with atomic electrons. Even if the collision is not exactly head-on, a substantial fraction of the neutron's kinetic energy will be transferred to the proton. If the struck nucleus is more massive than the neutron, a smaller amount of energy will be transferred and a large number of collisions will be necessary to slow the neutron from the MeV energy range to an energy near 1 eV.

The most effective neutron moderator is *hydrogen,* the only material whose nuclear mass is equal to that of the neutron. However, hydrogen has a serious drawback as a moderator: instead of always deflecting a neutron and carrying off some of its kinetic energy, hydrogen will sometimes *capture* a neutron, forming deuterium, ^2H, thereby preventing the neutron from inducing a fission event. A practical moderating substance is therefore one that has a small nuclear mass and a low probability for capturing a neutron. Nevertheless, the general practice is simply to accept the losses associated with the capture of neutrons by hydrogen and to use water (H_2O) as both the moderating substance and the coolant.

In addition to a moderator that slows down the fast fission neutrons, a reactor must be provided with a means for controlling the number of neutrons available to induce fission events so that each event contributes, on the average, exactly one neutron that triggers a new event. This function is performed by a material, such as boron, that has a high probability of capturing slow neutrons. By moving boron carbide *control rods* into or out of the reactor core, the number of effective neutrons per fission event can be maintained at the desired value. Furthermore, by dropping into the core several control rods, the chain reaction can be quickly stopped in the event that the reactor must be shut down for maintenance or in an emergency.

The central part of a reactor therefore consists of four main components: uranium or plutonium fuel, a moderator, control rods, and the heat transfer coils that carry

The reactor vessel head is being lowered over the control rod drive shafts after the core was first fully loaded with fuel assemblies. This Point Beach nuclear plant at Two Creeks, Wisconsin is a pressurized water nuclear reactor which began operating in 1970.

SOUTHERN CALIFORNIA EDISON CO.

The San Onofre nuclear generating station near San Clemente, California has an electrical capacity of 430 MW and began operation in 1967.

the water or other liquid to be heated. The uranium or plutonium fuel is in the form of long cylinders which are clad with a strong metal jacket. The fuel rods must be able to withstand the high temperatures at which all reactors operate. The fragments that result from the fission process are always highly radioactive. Therefore, another requirement on the cladding of fuel rods is that it not leak its radioactive contents into the reactor during operation nor into the environment when removed for replacement.

The Nuclear Power Industry

In 1957 the Duquesne Light Company began operating the world's first commercial power-producing nuclear reactor. This unit, located at Shippingport, Pennsylvania, produces 90 MW of electrical power. By present standards, Shippingport Unit 1 is a small power station. Most of the nuclear power plants recently opened or under construction have power ratings in the vicinity of 1000 MW.

By 1975 there were 53 nuclear power stations operating in the United States, with a total generating capacity of 34 000 MW. This figure represents about 7 percent of the total electrical generating capacity of the United States. More than 50 plants are under construction and over 100 are under firm order. It is anticipated that the fraction of the total electrical generating capacity contributed by nuclear power plants will be 21 percent in 1980 and 33 percent in 1985. By the year 2000, the Energy Research and Development Administration (ERDA) expects that nuclear energy will account for the majority of electrical power produced in this country.

Benefit versus Risk

The energy requirements of this country and of the world are increasing at a rapid rate (see Chapter 8). We are therefore faced with the necessity of providing more and more energy, particularly in the form of *electrical* energy. At the present time, most of the world's electrical energy is generated in power plants that operate by burning fossil fuels, primarily coal. The supply of coal for this purpose is probably secure for the next few hundred years, but eventually the reserves of coal, as for all fossil fuels, will be exhausted. Moreover, unless alternative methods of burning coal are developed (for example, by converting coal into cleaner-burning gases), the atmosphere will be increasingly burdened with fly ash, smoke, sulfur dioxide, and other noxious fumes from these plants.

Nuclear reactors offer the prospect of a greater available fuel supply as well as the advantage of operation without smoke and fumes. But reactors have their own peculiar set of disadvantages, mainly associated with the production of radioactivity in the fuel rods. We can divide the problems into several categories:

(1) *Explosions and melt downs.* The interior of an operating reactor is always radioactively "hot" because fission reactions produce radioactive fragments and because neutrons produce radioactive isotopes when they are captured by most reactor construction materials. One of the fears that has been expressed concerning reactors is that in the event of some sort of accident, radioactive material could be strewn about the surrounding countryside with catastrophic consequences. (See Section 21-3 for a discussion of the biological effects of radiation.) The likelihood of the occurrence of such a disaster is extremely small. The construction of a reactor is entirely different from that of a nuclear weapon, so that an uncontrolled chain reaction leading to the weaponlike explosion of a reactor is not possible.

Of far greater significance is the possibility of a failure of the cooling system; this could result in the *melt down* of the reactor core. Every reactor is equipped with a "backup" cooling system and so the probability that a melt down will ever occur is very small. But if a situation ever developed in which both cooling systems failed, the sudden temperature increase in the core would cause the core to melt. In such an accident the fuel rods would probably rupture and highly radioactive material would be released. But to forestall the possible spread of the dangerous radioactivity, every reactor core is surrounded by two containment vessels.

The only nuclear power plant to have suffered a melt down leading to the release of substantial amounts of radioactivity is the unit at Windscale, England. The graphite moderator of the Windscale reactor caught fire, causing some of the fuel elements to melt. A considerable amount of radioactive iodine, ^{131}I, was spread over the countryside and contaminated crops and milk supplies. The Windscale reactor did not have an outer protective container, but all of the reactors in U.S. nuclear electric stations do have containment vessels. The only other melt down to have occurred in a commercial power plant took place in 1966 at the Fermi reactor, 18 miles downriver from Detroit. The containment vessel was not breached and no serious leakage of radioactivity occurred.

Reactor engineers have been exceedingly conservative in the design of the safety features in nuclear reactors. All of the parts that are subject to high pressures or high temperatures are rated far in excess of the operating values. Every control circuit has at least one backup system and usually more than one. There is an emergency cooling system which comes into operation if the primary system fails or is overloaded. And, finally, in the event of some unforeseen difficulty, the reactor will *fail safe* and shut down. The likelihood of an explosion or a melt down has been reduced to a very low level, but, of course, these disasters are always *possibilities*. Critics have charged that insufficient attention has been given to the improvement of reactor safety measures, particularly those relating to possible melt downs. Nevertheless, the nuclear power industry has a better safety record than any other major industry.

(2) *Radioactive emissions.* Every reactor in normal operation releases small amounts of radioactivity into the atmosphere. Maximum limits have been set for the amount of radioactivity that any reactor can emit and most reactors release far less than the limit. But no amount of radioactivity moving freely through the air is "good," and efforts are continuing to reduce these emissions to the absolute minimum. At the present level of emission, persons living near nuclear power plants receive considerably less radiation from the plant then they do from other sources (cosmic rays, medical X rays, color television sets, and so forth). Critics contend that even this small increment in the level of radiation is unwarranted and leads to increased danger of leukemia and other radiation-induced cancers. Nuclear proponents admit that all radiation is dangerous to humans but that the small increases caused by reactor operations pose such a tiny additional health hazard that the benefits far outweigh the risks.

It is interesting to note that even a coal-burning power plant releases some radioactivity into the air due to the occurrence in the coal supply of minerals that contain radioactive elements, particularly radium. These emissions often exceed those of nuclear power plants in normal operation.

(3) *Fuel reprocessing.* In the course of normal operations, the fuel rods in a reactor undergo various changes and after a time must be replaced with new rods. When safety or reduced power output dictates the removal of a rod, it contains, in addition to the radioactive fission fragments, a substantial fraction of the original uranium or plutonium. After a "cooling-off" period, during which the short-lived radioactivity decays, the used fuel rods are shipped to a reprocessing plant where the remaining fuel is removed and incorporated into new rods. Those radioactive isotopes that are useful in medical, industrial, and research applications are separated and prepared in convenient forms. The remaining radioactive material is put into a form suitable for disposal (see below). All of this handling of the "hot" fuel rods must be carried out remotely behind thick shielding walls. Close controls are necessary to ensure that the amount of radioactivity released during reprocessing operations is held to minimum levels. In the United States, only pilot reprocessing operations have been carried out so far.

(4) *Disposal of radioactive wastes.* Although much of the material in used fuel rods can be recovered in the reprocessing operation, there remains a quantity of radioactive "garbage" that is not particularly useful. As more and more nuclear power plants become operational, these materials accumulate at an increasing rate. The safe disposal of radioactive wastes represents a serious problem because some of the isotopes have half-lives of hundreds or thousands of years. Various methods of disposal have been used. The earliest was simply to dump steel containers of the wastes at sea. But the containers corroded and eventually leaked radioactivity into the water. The practice of disposal at sea has now been halted.

Having found no really acceptable long-term solution to the disposal problem, radioactive wastes are now stored in liquid form in huge million-gallon stainless steel vats in concrete-shielded underground bunkers. Because of the corrosion problem, the storage sites are continually monitored for leaks and the highly radioactive material is transferred periodically to new containers. At the present time, nearly 100 million gallons of radioactive wastes from reactors are stored in this way.

It has been proposed that radioactive wastes be deposited in abandoned salt mines. One of the main problems in waste disposal is to ensure that the radioactivity does not enter a water system that eventually connects with the population's supply. Because salt is quite soluble in water, the existence of salt deposits indicates that little or no water seeps through the region. A salt-mine depository should therefore ensure that the radioactivity will not enter the underground water system. Although such a plan appears reasonable, there are many uncertainties: for example, oil wells or dry holes that penetrate the salt deposit might connect with water-bearing layers and could conceivably flood the mine. Radioactive material might then be carried away to the water supply of a nearby town or city. Consequently, the proposals for underground storage of radioactive wastes in salt mines as well as in rock layers are still under study and are directed toward finding a site with no possible connections to the local water system.

(5) *Thermal pollution.* Any electrical generating plant that uses steam to drive turbines must have a cooling system to condense the steam back into water. The cooling system necessarily exhausts heat into a water system or into the air. In this regard, a nuclear power plant is no different from a coal-burning plant: both systems release excess heat into the environment causing *thermal pollution.* Because nuclear power plants are, at present, less efficient than coal-burning plants (32 percent compared to 40 percent), a nuclear plant will exhaust about 1.4 times as much heat to the environment as will a coal-burning plant with the same power output.

If the heat is exhausted into a moving water system (a river or a bay), the water temperature will be increased measurably for some distance downstream. The amount of temperature increase depends on the power level of the plant, the energy conversion efficiency, and the flow rate of the water reservoir. Extensive studies have shown no drastic changes in the marine ecology downstream from reactor sites although some changes in the populations of marine life forms have been noted. As mentioned in Section 8-6, the effects of thermal pollution in static reservoirs, such as lakes, are potentially more serious than in moving water systems.

Instead of exhausting heat into a water reservoir, a *cooling tower* can be used to dissipate the heat into the atmosphere. One type of cooling tower is shown in Fig. 20-11. Air is pulled up through the tower by large fans and this continual flow of air removes heat from the water in the reactor's cooling loop. Although exhausting

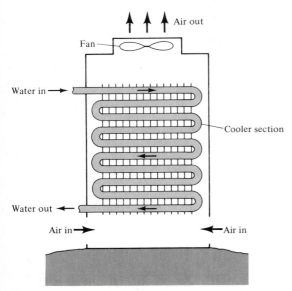

Figure 20-11   A cooling tower for removing heat from the water in the cooling loop of a steam power plant (either nuclear or coal-burning).

heat into the atmosphere does influence to some extent the local weather conditions downwind, it is generally believed that cooling tower systems perturb the environment less severely than systems that exhaust into water reservoirs.

The operation of nuclear power plants certainly involves risks. But almost every aspect of our modern technological society—airplane or automobile travel, handling electrical equipment, even crossing the street—involves a certain risk. The important issue is whether the benefits are worth the risk. The evidence that we now have appears to favor nuclear power, but this is no reason to ignore the possible risks nor to cease efforts to improve reactor safety.

The Calder Hall Nuclear Power Station in England. Four giant cooling towers exhaust the surplus heat into the atmosphere. The plumes are condensed water vapor (literally, clouds).

Breeder Reactors

The only naturally occurring isotope that undergoes fission with slow (moderated) neutrons is ^{235}U. Because ^{235}U constitutes only a small fraction (0.7 percent) of natural uranium, enormous quantities of uranium ore must be processed in order to provide fuel for slow-neutron reactors. The supplies of uranium ores are not unlimited. If we continue to use ^{235}U in the nuclear power plants that are projected until the year 2000, we will have exhausted all of the known reserves of high-quality ores and will then be using low-grade ores. The cost of separating ^{235}U will therefore increase and electrical energy will be considerably more expensive.

We know that the abundant uranium isotope, ^{238}U, can be converted into ^{239}Pu which is an excellent fuel in slow-neutron reactors. Furthermore, thorium is a plentiful element and the single stable isotope ^{232}Th can be converted into ^{233}U by a neutron capture reaction followed by a β decay, a process analogous to the ^{238}U $\rightarrow$ ^{239}Pu conversion. ^{233}U is radioactive but the half-life is long (162 000 years) so that bulk quantities of the isotope can be accumulated. ^{233}U is similar to ^{239}Pu in that it undergoes slow-neutron fission and can be used as a nuclear fuel.

Can we produce sufficient quantities of ^{239}Pu and ^{233}U to supply the increasing number of reactors with relatively inexpensive fuel? In any type of reactor, one neutron from each fission is required simply to maintain the chain reaction by inducing a new fission event. If one additional neutron is captured by ^{238}U or ^{232}Th, the fuel supply will remain constant—just as many fuel nuclei are produced as are used. A reactor which produces fuel nuclei as it operates is called a *breeder reactor*. The most important types of breeder reactors will be those that produce *more* fuel nuclei than they consume. These reactors, which have not yet been perfected to the degree of the present slow-neutron reactors, appear to offer the best hope for continued cheap electrical power during the unknown time interval before fusion reactors become operational (see the following section). A breeder reactor will not only produce fuel to compensate for its own consumption, but it will also provide fuel for new reactors.

Experimental breeder reactors have been in operation in the United States and elsewhere for a number of years. But the first breeder reactor to produce commercially useful quantities of electrical power was a Soviet unit. This reactor, on the eastern shore of the Caspian Sea, began operating in early 1972 and produces 350 MW of electrical power. Also in 1972 the United States announced a stepped-up program to make a commercially viable breeder reactors a reality by the mid-1980's.

20-6 NUCLEAR FUSION

Energy from Light Nuclei

When a heavy nucleus such as uranium undergoes fission, energy is released because the binding energy per particle is greater for the fission fragments than for the original nucleus. In fact, energy will be released in *any* type of nuclear process that results in an *increase* of the binding energy. How can we take advantage of this fact in a process other than fission? Referring to Fig. 20-6, we see that the binding energy per particle increases with mass number for A less than about 50. Therefore, if we bring together two light nuclei to form a more massive nucleus with $A < 50$, energy will be liberated in the reaction. For example, when two deuterium nuclei combine to form a helium nucleus, approximately 24 MeV of energy is released:

$$^2\text{H} + {}^2\text{H} \longrightarrow {}^4\text{He} + 24 \text{ MeV} \qquad (20\text{-}10)$$

Reactions in which two light nuclei combine and release energy are called *fusion reactions* (the nuclei *fuse* together).

Actually, when two deuterium nuclei collide and interact, the production of ^{4}He is relatively unlikely. It is much more probable that there will take place a reaction which produces either a proton or a neutron:

$$^2H + {}^2H \longrightarrow {}^3H + {}^1H + 4.0 \text{ MeV} \quad (20\text{-}11a)$$

$$^2H + {}^2H \longrightarrow {}^3He + n + 3.3 \text{ MeV} \quad (20\text{-}11b)$$

That is, in the $^2H + {}^2H$ reactions, approximately 1 MeV of energy is released for each of the four particles involved. This is about the same efficiency of mass-to-energy conversion that occurs in fission (approximately 200 MeV for the 236 particles involved in $^{235}U + n$ fission).

How does the availability of energy from fusion reactions influence the world energy picture? We need only look to the oceans for the answer. Deuterium constitutes 0.015 percent of natural hydrogen, and 1/9 of the mass of water is in the form of hydrogen. If we could extract 10 percent of the deuterium from the ocean waters, we could generate enough fusion energy to supply the entire world at the present rate of energy consumption for a billion years! The fusion of deuterium therefore represents an essentially inexhaustible supply of energy.

Thermonuclear Reactions

Although both release energy, the fission and fusion processes differ in a significant respect. In the fission case, the electrical repulsion that exists between the two parts of the nucleus which become fission fragments *assists* in breaking the nucleus apart. In a fusion reaction, on the other hand, the electrical repulsion between the two nuclei *resists* their combining into a single nucleus. Consequently, a fusion reaction between two deuterium nuclei will take place only if the nuclei are projected toward one another with high speeds.

How can we produce high-speed collisions between deuterium nuclei? One way would be to use some sort of accelerator (for example, a cyclotron) to project deuterium nuclei onto a deuterium target. In fact, this technique has been extensively used to study the $^2H + {}^2H$ reactions. But such a method is not practical if we expect to produce useful amounts of fusion energy. Another way is to take advantage of the fact that the atoms in a gas are continually in motion—if we need high speeds, we raise the temperature. However, in order to achieve the high speeds that are necessary to produce fusion reactions among deuterium atoms, a temperature of about 10 million degrees is needed! Reactions that require these extraordinarily high temperatures are called *thermonuclear reactions*. At these temperatures, *atoms* cannot exist. The violent collisions strip the electrons from the atoms and leave the gas in the *plasma* state, a sea of rapidly moving electrons and nuclei.

The interior of the Sun is at a sufficiently high temperature that fusion reactions take place. Indeed, the Sun's source of energy is the fusing together of hydrogen in the core to produce helium (see Section 20-7). On the Earth, thermonuclear temperatures can be generated in the explosions of nuclear fission devices. The *hydrogen bomb* operates on this principle: a fission device serves as a high-temperature trigger to induce the fusion of hydrogen isotopes (deuterium and tritium) with the release of enormous amounts of energy. A hydrogen bomb (or *thermonuclear bomb*) can be constructed to yield considerably more energy than would be practical with a device that uses only the fission of uranium or plutonium.

Fusion Reactors—Prospect
for the Future

The fantastic potential that fusion reactions have for the production of useful energy has been realized for many years, but the technical problems in building a practical fusion reactor are much more complex than those involved in fission reactors. How can a plasma at 10 000 000 °K be confined and controlled so that thermonuclear energy is made available at a steady rate? Several methods are being investigated. One is to confine the plasma in a magnetic field (see Section 12-4) while the nuclei interact. Another is to start with a small solid pellet of fusion material and to drive the nuclei toward one another by blasting the pellet from all sides with a powerful burst of laser radiation or high energy electrons. Perhaps one of these schemes will prove successful—perhaps by 1990 or 2000 we will have an operating fusion reactor.

There are many advantages to fusion-produced power. The fuel supply is plentiful and relatively inexpensive. (The main expense is in the separation of deuterium from water.) Moreover, the products of fusion reactions are either stable isotopes or they are only weakly radioactive. Radioactivity will also be produced by the neutrons released in the reactions when they are

Thermonuclear Fission—A New Twist on Nuclear Energy

Apart from the technological problems of constructing a fusion reactor, there are two main difficulties with the fusion process as a source of power. The first is that the fusion reactions, $^2H + ^2H$ and $^2H + ^3H$, produce neutrons. In order to utilize the kinetic energy of these neutrons, they must be slowed down in some material thereby causing the material to become heated; the extraction of this heat energy is an inefficient process. Second, the slow neutrons are absorbed by the reactor materials which then become radioactive. Radioactivity is also present in the form of tritium, ^3H, which will be produced in massive quantities in fusion reactors. There would be substantial advantages if a nuclear reaction were used in which only charged particles are emitted and which leaves no radioactive residue.

It has recently been proposed that the boron-plus-hydrogen reaction could be used to meet these criteria. In this reaction, the nucleus ^{11}B combines with a proton to produce three ^4He nuclei (α particles):

$$^{11}B + {}^1H \longrightarrow {}^4He + {}^4He + {}^4He + 8.7 \text{ MeV} \qquad (20\text{-}12)$$

This reaction is radically different from those that have been proposed for use in fusion reactors. Usually, it is possible to extract energy from nuclei only when a heavy nucleus undergoes fission or when two light nuclei undergo fusion. The boron-plus-hydrogen reaction, however, is really a fission process involving a light nucleus. Ordinarily such a process requires the input of energy. But because the end products of the $^{11}B + {}^1H$ reaction are tightly bound helium nuclei, this reaction actually releases energy.

Boron is a plentiful element (found in the oceans and in dry lake beds), and so there is an abundant fuel supply. The primary difficulty is that the $^{11}B + {}^1H$ reaction requires a substantially higher temperature for ignition (about 3×10^9 °K) than do the reactions involving deuterium. (This is because of the greater nuclear charge of the boron nucleus.) However, these extremely high temperatures can probably be developed eventually. The high-temperature fission of boron has been termed *thermonuclear fission*.

In a "conventional" fusion reactor, the neutrons are trapped and their kinetic energy is converted into heat for the purpose of boiling water to drive a steam generator. Because the products of the $^{11}B + {}^1H$ reaction are rapidly moving charged particles, they automatically represent an electrical current and this can be converted directly into useful output power without the necessity of a thermal cycle. Moreover, the products of boron fission are not radioactive.

Although the thermonuclear fission of boron may not be attempted until the deuterium systems are thoroughly explored, this new idea is potentially of great importance in the eventual generation of clean, inexpensive nuclear power.

captured in the materials of the reactor. But even so, the amount of radioactivity associated with the operation of a fusion reactor will be only a small fraction of that produced in the several phases of fission reactor operations.

20-7 NUCLEAR REACTIONS IN THE SUN

The Proton–Proton Chain

Our Sun is a rather typical star. Like most other stars, the Sun consists primarily of hydrogen, with some helium and a small amount of heavier elements. When a mass of gas in space begins to contract because of the mutual gravitational attraction among the atoms, some of the gravitational potential energy is converted into kinetic energy. That is, the gas atoms move more rapidly and the temperature of the gas increases. At the same time the density of the gas increases. In the 1920's it was believed that gravitational potential energy was the only source of energy available to a star. But it soon became apparent that a star cannot live by gravity alone. A typical star simply radiates too much energy during too long a time for gravity to be the sole source. It was soon realized that a star's primary source of energy (after the initial contraction phase) depends upon *nuclear reactions*.

We know from our discussion of fusion reactions (Section 20-6) that these reactions can take place only at high temperatures. Therefore, nuclear processes cannot begin in a condensing star until the gravitational forces have converted the core of the star into a hot, dense mass of hydrogen. Moreover, when the core has become sufficiently hot and dense, nuclear reactions *must* automatically begin. When the central temperature has reached about 10 million degrees (10^7 °K) and the central density has reached about 100 g/cm³, two hydrogen nuclei (protons) can fuse together, producing a nucleus of deuterium together with a positron (e^+) and a neutrino (ν_e):

$$^1\text{H} + {}^1\text{H} \longrightarrow {}^2\text{H} + e^+ + \nu_e + \text{energy} \quad (20\text{-}13)$$

Deuterium reacts readily with hydrogen to yield the light isotope of helium, ³He:

$$^2\text{H} + {}^1\text{H} \longrightarrow {}^3\text{He} + \gamma + \text{energy} \quad (20\text{-}14)$$

Finally, two ³He nuclei react in the following way:

$$^3\text{He} + {}^3\text{He} \longrightarrow {}^1\text{H} + {}^1\text{H} + {}^4\text{He} + \text{energy} \quad (20\text{-}15)$$

This reaction produces the normal isotope of helium (^{4}He) and completes the series of reactions.

Looking at the summary in Fig. 20-12 we see that a total of six hydrogen atoms have participated in these reactions and in the final reaction two hydrogen atoms have been returned. The net result of this series of reactions—called the *proton–proton* or *p–p chain*—is the conversion of four hydrogen atoms into one helium atom with the simultaneous release of energy.

The net output of energy in the *p–p* chain is approximately 26 MeV, or about 6.5 MeV per hydrogen atom consumed. (Compare this with the energy release per particle involved in the fission of uranium.) Each kilogram of hydrogen that is converted into helium produces approximately 6×10^{14} J. In the Sun, the *p–p* reactions convert hydrogen at a rate of about 6×10^{11} kg/s and generate about 4×10^{26} watts of power. [Recall that a 1000-MW (10^9-watt) power plant is considered a very large installation.]

When we look at the Sun, we see only the radiation from the surface layer of gases—we cannot *directly* see any of the results of processes taking place deep in the interior. How do we know, then, that the hydrogen fusion reactions are actually taking place in the Sun's core? Our evidence is based entirely on laboratory measurements and theoretical calculations of the ways in which hydrogen and helium nuclei behave. We study the interaction processes in controlled experiments designed so that the nuclei have the same energies that they have in the hot interior of the Sun. From the results of these investigations, we deduce that the hydrogen and helium nuclei in the Sun must react according to Eqs. 20-13–20-15. (Moreover, there appears to be no *other* way to account for the huge amount of energy required to keep the Sun shining.)

The Stable Star

Prior to the onset of the *p–p* chain of reactions, the generation of energy within a star takes place exclusively through the conversion of gravitational potential energy. The star necessarily continues to grow smaller in size during this phase. When thermonuclear reactions begin to take place in the core, the energy released in these reactions produces an outward pressure that soon becomes equal to the inward pressure due to gravity. Thus, when thermonuclear processes begin, a star no longer contracts but becomes stabilized in size (Fig.

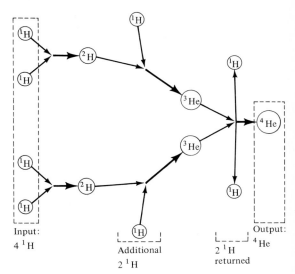

Figure 20-12 Schematic representation of the chain of reactions that leads from hydrogen to helium. The input is 4 atoms of ^{1}H and the output is 1 atom of ^{4}He. (Notice that 2 additional atoms of hydrogen enter the chain, but they are recovered when the ^{3}He nuclei interact.) The net energy release in this series of reactions is approximately 26 MeV.

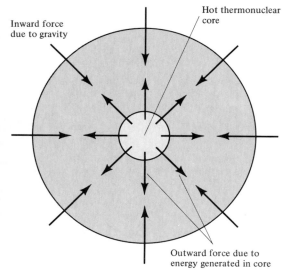

Inward force due to gravity

Hot thermonuclear core

Outward force due to energy generated in core

Figure 20-13 A condensing star ceases to contract and becomes stabilized in size when thermonuclear reactions begin in the core. In the equilibrium situation, the inward pressure due to gravity is just balanced by the outward pressure due to energy generation in the core.

20-13). Most stars (including our Sun) spend most of their lives in this condition—stable in size and generating energy through the conversion of hydrogen to helium.

The important stages in the formation of a star can be summarized as follows:

(1) Gravitational contraction and the conversion of gravitational energy into kinetic energy.

(2) Further increase in temperature due to contraction; infrared and red radiation; the proto-star becomes a star.

(3) The central temperature and density increase to about 10^7 °K and 100 g/cm³; the p–p chain of reactions begins, converting hydrogen to helium and releasing energy.

(4) The star becomes stabilized in size and in energy output.

What happens next in the history of a star? How are the elements heavier than helium formed?

The Formation of Carbon

After helium is prouduced in a star's core through the reactions in the p–p chain, we would expect that additional fusion reactions involving hydrogen plus helium and helium plus helium would take place. But consider the results of such reactions:

$$^1\text{H} + {}^4\text{He} \longrightarrow {}^5\text{Li} \qquad (20\text{-}16)$$

$$^4\text{He} + {}^4\text{He} \longrightarrow {}^8\text{Be} \qquad (20\text{-}17)$$

Here is an important point that puzzled astronomers and physicists for many years: neither ^5Li nor ^8Be is stable! Each of these nuclei disintegrates within a tiny fraction of a second into the original nuclei. Therefore, ^5Li and ^8Be cannot be parts of a continuing chain of fusion reactions. The building of nuclei heavier than helium appears to be blocked by the absence of stable nuclei with $A = 5$ and $A = 8$. (These are the only mass numbers between hydrogen and the heavy radioactive elements for which there are no stable nuclei.) This problem was not solved until the 1950's when laboratory experiments concerned with the properties of ^{12}C showed that a helium–helium reaction more complex than that in Eq. 20-17 is possible. The blockage caused by the instability of ^5Li and ^8Be is overcome by the combining of *three* helium nuclei to form ^{12}C:

$$^4\text{He} + {}^4\text{He} + {}^4\text{He} \longrightarrow {}^{12}\text{C} \qquad (20\text{-}18)$$

Because the nuclear force is effective only over a very short distance (see Section 6-6), it is extremely unlikely that in a collection of moving helium nuclei, *three* nuclei will come sufficiently close together at the same time so that a carbon nucleus can be formed. This process is so rare that it has never been observed in the laboratory. However, we *do* know enough about the properties of colliding helium nuclei and about the structure of ^{12}C to be able to predict with confidence that the reaction, $3\ ^4He \rightarrow\ ^{12}C$, *will* take place if the temperature and the density of the gas are sufficiently high. Within the core of a star, the conditions of high temperature and high density allow enough of these helium fusion reactions to take place so that essentially all of the carbon in the Universe is actually formed in this way.

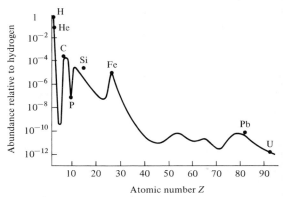

Figure 20-14   Relative abundances of the chemical elements in the Universe. The values are based primarily on spectroscopic observations of the light from stars and on the composition of meteorites. Notice the peak in the vicinity of iron. The shape of this graph is explained in terms of the nuclear reactions that take place within stars.

Heavier Nuclei

Once the great leap from ^4He to ^{12}C has taken place in a star, a whole host of nuclear reactions can occur. Fusion-type reactions involving charged particles appear to be responsible for synthesizing all of the elements up to iron. At this point the electrical repulsion between colliding nuclei becomes too strong to permit heavier nuclei to be formed by charged-particle reactions. Nuclei from iron to the heaviest that are known are formed by the capture of *neutrons*. By using laboratory measurements of nuclear properties and the current models of the conditions in the interiors of stars, it has been possible to account in a surprisingly accurate way for the observed abundances of the elements in the Universe (Fig. 20-14).

The Carbon–Nitrogen Cycle

Although energy production in the Sun is almost exclusively the result of the *p–p* chain of reactions, some types of stars utilize a different set of reactions. When the central temperature of a star exceeds about 2×10^7 °K, the most important energy source is the *carbon–nitrogen cycle* of nuclear reactions. The net result of the C–N cycle is exactly the same as that of the *p–p* chain, namely, the conversion of four atoms of hydrogen into one atom of helium with the release of about 26 MeV of energy.

In the C–N cycle, carbon acts as a kind of catalyst, participating in the conversion, $4\ ^1H \rightarrow\ ^4He$, but not itself being consumed in the process. The cycle of reac-

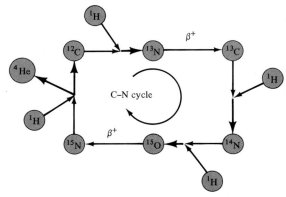

Figure 20-15   The carbon–nitrogen cycle of nuclear reactions which converts 4 ^1H atoms into 1 ^4He atom. About 26 MeV is released in the C–N cycle.

tions is shown schematically in Fig. 20-15. Notice that four hydrogen atoms are absorbed in these reactions, beginning with the capture of a proton by ^{12}C. In the final reaction, a helium atom is formed and ^{12}C is returned to participate in additional reactions.

The C–N cycle was first proposed in 1937 as a source of energy in stars by the American physicist Hans Bethe (1906–). The 1967 Nobel Prize in physics was awarded to Bethe for his work on energy generation processes in stars.

SUGGESTED READINGS

N. Feather, *Lord Rutherford* (Crane, Russak, New York, 1973).

L. Lamont, *Day of Trinity* (Atheneum, New York, 1965).

Scientific American articles:

W. A. Fowler, "The Origin of the Elements," September 1956.

O. Hahn, "The Discovery of Fission," August 1965.

QUESTIONS AND EXERCISES

1. The nucleus ^{212}Po decays by the emission of an α particle. What is the daughter nucleus? (Use Table 2-1 in order to find the atomic numbers.)

2. ^{10}Be exhibits β^- radioactivity. Into what nucleus does ^{10}Be decay?

3. Bismuth-214 (^{214}Bi) has the property that it can undergo either α decay or β decay. What are the two possible daughter nuclei that can remain when ^{214}Bi decays? (Use Table 2-1 in order to find the atomic numbers.)

4. When the Earth was formed (about 4.5 billion years ago) there was present some ^{232}Th and ^{238}Pu. Do you expect that any of this original thorium and plutonium still exists today? Explain your reasoning. (Refer to Table 20-1.)

5. An important method of determining the age of archeological items is by *radioactive carbon dating*. Radioactive ^{14}C is produced at a uniform rate in the atmosphere by the action of cosmic rays. This ^{14}C finds its way into living systems and reaches an equilibrium concentration of about 10^{-6} percent compared to normal, stable ^{12}C. When the organism dies, ^{14}C ceases to be taken up. Therefore, after the death of the organism, the ^{14}C concentration decreases with time according to the radioactive decay law with $\tau_{1/2} = 5730$ years. An archeologist working a *dig* finds an ancient firepit containing some crude pots and bits of partially consumed firewood. In the laboratory he determines that the wood contains only 12.5 percent of the amount of ^{14}C that a living sample would contain. What date does he place on the artifacts discovered in the dig?

6. A sample of β-radioactive material placed near a Geiger counter (a detector of β rays). The detector is found to count at a rate of 640 per second. Eight hours later, the detector counts at a rate of 40 per second. What is the half-life of the material?

7. The nucleus of ^{7}Li consists of 3 protons and 4 neutrons. Could ^{7}Li exist if its mass were equal to 3 proton masses plus 4 neutron masses? Explain.

8. Show that the amount of energy equivalent to 1 AMU is 931 MeV.

9. The mass of a helium nucleus is 4.0016 AMU. What is the binding energy per particle for ^{4}He?

10. When ^{15}N is bombarded by protons, a reaction takes place in which an α particle is emitted. What is the residual nucleus in this reaction?

11. Would energy be released by the fission of a nucleus with $A = 60$ into two equally massive fragments? (Refer to Fig. 20-6.)

12. Why do you suppose that Rutherford discounted the possibility of extracting useful amounts of energy from nuclear reactions?

13. Could a uranium nucleus fission into nuclei of iodine and zirconium? Into cerium and selenium?

14. What feature of the fission process makes possible a self-sustaining chain reaction?

15. Neptunium-239 (^{239}Np) undergoes fission with slow neutrons just as ^{235}U and ^{239}Pu do. Why is this isotope not useful as a reactor fuel?

16. The energy release in the detonation of 1 ton (2000 lb) of TNT is approximately 4×10^9 J. Express the energy released in the fission of 1 kg of ^{235}U in terms of tons of TNT. (It has become common practice to express the yields of fission weapons in terms of tons or kilotons of TNT; thermonuclear H-bombs with yields in excess of 100 megatons have been constructed.)

17. The element strontium is chemically similar to calcium. Why is the radioactive fission product ^{90}Sr particularly dangerous?

18. All elements with atomic number Z greater than 83 are radioactive. (Uranium, $Z = 92$, is radioactive, but the half-lives of the isotopes ^{235}U and ^{238}U are sufficiently long that uranium occurs naturally in the Earth.) Some of these high-Z elements are found in uranium ores. Are there likely to be any hazards associated with the residues of material (the *tailings*) that result from extracting uranium from its ores?

19. Why have nuclear-powered submarines been so successful and yet our only nuclear-powered freighter (the *Savannah*) has been retired while still in good condition? (Consider the mission of a submarine compared to that of a freighter. Which type of vessel is at sea for long periods of time?)

20. Iron has a relatively low probability for absorbing a slow neutron. Would iron make a satisfactory moderator for use in a reactor?

21. In Fig. 20-10 notice that the water which passes through the reactor core does not also pass through the turbine. Instead, the heat is transferred to a second water loop which is entirely outside the reactor. Why is this done?

22. In order to be practical, a fusion reactor must produce more energy than it uses. What are some of the ways in which energy must be used to operate a fusion reactor?

23. Explain how the energy released in the core of a star in the form of γ rays and rapidly moving particles is eventually radiated from the surface as *photons*.

24. The p–p chain of nuclear reactions is continually converting mass into energy in the Sun. Even though the mass of the Sun is steadily decreasing, we speak of *the* mass of the Sun. Why can we do this? (At present, the Sun's mass is approximately 2×10^{30} kg. The mass of 4 hydrogen atoms is greater than the mass of one helium atom by approximately 0.7 percent. Use this information plus the fact that about 6×10^{11} kg of hydrogen are con-

sumed each second in the Sun to compute the fraction of the Sun's mass that is converted to energy each year. Is this fraction sufficiently small so that we can consider the mass of the Sun to be essentially constant?)

25. Using Fig. 20-15, write down the series of nuclear reactions in the carbon–nitrogen cycle so that they appear similar to the equations for the proton–proton chain of reactions (Eqs. 20-13–20-15).

21

RADIATION – EFFECTS AND USES

The "nuclear age" began in 1945 with an awesome display of the destructive power that can be released from the nuclei of atoms. The development of nuclear weapons represents a *negative* aspect of the discovery and exploitation of nuclear energy sources. But what about the *positive* aspects? We have already seen that fission reactors are rapidly assuming the burden of producing the energy required to meet the world's needs and that fusion reactors hold the prospect for cheap and abundant power in the future. Although the generation of electricity in nuclear power plants is the best-known example of a positive contribution of nuclear energy, there are many other useful and important applications besides. The use of radioactive isotopes and artificially produced radiations in biology and medicine has had an enormous impact on these fields. By using radiation techniques, our knowledge of the functioning of living things has increased more in the last 20 years than in the previous two centuries. Radioisotopes are now routinely used in medicine to diagnose and to treat various ailments. New techniques in chemical processing using radiation permit, for example, the inexpensive production of polymers and the rapid curing of paints. Other applications are found in such diverse fields as archeology, art history, and the law. We certainly now live in a "nuclear age" – and the results are not all bad!

The Interaction of Radiation with Matter

In the various radioactive decay processes, α particles (^{4}He nuclei), β particles (electrons), and γ rays (high-energy photons) are emitted. What happens when these radiations strike and interact with matter? When an α or a β particle or a γ ray enters a piece of matter, energy is transferred to the material through collisions with the atoms in the material. These interactions lead to the ejection of electrons from the atoms and therefore produce ions in the material. This ionization, in turn, gives rise to chemical reactions and to a general heating of the absorbing material. It is the ionization produced in matter that makes these radiations useful in a variety of practical situations, and makes them dangerous if they enter the body.

When an α particle passes through matter, the double nuclear charge ($+2e$) causes intense ionization along its path. Furthermore, because an α particle is so much more massive than an electron, the ionization encounters (which involve electrons) do not appreciably deflect the α particle from its original direction of motion. As a result, an α particle plows almost straight through matter, leaving a high density of ions in its wake (Fig. 21-1a). The large-angle deflections observed in the Rutherford experiment (Section 2-2) are the result of *nuclear* encounters. Because of the extremely small size of a nucleus compared to that of an atom, ionization events are much more likely than nuclear collisions. An α particle traveling through matter will produce many millions of ions for each nuclear collision.

An electron, on the other hand, because of its small mass and single electrical charge, leaves behind far fewer ions per centimeter traveled and is frequently deflected in the electron–electron collisions. (Why?) The ionization produced by an electron is much more diffuse than that produced by an α particle (Fig. 21-1b). Consequently, an electron can penetrate much deeper into matter than can an α particle with the same energy. An α particle with 5 MeV (a typical energy for α particles from radioactive materials) will be stopped by a sheet of paper, but a 5-MeV electron will penetrate about an inch of biological material.

When a γ ray passes through matter, it can interact by the photoelectric effect (Section 17-1), in which the γ

(a)

(b)

Figure 21-1 (a) When an α particle passes through matter, it proceeds in an almost straight line, leaving a high density of ions along its path. (b) When a β particle (or an electron from any source) passes through matter, the ionization density is much less than in the case of an α particle. Moreover, the β particle suffers frequent deflections in its encounters and therefore proceeds along an erratic path. The electrons that result from direct ionization by the passage of charged particles can themselves produce further ionization. (The secondary ion pairs are not indicated in this diagram.)

ray is completely absorbed and an energetic electron is ejected from an atom. Or the γ ray can be deflected by an atomic electron, transferring to the electron some of its energy. The deflection of a γ ray without absorption is called the *Compton effect,* after A. H. Compton (1892–1962), an American physicist who studied this type of interaction in the 1920's. The γ ray is not absorbed in this process and it continues on to interact again with some other electron. Therefore, the ionization produced by a γ ray is due to the electrons that are released from atoms and has the characteristics of electron ionization described above. [Gamma rays with energies greater than 1 MeV can interact with matter to produce electron–positron (e^-–e^+) pairs, but we will not be concerned with this type of interaction here.]

α Particles and Other Nuclei

In addition to the emissions of radioactive substances, ionizing radiation is available from other sources as well. There is no difference between an α particle and the nucleus of a helium atom. Therefore, exactly the same effects will be produced by a 5-MeV α particle from a radioactive substance as by a helium nucleus that has acquired an energy of 5 MeV in some accelerator, such as a cyclotron. Depending on the application, particles from one source or another may be more convenient to use. Radioactive α sources emit particles at a rate that decreases only slowly with time (if the half-life of the isotope is long), require no maintenance, and are portable. A beam of helium nuclei from an accelerator, on the other hand, can be made much more intense than a radioactive source, the particles emerge all in the same direction, and the beam can be turned off when required.

Accelerators can be used to produce other energetic particles: for example, protons, deuterium nuclei, carbon nuclei, and so forth. High-energy machines can also provide beams of short-lived particles such as pions (see Section 2-4) which are useful in some biological applications, as we will see in Section 21-4. None of these other particles are emitted in radioactive decays, so accelerators represent unique sources for these radiations.

β Particles and Electrons

The negatively charged β particles that are emitted in radioactive decay are identical to ordinary atomic elec-

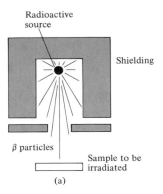

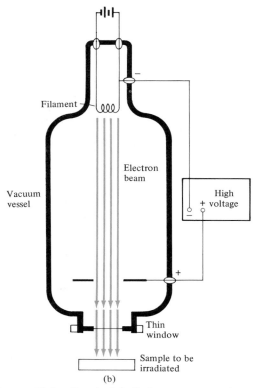

Figure 21-2 Simple irradiation systems using (a) a radioactive β source and (b) an electron accelerator (or electron *gun*). The thin window at the lower end of the accelerator permits the electrons to emerge from the vacuum container without appreciable energy loss.

trons. Therefore, any device that produces an electron beam can be substituted for a β source in a radiation application. A simple accelerator system is shown in Fig. 21-2b where it is contrasted with a radioactive source (Fig. 21-2a). The advantages of accelerators and radioactive sources for electrons are the same as those for α particles mentioned above.

γ Rays and X Rays

γ rays are high-energy electromagnetic radiations and, except for energy, are identical with X rays, light photons, and radio waves. For most radiation applications, high energy is required; therefore, in this chapter we will discuss only γ rays and X rays, and we will not be concerned with lower energy radiations.

The classification of a quantum as a γ ray or an X ray depends only upon its origin and not upon its energy. Any electromagnetic radiation that is emitted from a nucleus is called a γ ray. If the radiation originates in the atomic electron shells it is called an X ray. Thus, a 20-keV γ ray and a 20-keV X ray could be emitted from the same atom and the radiations would be exactly the same.

γ rays from radioactive decay processes result only in the deexcitation of a nucleus that is left in an excited energy state following α or β decay (see Fig. 20-2). The emission of γ radiation does not involve a nuclear transmutation. The nuclear isotope does not change during a γ-ray emission process.

Neutrons

Neutrons are not emitted in radioactive decay events, and, because they are electrically neutral particles, neutrons cannot be accelerated in machines as can electrons and nuclei. But neutrons can be produced in nuclear reactions initiated by high-energy particles in accelerator beams. A variety of target materials will yield neutrons when bombarded by high-speed particles. For example, the bombardment of lithium by protons produces neutrons according to the reaction,

$$^7Li + {}^1H \longrightarrow {}^7Be + n \qquad (21\text{-}1)$$

The absence of electrical charge makes the neutron an interesting and important particle. When a neutron strikes a piece of matter, it does not interact with the atomic electrons (this happens only with *charged* particles); instead, neutrons interact with the *nuclei*. These neutron–nucleus interactions can result in the transfer of energy from the neutron to the nucleus (see the discussion of neutron moderators in Section 20-5), or in a neutron-induced disintegration. The capture of a neutron by a nucleus often results in the formation of a radioactive isotope. (In the case of a heavy nucleus, the result can be fission.)

In traveling through a piece of matter, a neutron does not produce any ionization. When the neutron strikes a nucleus, the nucleus recoils as a result of the collision. As the nucleus moves through the surrounding atoms, some of the atomic electrons are stripped away. Thus, the collision produces ionization along the path of the recoiling nucleus. In a material that contains a large fraction of hydrogen (for example, biological tissue), neutrons interact primarily with the nuclear protons of the hydrogen atoms. The knocked-on protons are the particles that produce almost all of the ionization in such materials.

Supplies of Radioisotopes

Where do we obtain supplies of radioactive isotopes? Only a few radioisotopes, such as radium, are obtained from natural minerals, and huge amounts of material must be refined before useful quantities of the desired isotopes are separated. Most of the radioisotopes used today are produced in reactors. The spent fuel elements from reactors are prolific sources for many radioactive materials. The useful isotopes are chemically separated during the reprocessing of the fuel rods (see Section 20-5). Other radioisotopes are produced by purposely introducing into the reactor a material that will absorb neutrons and become radioactive. One of the most widely used radioisotopes is ^{60}Co, which is formed from natural cobalt, ^{59}Co, by the capture of a neutron. ^{60}Co is β radioactive and following each β decay two γ rays with energies of about 1 MeV are emitted. These energetic γ rays make ^{60}Co useful in the treatment of certain cancers and in the processing of plastics and foods. Finally, some radioisotopes do not appear as fission fragments nor are they formed by neutron capture reactions. These isotopes must be prepared by using charged-particle reactions in accelerators. For example, ^{7}Be is produced in this way using the reaction shown in Eq. 21-1.

Radiation Units

In order to specify the amount of radioactivity contained in a sample and the amount of radiation absorbed by an object, we make use of two units: the *curie* (Ci) and the *rad*. A *curie* of radioactivity represents 3.7×10^{10} decay events per second (regardless of the type or energy of the radiation). The *curie* is named for Marie and Pierre Curie who discovered radium (see Section 20-1). Originally, one curie (1 Ci) meant the number of disintegrations per second taking place in one gram of radium, but the definition has been broadened and standardized to mean *exactly* 3.7×10^{10} disintegrations per second of any radioactive material. Standard laboratory sources are usually near 10^{-6} Ci or 1 μCi (microcurie); sources used in industrial processing (for example, ^{60}Co) frequently are 10^3 Ci = 1 kCi (kilocurie) and sometimes are as large as 10^6 Ci = 1 MCi (megacurie). At the present time, industrial-size sources of ^{60}Co cost approximately \$0.50 per curie. The total *power* output of a 1000-curie source of ^{60}Co is approximately 15 watts, or only a fraction of the power supplied to an ordinary household lightbulb.

How big is a radioactive source? If we have a 1-gram sample of cobalt (a cube about $\frac{1}{4}$ inch on a side) in which every atom is radioactive ^{60}Co, the activity of the sample would be 1000 curies! The radioisotope ^{60}Co has a long half-life (5.24 years) and so decays rather slowly. The shorter the half-life, the greater is the activity of a given number of radioactive atoms. A 1-gram sample of ^{131}I (half-life = 8.05 days) would have an activity of about 10^5 curies. Samples consisting *entirely* of radioactive atoms of ^{60}Co or ^{131}I cannot actually be prepared (why?), and real samples ordinarily contain only a small fraction of radioactive atoms.

The *rad* is a unit which specifies the amount of radiation energy absorbed by an object. A dose of 1 rad corresponds to the absorption of 0.01 joule per kilogram of material:

$$1 \text{ rad} = 0.01 \text{ J/kg} \qquad (21\text{-}2)$$

Radiation doses up to 10^7 rad (10 Mrad) are commonly delivered to materials in industrial applications. A dose of 10 Mrad is an extremely large dose. For comparison, if a person were to stand 1 meter away from a 1-Ci laboratory source of ^{60}Co for 1 hour, he would receive a dose of approximately 1.2 rad at the front surface of his body and a dose of about half this amount at a depth of 10 cm because of the attenuation of the γ rays in passing through the body tissue. As we will see in the following section, there are no immediately detectable effects in humans of radiation doses below about 25 rad. However, even small doses of radiation are suspected of being harmful in some degree. Therefore, extreme caution should be exercised whenever a radiation source is in the vicinity.

21-2 RADIATION TECHNIQUES IN VARIOUS FIELDS

Chemical Reactions Induced by Radiation

What effects can radiation have in matter that represent desirable changes? First, consider a simple case. Suppose that a quantity of methane gas (CH_4) is irradiated with electrons that have energies of a few keV. What do we expect to happen? The electrons will produce ionization in the gas, so we expect to find the molecular ions, CH_4^+. Indeed, these ions are present in the sample during irradiation. But a molecule is bound together by electrons, and if a disruptive collision takes place, the molecule can easily be broken apart to produce such fragments as CH_3^+, CH_2^+, and CH^+, even H^+ and C^+. All of these ions are present in the gas during irradiation with abundances decreasing in the order given. Ions such as these are chemically active and can combine with methane molecules in reactions such as

$$CH_3^+ + CH_4 \longrightarrow C_2H_5^+ + H_2 \qquad (21\text{-}3)$$

A $C_2H_5^+$ ion could react with a hydrogen atom to produce a molecule of ethane, C_2H_6. Or, a $C_2H_5^+$ ion could combine with a CH_3^+ ion to produce a molecule of propane C_3H_8. That is, by irradiating methane with electrons, molecules are formed which have structures more complex than the original gas molecules. (Here is a hint as to how the first complex molecules of living matter probably originated on Earth.)

When the irradiation is terminated, the ions present during irradiation capture electrons and become electrically neutral. If we then analyze the composition of the gas, we find a host of new molecules, ranging from hydrogen gas, H_2, to long-chain hydrocarbons, up to 20 carbon atoms in length. The major products of the irradiation of methane are listed in Table 21-1.

TABLE 21-1 MAJOR PRODUCTS OF THE IRRADIATION OF CH_4

PRODUCT	FORMULA	NUMBER OF MOLECULES PER 100 eV ABSORBED
Hydrogen	H_2	5.7
Ethane	C_2H_6	2.2
Ethylene	C_2H_4	0.7
Propane	C_3H_8	0.36
n-Butane	C_4H_{10}	0.11
Iso-butane	C_4H_{10}	0.04
Pentanes	C_5H_{12}	0.001
Hexanes	C_6H_{14}	0.001
Longer-chain hydrocarbons	$(CH_2)_n$	2.1

The first industrial process to use radiation techniques was the synthesis of ethyl bromide by the Dow Chemical Company, beginning in 1963. A 3000-Ci source of ^{60}Co is used to process about 1 million pounds of ethyl bromide per year.

Radiation Processes in Space—Clues to the Origin of Life

Atoms and molecules can be identified by the characteristic radiations they emit. We usually think of these radiations in terms of visible light or, perhaps, infrared and ultraviolet radiations. But atoms and, particularly, molecules also emit radiation in the long-wavelength part of the electromagnetic spectrum. Such radiation is detected, not with ordinary optical instruments, but with special radio receivers that can be tuned to precise frequencies.

In recent years there has been an increasing interest in studying the radio-frequency emissions we receive from space. All stars emit radio signals to some degree, and some peculiar objects emit enormous amounts of radio energy. We have even found radio signals emanating from the gas and dust in the space between the stars. Among these signals are radiations with discrete frequencies that indicate the presence of particular types of molecules. More than 25 different molecular species have been identified in interstellar space. These range from simple compounds such as carbon monoxide (CO), cyanogen (CN), ammonia (NH_3), and water (H_2O), to more complex organic substances such as methyl alcohol (CH_3OH) and formaldehyde (CH_2O). One cloud of gas and dust that is especially rich in organic compounds is found in the constellation Sagittarius and is known as Sag B2. This cloud is approximately 30 L.Y. in diameter. The complex compound cyanoacetylene (H—C≡C—C≡N) has been identified in Sag B2 and it has been estimated that about 20 solar masses of this substance exists in the cloud.

Further evidence for the production of organic compounds in a space environment is found in the examination of meteorites that have crashed into the Earth and have been recovered. Careful chemical studies of these objects have revealed the presence of amino acids, the basic building blocks of the complex molecules of living matter.

What is the significance of the occurrence of organic compounds in space? Does this mean that there is living matter of some sort "out there"? Not at all. But we do have a strong hint as to how life began on Earth. The elements that are found in organic substances—hydrogen, carbon, nitrogen and oxygen—are among the most abundant elements in the Universe. When these atoms mingle together in space and are bathed in the various radiations from stars, the atoms can be joined together into molecules. At first, the simple diatomic molecules are formed, and then other atoms become attached to produce more complex molecules. Eventually, amino acids are formed—these are the basic building blocks of living matter.

These molecule-building chemical reactions can take place in interstellar clouds where the concentration of matter is extremely low and where the

where they pass through the electron beam from an accelerator (Fig. 21-3). The speed with which the conveyor belt moves is regulated so that each package remains in the beam for a time sufficient to kill any microbial activity within the package. When they emerge from the vault, the packages are removed and boxed for shipment. The energy absorbed in the irradiation raises the temperature of the product no more than 10 C deg while in the beam. Conventional thermal sterilization methods would require a 15-minute treatment at 120 °C.

It must be emphasized that in all such procedures, the irradiated material does *not* become radioactive. The only way in which an ordinary material can become radioactive is for some reaction to take place which alters the nuclei of the material. Irradiation with neutrons and charged particles can cause such reactions to occur, but low-energy electrons and γ rays cannot induce radioactivity.

Food Preservation

The length of time that a food product can be stored depends upon how rapidly bacteria and other microorganisms will attack the food and cause it to "spoil." Canned goods are heat-treated after canning in order to kill bacteria, and the shelf life of a can of vegetables, for example, is quite long—months or even years. Fresh foods, however, cannot be treated in this way and they generally deteriorate after a few days or weeks.

Experiments have demonstrated that radiation treatment will substantially prolong the shelf lives of many types of produce. Irradiated tomatoes, for example, will last 15 days before spoiling instead of the usual 8 days. And the shelf life of crab meat (at 33 °F) can be increased from 1 week to 5 weeks by irradiation. Unfortunately, the large doses of radiation that are necessary for food preservation often affect the odor, flavor, texture, and color of the food. For commercial purposes, these side-effects are undesirable. Even more important is the slim possibility that the irradiation of foods containing sugars will alter the chemical properties of the sugars and produce substances that cause cell or genetic damage in Man and in animals. The U.S. Army has been preserving selected foods by irradiation on an experimental basis for a number of years. Further research will be necessary before radiation techniques can be used in large-scale food preservation programs.

The effect of radiation as a food preservative technique is strikingly illustrated in these photographs. The potato at the bottom was irradiated with 20 000 rad of ^{60}Co γ rays; the top potato was not treated. After storage for 16 months, the untreated potato had developed an appreciable sprout structure, whereas the irradiated potato was still firm, fresh-looking, edible, and had no sprouts.

At the present time, only a few foods have been cleared by the U.S. Food and Drug Administration (FDA) for sterilization doses. Sub-sterilization doses have been approved to prevent sprouting in potatoes (see the photograph) and to kill insects in grains. Small doses of radiation could also be applied to certain foodstuffs to kill the salmonella bacteria that cause a disease characterized by diarrhea and vomiting. Used in this way, irradiation is similar to the *pasteurization* of milk to destroy harmful bacteria. This particular process has not yet been approved for use in the United States.

Gauging and Control

Radiation methods, particularly those involving radioisotopes, are used in a variety of situations for the gauging of thickness or density of items on production lines. Suppose that a pharmaceutical manufacturer wishes to determine whether medical capsules are being filled with the proper amount of material. The capsules are filled by machine in a continuously moving system, but occasionally a malfunction of the machinery will cause too much or too little material to be placed in a capsule or even in an entire batch of capsules. Prior to the introduction of radiation methods, *batch control* was accomplished by removing from the production line one capsule in 100 or 1000 and testing it for proper filling. Obviously, this sampling procedure cannot ensure that *every* capsule has been properly filled.

The modern method of production-line gauging of capsule filling (or other similar operations) involves the use of radioisotopes. A radioactive source of β particles or γ rays is placed under the conveyor system that carries the filled capsules and a detector is placed above the capsules (Fig. 21-4). Radiation passes through the capsules and enters the detector. If the capsule is properly filled, the detecting system will register a certain counting rate for the transmitted β particles or γ rays. If the capsule contains too much material, more of the radiation will be absorbed in the sample and the detector will register a counting rate less than normal. Similarly, if the capsule contains too little material, less radiation will be absorbed in the sample and the detector will register too high a counting rate. By adjusting a control device to accept capsules only if the counting rate falls within a certain narrow range, all of the faulty capsules can be automatically rejected. In this way it is possible

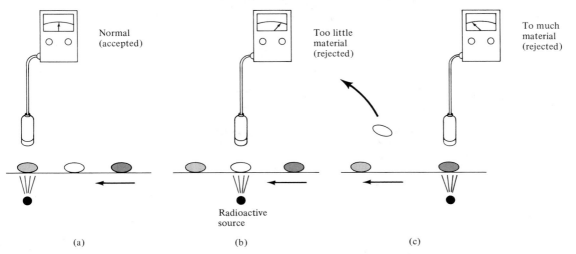

Figure 21-4   Control of capsule filling by radiation gaging. When the counting rate in the detector falls outside the acceptable limits (b, c), the capsule is rejected.

to exercise precise quality control over *every* item on the production line.

Radiography

When a casting is made, for example, of a valve to be used in a high-pressure system, it is important to know whether there are any voids or fissures in the casting that would weaken the valve and make it unsafe for use. X-ray (or *radiographic*) techniques are often used in such cases. The method is exactly the same as that used by a dentist when he takes an X-ray picture of your teeth. A piece of special photographic film is held inside the mouth and X rays from an X-ray tube are projected through the teeth. The differing absorption of the X rays by the teeth, fillings and inlays, the gums, and the jaw bone, produce an outline picture on the film when it is developed. But the peculiar shapes and thicknesses of many castings do not permit these X-ray methods to be used: either the geometry is not suitable or the large thickness of the material will cause all of the X rays (which are low-energy radiations) to be absorbed. In such cases, a radioactive source that emits high-energy γ radiation (which can penetrate even thick steel) is placed *inside* the casting, as shown in Fig. 21-5. The photographic film is attached to the outside of the casting and, upon being developed, will indicate the presence of any voids or fissures.

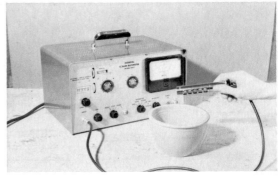

HARBRACE

A Geiger-counter system, one of the methods for detecting nuclear radiations.

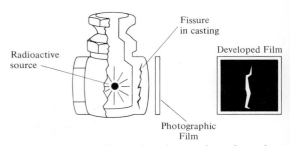

Figure 21-5   Radiography of a casting of a valve body by using a radioisotope.

Similar radiographic techniques employing radioisotopes (or X rays) are routinely used to inspect welded joints, ship parts, jet engine components, structural members of aircraft wings, and many other items whose structural failure could necessitate expensive repairs or could endanger human life.

Radiation Methods in Archeology

The pyramids of Egypt have intrigued mankind ever since their construction more than 4000 years ago. The largest, the Great Pyramid of Cheops is known to have several passageways and chambers. The nearby pyramid of Cheops' son, Chephren, has yielded only one interior passageway and a single chamber. But are there additional secret rooms that have so far escaped detection? Are there more undiscovered chambers that contain treasures from 2600 B.C.? One way to determine whether there are additional voids in the pyramid would be to perform radiographic tests. But how does one "X ray" a pyramid?

A clever scheme to investigate the interior of the Pyramid of Chephren was devised by Professor Luis Alvarez of the University of California and put into effect in cooperation with the Egyptian Department of Antiquities. The plan involved the use, not of radioactive sources, but of the natural radiations of cosmic rays. Among the particles that result from the reactions produced by cosmic rays in the atmosphere are muons (see Section 2-4). Muons do not interact strongly with matter and so they can penetrate to great depths in the Earth and can pass to the interior of a pyramid. Muon detectors were placed in the single known chamber in Chephren's pyramid (Fig. 21-6) and were arranged to give signals only for muons arriving from specific directions. If any voids existed in the pyramid, the absorption of muons passing through these voids would be less than normal and the detector counting rates would show higher than normal values in these directions. Alas, no indication of unknown chambers was found in the experiment. But at least it was shown that it is possible to "X ray" an object that stands 470 feet high!

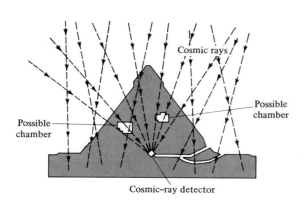

Figure 21-6 Schematic of the experiment to detect possible chambers in the Chephren Pyramid by using cosmic-ray muons to "X ray" the pyramid.

This famous prehistoric painting is located in the Lascaux Cave near Montignac, France. Radioactive ^{14}C dating of charcoal from fire sites in this cave has revealed that cave dwellers inhabited this region 9000 years ago and that the painting probably dates from this time. (By opening the cave to visitors, the air has become sufficiently contaminated that the painting is deteriorating; the cave is now closed to the public.)

GRAPHIC ARTS UNLIMITED, INC., N.Y. (BRAUN ET CIE)

Activation Analysis

Many of the stable isotopes, when bombarded with neutrons, capture a neutron and are converted into a radioactive isotope which subsequently undergoes β decay. For example, when silicon-30 (^{30}Si) captures a neutron, radioactive ^{31}Si is formed:

$$^{30}\text{Si} + n \longrightarrow \ ^{31}\text{Si}$$

$$^{31}\text{Si} \xrightarrow{\beta \text{ decay}} \ ^{31}\text{P} \qquad (\tau_{1/2} = 2.6 \text{ h})$$

The β decay of ^{31}Si to ^{31}P produces a γ ray with an energy of 1.27 MeV. This γ-ray energy and the half-life of 2.6 h are unique to the decay of ^{31}Si. In fact, every radioactive isotope can be identified by measuring the half-life and the type and energy of the emitted radiation. Therefore, if we bombard a sample of material with neutrons and find that there is present a radioactivity which produces a 1.27-MeV γ ray and which decreases in intensity according to a half-life of 2.6 h, then we know that the radioactivity is due to ^{31}Si and that ^{30}Si must be present in the sample. This technique for identifying the constituents of a sample is called *neutron activation analysis* (or simply, *activation analysis*).

There are two significant advantages of activation analysis over chemical methods. First, the activation technique is nondestructive—that is, the sample need not be changed by chemical reactions. Second, the sensitivity, at least for some elements, is enormously greater than is possible with chemical analysis. Quantities of some elements as small as 10^{-9} gram can be detected by activation analysis. This method is therefore particularly valuable in detecting *trace* quantities of specific elements, and applications have been found in a large number of fields. In law, for example, a suspect may be linked to the scene of a crime by activation analysis of a speck of material found in his clothing. If the speck has exactly the peculiar chemical composition of material from the crime scene, the speck must have come from the same site. In the area of historical research, activation analysis has been made of a strand of hair from Napoleon's body. The results indicate the presence of arsenic in sufficient quantity to demonstrate that the ex-Emperor did not die a natural death but was poisoned while in exile on St. Helena.

In this section we have discussed several of the diverse applications of radiation and radioisotope techniques. So varied are these methods that we have been able to give only a representative list. Rather than attempt to be more complete, we now proceed to a discussion of radiation effects in biological material and the uses of radiation in the biological and medical sciences.

21-3 BIOLOGICAL EFFECTS OF RADIATION

Radiation Damage

Every person on Earth is continually exposed to various kinds of radiation from many different sources. Ordinarily, these radiations do us no particular harm. But even the most familiar of radiations — solar radiation — can do damage to the skin or eyes if the exposure is too great. Infrared radiation from a heat lamp or ultraviolet radiation from a "sun lamp" can also cause uncomfortable burns (even serious burns) if used carelessly. However, when we use the term *radiation damage,* we usually mean the injurious effects that are caused by radiations of higher energy. In this category are X rays from medical or dental X-ray units and television sets, as well as α, β, and γ radiations from natural or artificial radioactive sources and from accelerators that produce nuclear radiations. The reason for this distinction is that radiations such as ultraviolet and infrared rays have very low penetrating power. Therefore, these radiations are stopped by the outer layers of skin and any damage that results from excessive exposure is superficial. On the other hand, high-energy X rays and, particularly, γ rays can easily penetrate the body and can damage the internal organs. Although a severe sunburn can be extremely painful, we do not ordinarily classify this annoyance as "radiation damage."

Almost all of the radiation that is capable of producing biological damage and to which the general public is exposed, is in the form of X or γ radiation. Persons who work with radioactivity or with accelerators are sometimes exposed to α and β particles or to other high-speed nuclear particles. All of these radiations produce ionization in matter and can therefore inflict damage on biological tissue.

When considering the biological effects of radiation, it is important to remember that the unit of absorbed dose—the *rad*—refers to the energy absorbed per kilogram. Therefore, the amount of radiation energy absorbed by a 100-kg man who receives a *whole-body* dose of 1 rad is much greater than if he receives a 1-rad dose only to his arm. On the other hand, if the same amount of *energy* is absorbed by the arm or by the body as a whole, the dose in rads is much less in the latter case.

Relative Biological Effectiveness
and the rem

It has been found that equal absorbed doses delivered by different types of ionizing radiations will produce varying amounts of biological damage. Thus, an individual who receives a whole-body dose of 1 rad due to high-speed α particles will suffer considerably more tissue damage than if he receives the same whole-body dose of 200-keV X rays. We say that α particles have a greater *relative biological effectiveness* (RBE) than low-energy X rays. Compared to 200-keV X rays (which are defined to have an RBE of 1), the RBE of α particles is approximately 20. Approximate RBE values for the more common radiations are given in Table 21-2. These values are only approximate because they depend to some extent on the energy of the radiation. Nevertheless, the tabulated values serve as useful guides to the effectiveness of the different radiations.

Fast neutrons produce radiation damage in tissue primarily through the protons that they set into motion because of collisions. Slow neutrons, on the other hand, have very little energy to impart to protons; nevertheless, they can produce high-energy secondary radiations by inducing nuclear reactions.

TABLE 21-2 RELATIVE BIOLOGICAL
EFFECTIVENESS OF VARIOUS RADIATIONS

RADIATION	RBE VALUE (approximate)
X or γ ray	1
Electrons (β particles)	1
α particles	20
Protons	10
Fast neutrons	10
Slow neutrons	5

Because of the differing biological effectiveness of different types of radiation, the *rad* (which measures only the total energy deposited per unit mass of the absorber) is not a useful unit for indicating radiation damage in living matter. Instead, a unit called the *rem* is used. This unit measures the energy deposited per unit mass multiplied by the RBE of the particular radiation — that is, the *equivalent dose:*

$$1 \text{ rem} = (1 \text{ rad}) \times (\text{RBE}) \qquad (21\text{-}4)$$

Thus, if a person receives a 0.2-rad dose of α particles (a substantial dose!), the exposure is measured as $(0.2 \text{ rad}) \times (20) = 4$ rem. If the exposure is entirely to X and γ radiation or electrons, the dose equivalent in rem is equal to the dose in rad.

Radiation Exposure

The largest contribution to the radiation received by an individual who is not a radiation worker is from natural sources — cosmic rays and the radioactivity that occurs in the Earth. The amount of natural radiation received during the course of a year by a particular individual depends upon his location and habits. Some parts of the country have more natural radioactivity than others; the intensity of cosmic radiation depends on altitude — the residents of Denver receive 50 percent more cosmic radiation than the residents of San Francisco; some wrist watches have luminous dials that contain radium; and so forth. The range of natural radiation doses received by individuals in the United States is approximately 90–150 mrem per year. (1 mrem = 1 millirem = 10^{-3} rem.)

The second most significant source of radiation exposure is medical and dental X rays. (We include here only routine diagnostic X rays; therapeutic treatments are special situations.) Again, there is a wide variation among individuals — some persons may have no X rays whereas others may require extensive sets of X rays for the diagnosis of particular medical problems. The normal range of exposure (in the U.S.) from this source is 50–100 mrem per year.

The radioactive fallout from weapons test amounts to about 5 mrem/y. *If* an agreement to stop all testing is reached, this figure will decrease gradually with time because of the decay of the radioactive residue still present in the atmosphere from previous tests.

The remaining source of radiation exposure — that due

CAUTION

RADIATION AREA

This symbol is universally used to indicate an area where radioactivity is being handled or artificial radiations are being produced.

to the operation of nuclear power reactors — is the most controversial of all. Averaged over the entire U.S. population, the individual exposure is about 0.003 mrem/y. But if a person were to live for the entire year on the down-wind boundary of one of the older nuclear plants (where the radiation containment is not as effective as for the new plants), the exposure could amount to 5 mrem/y. Of course in the unlikely event of a catastrophic accident (and this is the point of controversy), the exposure could be considerably higher. For comparison, it is interesting to note that a transcontinental trip by air typically exposes a passenger to a radiation dose greater than 0.01 mrem due to the effects of cosmic rays.

A summary of exposure figures for the U.S. population is given in Table 21-3.

Effects of Radiation Damage

What does radiation actually *do* to a person? Radiation effects can be divided into two categories: (a) effects on the individual exposed — these are called *somatic* effects, and (b) effects on the offspring of the individual exposed — these are called *genetic* effects. We can also divide the type of exposure into two categories: (a) long-term, or *chronic,* exposure at a relatively constant level, and (b) single-dose, or *acute,* exposure which is all received in a short time. Not all radiation damage is cumulative, so an individual may exhibit no somatic effects if exposed to 40 rem of radiation when the dose is distributed uniformly over a 40-year period. However, if a person received a 40-rem dose all at once, he would develop some of the symptoms of *radiation sickness,* but full recovery would be expected.

There are no immediately detectable somatic effects of acute exposure at dose levels below about 25 rem. However, there are *delayed* effects such as increased susceptibility to leukemia, bone cancer, and eye cataracts, as well as a shortened lifespan. In a sample of one million people, about 100 cases of leukemia will develop each year. If every person in this sample were to receive a 1-rem dose of radiation, an additional 1–2 cases of leukemia would be expected to develop during the following year. The shortening of lifespan due to radiation exposure is estimated to be 10 days/rem for acute exposure and 2.5 days/rem for chronic exposure. (Some estimates are even smaller.) Thus, the person referred to in the previous paragraph who received a single-

TABLE 21-3 RADIATION EXPOSURE OF INDIVIDUALS IN THE UNITED STATES

SOURCE	DOSE RANGE (mrem/y)	AVERAGE DOSE IN U.S. (mrem/y)
Natural (cosmic rays; radioactivity)	90–150	102
Medical and dental X rays (diagnostic only)	50–100	76
Weapons tests fallout	5	4
Nuclear power plant operation	<0.01–5	0.003[a]
Total:	145–260	182

[a] Increasing to about 3 mrem/y by the year 2000.

TABLE 21-4   SOMATIC EFFECTS OF RADIATION EXPOSURE

WHOLE-BODY DOSE (rem)	EFFECTS	REMARKS
0–25	None detectable	
25–100	Some changes in blood, but no great discomfort; mild nausea.	Some damage to bone marrow, lymph nodes, and spleen.
100–300	Blood changes, vomiting, fatigue, generally poor feeling.	Complete recovery expected; antibiotic treatment.
300–600	Above effects plus infection, hemorrhaging, temporary sterility.	Treatment involves blood transfusions and antibiotics, severe cases may require bone marrow transplants. Expected recovery about 50 percent at 500 rem.
>600	Above effects plus damage to central nervous system.	Death inevitable if dose >800 rem.

dose, whole body exposure of 40 rem, developed some symptoms of radiation sickness, and then recovered, would have a life expectancy up to 1 year shorter than normal. If the 40-rem dose resulted from chronic (instead of acute) exposure, the individual's life expectancy might be shortened by about 3 months.

Acute doses of more than a few hundred rem result in violent sickness and even death. The somatic effects of various levels of radiation exposure are summarized in Table 21-4.

The genetic effects of human exposure to radiation are much more subtle than somatic effects, and we still know relatively few details about the way in which the hereditary information carried in molecules of DNA is affected by radiation. Although many experiments have been carried out using insects and animals, the results are not directly applicable to the human case. Extensive studies have also been made on the survivors of the Hiroshima and Nagasaki blasts and their offspring, but too few generations have passed to assess the lasting genetic effects. Such observations are made especially difficult because some birth defects are due to causes other than radiation and because some of the radiation-induced mutations are due to natural radiation.

The science of radiation genetics was founded by the American biologist Hermann J. Muller (1890–1967), who began experimenting in the mid-1920's on the genetic effects produced in fruit flies (*Drosophila*) by X radiation. Muller was awarded the 1946 Nobel Prize in physiology for this work.

Public Exposure to Radiation

As far as most somatic effects are concerned (at least, for adults), the damage resulting from many small doses appears to be *less* than that resulting from a single large dose because the body has an opportunity to recover from small doses spaced in time. This may not be true for genetic effects and for some somatic effects such as the increased susceptibility to leukemia or other cancers. Thus, there is probably no level of radiation exposure below which there is zero damage to humans. All radiation is harmful to some extent. There is no argument on this point; but there is a continuing debate as to the amount of damage produced by small doses accumulated at low rates. The discussion centers around whether the benefits of radiation-producing devices, such as nuclear power plants, outweigh the increased risks of the resulting radiation exposure, even though this risk is small for the *average* citizen.

At the present time, the policy is to acknowledge that there are decided benefits in various operations that involve the production and use of radiation, so that *some* exposure of the general public is inevitable. It is also recognized that the exposure, especially to those who do not work with radiation and who may be unaware of any radiation in their environment, must be kept to the absolute minimum. Currently, the maximum permissible amount of radiation to which an individual may be exposed is set at 0.5 rem = 500 mrem per year. (Recall that the actual average exposure is less than half this figure; see Table 21-3.) Because very few individuals are actually monitored to determine the amount of exposure, the maximum permissible limit for a *typical* individual in a population sample has been set at 170 mrem over and above the dose due to natural radiations. It is estimated that this level of exposure would not unduly burden the population in terms of increased radiation risks.

There are opponents to this view, however, and they argue that steps should be taken to lower the amount of nonnatural radiation to which the population is exposed. This could be accomplished in a number of ways—for example, by stopping the testing of nuclear devices of any kind, by using diagnostic X rays only when absolutely required, by placing more stringent regulations on the allowable radiation from television sets and microwave cooking ovens, and by halting the proliferation of nuclear power plants. Some of these measures appear to

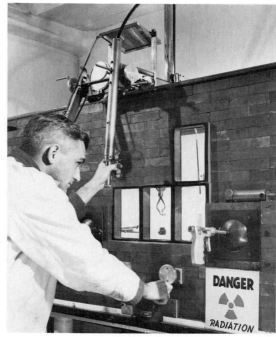

BROOKHAVEN NATIONAL LABORATORY

Laboratory experiments involving low-level radiation from encapsulated sources (in the microcurie range) can be performed without extensive shielding. But, when high-level sources or sources in liquid form are used, the manipulations must be carried out with remote-handling equipment behind elaborate radiation shields.

be desirable steps, whereas the population may be unwilling to accept the additional inconvenience and expense associated with others.

21-4 BIOLOGICAL AND MEDICAL USES OF RADIATION

Tracer Studies

One of the important problems in the biochemistry of life processes is to discover how an organism uses small molecules to build large molecules. How does a carbon atom in, for example, a CO_2 molecule become incorporated into a giant protein molecule? In the early days of biochemistry, complex questions such as this were extremely difficult, if not impossible to answer. But radioactivity methods have permitted life scientists to study in detail the flow of atoms of various types through the metabolic processes in plants and in animals. The technique is to substitute radioactive atoms for normal atoms of the same kind in certain molecules and then to follow the "tagged" atoms with radiation detectors as they move along the metabolic chain.

Many different types of radioactive atoms have been used in these *tracer* studies. Radioactive carbon (^{14}C) is one of the most useful tracers because of the occurrence of carbon in all organic molecules. An important application of ^{14}C has been in the investigation of photosynthesis in plants. Through what processes does carbon pass from the time that it enters the plant as CO_2 until it appears in glucose molecules? Suppose that a number of nearly identical plants are simultaneously provided with gaseous carbon dioxide that has been prepared from ^{14}C. At regular intervals of time after the "tagged" CO_2 has been introduced, the plants are one-by-one removed from the growing area and are pulverized so that all of the normal biochemical processes stop. A chemical separation is then carried out on each plant and the various types of molecules in the plant are isolated in separate batches. Next, radiation detectors are used to assay the amount of ^{14}C present in each of the molecular species. If the first plant were destroyed only a few minutes after the introduction of the "tagged" carbon dioxide, relatively little ^{14}C would be found in the glucose fraction. But the last plants to be destroyed would have a considerably greater concentration of radiocarbon in the molecular products of photosynthesis. Plants allowed to grow for intermediate periods of time would show the presence of ^{14}C in various types of molecules that are formed as parts of the sequence. Investigations such as this have provided detailed information on the series of chemical reactions that leads from carbon dioxide to carbohydrates in plants. The American biochemist Melvin Calvin (1911–), was awarded the 1961 Nobel Prize in chemistry for the leading role he played in the identification of photosynthesis reactions by using ^{14}C tracer methods.

Radioactive tracers can be used to study *mechanical* as well as *biochemical* processes in living organisms. Suppose, for example, that a person is suspected of having improper circulation of blood in one foot. A salt solution containing radioactive sodium (^{24}Na) is injected into the person's bloodstream and radiation detectors are placed near each foot. The detector near the foot with normal circulation will almost immediately begin to show the passage of ^{24}Na through the foot. If the other detector responds more slowly and with a lower counting rate, it indicates a smaller flow of blood through this foot and therefore impaired circulation. Such diagnostic techniques can be of great value to physicians in determining proper treatment for various ailments.

Medical Diagnostics and Therapeutics

Many different radioisotopes are now used on a routine basis for the diagnosis and treatment of a variety of illnesses. Hyperthyroidism (overactive thyroid gland), for example, can be identified by using the fact that iodine tends to concentrate in the thyroid gland. A patient drinks a solution containing radioactive iodine (^{131}I) and a radiation detector positioned near the thyroid gland is monitored. A normal thyroid gland will take up less than about 20 percent of the ingested iodine during the first hour, but a hyperactive gland will accumulate more than twice this amount. Therefore, the increase in the counting rate of the detector during the hour after the patient drinks the iodine solution will indicate whether the thyroid is hyperactive or not.

Cancerous tissue in the thyroid gland can also be detected by determining the distribution of radioiodine in the gland. In order to make such a measurement, a detector is moved over the thyroid area in a series of parallel sweeps, each displaced slightly from the preceding sweep. The counting rate is recorded at regular intervals

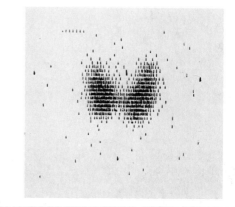

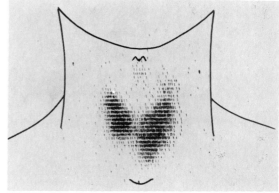

OAK RIDGE NATIONAL LABORATORY

The detector scan at the top shows the distribution of ^{131}I in a normal thyroid gland. The scan below shows the result obtained for a cancerous gland. The outline shows the patient's neck area.

during each sweep. In this way a two-dimensional plot of the concentration of ^{131}I is obtained, and from his information it is possible to deduce whether a cancerous condition is present. (See the photographs on the previous page.)

Treatment of a cancerous thyroid can be carried out by continuing the ingestion of radioiodine, using larger doses than required for diagnosis. The radiation emitted by ^{131}I tends to destroy the thyroid tissue, both normal and cancerous. But because the cancerous tissue takes up iodine at a rate that is greater than that for the normal tissue, more radioiodine is absorbed by the abnormal tissue and more of the cancerous cells are destroyed. Radioiodine treatments for cancerous thyroid glands have been found to be extremely effective.

Cancer Therapy with γ Rays and Nuclear Particles

Soon after the discovery of X rays it was found that certain types of skin diseases could be treated with these radiations. But X rays, because of their low energy and low penetrating power are not effective in treating internal cancers and tumors. The higher-energy γ rays from radioisotopes such as ^{60}Co are capable of penetrating deep into the body and destroying the diseased tissue associated with malignant growths. ^{60}Co therapy has now become standard practice for the treatment of several different kinds of cancers. Irradiation units similar to the one shown in the photograph on this page are to be found in many hospitals across the country. Also, small pellets of ^{60}Co can even be placed in the body by surgery at the sites of malignant growths; after the radiation arrests or destroys the growth, the pellets are removed.

Certain types of tumors cannot be effectively attacked by ^{60}Co γ rays; more intense and concentrated ionization is required in these cases. Beams of high energy particles, such as neutrons, protons, and heavier ions, as well as high-energy X rays, have been used in these cases with success.

One of the interesting short-lived particles that are produced when extremely high-energy protons interact with matter is the *pion* (see Section 2-4). Even though a pion lives, on the average, only about 10^{-8} s, intense beams of pions can be obtained from some of the large accelerators, such as the new facility at the Los Alamos Scientific Laboratory. When a negative pion (π^-) comes

An irradiation unit for the treatment of cancer with doses of γ rays from ^{60}Co.

to rest in matter, it is electrically attracted to and absorbed by a nucleus. In this process, the pion disappears and all of the mass-energy of the pion (140 MeV) is deposited in the nucleus. The nucleus cannot accommodate this amount of additional energy and it disintegrates, emitting charged particles in all directions. Thus, the stopping of a negative pion in matter produces a large energy release within a very small region of space. This is exactly what is needed to attack localized abnormal growths in the body. At the present time, *pion therapy* is still in the experimental stage, but it appears likely that this method of cancer "surgery" will receive increased attention in the future.

SUGGESTED READINGS

D. Harper, *Isotopes In Action* (Pergamon, New York, 1963).

J. Schubert and R. L. Lapp, *Radiation—What It Is and How It Affects You* (Viking, New York, 1958).

Scientific American articles:

P. J. Lovewell, "The Uses of Fission Products," June 1952.

S. Warren, "Ionizing Radiation and Medicine," September 1959.

QUESTIONS AND EXERCISES

1. A radiation detector is used to measure the activity of a sample of ^{90}Sr. It is found that the sample undergoes 8×10^8 decays in 1 hour. What is the activity of the sample (in μCi)?

2. The half-life of ^{131}I is 8 days. On a certain day, the activity of an ^{131}I sample is 6.4 mCi. What will be the activity of the sample 40 days later?

3. How many α particles are emitted per day by a sample of polonium which has an activity of 1 nCi (10^{-9} curie)?

4. Explain how radiation methods could be used to gage the level to which bottles are filled in a soft-drink bottling plant.

5. One of the operations in a steel mill is to form huge sheets of steel by passing the hot metal through a pair of rollers. Explain how radioisotopes could be used to determine whether the rollers are properly adjusted to produce sheets of uniform thickness.

6. One method of determining the amount of wear that pistons and piston rings experience in automobile engines is to use radioisotopes. Explain how such a measurement could be made. Points to consider: In what part should the radioisotope be incorporated? A radioisotope of what element should be used (iron, nickel, sulfur, lead, etc.)? Should the isotope be an α or a β emitter? What happens to metal in an engine when it is worn? Where does it go? Where would the radioisotope collect? (How is an engine lubricated?) How would the radioactivity measurements be made?

7. How much greater will be the radiation damage produced by a 3-MeV proton compared to that produced by a 1-MeV β particle?

8. One individual receives a whole-body dose of 10 rad of γ radiation and another individual receives a whole-body dose of 700 mrem of α particles. Which individual will suffer the greater radiation damage?

9. Speculate as to the long-term effects on the population if there were no controls on the use of radiation or the release of radioactivity into the atmosphere. How might the *gene pool* of the human race be affected?

10. Radium, thorium, and radioactive potassium (^{40}K) are found in small quantities in the materials from which bricks and concrete are made. Therefore, the radiation level in a brick-and-concrete house is generally higher than in a house constructed from wood. Do you believe that the radiation risk is worth the benefit of the increased insulation qualities and decreased fire hazard in a brick-and-concrete house compared to a frame dwelling?

11. The process of *evolution* (or *natural selection*) involves the carrying forward into future generations a characteristic or trait that results from a mutation and gives the individual some sort of competitive advantage over those not possessing this trait. Giraffes, for example, evolved long necks in order to reach leaves high on trees that could not be eaten by other animals. Because radiation is capable of producing mutations through ionizing changes in DNA, can it be argued that increased radiation is therefore *good?*

12. Radiation-induced mutations can be produced in grains, fruits, and other crops by irradiating the seeds from which the plants grow. What kind of mutations in, for example, wheat would be desirable and should be cultivated?

13. Some elements, when taken into the body, deposit selectively in certain organs or certain regions of the body. Iodine, for example, is concentrated in the thyroid gland and calcium is concentrated in the bones and teeth. Suppose that a certain radiation worker ingests a small speck of a radioisotope that is deposited exclusively and uniformly in the bone marrow. The total mass of the worker's bone marrow is 1 kg. The material emits 5-MeV α particles and the ingested sample has an activity of 10 nCi (10^{-8} curie). Assume that none of the sample is eliminated but instead resides permanently in the body. If the particular radioisotope has a long half-life (as does radium, for example), the activity will remain essentially constant for many years. What dose will the individual's bone marrow have received in one year? (Ans. 18 rem)

14. On one occasion, the author visited a dermatologist for treatment of a minor skin irritation. The dermatologist said he could treat the condition either by X radiation or with a salve. In order to decide which treatment to accept, I asked "What size dose of X radiation will you use?" Pointing to his X-ray apparatus, the dermatologist replied, "Oh, I'll use about 20 amps." Can you guess which treatment was accepted? Why?

APPENDIX

ESSENTIAL DEFINITIONS OF TRIGONOMETRY

Consider the right triangle shown in Fig. 1, in which the angle $\angle ACB$ is 90°. The lengths of the sides opposite the vertices A, B, and C are, respectively, a, b, and c. The side c is the *hypotenuse* of the triangle. We label by θ the angle $\angle BAC$. The *sine* of the angle θ is defined to be

$$\sin \theta = \frac{\text{length of side opposite } \theta}{\text{length of hypotenuse}} = \frac{a}{c} \qquad (1)$$

Similarly, the *cosine* of the angle θ is defined to be

$$\cos \theta = \frac{\text{length of side adjacent } \theta}{\text{length of hypotenuse}} = \frac{b}{c} \qquad (2)$$

Another quantity of interest is the *tangent* of the angle θ:

$$\tan \theta = \frac{\text{length of side opposite } \theta}{\text{length of side adjacent } \theta} = \frac{a}{b} \qquad (3)$$

The tangent of an angle is not independent of the sine and cosine because

$$\tan \theta = \frac{a}{b} = \frac{a/c}{b/c} = \frac{\sin \theta}{\cos \theta}$$

Two important right triangles are shown in Fig. 2. Notice that the lengths of the sides in each triangle are related by the Pythagorean theorem of plane geometry:

$$c^2 = a^2 + b^2 \qquad (4a)$$

or,

$$c = \sqrt{a^2 + b^2} \qquad (4b)$$

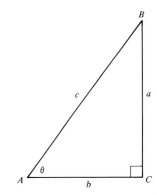

Figure 1

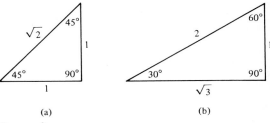

(a)   (b)

Figure 2

541

From Fig. 2a we can immediately write the values of the sine, cosine, and tangent of 45°:

$$\sin 45° = \frac{1}{\sqrt{2}} = 0.707$$

$$\cos 45° = \frac{1}{\sqrt{2}} = 0.707$$

$$\tan 45° = \frac{1}{1} = 1.000$$

From Fig. 2b we can obtain the values of the trigonometric functions of 30° and 60°:

$$\sin 30° = \frac{1}{2} = 0.500$$

$$\cos 30° = \frac{\sqrt{3}}{2} = 0.866$$

$$\tan 30° = \frac{1}{\sqrt{3}} = 0.577$$

and,

$$\sin 60° = \frac{\sqrt{3}}{2} = 0.866$$

$$\cos 60° = \frac{1}{2} = 0.500$$

$$\tan 60° = \sqrt{3} = 1.732$$

Notice that

$$\sin \theta = \cos(90° - \theta); \qquad \cos \theta = \sin(90° - \theta)$$

Values of $\sin \theta$, $\cos \theta$, and $\tan \theta$ are given in Table 1 for θ in 1° intervals from 0° to 90°.

The rectangular components of a vector can be expressed in terms of trigonometric functions in the following way (see Fig. 3):

$$A_x = A \cos \theta \tag{5a}$$
$$A_y = A \sin \theta \tag{5b}$$

or, since $\tan \theta = A_x/A_y$,

$$A_y = A_x \tan \theta \tag{5c}$$

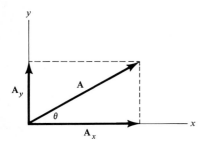

Figure 3

TABLE 1   SINES, COSINES, AND TANGENTS

ANGLE	SINE	COSINE	TANGENT	ANGLE	SINE	COSINE	TANGENT
0°	0.000	1.000	0.000				
1°	0.017	1.000	0.017	46°	0.719	0.695	1.036
2°	0.035	0.999	0.035	47°	0.731	0.682	1.072
3°	0.052	0.999	0.052	48°	0.743	0.669	1.111
4°	0.070	0.998	0.070	49°	0.755	0.656	1.150
5°	0.087	0.996	0.087	50°	0.766	0.643	1.192
6°	0.105	0.995	0.105	51°	0.777	0.629	1.235
7°	0.122	0.993	0.123	52°	0.788	0.616	1.280
8°	0.139	0.990	0.141	53°	0.799	0.602	1.327
9°	0.156	0.988	0.158	54°	0.809	0.588	1.376
10°	0.174	0.985	0.176	55°	0.819	0.574	1.428
11°	0.191	0.982	0.194	56°	0.829	0.559	1.483
12°	0.208	0.978	0.213	57°	0.839	0.545	1.540
13°	0.225	0.974	0.231	58°	0.848	0.530	1.600
14°	0.242	0.970	0.249	59°	0.857	0.515	1.664
15°	0.259	0.966	0.268	60°	0.866	0.500	1.732
16°	0.276	0.961	0.287	61°	0.875	0.485	1.804
17°	0.292	0.956	0.306	62°	0.883	0.469	1.881
18°	0.309	0.951	0.325	63°	0.891	0.454	1.963
19°	0.326	0.946	0.344	64°	0.899	0.438	2.050
20°	0.342	0.940	0.364	65°	0.906	0.423	2.145
21°	0.358	0.934	0.384	66°	0.914	0.407	2.246
22°	0.375	0.927	0.404	67°	0.921	0.391	2.356
23°	0.391	0.921	0.424	68°	0.927	0.375	2.475
24°	0.407	0.914	0.445	69°	0.934	0.358	2.605
25°	0.423	0.906	0.466	70°	0.940	0.342	2.747
26°	0.438	0.899	0.488	71°	0.946	0.326	2.904
27°	0.454	0.891	0.510	72°	0.951	0.309	3.078
28°	0.469	0.883	0.532	73°	0.956	0.292	3.271
29°	0.485	0.875	0.554	74°	0.961	0.276	3.487
30°	0.500	0.866	0.577	75°	0.966	0.259	3.732
31°	0.515	0.857	0.601	76°	0.970	0.242	4.011
32°	0.530	0.848	0.625	77°	0.974	0.225	4.331
33°	0.545	0.839	0.649	78°	0.978	0.208	4.705
34°	0.559	0.829	0.675	79°	0.982	0.191	5.145
35°	0.574	0.819	0.700	80°	0.985	0.174	5.671
36°	0.588	0.809	0.727	81°	0.988	0.156	6.314
37°	0.602	0.799	0.754	82°	0.990	0.139	7.115
38°	0.616	0.788	0.781	83°	0.993	0.122	8.144
39°	0.629	0.777	0.810	84°	0.995	0.105	9.514
40°	0.643	0.766	0.839	85°	0.996	0.087	11.43
41°	0.656	0.755	0.869	86°	0.998	0.070	14.30
42°	0.669	0.743	0.900	87°	0.999	0.052	19.08
43°	0.682	0.731	0.933	88°	0.999	0.035	28.64
44°	0.695	0.719	0.966	89°	1.000	0.017	57.29
45°	0.707	0.707	1.000	90°	1.000	0.000	

ANSWERS TO ODD–NUMBERED NUMERICAL EXERCISES

Chapter 1

3. 3280.8 ft/km
5. 5631.5 m
7. 28.3 kg; 62.4 lb
9. 2 cm
11. 21.46 g/cm^3
13. $V(Al)/V(Pb) = 4.2$

Chapter 2

3. ^{15}N
7. ^{208}Pb
9. 32 AMU
11. 18 cm^3
13. 1.6×10^{-19} m

Chapter 3

1. 18 km/h
5. 49 m/s
7. -5 (mi/h)/s; -7.3 ft/s^2
9. 4 s
15. 2.83 units, 45° above the x-axis
17. $A_x = 36.25$ units; $A_y = 16.90$ units
19. $C = 15.8$ units; **C** is directed 39.5° above the x-axis
21. 5.13 mi
23. 464 m/s
25. 24 ft/s; 2.36 s

Chapter 4

3. 6 m/s
5. 0.1 m/s^2; 0.5 m/s
9. 1176 N; 444 N
15. 46.3 N; 18.5°
17. 67.9 N
19. 1.28 m/s^2
21. 15.7 N
23. 42°

Chapter 5

1. $v = 2.5$ m/s
3. 3 m/s
9. $v = 5.29$ m/s at an angle of 15.0°
13. $x = 3$ m, $y = 4$ m
17. 295 N

Chapter 6

1. Venus: 35 km/s; Mars: 24 km/s
3. 10 kg
7. 1.62 m/s^2
11. 29.75 km/s
13. 87.5 min
17. 1.44×10^9 N
19. 43.2 N (attractive); 14.4 N (repulsive)
21. 4578 N at an angle of 10.6° above the x-axis

Chapter 7

1. 4900 J
3. 600 J
5. 41 N
7. 689 J
9. 25.2 J
13. 13 m/s
15. 4.2 J
17. 67 h.p.

Chapter 8

3. $60 billion; $100
7. 6250 mi²
9. 6.4×10^{13} kWh/y

Chapter 9

1. -28.9 °C; 26.7 °C; 37 °C; -10 °C
17. 19 Cal
21. 12 Cal
23. 0.11 C deg
29. 2×10^{17} W

Chapter 10

1. 137.5 N
3. 2.5 m
5. 10.3 m $= 33.9$ ft
7. ρ (balloon) $= 0.88$ kg/m³
13. 3.33 atm
19. increased by a factor of 2
23. 3822 °K
29. 12 000 Cal

Chapter 11

1. 1968 A
3. 0.72 kW; 121 kWh $= 4.36 \times 10^8$ J
5. $21
9. 3.25 Ω
15. nR; R/n
19. 8 Ω; 11 Ω
25. R
27. 66.7 W; 300 W
29. 24.3 AMU
31. 10 h

Chapter 12

3. 2000 V/m
5. 1.38×10^7 m/s; 1 MeV
15. 0.1 T
17. 1.08 m
19. 2.7 G
21. 2.9×10^{-14} N
27. 93.5%

Chapter 13

3. 12 Hz
5. 5, 10, 15 Hz
7. 8800 ft $= 1.67$ mi
9. 150
11. 3300 Hz
17. 10^{-4} W/m²
19. 1633 Hz; 1387 Hz

Chapter 14

3. 15 km
7. 1 cm to 1 m
9. 10 MHz
11. 187.5 m

Chapter 15

1. 3.17×10^4 rpm
3. 8.3 min
9. 2.25×10^8 m/s
11. 1.61
13. 48.6°
17. $x_i = -12$ cm
21. erect, enlarged, virtual image, 10 cm to the right of the first lens

Chapter 16

5. 0.6 y
9. 0.86 c
11. $\frac{5}{3}$
13. 3.6×10^{23} kW; 2×10^{-16}
15. $0.974c$; $0.65c$ away from the Earth

Chapter 17

1. 0.8 eV
3. 2.5 eV
5. 1.84×10^{-18} J
7. 1.2×10^4 eV
9. 4.1×10^{-9} eV
13. 2.2×10^{-34} m
17. 0.7 m/s

Chapter 18

5. 1.51 eV
7. -13.6 eV; -3.4 eV; -1.5 eV; -0.85 eV; -0.54 eV
9. 13.6 eV

13. $r_3 = 4.77 \times 10^{-10}$ m; $2\pi r_3 = 30 \times 10^{-10}$ m; $v = 7.29 \times 10^5$ m/s; $\lambda = 10 \times 10^{-10}$ m

Chapter 20

5. about 15 200 B.C.
9. 7.1 MeV

Chapter 21

1. 6 μCi
3. 3.2×10^6 per day
7. 30 times
13. 18 rem

GLOSSARY

A

Absolute zero is the temperature at which the volume of an ideal gas would be reduced to zero. All molecular motion (except that due to quantum effects) ceases at absolute zero, which is −273 °C. (Section 9-1)

An **absorption spectrum** (or *dark line* spectrum) is the series of lines of definite wavelength produced when a beam of white light (all colors) passes through a medium which selectively absorbs some of the light. The lines of an absorption spectrum are characteristic of the medium through which the light passes, not the source of the light. (Section 15-4)

Acceleration is the rate of change of velocity. An object is accelerated if its speed changes or if its direction of motion changes. (Sections 3-3, 4-2)

The **acceleration due to gravity** (g) is the acceleration experienced by an object that falls (without friction). Near the surface of the Earth, $g = 32$ ft/s² $= 9.8$ m/s². (Section 3-5)

An **alpha** (α) **particle** is the nucleus of a helium atom and consists of two protons and two neutrons bound together. α particles are spontaneously emitted by certain radioactive nuclei. (Sections 2-2, 20-1)

Alternating current is electrical current that reverses its direction of flow periodically. (Section 11-3)

The **ampere** (A) is the unit of electrical current. A current of 1 A flows in a wire when 1 C of charge passes a given point each second. 1 A = 1 C/s. (Section 11-1)

The **amplitude** of a wave or a vibration is the maximum amount of displacement of the medium from its normal condition. (Section 13-3)

An **amplitude modulated** (or **AM**) wave is a wave on which information has been impressed by varying the amplitude, with the frequency remaining the same. See also *frequency modulated* (FM) wave. (Section 14-1)

The **angular momentum** of an object is a measure of the object's rotation around a particular axis. As long as no torque is exerted on an object, its angular momentum will remain constant. (Section 5-4)

The **apogee** of an Earth satellite is the point in its elliptical orbit at which the satellite is farthest from the Earth. For the motion of a planet around the Sun, the corresponding term is **aphelion.** (Section 6-4)

Archimedes' principle states that the buoyant force exerted by a fluid on an object is equal to the weight of the fluid displaced by the object. (Section 10-2)

Artificial gravity can be produced in a spacecraft by rotating the spacecraft around an axis. The centripetal reaction experienced by an object or person within the spacecraft will be the equivalent of gravity. (Section 6-4)

The **astronomical unit** (A.U.) is a unit of length used in many astronomical distance measurements. 1 A.U. = distance from Earth to Sun = 1.5×10^{11} m.   (Chapter 1) '

The **atmospheric pressure** on a surface is the force per unit area exerted on the surface by the weight of the column of air above it. Under normal conditions at sea level, atmospheric pressure is 1.013×10^5 N/m²; this pressure is called 1 atmosphere (atm). (Section 10-1)

An **atom** is the smallest bit of matter that can be identified as a particular chemical element. (Section 2-1)

The **atomic mass unit** (AMU) is the unit used for expressing the masses of atoms and molecules. 1 AMU is defined to be the mass of a carbon atom whose nucleus consists of 6 protons and 6 neutrons (^{12}C).   (Section 2-3)

The **atomic number** of an element is equal to the number of electrons in a normal, electrically neutral atom of the element or the number of protons in the nucleus of the element. (Section 2-3)

Avogadro's hypothesis states that equal volumes of all gases (at the same temperature and pressure) contain equal numbers of molecules. (Section 2-5)

Avogadro's number is the number of molecules of a substance in one mole. $N_0 = 6.025 \times 10^{23}$ (Section 2-5)

B

A **beta** (β) **particle** is an electron that is emitted in the radioactive disintegration of certain nuclei. (Section 20-1)

The **binding energy** (or *ionization energy*) of an atom is the minimum energy required to free an electron from the atom. (Section 18-1)   The binding energy of a nucleus is the energy required to separate the nucleus into free protons and neutrons. (Section 20-2)

A **black hole** is an astronomical object with a gravitational field so intense that neither mass nor light can escape. (Section 16-5)

In the **Bohr model** of the hydrogen atom the electron is considered to move around the nuclear proton in a planetlike orbit. By allowing only discrete values for the angular momentum of the electron, Bohr succeeded in calculating the energies of hydrogen spectral lines that agreed with experiment. The Bohr model of atoms later gave way to quantum theory as a precise description of atomic matter. (Section 18-1)

The **boiling point** of a substance is the temperature at which the substance changes from the liquid to the gaseous state. The boiling point is usually stated for normal atmospheric pressure. (Section 10-5)

Boyle's law states that the volume of a gas at constant temperature is inversely proportional to the pressure. (Section 10-3)

A **breeder reactor** is a nuclear reactor that generates power and also produces more nuclear fuel than it consumes. (Sections 8-4, 20-5)

Buoyancy is the tendency of a fluid to exert an upward force on an immersed object due to the weight of the displaced fluid. (Section 10-2)

C

The **Calorie** (Cal) is a unit of heat. 1 Cal is the energy required to raise the temperature of 1 kg of water by 1 C deg. (Section 9-1)

The **carbon–nitrogen** (or **CN**) **cycle** is a series of nuclear reactions which take place in certain stars and convert hydrogen to helium, thereby generating energy. (Section 20-7)

The **cathode** is the negative electrode in an electrical circuit. **Cathode rays** are negatively charged particles (electrons) that are repelled from the cathode and stream through the partially evacuated space of a **cathode-ray tube.** (Section 2-1)

The **center of mass** of an object or system is the point at which all of the mass may be considered to be concentrated when making dynamical calculations. (Section 5-3)

Centrifugal force is the outward directed reaction to centripetal force. (Sections 3-7, 4-3)

Centripetal acceleration is the inward acceleration experienced by an object moving in a curved path. (Section 3-7)

Centripetal force is that inward directed force which is required to maintain an object moving in a curved path. (Sections 3-7, 4-3, 12-4)

The law of **Charles and Gay-Lussac** states that the volume of a gas at constant pressure is directly proportional to the absolute temperature. (Section 10-3)

A chemical **compound** is a substance in which the molecules are composed of two or more different elements. The smallest bit of matter that retains the properties of a compound is a molecule. (Section 2-1)

Compressional wave. See *longitudinal wave.*

A **conductor** is a material through which electrical current will readily flow. Metals are generally good conductors. (Sections 6-5, 11-1)

Convection is the flow that takes place within a fluid due to density differences arising from temperature differences. (Sections 9-2, 9-4)

Cosmic rays are high-speed particles of various types that are produced in violent events in stars, travel through space, and enter the Earth's atmosphere. (Section 2-4)

The **coulomb** (C) is the unit of electrical charge. The charge on an electron is $e = 1.6. \times 10^{-19}$ C. (Sections 2-1, 6-5)

Coulomb's law. See *electrical force.*

A **couple** is a pair of forces which have the same magnitudes but opposite directions and which are applied off-center to an object, thereby producing a *torque.* (Section 5-2)

Covalent bonding is the joining together of atoms to form a molecule by sharing electrons in their outermost shells. (Section 19-2)

A **crystal** is a solid composed of atoms that are arranged in a regular geometrical pattern. (Sections 2-5, 19-1, 19-2)

The **curie** (Ci) is a unit that is used to specify the amount of radioactivity in a sample. 1 Ci of radioactivity represents 3.7×10^{10} decay events per second. (Section 21-1)

Electrical **current** is the flow of electrical charge (usually electrons) around a circuit. The unit of current is the ampere (A). (Section 11-1)

D

The **de Broglie wavelength** of a particle is inversely proportional to its momentum. $\lambda = h/p$. A particle with a wavelength λ will exhibit wavelike properties that are the same as those of electromagnetic radiation with wavelength λ. (Section 17-2)

The **decibel** (dB) is a unit of sound intensity. An increase of 10 dB corresponds to an increase of sound intensity by a factor of 10; an increase of 20 dB corresponds to an intensity increase by a factor of 100; and so forth. (Section 13-4)

The law of **definite proportions** states that in every chemical compound there is always a definite proportion by mass of each constituent element. (Section 2-1)

The **density** of a substance is the mass per unit volume. $\rho = M/V$ (Chapter 1, Section 10-2)

Deuterium is the name given to hydrogen if the nucleus consists of one neutron in addition to the one proton of ordinary hydrogen. Deuterium is an *isotope* of hydrogen. (Section 2-3)

Diffraction is the bending of a wave disturbance around an obstacle in the medium. Sound waves readily diffract around corners. (Section 13-5)

A **diode** is a device that allows electrical current to pass in only one direction. Diodes are often constructed from semiconductor materials. (Section 19-4)

The **Doppler effect** is the change in frequency (or wavelength) of a wave disturbance sensed by an observer due to relative motion of the source and the observer. (Section 13-4)

E

The **Earth's magnetic field** is similar to that of a giant bar magnet and is believed to be due to intense electrical currents circulating in the molten iron core. (Section 12-3)

The **Earth's radiation belts** are due to the trapping of electrons and protons in the Earth's magnetic field. (Section 12-4)

The **efficiency** of a device or process is the ratio of the amount of work or energy delivered to the input amount of work or energy. Efficiency = (work done)/(energy used). (Section 7-4)

An **electric field** is a condition in space set up by electrical charges to which other electrical charges react. A free charged particle will experience a force and will be accelerated in an electric field. (Section 12-7)

The **electric field strength** (E) gives the force per unit charge exerted on an electrical charge in the field. $E = F_E/q$. Electric field strength is measured in V/m. (Section 11-1)

Electrical force is one of the four basic forces in Nature. The electrical force between two objects is directly proportional to the product of their electrical charges and inversely pro-

portional to the square of the distance between them. This is *Coulomb's law:* $F_E = Kq_1q_2/r^2$. The force is repulsive if the charges have the same sign and is attractive if the charges have opposite signs. (Section 6-5)

The **electrical power** delivered to a particular part of an electrical circuit is equal to the product of the voltage across the element and the current flowing through it. $P = VI$. The unit of electrical power is the watt (**W**). See also *power.* (Section 11-2)

Electrolysis is the process by which free elements are liberated from an ionic solution as the result of an electrical current passing through the solution. (Section 11-5)

An **electromagnet** is a magnet whose temporary magnetism is due to an electrical current flowing in a wire that is wound around a part of the magnet. (Section 12-6)

Electromagnetic induction is the generation of an electrical current by a changing magnetic field. (Section 12-6)

Electromagnetic radiation is produced by accelerated charges and takes the form of propagating waves. (Section 9-4, Chapter 14)

The **electromagnetic spectrum** consists of radiations of all frequencies, from low-frequency radio waves to high-frequency γ rays, and includes microwaves, radar waves, X rays, as well as visible light. (Section 14-2)

An **electromagnetic wave** is a propagating disturbance in the electromagnetic field. Such waves require no material medium and can propagate through empty space. Light, radio waves, and X rays are all electromagnetic waves. (Section 14-1)

An **electromotive force** (or **EMF**) is any force that can cause electric charges to move and thereby give rise to an electrical current. A battery is a source of EMF. (Section 11-2)

Electrons are elementary particles which constitute the outer portions of all atoms. Electrons are the basic carriers of negative electrical charge. (Sections 2-1, 6-5)

The **electron charge** (*e*) is the basic unit of electrical charge. The charge carried by any particle or object is an integer number of electron charges. The electron charge is $-e$ and the proton charge is $+e$. $e = 1.6 \times 10^{-19}$ C. (Sections 2-1, 6-5, 12-2)

The **electron volt** (eV) is a unit of energy. If a particle carrying a charge *e* accelerates from rest through a potential difference of 1 volt, it acquires a kinetic energy of 1 eV $= 1.6 \times 10^{-19}$ J. (Section 12-2)

Electroplating is the deposition of one metal upon another at the cathode of an electrolytic solution when an electrical current is passed through the solution. (Section 11-5)

Elements are substances that cannot be decomposed or transformed into one another by chemical means. Just over one hundred chemical elements are known. (Section 2-1)

An **elementary particle** is one that cannot be broken down into smaller components. Over one hundred elementary particles are known, most of which have extremely short half-lives. The common elementary particles are electrons, protons, neutrons, and photons. (Section 2-4)

An **emission spectrum** (or *bright-line* spectrum) is the series of lines of definite wavelength produced when light from a source is passed through a prism. The lines in an emission spectrum are characteristic of the source of the light. (Section 15-4)

The **emissivity** of a material determines how much electromagnetic energy will be radiated by the material at a particular temperature. A good emitter is also a good absorber. (Section 9-4)

Energy is the quality possessed by an object that enables it to do work. We identify the energy due to motion (*kinetic energy*) and the energy due to position in a field of force (*potential energy*). The metric unit of energy is the joule (J). (Energy is often measured in kilowatt-hours, kWh.) (Chapters 7, 8)

An object or system is said to be in **equilibrium** if there is no net force acting. Often, a condition of equilibrium is one of rest, but it can be one of uniform motion. (Section 4-4)

The **equivalence principle** states that effects due to gravity cannot be distinguished from effects due to an accelerated reference frame. (Section 16-5)

Evaporation is the process by which molecules escape from the surface of a liquid, thereby gradually converting the liquid into a gaseous vapor. (Section 10-5)

The **exclusion principle** states that no two electrons in an atom can have exactly the same four quantum numbers. The exclusion principle accounts for the occurrence of electron *shells* in atoms. (Section 18-3)

F

Field lines represent the map of a field, such as an electric field or a gravitational field. The direction of a field line through a point indicates the direction of the force that a particle will experience at that point due to the field. The density of field lines indicates the magnitude of the force. (Section 12-1)

Fission is the splitting of a nucleus into two more-or-less equal fragments with the release of a substantial amount of energy. (Sections 8-4, 20-4)

Force is a push or a pull that can alter the state of motion of an object (produce an acceleration). The unit of force is the newton (N). (Chapter 4)

Fossil fuels are those natural fuels that are derived from previously living matter: coal, oil, natural gas. (Sections 8-3, 8-7)

The **frequency** of a wave is the number of times per second that the wave motion repeats itself. The unit of frequency is the hertz (Hz). (Section 13-2)

A **frequency modulated** (or **FM**) wave is a wave on which information has been impressed by varying the frequency, with the amplitude remaining the same. See also *amplitude modulated* (AM) wave. (Section 14-1)

Friction is the resistance to motion that an object experiences because it moves through, moves over, or rests on another object or substance. We distinguish moving (or kinetic) friction and static friction. (Sections 3-5, 4-5, 7-1)

A **fuel cell** is a device that produces electrical power through the flameless oxidation of a fuel. (Section 11-5)

The **fundamental frequency** of a standing wave is the lowest frequency at which wave motion can be set up in a particular situation. The higher frequencies that are possible are called *harmonics* or *overtones*. (Section 13-3)

Fusion is the combining of two light nuclei into a more massive nucleus with the release of a substantial amount of energy. (Section 20-6)

G

A **galvanometer** is an instrument used to measure electrical currents. By combining various resistances with a galvanometer, a voltmeter or an ammeter can be constructed. (Section 12-6)

A **gamma** (γ) **ray** is a bundle (or *photon*) of very high frequency electromagnetic radiation. Gamma rays are often emitted by nuclei following radioactive α or β decay. (Section 20-1)

The **gauss** (G) is a metric unit of magnetic field strength. (Section 12-4)

Genetic radiation damage is damage to human genes caused by exposure to radiation. The effects of such damage can be manifest in the progeny of the exposed person. (Section 21-3)

The **geomagnetic poles** are the N and S poles of the Earth's magnetic field. The S pole is located in the Northern Hemisphere, about 800 miles from the geographic north pole. (Section 12-3)

Geothermal energy is thermal energy within the Earth's crust which is due to heating caused by the decay of radioactive materials. This energy can be tapped by drilling wells to release steam or hot water. (Section 8-5)

The **gravitational field** is a condition in space set up by a mass to which any other mass will react. (Section 6-3)

Gravitational force is one of the four basic forces in Nature. Newton's *law of universal gravitation* states that the gravitational force between two objects is directly proportional to the product of their masses and inversely proportional to the square of the distance separating their centers. $F_G = Gm_1m_2/r^2$. Gravitational force is always attractive. (Section 6-3)

H

The **half-life** of a radioactive substance is the time required for one-half of the atoms in any sample of the substance to undergo decay. (Section 20-1)

Harmonics. See *fundamental frequency*.

Heat is thermal energy in transit. If a hot object is placed in contact with a cold object, heat will flow from the hot to the cold object; some of the molecular motion of the hot object will be transferred to the cold object. The unit of heat is the Calorie (Cal). (Section 9-1)

The **heat of fusion** of a substance is the amount of energy required to convert the substance from the solid state to the liquid state at the freezing temperature. The heat of fusion of water is 80 Cal/kg. (Section 10-5)

The **heat of vaporization** of a substance is the amount of energy required to convert the substance from the liquid state to the gaseous state at the boiling temperature. The heat of vaporization of water is 540 Cal/kg. (Section 10-5)

The **hertz** (Hz) is the unit of frequency. 1 Hz means one cycle or vibration per second. Also used are kHz (10^3 Hz) and MHz (10^6 Hz). (Section 13-2)

Holography is the process by which three-dimensional optical images are produced by laser beams. (Section 18-5)

Hooke's law states that the extension of an elastic object is directly proportional to the stretching force that is applied. $F = kx$. (Section 7-2)

Humidity. See *relative humidity*.

Hybrid bonding is the way in which carbon atoms enter into compounds by using their S and P electrons in an equivalent way. (Section 19-2)

Hydrogen bonds are strong electrical bonds between polar molecules that are due to the exposed nuclear proton in molecules containing hydrogen. (Section 19-3)

The **hydronium ion** (H_3O^+) is formed when a hydrogen ion (H^+) attaches itself to a water molecule in solution. The hydronium ion is the principal hydrogen-containing positive ion in water solutions. (Section 11-5)

I

The **ideal gas law** combines the laws of Boyle and of Charles and Gay-Lussac, and states that the pressure, volume, and absolute temperature of a gas are related according to PV/T_K = constant. (Section 10-3)

The **index of refraction** (*n*) of a transparent medium is a measure of the ability of the medium to refract a light wave. *n* is equal to the ratio of the speed of light in vacuum to the speed of light in the medium. (Section 15-1)

The **inert gases** (or *noble* gases) are those gases that have completely filled outer shells and therefore are chemically inactive. (Section 2-1)

The **inertia** of an object is that quality of the object that resists a change in its state of motion. The measure of an object's inertia is its mass. (Section 4-1)

An **inertial reference frame** is an unaccelerated frame, one in which Newton's laws are valid. Any frame that moves with constant velocity with respect to an inertial frame is also an inertial frame. (Section 4-2)

An **insulator** is a material through which electrical current will not readily flow. Glass and plastics are good insulators (poor conductors). (Sections 6-5, 11-1)

Two (or more) waves exhibit constructive **interference** if they combine in phase. The interference is destructive if the waves are out of phase. (Section 13-5)

The **internal reflection** of a light ray results when a ray is incident on the boundary between two media from the side of the more dense medium. If the angle of incidence exceeds the critical angle, the ray will be totally reflected and no light will be transmitted into the less dense medium. (Section 15-1)

An **ion** is an atom that has lost or gained one or more electrons compared with the normal, electrically neutral atom. (Sections 2-1, 11-1, 11-5)

Ionic binding is the joining together of atoms in the form of positive and negative ions to produce a chemical compound. (Section 19-1)

The **ionization energy** of an atom or molecule is the minimum energy required to remove an electron and produce an ion. See also *binding energy*. (Section 18-1)

Ionizing radiations are those particles or rays that produce ionization in matter. Usually, this term includes all charged particles (such as electrons, protons, and α particles) and high energy electromagnetic radiations (such as X rays and γ rays). (Section 21-1)

The **isotopes** of a particular chemical element all have nuclei with the same number of protons but with different numbers of neutrons.

Such isotopes have identical chemical properties but different nuclear properties. (Sections 2-3 and 20-3)

J

The **joule** (J) is the metric unit of work or energy. 1 J = 1 N-m. (Section 7-1)

K

Kepler's laws are empirical laws formulated to describe the motions of planets. The most significant of these laws states that planets move around the Sun in elliptical orbits. (Section 6-2)

The **kilowatt-hour** (kWh) is a common unit of energy or work. 1 kWh = 3.6×10^6 J. (Section 7-3)

The **kinetic energy** of an object is the energy possessed by the object because of its motion. K.E. = $\frac{1}{2}mv^2$. The metric unit of kinetic energy is the joule (J). (Section 7-4)

The **kinetic theory of gases** relates the bulk properties of a gas to the microscopic motions of the constituent molecules. (Section 10-4)

L

A **laser** is a device that emits a narrow beam of light with a pure frequency and with all of the photons in phase. (Section 18-5)

Length contraction takes place when two observers are in relative motion. An observer in motion with respect to an object will see that object with its length contracted compared to the length seen by an observer at rest with respect to the object. This effect is usually important only at relativistic speeds. (Section 16-3)

A **lens** is a shaped piece of optically transparent material that is capable of focusing a beam of light or of diverging the light from an apparent source. (Section 15-2)

Lenz's Law states that if any electromagnetic change causes an effect, then that effect will always induce a reaction that tends to oppose the original change. (Section 12-6)

Light is electromagnetic radiation that is visible to the eye. (Chapter 15)

Lightning is an electrical discharge between two points (cloud-to-cloud or cloud-to-ground) that have a large potential difference. Currents up to 200 000 A flow in lightning strokes. (Section 11-6)

A **light year** (L.Y.) is the distance that light will travel in one year. 1 L.Y. = 9.5×10^{15} m. (Chapter 1)

Lines of force. See *field lines*.

A **longitudinal wave** is a wave in which the motion or disturbance of the medium is in the same direction as the direction of propagation of the wave. Sound waves are longitudinal waves. See also *transverse wave*. (Section 13-4)

M

Macroscopic matter is large-scale matter, objects that are visible to the unaided eye and ranging in size to the largest astronomical objects.

Every **magnet** has N and S poles. Like poles repel and unlike poles attract. *Permanent magnets* retain their magnetism; *electromagnets* are magnetic only as long as the exciting electrical current flows in the windings. (Section 12-3)

A **magnetic domain** is a tiny crystal that has permanent magnetic properties. If the domains of a piece of iron are aligned, a net magnetism results; if the domains are oriented at random, the sample as a whole is not magnetic. (Section 12-3)

The **magnetic field** is the condition in space set up by a magnet to which other magnets or moving charged particles react. (Sections 12-3, 12-5)

Magnetic field strength (B) is given in terms of the magnetic force on a charged particle moving in the field. $B = F_M/qv$. The unit of magnetic field strength is the tesla (T) or the gauss (G). (Section 12-4)

The **magnetic poles** of a magnet are labeled N and S. The N pole of a compass magnet is north-seeking and points toward the Earth's geomagnetic pole in the Northern Hemisphere (which is actually an S pole). (Section 12-3)

The **mass** of an object is a measure of the object's inertia (its resistance to change in state of motion). Mass may be considered to be the quantity of matter in an object. (Section 4-2)

Mass energy is the energy associated with a quantity of matter according to the Einstein equation $\mathscr{E} = mc^2$. (Sections 7-5, 16-4)

The **mass number** of a nucleus is equal to the sum of the number of protons and neutrons in the nucleus. (Section 2-3)

The **mechanical equivalent of heat** is the conversion factor connecting mechanical work and heat energy: 1 Cal = 4186 J. (Section 9-3)

The **melting point** of a substance is the temperature at which the substance changes from the solid state to the liquid state. (Sections 2-5, 10-5)

Microscopic matter is matter too small to be seen by the unaided eye, ranging in size down to atomic and subatomic particles.

One **mole** of a substance is an amount with a mass in grams equal to the molecular mass of the substance expressed in atomic mass units (AMU). (Section 2-5)

A **molecule** is the smallest unit of a particular substance that exists in Nature. Some elements exist naturally as molecules (for example, H_2, N_2, O_2), and all compounds exist as molecules. (Section 2-1)

The linear **momentum** of an object is the product of the object's mass and its velocity,

a vector quantity. $\mathbf{p} = m\mathbf{v}$. The linear momentum of an object can be changed only by the application of a force. See also *angular momentum*. (Section 5-1)

Muons are short-lived elementary particles that result from the decay of positively and negatively charged pions. Muons decay into electrons and neutrinos. (Section 2-4)

N

Neutrinos are elementary particles that are produced and emitted in certain radioactive decay processes, such as β decay and the decay of muons. Neutrinos have no mass and carry no electrical charge. There are four distinct types of neutrinos. (Section 2-4)

Neutrons are electrically neutral elementary particles found in the nuclei of all atoms (except the lightest isotope of hydrogen). Neutrons are very similar to protons, the main difference being the lack of any electrical charge on a neutron. (Section 2-3)

The **newton** (N) is the metric unit of force. 1 N = 1 kg-m/s². (Section 4-2)

Newton's laws of motion: I. An object will maintain its state of rest or motion unless acted upon by a force. (Section 4-1) II. An object will accelerate in the direction of a force applied to it. $\mathbf{F} = m\mathbf{a}$. (Section 4-2) III. For every force there is an equal and opposite reaction force. (Section 4-3)

A **node** is a position of no vibration in a standing wave. (Section 13-3)

The **nuclear force** is the strongest of the four basic forces in Nature. The nuclear force is responsible for binding together protons and neutrons in nuclei. (Section 6-6)

A **nuclear reactor** is a device in which fission reactions are self-sustaining. The energy released in the continuing fission reactions can be converted into useful electrical energy. (Section 20-5)

The **nucleus** of an atom is the tiny central core which carries most of the mass and all of the positive charge of the atom. All nuclei consist of protons and neutrons. (Sections 2-2, 20-1)

O

The **Ohm** (Ω) is the unit of electrical resistance. $1\ \Omega = 1$ V/A. (Section 11-3)

Ohm's law states that the current flow (I) through a conductor is directly proportional to the potential difference (V) across the conductor. If the resistance of the conductor is R, then $I = V/R$. Ohm's law is approximately valid for many conducting materials. (Section 11-3)

Overtones. See *fundamental frequency.*

P

An object that is projected with some horizontal velocity component will undergo **parabolic motion.** (Section 3-7)

Pascal's principle states that a pressure applied to one part of a fluid is transmitted undiminished to every other part of the fluid. (Section 10-1)

The **perigee** of an Earth satellite is the point in its elliptical orbit at which the satellite is closest to the Earth. For the motion of a planet around the Sun, the corresponding term is **perihelion.** (Section 6-4)

The **period** of a motion or a wave is the time required to complete one revolution or cycle and to return to the initial condition. (Sections 3-7, 13-2)

The **periodic table** of the chemical elements is a way of ordering the elements to show the periodicity of similar chemical properties. (Section 18-3)

The **photoelectric effect** is the emission of electrons from a material when light of sufficiently high frequency is incident on the surface. (Section 17-1)

Photons are bundles of electromagnetic wave energy. The energy of a photon is propor-

tional to its frequency. $\mathscr{E} = h\nu$. (Sections 14-4, 17-1)

Pions are short-lived elementary particles that are produced in many types of high-speed collisions between nuclear particles. There are three types of pions, identified by their electrical charge: $+e$, $-e$, 0. (Section 2-4)

Planck's constant (h) is the proportionality factor between the energy and frequency of a photon. $\mathscr{E} = h\nu$. $h = 6.6 \times 10^{-34}$ J-s. (Section 17-1)

A **plasma** is matter at such a high temperature that all of the atoms are ionized—an intimate mixture of nuclei and electrons. (Section 2-5)

A **polar molecule** is one in which there is a preponderance of negative charge at one end and positive charge at the other end. Polar molecules usually contain hydrogen and can be joined together by hydrogen bonds. (Section 19-3)

Polarization is the act of forcing the charge to separate in an object so that one part bears a positive charge and another part bears an equal negative charge. (Section 12-1)

Polarized light is light in which the electric field vector points preferentially in one direction. (Section 14-3)

Polymerization is the process by which long-chain molecules are constructed from small molecules. (Section 21-2)

A **positron** is an elementary particle that is the same as an ordinary electron except that it carries a positive charge. (Section 2-4)

Potential difference. See *voltage.*

The **potential energy** of an object is the energy possessed by the object by virtue of its position in a field of force. If an object is at a height h above the surface of the Earth, an amount of potential energy equal to mgh can be released if the object falls to the Earth. The metric unit of potential energy is the joule (J). (Section 7-4)

Power is the rate at which work is done or energy is expended. The metric unit of power is the watt (W). See also *electrical power*. (Section 7-3)

Pressure is the force per unit area exerted on an object. $P = F/A$. Pressure is measured in N/m^2. See also *atmospheric pressure*. (Section 10-1)

A **prism** is a triangular piece of glass which, by refracting light passing through it, separates the light into its various component colors. (Section 15-4)

Protons are positively charged elementary particles found in the nuclei of all atoms. The magnitude of the charge carried by a proton is exactly equal to that carried by an electron. (Section 2-3)

The **proton-proton chain** is the sequence of nuclear reactions that converts hydrogen into helium in stars. (Section 20-7)

Q

Quanta. See *photons*.

The **quantum numbers** that are necessary to specify completely the state of an electron in an atom are: n, the principal quantum number; l, the angular momentum quantum number; m_l, the magnetic or angular momentum projection quantum number; and m_s, the spin quantum number. (Section 18-1)

Quantum theory is the mathematical theory of the behavior of microscopic matter. It is a completely wave theory and does not invoke any mechanical models. (Sections 17-3, 18-2)

R

The **rad** is a unit that is used to specify the amount of radiation energy absorbed by an object. 1 rad = 0.01 J/kg. (Section 21-1)

Radiation damage is the damage done to living biological matter by the action of ionizing radiations. (Section 21-3)

Radioactivity is the property possessed by certain substances to become transformed spontaneously into other substances by the emission of α particles (helium nuclei) or β particles (electrons) from their nuclei. (Section 20-1)

The **RBE** (or *relative biological effectiveness*) of an ionizing radiation is a measure of how much biological damage that radiation produces compared to the damage produced by electrons or X rays of the same energy. (Section 21-3)

The image formed by an optical system is a **real image** if that image can be projected onto a screen. (Section 15-2)

Reflection is the turning back of a wave when it is incident on a surface. (Sections 14-2, 15-1)

Refraction is the bending or changing of direction of a wave when it passes from one medium into another in which the speed of wave propagation is different. All types of waves can exhibit refraction. (Sections 13-5, 15-1)

The **relative humidity** of a mass of air is the amount of water vapor in the air expressed as a percentage of the amount that the air is capable of holding at the particular temperature. A relative humidity of 100 percent means that the air is saturated. (Section 10-5)

The theory of **relativity** improves upon Newtonian theory, especially in the description of phenomena that take place at high speeds. The *special* theory relates to situations in which two reference frames move relative to one another at constant velocity. The *general* theory treats cases of acceleration and gravitation. (Chapter 16)

The **rem** is a unit that is used to measure radiation dosage in living tissue. (Section 21-3)

The **rest mass** of an object is the mass as measured by an observer at rest with respect to the object. (Section 16-4)

The **right-hand rule** for determining the direction of the magnetic field lines surrounding a current-carrying wire is as follows: grasp the wire with the right hand, thumb pointing in the direction of current flow; the fingers will then encircle the wire in the direction of the magnetic field lines. (Section 12-3) There are similar right-hand rules for determining the direction of the angular momentum vector (Section 5-4) and the direction of the magnetic force on a moving charged particle. (Section 12-4)

S

A **satellite** is an object that orbits around an astronomical body. The planets are satellites of the Sun; the Moon is a satellite of the Earth; and artificial satellites have been placed into orbit around the Earth. (Sections 3-7, 6-4)

Semiconductors are materials with electrical conductivity properties intermediate between those of conductors and insulators. (Section 19-4)

A **shock wave** is the concentration of wave motion along a surface due to the motion of the source through the medium at a speed greater than the speed of the wave in the medium. A *sonic boom* is an example of a shock wave. (Section 13-4)

Snell's law of optics relates the angle of refraction of a light ray to the angle of incidence and to the indexes of refraction of the two materials. (Section 15-1)

The **solar wind** consists of rapidly moving charged particles ejected from the Sun. These particles influence the magnetic field of the Earth in space. (Section 12-4)

Somatic effects of radiation are those effects that appear in an individual exposed to a large dose of radiation. (Section 21-3)

Sonic boom. See *shock wave.*

Sound waves are longitudinal waves in air or some other medium that are in the audible range of frequencies. (Section 13-4)

The **specific heat** of a substance is the amount of heat required to change the temperature of 1 kg of the substance by 1 C deg. (Section 9-3)

A **spectrum** is the series of colors or radiations with various wavelengths from a source of waves. See also *absorption spectrum* and *emission spectrum.* (Sections 14-2, 15-4)

Speed is the rate at which an object moves, regardless of the direction of motion. See also *velocity.* (Section 3-1)

Spin is the common name for the intrinsic angular momentum of an elementary particle. (Section 18-1)

A **standing wave** is a periodic disturbance set up between two boundaries such that reflections cause a regular pattern of reinforcements and cancellations. (Section 13-3)

The **stimulated emission** by an atom of a photon with a certain frequency occurs when another photon with the same frequency passes close to the excited atom. Stimulated emission is the basic process that takes place in lasers. (Section 18-5)

A **superconductor** is a metallic element or alloy that loses all resistance to the flow of electrical current at some temperature near absolute zero. (Section 19-5)

A **synchronous satellite** is one that revolves in its orbit around the Earth at the same rate that the Earth rotates on its axis and therefore maintains a fixed position relative to the Earth. (Section 6-4)

T

Temperature is a measure of the internal motion of an object's constituent molecules. The greater the motion, the greater is the internal energy and the higher is the temperature. (Section 9-1)

The **tesla** (T) is a metric unit of magnetic field strength. (Section 12-4)

Thermal energy is the internal energy of an object due to the motion of the constituent molecules. The greater the thermal energy of an object, the higher is its temperature. (Section 9-1)

Thermal pollution is the result of waste heat exhausted by power plants into bodies of water or into the air. (Sections 8-7, 20-5)

The *first law* of **thermodynamics** is the law of energy conservation when internal or thermal energy is explicitly taken into account. The *second law* states that heat can flow spontaneously only from a hotter body to a cooler body. (Section 9-1)

A **thermonuclear reaction** is a fusion reaction that will proceed only if the reactants are at an extremely high temperature. (Section 20-6)

Time dilation takes place when time is measured in two reference frames that are in relative motion. An observer will see that a clock in motion will run more slowly than will an identical clock that is at rest in his reference frame. (Section 16-3)

An object experiences a **torque** when a force is applied in such a way that the object tends to rotate. (Section 5-2)

A **transformer** is a device for increasing or decreasing the voltage in an alternating current circuit. If the voltage is increased, the current is decreased (or vice versa) so that the power $P = IV$ is constant. (Section 12-6)

A **transistor** is a three-element device consisting of p–n–p or n–p–n semiconductor layers which can control and amplify electrical signals. (Section 19-4)

A **transverse wave** is a wave in which the motion or disturbance of the medium is perpendicular to the direction of propagation of the wave. Electromagnetic waves are always transverse waves. See also *longitudinal wave*. (Section 13-2)

A **traveling wave** is a periodic disturbance that moves forward in a medium. (Section 13-2)

Tritium is the name given to the radioactive form of hydrogen in which the nucleus con-

sists of two neutrons in addition to the one proton of ordinary hydrogen. Tritium is an *isotope* of hydrogen. (Section 2-3)

U

The **uncertainty principle** states that it is not possible to measure simultaneously the position and the momentum of a particle or photon with unlimited precision. The uncertainty principle is a key ingredient of quantum theory. (Section 17-3)

V

The **vapor pressure** of a substance at a given temperature is the pressure of the vapor in a confined space above the substance. At the boiling point of an unconfined liquid, the vapor pressure is equal to the atmospheric pressure. (Section 10-5)

A **vector** is a quantity that requires both magnitude and direction for its complete specification. Examples of vectors are velocity, force, momentum, electric field, and so forth. (Section 3-6)

Velocity combines the rate at which an object moves (the *speed*) with the direction of motion. Velocity is a vector quantity. (Section 3-6)

An optical system forms a **virtual image** if that image can only be perceived by the eye but cannot be projected onto a screen. (Section 15-2)

The **volt** (V) is the unit of measure of potential difference. 1 V = 1 J/C. (Section 11-2)

Voltage is the electrical "pressure" or potential difference which causes current to flow. The unit of potential difference is the volt (V). (Section 11-2)

W

The **watt** (W) is the metric unit of power. 1 W = 1 J/s. (Section 7-3)

A **wave** is a propagating disturbance in a medium. All waves carry energy and momentum, but not matter. (Chapter 13)

The **wavelength** of a wave is the distance between successive crests or troughs of the wave. (Section 13-2)

A **wave packet** is a bundle of waves that retains its oscillatory character but is localized in space. (Section 17-3)

The **weak force** is one of the four basic forces in Nature. The weak force is responsible for radioactive β decay and for various processes involving elementary particles. (Section 6-6)

The **weight** of an object is the gravitational force acting on the object. Weight is proportional to mass. (Section 4-2)

Work is the product of force and the distance through which the force acts. The metric unit of work is the joule (J). (Section 7-1)

The **work function** of a material is the minimum energy required to release an electron from the surface of the material by the photoelectric effect. (Section 17-1)

X

X rays are electromagnetic radiations that are emitted by atoms when transitions occur in the inner electron shells. (Section 18-4)

INDEX

in stars, 508
states of hydrogen, 427
storage of, 175
thermal, 194
unit of, 151
usage of, 161
of waves, 309
Equilibrium of forces, 79
Equivalence principle, 400
Evaporation, 229
Exclusion principle, 441

F

Fahrenheit, G. D., 196
Fahrenheit temperature scale, 196
Faraday, M., 260, 273, 339
Faraday (unit), 260
Faraday's law of electrolysis, 259
Fields
electrical, 271
gravitational, 118
magnetic, 281
time-varying, 294
Fission, 158, 489
Force, 63
centrifugal, 78
centripetal, 78, 101
electrical, 129, 134
gravitational, 115
lines of, 271
magnetic, 286
nuclear, 136
unit of, 67
weak, 137
Fossil fuels, 165
environmental effects of, 182
supplies of, 167
Frames of reference, 69
inertial, 70, 382
Fraunhofer, J. von, 370
Fraunhofer lines, 370
Frequency, 307
of light, 410
Friction, 81
effect on free fall, 49
kinetic, 81
static, 83
Frisch, O., 489
Fuel cells, 263
Fusion, nuclear, 504
Fusion reactors, 506

G

Galaxies, formation of, 102

Galileo, 45, 47, 348, 366
Galle, J., 118
Galvanometer, 299
Gamma rays, 25, 479, 518
Garlits, D., 46
Gauss (unit), 287
Gay-Lussac, J. L., 221
Geiger, H., 20
Genetic radiation effects, 533
Geothermal energy, 171
Germer, L. H., 415
Goudsmit, S., 436
Gravitational constant, 116
Gravity, 112
acceleration due to, 45
artificial, 125
force law for, 116
in relativity theory, 400

H

Hahn, O., 489
Half-life, 482
of elementary particles, 28
Harmonics, 311
Heat, 193
of fusion, 234
mechanical equivalent of, 206
transfer, 207
unit of, 198
of vaporization, 233
Heisenberg, W., 421, 439
Helmholtz, H., 141
Hertz, H., 307
Hertz (unit), 307
Holography, 453
Hooke, R., 147
Hooke's law, 147
Horsepower (unit), 149
Humidity, 232
Hybrid binding, 463
Hydrogen atom, 426
Bohr model of, 427
quantum theory of, 437
Hydrogen bomb, 506
Hydrogen bonds, 465
Hydrogen molecule, 462
Hydropower, 164

I

Ideal gas law, 223
Induction, 294
electrostatic, 132
Inertia, 65
law of, 65
Inertial reference frame, 70, 382